www.wadsworth.com

wadsworth.com is the World Wide Web site for Wadsworth and is your direct source to dozens of online resources.

At *wadsworth.com* you can find out about supplements, demonstration software, and student resources. You can also send email to many of our authors and preview new publications and exciting new technologies.

wadsworth.com
Changing the way the world learns®

The Emergence
of Sociological Theory

Fifth Edition

JONATHAN H. TURNER
University of California, Riverside

LEONARD BEEGHLEY
University of Florida

CHARLES H. POWERS
Santa Clara University

WADSWORTH

™

THOMSON LEARNING

Australia • Canada • Mexico • Singapore • Spain • United Kingdom • United States

WADSWORTH
THOMSON LEARNING
™

Sociology Editor: Lin Marshall
Assistant Editor: Analie Barnett
Editorial Assistant: Reilly O'Neal
Technology Project Manager: Dee Dee Zobian
Marketing Manager: Matthew Wright
Project Manager, Editorial Production:
  Jerilyn Emori
Print/Media Buyer: Tandra Jorgensen

Permissions Editor: Stephanie Keough-Hedges
Production Service: Scott Rohr/Buuji, Inc.
Copy Editor: Robin Gold
Cover Designer: Carole Lawson
Compositor: Buuji, Inc.
Text and Cover Printer: R.R. Donnelley & Sons
  Company/Crawfordsville

Wadsworth Thomson Learning
10 Davis Drive
Belmont, CA 94002-3098
USA

For more information about our products,
contact us:
Thomson Learning Academic Resource Center
1-800-423-0563
http://www.wadsworth.com

International Headquarters
Thomson Learning
International Division
290 Harbor Drive, 2nd Floor
Stamford, CT 06902-7477
USA

UK/Europe/Middle East/South Africa
Thomson Learning
Berkshire House
168-173 High Holborn
London WC1V 7AA
United Kingdom

Asia
Thomson Learning
60 Albert Street, #15-01
Albert Complex
Singapore 189969

Canada
Nelson Thomson Learning
1120 Birchmount Road
Toronto, Ontario M1K 5G4
Canada

Library of Congress Cataloging-in-Publication Data
Turner, Jonathan H.
    The emergence of sociological theory / Jonathan H. Turner, Leonard Beeghley,
Charles H. Powers. —5th ed.
    p. cm.
    Includes index.
    ISBN 0-534-51967-9 (alk. paper)
    1. Sociology—History. 2. Sociology—United States—History. I. Beeghley, Leonard. II.
Powers, Charles H. III. Title.
    HM19 .T97 2001
    301'.01—dc21   2001026521

Contents

Preface

The first edition of *The Emergence of Sociological Theory* was published in early 1981. At that time, our goal was to examine the first one hundred years of sociological theorizing—roughly the period between 1830 and 1930. In particular, we sought to communicate the explanatory power of each theorist's ideas. We visualized the early masters as scientists who sought to understand the operative dynamics of human action, interaction, and organization. Of course, not all the masters saw themselves as scientists but, except for Karl Marx and Max Weber, most founders of sociology believed in the epistemology of science. And, despite differences in their respective commitments to science, all of the theorists of sociology's first one hundred years discovered some of the fundamental properties and processes of the social universe. Our intent back in the 1970s was to highlight their respective discoveries of these properties and processes.

Over the years, the goals of the book have remained unchanged: to summarize the basic works of each theorist and to pull the explanatory models and principles from these works. This book has always summarized each theorist's ideas in great detail. We have never "watered down" the reviews of a theorist's basic works; rather, we have tried to present ideas in their full complexity, although we have also sought to do so in simple language. Ironically, when ideas are watered down or summarized too briefly, they become less understandable than when they are reviewed in their most robust form. In this new edition, we state the theorists' ideas in even more straightforward prose and provide thorough summaries of core ideas because we feel that this is the best way to understand the argument of a theorist. What has always distinguished *The Emergence of Sociological Theory* is a commitment to present the ideas of theorists in detail. When the details of arguments are reviewed, it becomes possible to see the ideas of the early masters as real theory—that is, as explanatory models and propositions. This mode of analysis is not everyone's cup of tea, of course, but our view is that these classical thinkers are read and re-read today because we sense the explanatory power of their ideas. And so, in our minds, it is appropriate to make the explanations more explicit.

New Tri-Chapter Organization

Classical theory can be taught in many ways. One method is to review the texts of the early masters in their intellectual contexts. Another is, to isolate the enduring theoretical ideas of a thinker and state these in more modern terms as models and principles. In this fifth edition of the *Emergence of Sociological Theory*, we try to accomplish all these goals, but in separate chapters so that readers can pick and choose which approach they wish to take. With the exception of Auguste Comte, who articulated a vision for sociology but little real theory, we devote three chapters to each theorist. The first of the three chapters is the biography of the theorist and the intellectual influences on that theorist's thinking. The second chapter summarizes the theorist's core works; as we have emphasized, this summary is detailed so that readers can understand the full complexity of the argument. We also use graphics to help the reader visualize complex ideas. The third chapter—new to this edition—goes beyond a summary of the theorist's works. This additional chapter for each theorist presents the underlying theory of a master as a model in which the key forces and their causal relations to each other are diagrammed. This kind of modeling is complex, especially when we present a scholar's entire scheme as one model. Still, it is useful to see a theory in its most robust form, and by presenting the overall theory, we can appreciate the entire theoretical scheme of a scholar as it evolved over time. For many theorists, we present separate causal models for particular aspects of their overall schemes, and these more focused models are generally much simpler. In all, we have developed twenty-four causal models that, we believe, allow us to see the complex causal arguments developed by sociology's early masters. These models have stood the test of time, and indeed, they are still relevant. At the close of this new third chapter, we present a master's theoretical ideas as a series of elementary propositions that are, in principle, testable. By presenting a theory in these two formats—complex causal models and abstract principles—the power of the theory is revealed. We recognize that, to many, this kind of exercise imposes the epistemology of science and more modern theoretical formats onto older discursive texts. While this charge is true, we believe that the enduring power of a scholar's ideas resides in the elegance of the theory as it can be expressed as a causal scheme or as a series of principles. This is why these scholars are still important as theorists rather than solely as historical figures who founded the discipline. These thinkers speak to us today because they developed scientific theories that are easily extracted from their discursive texts.

This tri-chapter organization of the book presents several options for teaching a course on the history of sociological theory. If a "text-in-historical context" approach is desired, then the first two chapters on a theorist can be used and the third chapter omitted. If a purely textual approach is desired, then the second chapter on each theorist can be emphasized, perhaps excluding the first and third chapters. If a more scientific emphasis is desired, then the third chapter on a theorist can be stressed, typically after reading the second chapter summarizing the arguments presented by the theorist's basic texts. And, of course, all these options can be combined into a very complete analysis of early theorizing in sociology. We have made the chapters modular, so each can stand alone. As a result, an instructor can pick and choose how to present the materials to students.

Changes in this Edition

To summarize the essential changes in this edition:

- A new third chapter on each major theorist has been added. This chapter presents the theories of the early masters as causal models, depicted visually and as abstract theoretical principles.

- Many new diagrams, tables, and other visual aids have been added, not only in the third chapter on each theorist but also throughout the text.

- The book is now more modular, presenting instructors with various options for what they want to emphasize: history of ideas in context, summaries of basic texts, or underlying theoretical models and principles.

- A new chapter on the nature of science and scientific theory has been added, particularly for those who want to pursue the analysis of the first masters as theorists who articulated enduring theoretical models and principles.

- Each chapter has been somewhat reorganized, particularly those on Marx and Weber. Other chapters have been shortened somewhat, especially those on Spencer and Comte.

- The writing style has been simplified, but not to the point of oversimplifying complex arguments by the early theorists.

- A new companion Web site has been added. Additional materials can be accessed via this site, including a discussion of how the early masters' ideas have been carried forward into contemporary theory. Moreover, the Web site will be constantly updated.

This is the most complete revision of the book since it was written in the 1970s. We have added many new materials and features but not abandoned the original goal of reviewing the theoretical ideas of the first sociologists in the context of their times and for their contributions to the development of cumulative scientific theory.

ACKNOWLEDGMENTS

We would like to thank the following reviewers for their helpful suggestions for revising this fifth edition. Even though we did not follow all their recommendations, we are grateful for the time and effort they put into making this a better book: Evandro Camara, Emporia State University; Walter Carroll, Bridgewater State College; Richard A. Garnett, Marshall University; Paul Johnson, Texas Tech University; David Monk, California State University, Stanislaus; and Will Wright, University of Southern Colorado.

About the Authors

Jonathan H. Turner
(Ph.D., Cornell University) is Distinguished Professor of Sociology at the University of California, Riverside. He is the author of more than two dozen books, including *The Structure of Sociological Theory*, which is published by Wadsworth as a companion volume to this one. He is the author of well over one hundred research articles in journals and books. He has served on the editorial boards of many journals as well as several publishers, including the University of California Press. He received the Faculty Research Lecturer award at the University of California for his research on social theory.

Leonard Beeghley
(Ph.D., University of California at Riverside) is professor of sociology at the University of Florida. He is the author of a number of books, primarily in the area of stratification and social policy issues. He has written many articles in research journals and has served in editorial positions for several publishers. He has served on committees within the American Sociological Association.

Charles H. Powers
(Ph.D., University of California at Riverside) is professor of sociology at Santa Clara University. Under his leadership the sociology program at Santa Clara won the American Sociological Association's Distinguished Contributions

to Teaching Award in 1998. He is the author of several books and research articles focusing on sociological theory and on change management in organizations.

1

✳

The Enlightenment
and New Ways
of Thinking

Humans have, no doubt, always thought about their lives and the conditions of their existence. Such thoughts are the life-blood of religion, philosophy, ideology, and the many other ways that humans can think about themselves and their world. There is, therefore, nothing new in the basic impulse that eventually led to sociology emerging as a discipline concerned with understanding human behavior, interaction, and organization. Sociology is, after all, only the more *systematic study* of what people do in their daily lives and routines. Sociology did not emerge as an inevitable extension of what people typically do; rather, it arose from the rebirth or Renaissance in Europe after centuries of apparent stagnation and misery. These "dark ages" were the aftermath of the collapse of the Roman Empire's last remnants, and they were only dark in retrospective comparison with the perceived accomplishments of the Greeks and Romans. Life was not so stagnant, however: New inventions and ideas were slowly accumulating, despite the oppressive poverty of the masses, the constant warfare among feudal lords, and the rigid dogma of religion. New systems of commerce were slowly emerging. New forms and experiments in political organization were emerging from the patterns of war and conquest. New religious ideas were subtly working their way around the dogmas of the dominant church. The great awakening in intellectual thought, art, commerce, politics, and other human pursuits was built on small achievements and advances that were slowly accumulating between the fifth and thirteenth centuries in Europe. Once a critical threshold was reached, human thinking took sudden leaps, recapturing much that had been lost from the Greeks and

Romans and, more significantly, re-creating systematic scientific thought about the universe.

Sir Francis Bacon (1561–1626) was the first to articulate clearly the new mode of inquiry: Conceptualizations of the nature of the universe should always be viewed with skepticism and tested against observable facts. This sounds like a commonplace idea today, but it was radical in its time. This idea both legitimated and stimulated the great achievements of the sixteenth and seventeenth centuries in astronomy, culminating in Sir Isaac Newton's famous law of gravity. Thinking about the universe took on a systematic character, but more than just systematic: Thinking also became abstract, articulating basic and fundamental relationships in highly general terms and, then, seeing if concrete events in the empirical world conformed to these general statements. Such is the essence of science, and it changed the world.

THE ENLIGHTENMENT

Sociology emerged as a discipline early in the nineteenth century, but it was not so much a dramatic breakthrough in human reasoning as an extension of what is often termed *the Enlightenment.* Perhaps the Enlightenment can be considered an intellectual revolution because it turned thinking about the human condition toward the view that progress was not only possible but also inevitable.

In England and Scotland, the Enlightenment was dominated by a group of thinkers who argued for a vision of human beings and society that both reflected and justified the industrial capitalism that first emerged in the British Isles. Scholars such as Adam Smith believed individuals should be free of external constraint and allowed to compete, thereby creating a better society. In France, the Enlightenment is often termed the *Age of Reason,* and it was dominated by a group of scholars known as the *philosophes.* Sociology was born from the intellectual ferment generated by the French philosophes.

Although the Enlightenment was fueled by the political, social, and economic changes of the eighteenth century, it derived considerable inspiration from the scientific revolution of the sixteenth and seventeenth centuries. The scientific revolution reached a symbolic peak, at least to eighteenth-century thinkers, with Newtonian physics. The post-Newtonian view of science was dramatically different from previous views. The old dualism between reason and the senses had broken down, and for the first time, it could be confidently asserted that the world of reason and the world of phenomena formed a single unity. Through concepts, speculation, and logic, the facts of the empirical world could be understood, and through the accumulation of facts, reason could be disciplined and kept from fanciful flights of speculation.

The world was thus viewed as orderly, and people believed it was possible to understand the world's complexity through reason and facts. Newton's principle of gravity was hailed as the model for this reconciliation between reason and senses. Physics became the vision of how scientific inquiry and theory should be conducted. The individual and society were increasingly drawn into the orbit of the new view of science. This gradual inclusion of the individual and society into the realm of science represented a break with the past because heretofore these phenomena had been considered the domain of morals, ethics, and religion. Indeed, much of the philosophes' intellectual effort involved the emancipation of thought about humans from religious speculation, and although the philosophes were far from scientific, they performed the essential function of placing speculation about the human condition in the realm of reason. Indeed, as can be seen in their statements on universal human rights, laws, and the natural order, much of their work consisted of attacks on established authority in both the church and state. From notions of "natural law," it was but a short step to consideration of the laws of human organization. As we will see in the next chapter, many of the less shrill and polemical philosophes—first Charles Montesquieu, then Jacques Turgot and Jean Condorcet—actually made this short step and sought to understand the social order through principles they felt were the equivalent in the social realm of Newton's law of gravitation.

The philosophes' view of human beings and society was greatly influenced by the social conditions around them. They were vehemently opposed to the Old Regime in France and supported the bourgeoisie in free trade, free commerce, free industry, free labor, and free opinion. The large and literate bourgeoisie formed the reading public that bought the books, papers, and pamphlets of the philosophes. These philosophes' concerns with the "laws of the human condition" were as much, and probably more, influenced by their moral, political, and ideological commitments as by a dispassionate search for scientific laws. We should not ignore, however, the extent to which the philosophes heralded a science of society molded in the image of physics or biology.

The basic thesis of all philosophes, whether Voltaire, Jean Jacques Rousseau, Condorcet, Denis Diderot, or others, was that humans had certain "natural rights," which were violated by institutional arrangements. It would be necessary, therefore, to dismantle the existing order and substitute a new order considered more compatible with the essence and basic needs of humankind. The transformation was to occur through enlightened and progressive legislation; ironically, the philosophes watched in horror as their names and ideas were used to justify the violent French Revolution of 1789.

In almost all of the philosophes' formulations was a vision of human progress. Humanity was seen to be marching in a direction and was considered to be governed by a "law of progress" that was as fundamental as the law of gravitation was in the physical world. In particular, those who exerted the most influence on Auguste Comte—Turgot, Condorcet, and Claude-Henri de Saint-Simon—built their intellectual schemes around a law of progress. Thus, the philosophes were decidedly unscientific in their moral advocacy, but they offered at least the rhetoric of post-Newtonian science in their search for the natural laws of human order and in their formulation of the law of progress.

From these somewhat contradictory tendencies sociology emerged in the work of Comte, who sought to reconcile the seeming contradiction between moral advocacy and detached scientific observation.

The more enlightened of the philosophes, men such as Montesquieu, Turgot, and Condorcet, presented the broad contours of this reconciliation to Auguste Comte: The laws of human organization, particularly the law of progressive development, can be used as tools to create a better society. With this mixture of concerns—moral action, progress, and scientific laws—the Age of Reason ended and the nineteenth century began. From this intellectual milieu, as it was influenced by social, economic, and political conditions, Comte pulled diverse and often contradictory elements and forged a forceful statement about the nature of a science of society, as we will see in the next two chapters.

Systems of ideas do not suddenly appear, even ones as powerful and influential as those advocating science and its use for human betterment. Important ideas almost always reflect more fundamental changes in the distribution of power and the organization of production. Once created, however, ideas have the capacity to stimulate new forms of politics and new modes of production. But the Enlightenment was not just an intellectual revolution. Its emergence and persistence was a response to changes in economic and political organization.

THE POLITICAL ECONOMY
OF THE ENLIGHTENMENT

During most of the eighteenth century, the last remnants of the old economic order were crumbling under the impact of the commercial and industrial revolutions. Much of the feudal order had been eliminated by trade expansion during the seventeenth century. Yet economic activity in the eighteenth century had become greatly restricted by guilds, which controlled labor's access to skilled occupations, and by chartered corporations, which restrained trade and production.

The eighteenth century saw the growth of free labor and more competitive manufacturing. The cotton industry was the first to break the hold of the guilds and chartered corporations, but with each decade, other industries were subjected to the liberating effects of free labor, free trade, and free production. By the time large-scale industry emerged—first in England, then in France, and later in Germany—the economic reorganization of Europe had been achieved. Large-scale industry and manufacture simply accelerated the transformations in society that had been occurring for decades.

These transformations involved a profound reorganization of society. Labor was liberated from the land; wealth and capital existed independently of the large noble estates; large-scale industry accelerated urbanization of the population; the extension of competitive industry hastened the development of new technologies; increased production encouraged the expansion of markets and world trade for securing raw resources and selling finished goods; religious organizations lost much of their authority because of secular economic activities; family structure was altered as people moved from rural to urban areas; law became as concerned with regularizing the new economic processes as with preserving the privilege of the nobility; and the old political regimes legitimated by "divine right" became less tenable.

Thus, the emerging capitalist economic system inexorably destroyed the last remnants of the feudal order and the transitional mercantile order of restrictive guilds and chartered corporations. Such economic changes greatly altered the way people lived, created new social classes (such as the bourgeoisie and urban proletariat), and led not only to a revolution of ideas but also to a series of political revolutions. These changes were less traumatic in England than in France, where the full brunt of these economic forces clashed with the Old Regime. This volatile mixture of economic changes, coupled with the scientific revolution of the sixteenth and seventeenth centuries, spawned political and intellectual revolutions. From these combined revolutions, sociology emerged.

The Revolution of 1789 marked a dramatic transformation in French society. The Revolution and the century of political turmoil that followed provided early French sociologists with their basic intellectual problem: how to use the "laws of social organization" to create a new social order. Yet, in many ways, the French Revolution was merely the violent culmination of changes that had been occurring in France and elsewhere in Europe for the entire eighteenth century.

By the time of the French Revolution, the old feudal system was merely a skeleton. Peasants were often landowners, although many engaged in the French equivalent of tenant farming and were subject to excessive taxation. The old landed aristocracy had lost much of its wealth through indolence, incompetence, and unwillingness to pursue lucrative, yet low-status, occupations. Indeed, many of the nobility lived in genteel poverty behind the walls of their disintegrating estates. As the nobility fell into severe financial hardship, the affluent bourgeoisie were all too willing to purchase the land. Indeed, by 1789, the bourgeoisie had purchased their way into the ranks of the nobility as the financially pressed monarchy sold titles to upwardly mobile families. Thus, by the time of the Revolution, the traditional aristocracy was in a less advantageous position, many downtrodden peasants were landholders, the affluent bourgeoisie were buying their way into the halls of power and prestige, and the monarchy was increasingly dependent on the bourgeoisie for financial support.

The structure of the state best reflects these changes in the old feudal order. By the end of the eighteenth century, the French monarchy had become almost functionless. It had centralized government by suppressing old centers of feudal power, but its monarchs were now lazy, indolent, and incompetent. The real power of the monarchy increasingly belonged to the professional administrators in the state bureaucracy, most of whom had been recruited from the bourgeoisie. The various magistrates were virtually all recruited from the bourgeoisie, and the independent financiers, particularly the Farmers General, had assumed many of the tax-collecting functions of government. In exchange for a fixed sum of money, the monarchy had contracted to the financiers the right to collect taxes, with the result that the financiers collected all that the traffic could bear and, in the process, generated enormous resentment and hostility in the population. With their excessive profits, the financiers became the major bankers of the monarchy; the king, nobility, church, guild master, merchant, and monopolistic corporate manufacturer often went to them for loans.

Thus, when the violent revolution came, it hit a vulnerable political system that had been in decline for most of the eighteenth century. The ease with which the system crumbled highlighted its vulnerability, and the political instability that followed revealed the extent to which the ascendance of the bourgeoisie and large-scale industrialists had been incomplete. In other societies where sociology also emerged, this transition to industrial capitalism and new political forms was less tumultuous. Particularly in England, the political revolution was more evolutionary than revolutionary, creating a sociology distinctly different from that in France.

THE EMERGENCE OF SOCIOLOGY

Change forces new ways of thinking to emerge, and so it was with the intellectual currents of the Enlightenment, the transformation of feudalism into capitalism, and the political upheaval that came with the old feudal order's demise and the state's rise. New material conditions force both ordinary people in their daily routines and scholars in their more systematic pursuit of understanding to reconceptualize the world. For much of the eighteenth century, scholars had been grappling with changes in the old order, trying to find comfort and promise in what was occurring.

By the nineteenth century, the time was right for a new discipline, sociology. Auguste Comte, who proposed this name for the new discipline, actually preferred the title "social physics" to sociology because it captured the essence of the Age of Science and the Enlightenment. At the time, the term *physics* was not so welded to the current discipline by this name; rather, it denoted an effort to understand the nature of phenomena. Thus, Comte believed that social physics would understand the nature of human social organization through the epistemology of science. This was not a radical idea by the time that Comte presented it, but he was the first to advocate in a forceful way that thinking about the social world could become as scientific as efforts to understand the physical universe.

2

✳

Auguste Comte
and the Emergence
of Sociological Theory

Auguste Comte gave sociology its name, so he is generally credited with being the founder of sociology. Yet, new ideas are often extensions and codifications of ideas developed by a scholar's predecessors and contemporaries. Such is certainly the case for Comte, for he was working within a long French tradition of thinking that gave him the critical elements for his pronouncement that the era of sociology had arrived. Comte was an odd man, and indeed, he went rather insane in his later years, but he was young and well-situated at the time he began to visualize a "system of positive philosophy," with sociology being its culminating science. Before exploring the intellectual influences on Comte, let us briefly review how Comte's biography influenced his thinking.

THE STRANGE BIOGRAPHY
OF AUGUST COMTE[1]

Auguste Comte was born in 1798 in Montpellier, a city in the south of France. The period from Comte's birth to his formative years as a student in Paris was socially and politically tumultuous, punctuated by revolution and revolt. At his birth, the Directory ruled France after the Reign of Terror imposed by Robespierre, but within a couple of years Napoleon had led a coup and become first among equals in the council. Before Comte entered school five years later, Napoleon had been crowned emperor of France. After Napoleon's defeat in 1814, the throne was restored to the brother of the former king, retaken briefly by Napoleon on his escape from the island of Elba, and restored again to Louis XVIII in 1815 after Napoleon's defeat at Waterloo. For the next fifteen years, Louis XVIII, and later his brother, Charles X, ruled France. In 1830, the publication date for the first installment of Comte's *The Course of Positive Philosophy,* yet another revolution established Louis Philippe, duke of Orléans as king, and between 1848 and 1852, a series of popular revolts and military coups reestablished the Empire with Napoleon III as emperor.

This constantly changing political landscape was accompanied by the rapid industrialization of France. Industrialization brought new classes of individuals, including urban wage earners and expanded numbers of bourgeoisie, as well as new structural systems like the factory, bureaucracy, and open market. During these transformations Comte's career began, reached a promising beginning, and then faded to embarrassment and ridicule. He had been an impressive student in the *lycée* of his hometown, but he had also been a somewhat rebellious and difficult person—traits that, in the end, caused his ruin. On the basis of competitive examinations, in 1814 he secured a place in the École Polytechnique, the elite technical school of France. He soon established himself as a brilliant, though difficult, student. His generation of Parisian students and intellectuals had lost much of their religiosity; yet they retained the desire to construct a new, more stable order in terms of some faith. Many believed science was this new faith and could be used to make a better society. Moreover, there was a growing feeling that the sciences—both natural and social—could be unified to reconstruct the world. The authority of scientific laws, and their engineering applications, were to be the substitute for religion. Comte was captured by these ideas, especially because during his years in the *lycée* he had lost his religious faith and abandoned Catholicism.

But the École Polytechnique of Comte's time closed temporarily in a dispute between its students and faculty, on the one side, and its financial benefactor, the government, on the other side. The dispute was over the mission of the school: Was it to be devoted to pure science, to engineering, or to military training? During the university's closure, Comte briefly returned home but was soon back in Paris, giving private lessons. He even sought to move to America, but his hopes went unrealized when the U.S. government did not create an equivalent technical university. When the École Polytechnique reopened, he did not seek readmission, perhaps because he had made too many enemies.

At this time, in 1817, Comte began his association with Claude-Henri de Saint-Simon, at first as secretary and later as a collaborator. They worked closely together, and most of Comte's major ideas were developed during this period. But he also began to resent Saint-Simon's dominance, especially because Comte had become the intellectual force behind the work that was making Saint-Simon a leading thinker and reformer. Moreover, because

Comte was interested in developing theory before ameliorative action, he constantly came into conflict with the activist Saint-Simon and his followers, the "Saint-Simonians." By 1824, these tensions had built to the point where Comte acrimoniously broke from Saint-Simon, a move that destroyed Comte's career.

In 1825, Comte married, but the marriage was problematic; hence, instability in his personal and intellectual lives was about to overwhelm him. Yet he began to write the ideas for his most famous work, *The Course of Positive Philosophy,* in which he explicitly created the discipline that became sociology. During this period of creativity, however, he still needed to earn a living, and he was forced to tutor and perform marginal academic tasks that were beneath his abilities and aspirations. In a bold move to recapture some of his fading esteem, he proposed a series of public lectures to communicate his ideas. Forty eminent scientists and intellectuals subscribed to the lectures, but he gave only three before the pressure and tension of the enterprise made him too ill to continue. Three years later, he was sufficiently well to restart the lectures, with many notables still in attendance.

Yet his support was fragile, and his difficult personality drove people away. His grand goal, to incorporate the development of all the sciences in one scheme, was attacked by specialists in each science, so that, as is often the case in academia, a pretentious and ambitious nonacademic soon became an object of derision and ridicule.

Thus, by the time that the first installment of *Positive Philosophy* was published, Comte was becoming more isolated. Yet the first volume received critical acclaim, and his ideas did attract considerable attention. But this acclaim was short-lived. Within a few years, he had fully antagonized the last of his important scientific admirers; he had lost his academic colleagues; he was the confirmed enemy of the Saint-Simonians; he began to lose his old personal friends; he continued to have marital problems; and he was reduced to ever more menial and marginal academic work as a reader, tutor, and examiner. When the six volumes of *Positive Philosophy* were completed in 1842, there was not a single review in the French press. At this point, moreover, his wife left him. Many of Comte's problems were of his own making: He was obnoxious to friends and critics, he was defensive and dogmatic, and he was so arrogant as to cease reading others' works in an act of defiance to his critics (what he termed "cerebral hygiene").

The titular founder of sociology had gone from promising brilliance to intellectual isolation, at least in France. In England, Comte did exert some influence on both John Stuart Mill and Herbert Spencer, and he influenced subsequent generations of French thinkers, most particularly Émile Durkheim.

In a desperate search for an audience, Comte was reduced to lecturing to a ragtag collection of workers and other interested parties. He had lost, forever in his lifetime, the respect of the scientific, academic, and intellectual community. He began to see himself as the "High Priest of Humanity," making pronouncements to his followers, and in his *System of Positive Polity* his science had taken a back seat to his advocacy—ironically, the very point that had led to his break with Saint-Simon decades earlier. He was a pathetic figure, lecturing to his intellectual inferiors and seeking to create a semi-religious cult of followers. The founder of sociology, the person with the first clear and still relevant vision of what sociology could be, died as a demented fool. The promise and power of his early vision for sociology had been lost—only to be picked up again after his death in 1857.

THE INTELLECTUAL ORIGINS OF COMTE'S THOUGHT

Comte's sociology emerged from the economic, political, and social conditions of postrevolutionary France. No social thinker could ignore the oscillating political situation in France during the first half of the nineteenth century or the profound changes in social organization that accompanied the growth of large-scale industry. Yet despite the influence of these forces, the content of Comte's sociology represents a selective borrowing of ideas from the Enlightenment of the eighteenth century. Comte absorbed, no doubt, the general thrust of the philosophes' advocacy, but he appears to have borrowed and then synthesized concepts from four major figures: Charles Montesquieu (1689–1755), Jacques Turgot (1727–1781), Jean Condorcet (1743–1794), and Claude-Henri de Saint-Simon (1760–1825). In addition, Comte seems to have been influenced by the liberal tradition of Adam Smith (1723–1790) as well as by the reactionary traditionalism of such scholars as Joseph-Marie de Maistre and Louis de Bonald.[2] In reviewing the origins of Comte's thought, we will focus primarily on the influence of Montesquieu, Turgot, Condorcet, and Saint-Simon, with a brief mention of the traditional and liberal elements in Comte's thinking.

Montesquieu and Comte

In Chapter 17, when we examine the culmination of French sociology in the work of Émile Durkheim, Montesquieu's ideas will be explored in more detail. For the present, we will stress those key concepts that Comte borrowed from Montesquieu. Written in the first half of the eighteenth century, Montesquieu's *The Spirit of Laws* can be considered one of the first sociological works in both style and tone.[3] Indeed, if we wanted to push back by seventy-five years the founding of sociology, we could view *The Spirit* as the first distinctly sociological work. There are, however, too many problems with Montesquieu's great work for it to represent a founding effort. Its significance resides more in the influence it had on scholars of the next century, particularly Comte and Durkheim.

In *The Spirit,* Montesquieu argued that society must be considered a "thing." As a thing, its properties could be discovered by observation and analysis. Thus, for Montesquieu, morals, manners, and customs, as well as social structures, are amenable to investigation in the same way as are things or phenomena in physics and chemistry. Comte's concern with "social facts" and

Durkheim's later proclamation that sociology is the study of social facts, can both be traced to Montesquieu's particular emphasis on society as a thing.

As a thing, Montesquieu argued, society can be understood by discovering the "laws" of human organization. Montesquieu was not completely clear on this point, but the thrust of his argument appears to have been that the laws of society are discoverable in the same way that Newton had, in Montesquieu's mind, uncovered the laws of physical matter. This point became extremely important in Comte's sociology. Indeed, Comte preferred the label *social physics* to *sociology*. In this way he could stress that social science, like the physical sciences, must involve a search for the laws of social structure and change.

Montesquieu also viewed scientific laws as a hierarchy—a notion that, along with Saint-Simon's emphasis, intrigued Comte. Sciences low in the hierarchy, such as physics, will reveal deterministic laws, as did Newton's principle of gravitation. Sciences higher in the hierarchy will, Montesquieu argued, be typified by less determinative laws. The laws of society, therefore, will be more probabilistic. In this way, Montesquieu was able to retain a vision of human freedom and initiative within the context of a scientific inquiry. Comte appears to have accepted much of this argument, for he stressed that the complexity of social phenomena renders strictly determinative laws difficult to discover. For Comte, sociological laws would capture the basic tendencies and directions of social phenomena.

Montesquieu's *The Spirit* also developed a typology of governmental forms. Much of this work is devoted to analyzing the structure and "spirit" (cultural ideas) of three basic governmental forms: republic, monarchy, and despotism. The details of this analysis are less important than is the general thrust of Montesquieu's argument. First, his analysis implies a developmental sequence, although not to the degree evident in the next generation of social thinkers, such as Turgot and Condorcet. Second, Montesquieu's abstract typology was constructed to capture the diversity of empirical systems in the world and throughout history. Thus, by developing a typology with an implicit developmental sequence, he believed he was achieving a better sense of the operation of phenomena, a point central to Comte's scheme. And third, Montesquieu's separate analysis of the "spirit of a nation" and its relation to structural variables, especially political structures, was adopted by Comte in his analysis of societal stages that reveal both "spiritual" (ideas) and "temporal" (structural) components.

In sum, then, Montesquieu laid much of the intellectual foundation on which Comte built his scheme. The emphasis on society as a thing, the concern for laws, the stress on hierarchies of laws, the implicit developmental view of political structures, the belief that empirical diversity can be simplified through analytical typologies, and the recognition that the social world is composed of interdependent cultural and structural forces all found their way into Comtean sociology, as well as into the sociology of Comte's intellectual successors, such as Durkheim. Montesquieu's ideas were transformed, however, in Comte's mind under the influence of other eighteenth-century scholars, particularly Turgot, Condorcet, and Adam Smith.

Turgot and Comte

Jacques Turgot was one of the more influential thinkers of the eighteenth-century Enlightenment. As a scholar, and for a short time as the finance minister of France, Turgot exerted considerable influence within and outside intellectual circles. His work, like that of many scholars of his time, was not published in the conventional sense but initially appeared as a series of lectures or discourses that were, no doubt, informally distributed. Only later, in the early nineteenth century, were many of his works edited and published.[4] Yet his ideas were well known to his contemporaries and his successors, particularly Condorcet, Saint-Simon, and Comte.

In 1750, Turgot presented two discourses at the Sorbonne and established himself as a major social thinker. The first discourse was delivered in July and was titled "The Advantages Which the Establishment of Christianity Has Procured for the Human Race."[5] The second discourse was given on December 11, and was titled "Philosophical Review of the Successive Advances of the Human Mind."[6] Although the first discourse is often discounted in sociological circles, it presented a line of reasoning that would be reflected in Comte's writings. Basically, Turgot argued that religion performed some valuable services for human progress and, although no longer an important ingredient in human development, made subsequent progress possible. Had Christianity not existed, basic and fundamental events such as the preservation of classical literature, the abolishment of cruel treatment of children, the eradication of extreme and punitive laws, and other necessary conditions for further progress would not have been achieved. Comte took this idea and emphasized that each stage of human evolution, particularly the religious or theological, must reach its zenith, thereby laying down the conditions necessary for the next stage of human development.

The second 1750 discourse influenced Comte and other sociologically inclined thinkers more directly. In this discourse, Turgot argued that because humans are basically alike, their perceptions of, and responses to, situations will be similar, and hence they will all evolve along the same evolutionary path. Humanity, he argued, is like an individual in that it grows and develops in a similar way. Thus, the human race will be typified by a slow advancement from a less developed to a more developed state. Naturally, many conditions will influence the rate of growth, or progress, for a particular people. Hence progress will be uneven, with some peoples at one stage of growth and others at a more advanced stage. But in the end, all humanity will reach a "stage of perfection." Comte saw much in this argument because it implicitly accounted for variations and diversity among the populations of the world. Populations differ because their societies are at different stages in a single and unified developmental process.

In this second discourse, Turgot also presented a rather sophisticated economic analysis in which parts are seen as connected in a system or structure. Change occurs as a result of economic forces that inevitably produce alterations in various parts and, hence, in society as a whole. For example, the emergence of agriculture produces an economic surplus, which, in turn, allows for the expansion of the division of labor. Part of this expansion involves commercial activity, which encourages innovations in shipbuilding, and the extensive use of ships causes advances in navigation, astronomy, and geography. The expansion of trade creates towns and cities, which preserve the arts and sciences, thereby encouraging the advance of technologies. Thus, Turgot saw progress in more than a moralistic or metaphorical sense; he recognized that structural changes in one area of a social system, especially in economic activity, create pressures for other changes, with these pressures causing further changes, and so on. This mode of analysis anticipated Marx's economic determinism by a hundred years, and it influenced evolutionary theorizing in France for 150 years.

Turgot's next works made more explicit the themes developed in these two early discourses. *On Universal History*[7] and *On Political Geography*[8] were apparently written near the end of Turgot's stay at the Sorbonne, perhaps around 1755. *On Political Geography* is most noteworthy for its formulation of the three stages of human progress, an idea that became a central part of Comte's view of human evolution. Moreover, Turgot used the notion of universal stages not only to explain human development but also to account for the diversity of human societies, an analytical tactic similar to that used by Montesquieu. All societies of the world are, Turgot argued, at one of three stages, "hunters, shepherds, or husbandmen." In *On Universal History,* he developed the notion of three stages even further, presenting several ideas that became central to Comte's sociology. First, Turgot divided evolution into "mental" and "structural" progression so that development involves change in economic and social structures as well as in idea systems. Second, the progress of society is explicitly viewed as the result of internal structural and cultural forces rather than as a result of intervention by a deity. Third, change and progress can be understood as abstract laws that depict the nature of stasis and change in social systems. And fourth, Turgot's empirical descriptions of what we would now call hunting, horticultural, and agrarian societies are highly detailed and filled with discussion of how structural and cultural conditions at one point create pressures for new structures and ideas at the next point.

Later in his career, probably during the 1760s, Turgot turned his analytical attention increasingly to economic matters. Around 1766 he formulated *Reflections on the Formation and Distribution of Wealth,*[9] which parallels and, to some extent, anticipates the ideas developed by the classical economists in England. Turgot's advocacy of free trade, his analysis of how supply and demand influence prices, and his recognition of the importance of entrepreneurs to economic development are extremely sophisticated for his time. From this analysis, an implicit fourth stage of development is introduced: As capital increasingly becomes concentrated in the hands of entrepreneurs in advanced agricultural societies, a commercial type of society is created. Much of Turgot's description of the transition to, and the arrangements in, this commercial stage were restated by Marx, Spencer, and Comte in the nineteenth century, although only Comte would be directly influenced by Turgot's economic analyses. Yet Comte never expanded Turgot's great insights into the importance of economic variables on the organization and change of society. Only the emphasis on entrepreneurial activity in industrial societies was retained, which, in the end, made Comte's analysis of structural change superficial compared with that of Turgot, Spencer, and Marx.

In sum, then, Turgot dramatically altered the course of social thought in the eighteenth century. Extending Montesquieu's ideas in subtle but nevertheless important ways, Turgot developed a mode of analysis that influenced Comte both directly and indirectly. The idea of three stages of progress, the notion that structures at one stage create the necessary conditions for the next, and the stress on the lawlike nature of progress became integral parts of Comte's sociology. Much of Turgot's influence on Comte, however, could have been indirect, working its way through Condorcet, whose work was greatly affected by Turgot.[10]

Condorcet and Comte

Jean Condorcet was a student, friend, and great admirer of Turgot, so it is not surprising that his work represents an elaboration of ideas developed by Turgot. Throughout Condorcet's career, which flowered during the French Revolution and then foundered, Condorcet concerned himself with the relation of ideas to social action. In particular he emphasized the importance of science as a means to achieve the "infinite perfectability" of the human race. The culmination of his intellectual career was the short and powerful *Sketch for a Historical Picture of the Progress of the Human Mind*,[11] which was written while he was in hiding from an unfavorable political climate.

Progress of the Human Mind was written in haste by a man who knew he would soon die, yet it is by far his best work. In its hurried passages, Condorcet traced ten stages of human development, stressing the progression of ideas from the emergence of language and simple customs to the development and elaboration of science. Condorcet felt that with the development of science and its extension to the understanding of society, humans could now direct their future toward infinite perfectibility. Human progress, Condorcet argued, "is subject to the same general laws that can be observed in the development of the faculties of the individual," and once these faculties are fully developed, "the perfectability of man is truly indefinite; and . . . the progress of this perfectability, from now onwards independent of any power that might wish to halt it, has no other limit than the duration of the glove upon which nature has cast us?"[12]

The historical details of Condorcet's account are little better than Turgot's, but several important changes in emphasis influenced Comte's thinking. First, Condorcet's stress on the movement of ideas was retained in Comte's view of progress. Second, the emphasis on science as representing a kind of intellectual takeoff point for human progress was reaffirmed by Comte. And third, the almost religious faith in science as the tool for constructing the "good society" became central to Comte's advocacy. Thus, Comte's great synthesis took elements from Turgot's and Condorcet's related schemes. Turgot's law of the three stages of progress was used instead of Condorcet's ten stages, but Condorcet's emphasis on ideas and on the use of science to realize the laws of progress was preserved.

Yet Comte's synthesis was, in some respects, merely an extension of ideas that his master, Saint-Simon, had developed in rough form. And to understand fully the origins of Comte's thought, and hence the emergence of sociology, we must explore the volatile relationship between Saint-Simon and Comte. Sociology was born as a self-conscious field of inquiry from their interaction.

Saint-Simon and Comte

In many ways, Claude-Henri de Saint-Simon represented a bridge between the eighteenth century and the early nineteenth century. Born into an aristocratic family, Saint-Simon initially pursued a nonacademic career. He fought with the French in the American Revolution; traveled the world; proposed a number of engineering projects, including the Panama Canal; was politically active during the French Revolution; became a land speculator in the aftermath of the 1789 Revolution; and amassed and then lost a large fortune. Only late in life, at the turn of the century, did he become a dedicated scholar.[13]

The relationship between Saint-Simon's and Comte's ideas has been debated ever since their violent quarrel and separation in 1824. Just what part of Saint-Simon's work is Comte's, and vice versa, will never be completely determined. But it is clear that between 1800 and 1817, Saint-Simon's ideas were not influenced by Comte because the young Comte did not join the aging Saint-Simon as a secretary, student, and collaborator until 1817. In the seven years between 1817 and 1824, Comte's and Saint-Simon's ideas were intermingled, but we can see in the pre–1817 works of Saint-Simon many of the ideas that became a part of Comte's sociology.[14] The most reasonable interpretation of their collaboration is that Comte took many of the crude and unsystematic ideas of Saint-Simon, refined and polished them in accordance with his greater grasp of history and science, and extended them in small but critical ways as a result of his exposure to Montesquieu, Turgot, Condorcet, Adam Smith, and the traditionalists. To appreciate Saint-Simon's unique contribution to the emergence of sociology, then, we must examine first the period between 1800 and 1817 and then the post-1817 work, with speculation on the contribution by Comte to this later work.

Saint-Simon's Early Work Saint-Simon read Condorcet carefully and concluded that the scientific revolution had set the stage for a science of social organization.[15] Saint-Simon argued in his first works that the study of humankind and society must be a "positive" science, based on empirical observation. Like many others of this period, Saint-Simon saw the study of society as a branch of physiology because society is a type of organic phenomenon. Like any organic body, society is governed by natural laws of development, which are to be revealed by scientific observation. As an *organism,* then, society would be studied by investigating social organization, with particular emphasis on the nature of growth, order, stability, and abnormal pathologies.[16]

Saint-Simon saw that such a viewpoint argued for a three-part program: (1) "a series of observations on the course of civilization" must be the starting point of the new science; (2) from these observations, the laws of social organization would be revealed; and (3) on the basis of these laws, humans could construct the best form of social organization. From a rather naive and ignorant view of history,[17] Saint-Simon developed a law of history in which ideas move from a polytheistic stage to a Christian theism and then to a positivistic stage. In his eye, each set of ideas in human history had been essential to maintaining social order, and with each transition came a period of crisis. The transition to positivism, therefore, revolved around the collapse of the feudal order and its religious underpinnings and the incomplete establishment of an industrial order in European societies, with its positivistic culture of science.

In analyzing this crisis in European society, Saint-Simon noted that scientific observations had first penetrated astronomy, then physics and chemistry, and finally physiology, including both biological and social organs.[18] With the application of the scientific method to social organization, the traditional order must give way to a new system of ideas. Transitional attempts to restore order, such as the "legal-metaphysical" ideas of the eighteenth century, must yield to a "terrestrial morality" based on the ideas of positivism—that is, the use of observations to formulate, test, and implement the laws of social organization.[19]

Founded on a terrestrial morality, this new order resulted from a collaboration of scientists and industrialists. In Saint-Simon's early thought, scientists were to be the theoreticians, and industrialists were to be the engineers who performed many of the practical tasks of reconstructing society. Indeed, scientists and industrialists were to be the new priests for the secular religion of positivism. Saint-Simon's thought on social reorganization, however, underwent considerable change between 1814 and 1825, when he fell ill and died. The increasingly political and religious tone of his writings alienated the young Comte, who saw the more detailed study of history and the movement of ideas as necessary for the formulation of the scientific laws of social organization. Ironically, Comte's own work, later in his career, took on the same religious fervor and extremes as Saint-Simon's last efforts.

Comte's early sociological work owed much to Saint-Simon's initial period of intellectual activity. The law of the three stages became even more prominent; the recognition of the successive penetration of positivism into astronomy, physics, chemistry, and biology was translated into a hierarchy of sciences, with physics at the bottom and sociology at the top; and the belief that sociology could be used to reconstruct industrial society became part of Comte's program. Comte, however, rejected much of Saint-Simon's argument. In particular, Comte did not accept the study of social organization as a part of physiology; rather, he argued that sociology was a distinct science with its own unique principles. In this vein, he also rejected Saint-Simon's belief that one law of all the universe could be discovered; instead, Comte recognized that each science had its own unique subject matter, which could be fully understood only through its own laws and principles. These objections aside, much of Comte's early work represented the elaboration of ideas developed and then abandoned by Saint-Simon as the aging scholar became increasingly absorbed in the task of reconstructing society.

Saint-Simon's Later Work After 1814, Saint-Simon turned increasingly to political and economic commentary. He established and edited a series of periodicals to propagate his ideas on the use of scientists and industrialists to reconstruct society.[20] His terrestrial morality thus became elaborated into a plan for political, economic, and social reform.

The emphasis was on *terrestrial* because Saint-Simon argued that the old supernatural basis for achieving order could no longer prevail in the positivistic age. Yet by his death, he recognized that a "religious sense" and "feeling" are essential to the social order. People must have faith and believe in a common set of ideas, a theme that marked all French sociology in the nineteenth century. The goal of terrestrial morality, therefore, is to create the functional equivalent of religion with positivism. Scientists and artists are to be the priests and the "spiritual" leaders,[21] and industrialists are to be the "temporal" leaders and are to implement the spiritual program by applying scientific methods to production and the organization of labor.

For Saint-Simon, terrestrial morality had both spiritual and temporal components. Spiritual leaders give a sense of direction and a new religious sense to societal activity. Temporal leaders ensure the organization of industry in ways that destroy hereditary privilege and give people an equal chance to realize their full potentials. The key to Saint-Simon's program, then, was to use science as the functional equivalent of religion and to destroy the idle classes so that each person worked to his or her full potential. Although he visualized considerable control of economic and social activity by government (to prevent exploitation of workers by the "idle"), Saint-Simon also believed that people should be free to realize their potentials. Thus, his doctrine is a mixture of free enterprise economics and a tempered but heavy dose of governmental control (an emphasis that has often led commentators to place him in the socialist camp).

Saint-Simon's specific political, educational, and social programs were, even for his time, naive and utopian, but they nevertheless set into motion an entire intellectual movement after his death. Comte, however, was highly critical of Saint-Simon's later writings, and Comte waged intellectual war with the Saint-Simonians after 1825. Although Comte wrote much of Saint-Simon's work between 1817 and 1824, Comte's contribution is recognizable because it is more academic and reasoned than is Saint-Simon's advocacy.[22] Later, Comte's own work took on the same religious extremes as Saint-Simon's. Thus Comte clearly accepted in delayed and subliminal form much of Saint-Simon's advocacy of science as a functional substitute for religion.

Comte's real contribution comes not from Saint-Simon's political commentary but, rather, from his systematization of Saint-Simon's early historical

and scientific work because from this effort, sociology as a self-conscious discipline emerged.

LIBERAL AND CONSERVATIVE TRENDS
IN COMTE'S THOUGHT

We can see that Saint-Simon's work encompassed both liberal and conservative elements. He advocated change and individual freedom, yet he desired
that change produce a new social order and that individual freedom be subordinated to the collective interests of society. Comte's work also revealed this
mixture of liberal and conservative elements, in that the ideas of economic liberals, such as Adam Smith, and conservatives or traditionalists, like de Maistre
and de Bonald, all played a part in his intellectual scheme.

Liberal Elements in Comte's Thought

In England, Adam Smith (1723–1790) had the most decisive effect on social
thought in his advocacy of an economic system consisting of free and competitive markets. His *Wealth of Nations* (1776), however, is more than a simple
model of early capitalism; the fifth book reveals a theory of moral sentiments
and raises a question that concerned Comte and, later, Durkheim: How can
society be held together at the same time that the division of labor compartmentalizes individuals? For Smith, this dilemma was not insurmountable,
whereas for French sociologists who had experienced the disintegrative effects
of the Revolution and its aftermath, splitting society into diverse occupations
posed a real intellectual problem. For French sociologists, the solution to this
liberal dilemma involved creating a strong state that coordinated activities, preserved individual liberties, and fostered a set of unifying values and beliefs.

Comte also absorbed liberal ideas from the French followers of Smith, particularly Jean-Baptiste Say, who had seen the creative role played by entrepreneurs in the organization of other economic elements (land, labor, capital).
Saint-Simon appears to have had a notion of entrepreneurship in his proposal
that the details of societal reconstruction be left to "industrialists," but Say's
explicit formulation of the creative coordination of labor and capital by those
with "industry" explicitly influenced Comte's vision of how a better society
could be created by entrepreneurs.[23]

Traditional Elements in Comte's Thought

Saint-Simon had attacked those who, in the turmoil of the Revolution,
wanted to return to the Old Regime. Writing from outside France, Catholic
scholars such as de Maistre and de Bonald argued that the Revolution had
destroyed the structural and moral fiber of society.[24] Religious authority had
not been replaced by an alternative, the order achieved from the old social
hierarchies had not been reestablished, and the cohesiveness provided by local

communities and groups had been allowed to disintegrate. Both Saint-Simon and Comte, as well as an entire generation of French thinkers, agreed with the traditionalist's diagnosis of the problem but disagreed with the proposed solution. The traditionalists believed the solution was the reinstatement of religion, hierarchy, and traditional local groupings (on the feudal model).

Although Comte became an atheist in his early teens, he had been reared as a Catholic; hence, he shared with many of the traditionalists a concern about order and spiritual unity. The Enlightenment and liberal economic doctrines had also influenced him, and thus, he saw that a return to the old order was not possible. Rather, it was necessary to create the functional equivalent of religious authority and to reestablish nonascriptive hierarchies and communities that would give people an equal chance to realize their full potentials. For Comte, then, the religious element is to be secular and positivistic; hierarchies are to be based on ability and achievement rather than on ascriptive privilege; and community is to be re-created through the solidarity of industrial groups. He thus gave the traditionalists' concerns a liberal slant, although his last works were decidedly authoritarian in tone, perhaps revealing the extent to which the traditionalists' ideas had remained with him.

In reviewing the specific thinkers who preceded Comte, we can see that the emergence of sociology was probably inevitable. Science had become too widespread to be suppressed by a return to religious orthodoxy, and the economic and political transformations of society caused by industrialization and urbanization needed explanation. All that was necessary was for one scholar to take that final step and seek to create a science of society. Drawing from the leads provided by his predecessors, Auguste Comte took this final step. In so doing, he gave the science of society a name and a vision of how it should construct theory. Now, let us turn to the substance of Comte's vision.

THE SOCIOLOGY OF AUGUSTE COMTE

Auguste Comte's works can be divided into two distinct phases: (1) the early scientific stage, between 1820 and 1842 and (2) the moralistic and quasi-religious stage, which culminated between 1851 and 1854. The scientific phase involved the publication of several important articles and then, between 1830 and 1842, the five volumes of *The Course of Positive Philosophy,*[25] where the science of society was formally established. The second period in Comte's life is marked by personal tragedy and frustration; during this period he wrote *System of Positive Polity,*[26] which represented his moralistic view of how society should be reconstructed. Despite the excessive moral preachings of this work, its vision of science as the tool for reconstructing society was an important element in sociology's mission as seen by later generations of French sociologists.

In our review of Comte's work, we will focus primarily on his purely sociological works—that is, on those from his scientific phase. We will not ignore his more moralistic efforts, but they did not contribute to the emergence of sociological theory. Thus, we will examine Comte's most important early essay, "Plan of the Scientific Operations Necessary for Reorganizing Society,"[27] which was written in 1822, just before his break with Saint-Simon. This essay contains the germs of both Comte's scientific and moralistic phases. Then, and for the bulk of our examination of his work, we will explore his greatest treatise, *The Course of Positive Philosophy.* Finally, we will look briefly at *Positive Polity.*

Comte's Early Essays

It is sometimes difficult to separate Comte's early essays from those of Saint-Simon, because the aging master often put his name on works written by the young Comte. Yet the 1822 essay, "Plan of the Scientific Operations Necessary for Reorganizing Society," is clearly Comte's and represents the culmination of his thinking while working under Saint-Simon. This essay also anticipates, and presents an outline of, the entire Comtean scheme as it was to unfold over the succeeding decades.

In this essay, Comte argued that it was necessary to create a "positive science" based on the model of other sciences. This science would ultimately rest on empirical observations, but, like all science, it would formulate the laws governing the organization and movement of society, an idea implicit in Montesquieu's *The Spirit of Laws.* Comte initially called this new science *social physics.* Once the laws of human organization have been discovered and formulated, Comte believed that these laws could be used to direct society. Scientists of society are thus to be social prophets, indicating the course and direction of human organization.

Comte felt that one of the most basic laws of human organization was the "law of the three stages," a notion clearly borrowed from Turgot, Condorcet, and Saint-Simon. He termed these stages the *theological-military, metaphysical-judicial,* and *scientific-industrial* or "positivistic." Each stage is typified by a particular "spirit"—a notion that first appeared with Montesquieu and was expanded by Condorcet—and by temporal or structural conditions. Thus, the theological-military stage is dominated by ideas that refer to the supernatural while being structured around slavery and the military. The metaphysical-judicial stage, which follows from the theological and represents a transition to the scientific, is typified by ideas that refer to the fundamental essences of phenomena and by elaborate political and legal forms. The scientific-industrial stage is dominated by the "positive philosophy of science" and industrial patterns of social organization.

Several points in this law were given greater emphasis in Comte's later work. First, the social world reveals both cultural and structural dimensions, with the nature of culture or idea systems being dominant—an idea probably taken from Condorcet. Second, idea systems, and the corresponding structural arrangements that they produce, must reach their full development before the next stage of human evolution can occur. Thus, one stage of development creates the necessary conditions for the next. Third, there is always a period of crisis and conflict as systems move from one stage to the next because elements of the previous stage conflict with the emerging elements of the next stage. Fourth, movement is always a kind of oscillation, for society "does not, properly speaking, advance in a straight line."

These aspects of the law of three stages convinced Comte that cultural ideas about the world were subject to the dictates of this law. All ideas about the nature of the universe must move from a theological to a scientific, or positivistic, stage. Yet some ideas about different aspects of the universe move more rapidly through the three stages than others do. Indeed, only when all the other sciences—first astronomy, then physics, later chemistry, and finally physiology—have successively reached the positive stage will the conditions necessary for social physics have been met. With the development of this last great science, it will become possible to reorganize society by scientific principles rather than by theological or metaphysical speculations.

Comte thus felt that the age of sociology had arrived. It was to be like Newton's physics, formulating the laws of the social universe. With the development of these laws, the stage was set for the rational and scientific reorganization of society. There is much of Saint-Simon in this advocacy, but Comte felt that Saint-Simon was too impatient in his desire to reorganize society without the proper scientific foundation. The result was Comte's *Course of Positive Philosophy,* which sought to lay the necessary intellectual foundation for the science of society.

Comte's Course of Positive Philosophy

Comte's *Course of Positive Philosophy* is more noteworthy for its advocacy of a science of society than for its substantive contribution to understanding how patterns of social organization are created, maintained, and changed. *Positive Philosophy* more nearly represents a vision of what sociology can become than a well-focused set of theoretical principles. In reviewing this great work, then, we will devote most of our attention to how Comte defined sociology and how he thought it should be developed. Accordingly, we will divide our discussion into the following sections: (1) Comte's view of sociological theory, (2) his formulation of sociological methods, (3) his organization of sociology, and (4) his advocacy of sociology.

Comte's View of Sociological Theory

As a descendant of the French Enlightenment, Comte was impressed, as were many of the philosophes, with the Newtonian revolution. Thus, he argued for a particular view of sociological theory: All phenomena are subject to invariable natural laws, and sociologists must use their observations to uncover the laws governing the social universe, in much the same way as Newton had formulated the law of gravity. As Comte emphasized in the opening pages of *Positive Philosophy,*

The first characteristic of Positive Philosophy is that it regards all phenomena as subject to invariable natural *Laws*. Our business is—seeing
how vain is any research into what are called *Causes* whether first or
final—to pursue an accurate discovery of these Laws, with a view to
reducing them to the smallest possible number. By speculating upon
causes, we could solve no difficulty about origin and purpose. Our real
business is to analyse accurately the circumstances of phenomena, and to
connect them by the natural relations of succession and resemblance. The
best illustration of this is in the case of the doctrine of Gravitation.[28]

Several points are important in this view of sociological theory. First, sociological theory is not to be concerned with causes per se but, rather, with the laws
that describe the basic and fundamental relations of properties in the social
world. Second, there is an explicit rejection of "final causes"—that is, analysis of
the results of a particular phenomenon for the social whole. There is a certain
irony in this disavowal because Comte's more substantive work helped found
sociological functionalism, a mode of analysis that often examines the functions
or final causes of phenomena. Third, there is a clear recognition that the goal of
sociological activity is to reduce the number of theoretical principles by seeking
only the most abstract and only those that pertain to understanding fundamental properties of the social world—a point that, unfortunately, has been lost in
modern sociology's concern with introducing multiple and manifold variables
into theoretical activity. Comte thus held a vision of sociological theory as based
on the model of the natural sciences, particularly the physics of his time. For this
reason he preferred the term *social physics* to *sociology*.[29]

The laws of social organization and change, Comte felt, will be discovered,
refined, and verified through a constant interplay between theory and empirical organization. For, as he observed in the opening pages of *Positive
Philosophy*, "if it is true that every theory must be based upon observed facts, it
is equally true that facts cannot be observed without the guidance of some theory."[30] In later pages, Comte became even more assertive and argued that what
we might now term *raw empiricism*, or the collection of data for its own sake,
runs counter to the goals of science. He saw strict empiricism as an absolute
hindrance to the development of sociological theory. In a passage that sounds
distinctly modern, he noted:

> The next great hindrance to the use of observation is the empiricism
> which is introduced into it by those who, in the name of impartiality,
> would interdict the use of any theory whatever. No other dogma could be
> more thoroughly irreconcilable with the spirit of the positive philosophy.
> . . . No real observation of any kind of phenomena is possible, except in as
> far as it is first directed, and finally interpreted, by some theory.[31]

And he concluded:

> Hence it is clear that, scientifically speaking, all isolated, empirical obser-
> vation is idle, and even radically uncertain; that science can use only those
> observations which are connected, at least hypothetically, with some law.[32]

For Comte, then, sociology's goal was to seek to develop abstract theoretical principles. Observations of the empirical world must be guided by such principles, and abstract principles must be tested against the empirical facts. Empirical observations that are conducted without this goal in mind are not useful in science. Theoretical explanation of empirical events thus involves seeing how they are connected in lawlike ways. For social science "endeavors to discover . . . the general relations which connect all social phenomena; and each of them is *explained*, in the scientific sense of the word, when it has been connected with the whole of the existing situation."[33]

Comte held a somewhat ambiguous view of how such an abstract science should be "used" in the practical world of everyday affairs. He clearly intended that sociology must initially establish a firm theoretical foundation before making efforts to use the laws of sociology for social engineering. In volume 1 of *Positive Philosophy,* he stressed,

> We must distinguish between the two classes of Natural science—the abstract or general, which have for their object the discovery of the laws which regulate phenomena in all conceivable cases, and the concrete, particular, or descriptive, which are sometimes called Natural sciences in a restricted sense, whose function it is to apply these laws to the actual history of existing beings. The first are fundamental, and our business is with them alone; as the second are derived, and however important, they do not rise to the rank of our subjects of contemplation.[34]

Comte believed that sociology must not allow its scientific mission to be confounded by empirical descriptions or by an excessive concern with a desire to manipulate events. Once sociology is well established as a theoretical science, its laws can be used to "modify" events in the empirical world. Indeed, such was the historic mission of social physics. As Comte's later works testify, he took this mission seriously, and at times to extremes. But his early work is filled with more reasoned arguments for using laws of social organization and change as tools for creating a variety of new social arrangements. He stressed that the complexity of social phenomena gives them more variation than either physical or biological phenomena have, and hence it would be possible to use the laws of social organization and change to modify empirical events in a variety of directions.[35]

In sum, then, Comte believed that sociology could be modeled after the natural sciences. Sociology could seek and discover the fundamental properties and relations of the social universe, and like the other sciences, it could express these in a small number of abstract principles. Observations of empirical events could be used to generate, confirm, and modify sociology's laws. Once well-developed laws had been formulated, they could be used as tools or instruments to modify the social world.

Comte's Formulation of Sociological Methods

Comte was the first social thinker to take methodological questions seriously—that is, how facts about the social world are to be gathered and used to develop, as well as to test, theoretical principles. He advocated four methods in the new science of social physics: (1) observation, (2) experimentation, (3) comparison, and (4) historical analysis.[36]

Observation For Comte, positivism was based on use of the senses to observe *social facts*—a term that the next great French theorist, Émile Durkheim, made the center of his sociology. Much of Comte's discussion of observation involves arguments for the "subordination of Observation to the statical and dynamical laws of phenomena"[37] rather than a statement on the procedures by which unbiased observations should be conducted. He argued that observation of empirical facts, when unguided by theory, will prove useless in the development of science. He must be given credit, however, for firmly establishing sociology as a science of social facts, thereby liberating thought from the debilitating realm of morals and metaphysical speculation.

Experimentation Comte recognized that artificial experimentation with whole societies, and other social phenomena, was impractical and often impossible. But, he noted, natural experimentation frequently "takes place whenever the regular course of the phenomenon is interfered with in any determinate manner."[38] In particular, he thought that, much as is the case in biology, pathological events allowed "the true equivalent of pure experimentation" in that they introduced an artificial condition and allowed investigators to see normal processes reasserting themselves in the face of the pathological condition. Much as the biologist can learn about normal bodily functioning from the study of disease, so social physicists can learn about the normal processes of society from the study of pathological cases. Thus, although Comte's view of "natural experimentation" was certainly deficient in the logic of the experimental method, it nonetheless fascinated subsequent generations of scholars.

Comparison Just as comparative analysis had been useful in biology, so the comparison of social forms with those of lower animals, with coexisting states, and with past systems could generate considerable insight into the operation of the social universe. By comparing elements that are present and absent, and similar or dissimilar, knowledge about the fundamental properties of the social world can be achieved.

Historical Analysis Comte originally classified historical analysis as a variation of the comparative method (that is, comparing the present with the past). But his "law of the three stages" emphasized that the laws of social dynamics could ultimately be developed only with careful observations of the historical movement of societies.

In sum, then, Comte saw these four basic methods as appropriate to sociological analysis. His formulation of the methods is quite deficient by modern standards, but we should recognize that before Comte, little attention had been paid to how social facts were to be collected. Thus, although the specifics of Comte's methodological proposals are not always useful, their spirit and intent are important. Social physics was, in his vision, to be a theoretical science capable of formulating and testing the laws of social organization and change. His formulation of sociology's methods added increased credibility to this claim.

Comte's Organization of Sociology

Much as Saint-Simon had emphasized, Comte saw sociology as an extension of the study of "organisms" in biology to "social organs." Hence, sociology was to be the study of social *organization*. This emphasis forces the recognition that society is an "organic whole" whose components stand in relation to one another. To study these parts in isolation is to violate the essence of social organization and to compartmentalize inquiry artificially. As Comte emphasized, "there can be no scientific study of society, either in its conditions or its movements, if it is separated into portions, and its divisions are studied apart."[39]

Implicit in this emphasis is a mode of analysis that later became known as *functionalism*. As biology's prestige grew during the nineteenth century, attempts at linking sociological analysis to the respected biological sciences increased. Eventually scholars began asking, What is the function of a structure for the body social? That is, what does a structure "do for" the social whole? Comte implicitly asked such questions and even offered explicit analogies to encourage subsequent organismic analogizing. For example, his concern with social pathology revealing the normal operation of society is only one illustration of a biological mode of reasoning. In his later work, Comte explicitly argued in biological terms when he viewed various structures as analogous to "elements, tissues, and organs" of biological organisms.[40] In his early works, however, this organismic analogizing is limited to dividing social physics into statical and dynamical analysis.

This division, we suspect, represents a merger of Comte's efforts to build sociology on biology and to retain his heritage from the French Enlightenment. As a scholar who was writing in the tumultuous aftermath of the French Revolution, he was concerned with order and stability. The order of biological organisms, with their interdependent parts and processes of self-maintenance, offered him a vision of how social order should be constructed. Yet the Enlightenment had emphasized "progress" and movement of social systems, holding out the vision of better things to come. For this reason Comte was led to emphasize that "ideas of Order and Progress are, in Social Physics, as rigorously inseparable as the ideas of Organization and Life in Biology: from whence indeed they are, in a scientific view, evidently derived."[41] And thus he divided sociology into (1) social statics (the study of social order) and (2) social dynamics (the study of social progress and change). We will explore these two aspects of his sociology in more detail.

Social Statics Comte defined social statics as the study of social structure, its elements, and their relations. He first analyzed "individuals" as elements in the analysis of social structure. Generally, he viewed the individual as a series of capacities and needs, some innate and others acquired through participation in society.[42] He did not view the individual as a "true social unit";

indeed, he relegated the study of the individual to biology—an unfortunate oversight because it denied the legitimacy of psychology as a distinct social science. The most basic social unit, he argued, is "the family." It is the most elementary unit, from which all other social units ultimately evolved:

> As every system must be composed of elements of the same nature with itself, the scientific spirit forbids us to regard society as composed of individuals. The true social unit is certainly the family—reduced, if necessary, to the elementary couple which forms its basis. This consideration implies more than the physiological truth that families become tribes, and tribes become nations: so that the whole human race might be conceived of as the gradual development of a single family. . . . There is a political point of view from which also we must consider this elementary idea, inasmuch as the family presents the true germ of the various characteristics of the social organism.[43]

Comte took a strong sociologistic position that social structures could not be reduced to the properties of individuals. Rather, social structures are composed of other structures and can be understood only as the properties of, and relations among, these other structures. Comte's analysis of the family then moves to descriptions of its structure—first the sexual division of labor and then the parental relation. The specifics of his analysis are not important because they are flawed and inaccurate. Far more important is the view of structure that he implied: Social structures are composed of substructures and develop from the elaboration of simpler structures.

After establishing this basic point, Comte moved to the analysis of societal structures. His opening remarks reveal his debt to biological analysis and the functional orientation it inspired:

> The main cause of the superiority of the social to the individual organism is according to an established law; the more marked is the specialization of the various functions fulfilled by organs more and more distinct, but interconnected; so that unity of aim is more and more combined with diversity of means.[44]

Thus, as social systems develop, they become increasingly differentiated, and yet like all organisms, they maintain their integration. This view of social structure led Comte to the problem that Adam Smith had originally suggested with such force: How is integration among parts maintained despite increasing differentiation of functions? This question occupied French sociology in the nineteenth century, culminating in Durkheim's theoretical formulations. As Comte emphasized,

> If the separation of social functions develops a useful spirit of detail, on the one hand, it tends on the other, to extinguish or to restrict what we may call the aggregate or general spirit. In the same way, in moral relations, while each is in close dependence on the mass, he is drawn away from it by the expansion of his special activity, constantly recalling him to

his private interest, which he but very dimly perceives to be related to the public.[45]

Comte's proposed solution to this problem reveals much about how he viewed the maintenance of social structure. First, the potentially disintegrating impact of social differentiation is countered by the centralization of power in government, which will then maintain fluid coordination among system parts. Second, the actions of government must be more than "material"; they must also be "intellectual and moral."[46] Hence, human social organization is maintained by (1) mutual dependence of system parts on one another, (2) centralization of authority to coordinate exchanges of parts, and (3) development of a common morality or spirit among members of a population. To the extent that differentiating systems cannot meet these conditions, pathological states are likely to occur. Figure 2.1 shows Comte's implicit model of social statics.

In presenting this analysis, Comte felt that he had uncovered several laws of social statics because he believed that differentiation, centralization of power, and development of a common morality were fundamentally related to the maintenance of the social order. Although he did not carry his analysis far, he presented both Herbert Spencer and Durkheim with one of the basic theoretical questions in sociology and the broad contours of the answer.

Social Dynamics Comte appeared far more interested in social dynamics than in statics, for "the dynamical view is not only the more interesting . . . , but the more marked in its philosophical character, from its being more distinguished from biology by the master-thought of continuous progress, or rather of the gradual development of humanity."[47] Social dynamics studies the "laws of succession," or the patterns of change in social systems over time. In this context Comte formulated the details of his law of the three stages, in which idea systems, and their corresponding social structural arrangements, pass through three phases: (1) the theological, (2) the metaphysical, and (3) the positivistic. The basic cultural and structural features of these stages are summarized in Table 2.1.

Table 2.1 ignores many details that have little relevance to theory,[48] but the table communicates, in a rough fashion, Comte's view of the laws of succession.

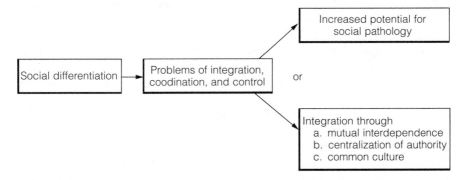

FIGURE 2.1 Comte's Implicit Model of Social Statics

Table 2.1 Comte's "Law of the Three Stages"

SYSTEM	STAGES		
	Theological	**Metaphysical**	**Positivistic**
1. Cultural (moral) system			
a. Nature of ideas	Ideas are focused on nonempirical forces, spirits, and beings in the supernatural realm.	Ideas are focused on the essences of phenomena and rejection of appeals to supernatural.	Ideas are developed from observation and constrained by the scientific method; speculation not based on observation of empirical facts is rejected.
b. Spiritual leaders	Priests	Philosophers	Scientists
2. Structural (temporal) system			
a. Most prominent units	Kinship	State	Industry
b. Basis of integration	Attachment to small groups and religious spirit; use of coercive force to sustain commitment to religion	Control by state, military, and law	Mutural dependence; coordination of functions by state and general spirit.

Several points should be noted: First, each stage sets the conditions for the next. For example, without efforts to explain references to the supernatural, subsequent efforts at more refined explanations would not have been possible; or, without kinship systems, subsequent political, legal, and military development would not have occurred, and the modern division of labor would not have been possible. Second, the course of evolution is additive: New ideas and structural arrangements are added to, and build on, the old. For instance, kinship does not disappear, nor do references to the supernatural. They are first supplemented, and then dominated, by new social and cultural arrangements. Third, during the transition from one stage to the next, elements of the preceding stage conflict with elements of the emerging stage, creating a period of anarchy and turmoil. Fourth, the metaphysical stage is a transitional stage, operating as a bridge between theological speculation and positivistic philosophy. Fifth, the nature of cultural ideas determines the kinds of social structural (temporal) arrangements, circumscribing what social arrangements are possible. And sixth, with the advent of the positivistic stage, true understanding of how society operates is possible, allowing the manipulation of society in accordance with the laws of statics and dynamics.

Although societies must eventually pass through these three stages, they do so at different rates. Probably the most important of the variable empirical conditions influencing the rate of societal succession is population size and density, an idea taken from Montesquieu and later refined by Durkheim. Thus, Comte felt that he had discovered the basic law of social dynamics in his analysis of the three stages, and, coupled with the laws of statics, a positivistic science of society—that is, social physics or sociology—would allow for the reorganization of the tumultuous, transitional, and conflictual world of the early nineteenth century.

Comte's Advocacy of Sociology

Comte's *Positive Philosophy* can be viewed as a long and elaborate advocacy for a science of society. Most of the five volumes review the development of other sciences, showing how sociology represents the culmination of positivism. As the title, *Positive Philosophy,* underscores, Comte was laying a philosophical foundation and justification for all science and then using this foundation as a means for supporting sociology as a true science. His advocacy took two related forms: (1) to view sociology as the inevitable product of the law of the three stages and (2) to view sociology as the "queen science," standing at the top of a hierarchy of sciences. These two interrelated forms of advocacy helped legitimate sociology in the intellectual world and should, therefore, be examined briefly.

Comte saw all idea systems as passing through the theological and metaphysical stages and then moving into the final, positivistic, stage. Ideas about all phenomena must pass through these phases, with each stage setting the conditions for the next and with considerable intellectual turmoil occurring during the transition from one stage to the next. Ideas about various phenomena, however, do not pass through these stages at the same rate, and, in fact, a positivistic stage in thought about one realm of the universe must often be reached before ideas about other realms can progress to the positivistic stage. As the opening pages of *Positive Philosophy* emphasize,

> We must bear in mind that the different kinds of our knowledge have passed through the three stages of progress at different rates, and have not therefore arrived at the same time. The rate of advance depends upon the nature of knowledge in question, so distinctly that, as we shall see hereafter, this consideration constitutes an accessory to the fundamental law of progress. Any kind of knowledge reaches the positive stage in proportion to its generality, simplicity, and independence of other departments.[49]

Thus, thought about the physical universe reaches the positive stage before conceptions of the organic world do because the inorganic world is simpler and organic phenomena are built from inorganic phenomena. In Comte's view, then, astronomy was the first science to reach the positivistic stage, then came physics, next came chemistry, and after these three had reached the positivistic (scientific) stage, thought about organic phenomena could become more positivistic. The first organic science to move from the metaphysical to the positivistic stage was biology, or physiology. Once biology became a positivistic doctrine, sociology could move away from the metaphysical speculations of the seventeenth and eighteenth centuries (and the residues of earlier theological thought) toward a positivistic mode of thought.

Sociology has been the last to emerge, Comte argued, because it is the most complex and because it has had to wait for the other basic sciences to reach the positivistic stage. For the time, this argument represented a brilliant advocacy for a separate science of society, while it justified the lack of scientific rigor in social thought when compared with the other sciences. Moreover, though dependent on, and derivative of, evolutionary advances in the other sciences, sociology will study phenomena that distinguish it from the lower inorganic phenomena as well as from the higher organic science of biology. Although it is an organic science, sociology will be independent and study phenomena that "exhibit, in even a higher degree, the complexity, specialization, and personality which distinguish the higher phenomena of the individual life."[50]

This notion of hierarchy[51] represented yet another way to legitimate sociological inquiry: It explained why sociology was not as developed as the other highly respected sciences, and it placed sociology in a highly favorable spot (at the top of a hierarchy) in relation to the other "positive sciences." If sociology could be viewed as the culmination of a long evolutionary process and as the culmination of the positive sciences, its legitimacy could not be questioned. Such was Comte's goal, and although he was only marginally successful in his efforts, he was the first to see clearly that sociology could be like the other sciences and that it would be only a matter of time until the old theological and metaphysical residues of earlier social thought were cast aside in favor of a true science of society. This advocacy, which takes up the majority of pages in *Positive Philosophy,* rightly ensures Comte's claim to being the founder of sociological theory.

CRITICAL CONCLUSIONS

Comte gave sociology its name, however reluctantly, because he preferred the label "social physics," but he did much more: He gave the discipline a vision of what it could be. Few have argued so forcefully about the kind of science sociology should be, and he provided an interesting if somewhat quirky explanation for why this discipline should emerge and become increasingly important in the realm of science. Not all who followed Comte during the last two centuries would accept his positivism—that of a theoretically driven social science that could be used in the reconstruction of society—but he made several important points. First, theories must be abstract, seeking to isolate and explain the nature of the fundamental forces guiding the operation of society. Second, theories must be explicitly and systematically tested against the empirical world, using a variety of methods. Third, collecting data without the guidance of theory will not contribute greatly to the accumulation of knowledge about how the social universe operates. Finally, sociology should be used to rebuild social structures, but these applications of sociology must be guided by theory rather than by ideologies and personal biases.

Comte also anticipated the substantive thrust of much early sociology, especially that of Herbert Spencer and Émile Durkheim. Comte recognized that as societies grow, they become more differentiated, and the differentiation requires new bases of integration revolving around the concentration of power and around mutual interdependence. He did not develop these ideas very far, but he set an agenda. Comte also reintroduced the organismic analogy to social thinking, although many would not see this as a blessing. At the very least, however, he alerted subsequent sociologists that society is a system whose parts are interconnected in ways having consequences for the maintenance of the social whole. This basic analogy to organisms evolved into the functionalism of Spencer and Durkheim.

Still, there is much to criticize in Comte. He never really developed any substantive theory, aside from the relationship between social differentiation and new modes of integration. Most of Comtean sociology is a justification for sociology, and a very good one at that, but he did not explain how the social universe operates. He thought that his "law of the three stages" was the equivalent of Newton's law of gravity, but Comte's law is not so much a law as a rather simplistic view of the history of ideas. It made for an interesting way to justify the emergence of positivism and its queen science, sociology, but it did not advance sociology's understanding of the dynamics of the social universe.

Add to this lack of substantive contribution Comte's personal pathologies, which made him a truly bizarre and pathetic figure by the time of his death, and we are perhaps justified in ignoring Comte as a theorist who contributed to our understanding of the social universe. We should remember him for his forceful advocacy for scientific sociology. No one has done better since Comte first began to publish his positive philosophy.

NOTES

1. This brief review borrows heavily from Lewis A. Coser's *Masters of Sociological Thought* (New York: Harcourt Brace Jovanovich, 1978), pp. 13–40, as well as from Frank E. Manuel's *The Prophets of Paris* (Cambridge, MA: Harvard University Press, 1962).

2. Coser's *Masters of Sociological Thought,* pp. 25–27, is the first work to bring this line of influence to our attention.

3. Charles Montesquieu, *The Spirit of Laws,* vols. 1 and 2 (London: Colonial, 1900; originally published in 1748).

4. Du Pont de Nemours, for example, published a nine-volume edition of Turgot's work between 1808 and 1811, which, though deficient in many respects, brought Turgot's diverse pamphlets, discourses, letters, anonymously published articles, private memoranda, and so on together for the first time. Comte certainly must have read this work, although it is likely that he also read many of the original articles and discourses in their unedited form. See also W. Walker Stephens, ed., *The Life and Writings of Turgot* (London: Green, 1895).

5. See Du Pont de Nemours, *Collected Works of Baron A. R. J. Turgot,* 9 vols. (Paris, 1808–1811).

6. Reprinted in English in Ronald L. Meek, ed. and trans., *Turgot on Progress, Sociology, and Economics* (Cambridge: Cambridge University Press, 1973).

7. Meek, *Turgot on Progress.*

8. Du Pont, *Collected Works.*

9. See Meek, *Turgot on Progress,* for an English translation.

10. Indeed, Condorcet wrote *Life of Turgot* in 1786, which Comte, no doubt, read with interest.

11. Marquis de Condorcet, *Sketch for a Historical Picture of the Progress of the Human Mind* (London: Weidenfeld and Nicolson, 1955; originally published in 1794; translated into English in 1795).

12. Condorcet, *Sketch for a Historical Picture,* p. 4. The term *glove* here is ambiguous. It appears to refer to the constraints that the natural world imposes, although these constraints are but minimal—a gloved and gentle hand as opposed to a heavy hand.

13. See Keith Taylor, *Henri Saint-Simon* (London: Croom Helm, 1975), pp. 13–29, for a concise biographical sketch of Saint-Simon.

14. The most important of these works are *Letters from an Inhabitant of Geneva* (1803), *Introduction to the Scientific Studies of the Nineteenth Century* (1807–1808), *Essays on the Science of Man* (1813), and *The Reorganization of the European Community* (1814). Unfortunately, much of Saint-Simon is unavailable in convenient English translations. For convenient secondary works where portions of these appear, see F. M. H. Markham, *Henri Comte de Saint-Simon* (New York: Macmillan, 1952); and Taylor, *Henri Saint-Simon.* For interesting commentaries, see G. G. Iggers, *The Political Philosophy of Saint-Simon* (The Hague: Mouton, 1958); F. E. Manuel, *The New World of Henri de Saint-Simon* (Cambridge: Cambridge University Press, 1956); and Alvin Gouldner, *Socialism and Saint-Simon* (Yellow Springs, OH: Collier, 1962).

15. Saint-Simon gave Condorcet explicit credit for many of his ideas.

16. The French word *organization* means both "organization" and "organic structure." Saint-Simon initially used the term to refer to the organic structure of humans and animals and then extended it to apply to the structure of society.

17. For an interesting commentary, see Walter M. Simon, "Ignorance Is Bliss: Saint-Simon and the Writing of History," *International Review of Philosophy* 14 (nos. 3 and 4, 1960), pp. 357–383.

18. See Peyton V. Lyon, "Saint-Simon and the Origins of Scientism and Historicism," *Canadian Journal of Economics and Political Science* 27 (February 1961), pp. 55–63.

19. These ideas begin to overlap with Saint-Simon's collaboration with Comte.

20. All of these journals were short-lived, but they eventually gave Saint-Simon some degree of recognition as a publicist. These journals included *The Industry* (1816–1818); *The Political* (1819); *The Organizer* (1819–1820); *On the Industrial System* (1821–1822), actually a series of brochures; *Disasters of Industry* (1823–1824); and *Literary, Philosophic, and Industrial Opinions* (1825). From the latter, a portion on religion was published separately in book form as *New Christianity,* Saint-Simon's last major statement on science and the social order. Another important work of this last period was *On Social Organization.* In all these works there is a clear change in tone and mood; Saint-Simon is now the activist rather than the detached scholar.

21. Saint-Simon was initially anti-Christian, but with *New Christianity,* he changed his position so that the new spiritual heads of society were "true Christians" in that they captured and advocated the implementation of the "Christian spirit."

22. For example, Comte wrote much of *The Organizer* (1819–1820), especially the historical and scientific sections.

23. Naturally, Say and Saint-Simon did not explicitly use the concept of *entrepreneurs,* but they clearly grasped the essence of this economic function.

24. See Robert A. Nisbet, *Tradition and Revolt* (New York: Random House, 1968).

25. We will use and reference Harriet Martineau's condensation of the original manuscript. This condensation received Comte's approval and is the most readily available translation. Martineau changed the title and added useful margin notes. Our references will be to the 1896 edition of Martineau's original 1854 edition: Auguste Comte, *The Positive Philosophy of Auguste Comte,* vols. 1, 2, and 3, trans. and cond. H. Martineau (London: George Bell and Sons, 1896; originally published in 1854).

26. Auguste Comte, *System of Positive Polity,* vols. 1, 2, 3, and 4 (New York: Burt Franklin, 1875; originally published 1851–1854).

27. Auguste Comte, "Plan of the Scientific Operations Necessary for Reorganizing Society," reprinted in Gertrud Lenzer, ed., *Auguste Comte and Positivism: The Essential Writings* (New York: Harper Torchbooks, 1975), pp. 9–69.

28. Comte, *Positive Philosophy,* vol. 1, pp. 5–6 (emphasis in original).

29. In Comte's time, the term *physics* meant to study the "nature of'" phenomena; it was not merely the term for a particular branch of natural science. Hence, Comte's use of the label *social physics* had a double meaning: to study the "nature of'" social phenomena and to do so along the lines of the natural sciences. He abandoned the term *social physics* when he realized that the Belgian statistician Adolphe Quételet was using the same term. Comte was outraged that his original label for sociology had been used in ways that ran decidedly counter to his vision of theory. Ironically, sociology has become more like Quételet's vision of social physics, with its emphasis on the normal curve and statistical manipulations, than like Comte's notion of social physics as the search for the abstract laws of human organization—an unfortunate turn of events.

30. Comte, *Positive Philosophy,* vol. 1, p. 4.

31. Comte, *Positive Philosophy,* vol. 2, p. 242.

32. Ibid., p. 243.

33. Ibid., p. 240 (emphasis in original).

34. Comte, *Positive Philosophy,* vol. 1, p. 23.

35. See, for example, the following passages in *Positive Philosophy,* vol. 2: pp. 217, 226, 234, 235, and 238.

36. Comte, *Positive Philosophy,* vol. 2, pp. 241–257.

37. Ibid., p. 245.

38. Ibid., p. 246.

39. Ibid., p. 225.

40. See, in particular, his *System of Positive Polity,* vol. 2, pp. 221–276, on "The Social Organism."

41. Comte, *Positive Philosophy,* vol. 2, p. 141.

42. Ibid., pp. 275–281.

43. Ibid., pp. 280–281.

44. Ibid., p. 289.

45. Ibid., p. 293.

46. Ibid., p. 294.

47. Ibid., p. 227.

48. Most of *Positive Philosophy,* vol. 3, is devoted to the analysis of the three stages. For an abbreviated overview, see vol. 2, pp. 304–333.

49. Comte, *Positive Philosophy,* vol. 1, pp. 6–7.

50. Comte, *Positive Philosophy,* vol. 2, p. 258.

51. The hierarchy, in descending order, is sociology, biology, chemistry, physics, and astronomy. Comte added mathematics at the bottom because all sciences are ultimately built from mathematical reasoning.

3

✳

The Early Masters
and the Prospects
for Scientific Theory

uguste Comte's advocacy of a science of society was not accepted by
all of sociology's founding figures. Even Comte himself was caught
between trying to be a scientist who would articulate the laws of social
organization and one who would advocate the reconstruction of society. The
problem with such reasoning is this: Whose definition of the "good society"
should direct efforts at reconstruction? Even if scholars, politicians, and the
public could agree on what "the good society" should be, this consensus would
not be determined by scientific principles but by ideology, or what people
believe should exist. Thus, there is always a tension between science and advocacy. The best way to clarify the sources of this tension is to understand the
nature of science in relation to other systems of belief.

SCIENCE AS A BELIEF SYSTEM

Before examining the theories of Herbert Spencer, Karl Marx, Max Weber,
Émile Durkheim, Georg Simmel, Vilfredo Pareto, and George Herbert Mead,
as well as the many figures who influenced their thought, we should pause to
assess the relation of science to sociological theory. Let us begin with a general scheme represented in Figure 3.1, which shows two basic criteria among
belief systems: (1) Is the idea system evaluative, stating what should and ought
to exist, or is it neutral and objective?; and (2) is the idea system empirical,

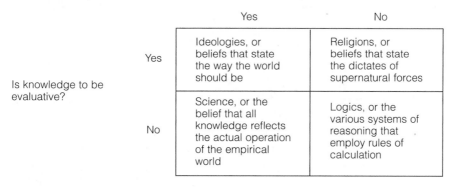

		Is knowledge to be empirical?	
		Yes	No
Is knowledge to be evaluative?	Yes	Ideologies, or beliefs that state the way the world should be	Religions, or beliefs that state the dictates of supernatural forces
	No	Science, or the belief that all knowledge reflects the actual operation of the empirical world	Logics, or the various systems of reasoning that employ rules of calculation

FIGURE 3.1 Types of Knowledge

making reference to use of empirical facts to assess the plausibility of statements, or is it nonempirical? Using these two criteria, a four-fold scheme is revealed. Science is a system of beliefs that argues for the assessment of theoretical statements about how the world works against systematically gathered empirical facts.

Science is a unique strategy for understanding the world because of this emphasis on explaining things through theories that are rigorously assessed against the reality of empirical observations. Compared with other types of belief systems, we can see what a revolution science was in a world dominated by religion. Indeed, religion does just the opposite of science; it makes statements about nonempirical realms (the sacred and supernatural) and does not test the validity of its statements with empirical facts. It is no wonder, then, that science and religion have clashed and continue to confront each other. Comte's belief that the "scientific-industrial" age would replace the theological was perhaps premature. Although most sociologists do not try to explain the world by appeals to the supernatural, other nonscientific belief systems such as political ideologies are still implicated in many sociological statements.

Ideologies are statements about what should occur in the empirical world. In the context of sociology, these statements indicate the kind of society that should, or ought, to exist. Sociology has never exorcised ideology from its statements, for several reasons. One is that many sociologists want to be critical of oppressive social arrangements, and they think it their duty to expose these arrangements and to propose more liberating alternatives. Karl Marx's entire theoretical scheme, for example, was devoted to these goals; and although it has elements of science in it, his scheme is heavily impregnated with ideology. Another reason for the difficulty in avoiding ideology is that sociologists are people who occupy certain places in society, who have beliefs and biases that distort their perceptions and analyses of the social world. To be human is to be biased, and it is often difficult to be completely objective, even when one is deliberately trying to be so. Still another reason for including ideology, often emphasized by critical theorists, is that in studying the way things are, it is easy to assume that this is also the way that they inevitably must be. This problem is compounded when laws of social organization are extracted from the way a society is organized at a particular time; these laws run the risk of becoming a legitimating ideology justifying the status quo with the language of science. Another source of ideology in sociology lies in the funding of science because those with money will pay for research and theory. As a result, leading scientists will often ask questions in line with the priorities of those who fund their work, whether this be government or private interests. No sector of society is unbiased and interest-free. For at least these reasons, then, the boundary between science and ideology is always tense and often permeable. Most social scientists try to eliminate their ideology from theory, but it is not always clear that they have done so.

The last box in Figure 3.1 denotes systems of ideas, termed "logics," that are neither evaluative nor empirical. These are formal languages, methods for calculation, and procedures for deductions, inferences, and other statements of relationships. Mathematics, symbolic logic, and computer languages all fall into this category. These "logics" are frequently the way theory makes statements about how the world operates, although ordinary language is used equally often. Comte felt that mathematics was the language of science, and at its highest level, this is true. Yet, many sciences also use ordinary language. Thus, like ideology, the boundaries between logics and science are often permeable, with theories being stated in a logic but equally often with problems in science stimulating the development of new logics.

What, then, can this simple scheme tell us about sociology's founding tradition? As we will see, some scholars—particularly Auguste Comte, Herbert Spencer, and Émile Durkheim—advocated a view of sociology as a scientific discipline, and they made significant contributions to developing the laws of human organization. Yet, their work was very much influenced by ideologies, even as they sought to remain objective. Karl Marx reveals the opposite profile in a theoretical scheme that is explicitly ideological. He made a number of important scientific statements even though his work was guided by a desire to construct a rationale for the inevitable destruction of capitalism. Max Weber was ambivalent about science; he thought that objectivity or what he termed "value neutrality" was a good thing, but he did not see patterns of social organization as revealing universal and invariant properties that could be subject to timeless laws of science. For Weber, much of what occurs is contingent on circumstances that do not follow laws and, hence, cannot be predicted; still, he did not want to abandon objectivity. Among the early theorists, then, Weber is perhaps the most tortured about the prospect for scientific sociology. Georg Simmel, the last of the three German sociologists we will examine, was much more upbeat about the prospects for science. He argued that if sociology studies the basic forms of social relationships among actors, whether individuals or collective units, it would be possible to develop scientific explanations of these forms. Vilfredo Pareto similarly felt that a science of human organization could be developed. After a long career in several fields, including economics where "Pareto curves" are today standard techniques, he turned to sociology because he felt that economics was too narrow and did not ask scientific questions about the social and cultural contexts of human action. Finally, George Herbert Mead was a philosopher rather than a sociologist, and although he did not make a strong case for science, clearly his goal was to develop statements about the basic processes of face-to-face interaction.

Thus, most early sociologists that we will examine were, to varying degrees, committed to science—that is, to developing statements about the universal properties and processes of human action, interaction, and organization. Others were more ambivalent, but despite their reservations, they nonetheless developed statements that look like laws of human organization. As we examine each of these theorists, we will first let them speak for themselves, but in the third chapter on each, we will put words in their mouths as we try to convert their ideas into more law-like theoretical statements. Some like Spencer, Durkheim, Simmel, and Pareto would probably not object; others such as Marx and Weber would consider the exercise inappropriate; and George Herbert Mead would probably not lodge a strong objection. The reason that we propose to convert these founding figures' ideas into abstract theoretical statements is that the enduring impact of their work ultimately resides, we believe, in the theories that they generated.

WHAT IS THEORY?

Scientific theories are statements that seek to explain how the universe operates. Hence, scientific sociological theory attempts to explain how the universe of human social action, interaction, and organization operates. Theories in all sciences reveal certain common features: (1) concepts, (2) statements linking concepts, and (3) formats for organizing statements.

Concepts

Concepts denote some property of the universe and, in the case of sociology, a property of the social world. A concept like "differentiation" denotes that the activities of people and collective actors become more specialized, distinctive, and different. Or, a concept such as the "bourgeoisie" alerts us to the existence of a class of individuals and organizations engaged in making, buying, and selling goods and services for profit. And, a concept like "stratification" signals that the social world reveals systematic inequalities in which people possess different amounts and kinds of valued resources. A concept is constructed from definitions that simply identify some phenomenon of interest.

It is, of course, desirable that concepts have common meanings; hence, the more precise the definition of a concept is, the more we can be confident that there is no ambiguity about what the concept is denoting. For example, the concept of bourgeoisie is a bit vague because it is not entirely clear who or what would be part of the bourgeoisie; even if the definition were elaborated, some ambiguity would remain. Part of this lack of precise clarity is the result of using ordinary language, which has its own built-in sources of uncertainty, but much of the problem centers on the phenomenon itself, which is somewhat amorphous. We can get only a suggestive vague sense of about where the bourgeoisie begins and ends, so the concept of "bourgeoisie" is not as clear as it should be. Thus, in science, the best concepts denote precisely the phenomenon in question, and what is true of science in general is also the case for sociology in particular.

Concepts of most use in scientific theory are *abstract;* they are not tied to the particulars of time and place, as is a notion like bourgeoisie. Rather, concepts are about general and generic classes of phenomena. For instance, the concept of differentiation is universal because all known human populations reveal some degree of this phenomenon—even if a simple society only distinguishes between the age and sex of people. Scientific theory always attempts to explain as much as is possible, and unless concepts are abstract, rising above the particulars of an empirical or historical situation, a theory can never explain very much beyond the particulars of an empirical case.

Concepts that emphasize *the variable states* of a phenomenon are the most useful in theory. For example, the concept of differentiation denotes the complexity of the social universe whereby people or collective units are distinctive, whereas the concepts of *level or degree* of differentiation makes the concept a variable, focusing our attention on how much differentiation is evident. The reasons that concepts denoting variable properties of the universe are so useful is that a theory tries to explain how variable states of one phenomenon are connected to variable states of another. For example, a simple hypothesis from Auguste Comte's work might be this: *The greater the degree of differentiation, the greater will be the degree of interdependence among members of the population and the more power will be used to integrate members of the population.* In the last chapter, we saw that Comte implied that the degree of regulation of a society by ties of mutual interdependence and centralized power is related to the level of differentiation.

Statements

Statements connect concepts, as is illustrated by Comte's arguments about the relations among differentiation, interdependence, and centralized power. Concepts by themselves are not very useful in theory, although they do call attention to important dimensions of the world. But concepts become particularly useful when they are connected. Recall Einstein's famous formula of $E=mc^2$. This elegant equation is a statement on the relationship among energy, matter, and the speed of light. Similarly, a sociological statement implied by Comte and developed by Spencer and Durkheim about the *relationship* between size of a population and the degree of differentiation among members of this population is more useful than is the case when these concepts stand alone and are not linked together in a theoretical statement. When the two are linked, we begin to see that two important phenomena—size and degree of differentiation—are related. If we add to this insight and note that, as differentiation increases, so does the degree of integration through mutual interdependence and centralization of authority, then we have an even more powerful set of theoretical statements.

Like concepts, highly abstract statements are the most useful because they are likely to rise above particulars of time and place. As a result, they can explain a wider range of phenomena across longer reaches of time. The more concepts in abstract statements are about generic phenomena that are always present when humans act, interact, and organize, the more powerful the theory is.

Theoretical Formats

A format is simply a way to array and arrange statements, incorporating concepts. In general terms, there are several basic kinds of formats. One is simple discursive statements in which a theory implicitly develops concepts, uses ordinary language to make statements, and loosely describes the relationships among statements. Most of the theories that we will examine in this book are of this nature. They are stated discursively, and although the train of thought might be well organized, the relationships among concepts and statements are imprecisely connected. Among the early theorists, Spencer and Durkheim probably came the closest to making more precise statements, and in places Marx, Simmel, and Pareto make more formal statements. But, by and large, the preferred format among early theorists and, for that matter, theorists today, is discursive text. As we will see, however, converting discursive statements into more precise formats is often rather easily done, as we will often do to highlight the power of these early thinkers' ideas. But what is the nature of these more precise formats?

Another type of format is an *analytical scheme* in which concepts and statements are arrayed into a typology. This typology simply classifies and categorizes what are considered important properties of the social universe. As we will see, Max Weber often employed such schemes, and as we saw in Chapter 2, Comte's portrayal of distinctions among the theological, metaphysical, and positivistic stages of evolution is, in many ways, an analytical scheme that simply classifies types of idea systems. Although the scheme posits the movements of ideas, each stage is really a typology like the one we constructed in Figure 3.1. Analytical schemes are useful in seeing the place of phenomena, but they rarely specify how the phenomena influence each other. Schemes are most useful when one is beginning to theorize, but they rarely are good theory in themselves.

Another kind of format is a *simple causal model,* in which concepts are linked together in a visual model that records how change in one phenomenon causes changes in others over time. For example, we might draw arrows indicating increasing size of a population as causing increased differentiation of the population, and this diagram would represent a simple causal model. The problem with simple causal models, however, is that they only go in one direction, from one event to an outcome event. Much of the social world is recursive, however, folding back on itself such that an outcome becomes, itself, a causal force that feeds back and effects the causal forces that brought this outcome about. For instance, population size may cause differentiation, but the relationship is two-way because as a population becomes more differentiated, the increased complexity of structure encourages the population to grow. So our causal diagram might draw a reverse causal arrow back from differentiation to population size (See Figure 3.2). When many causal chains are linked in this way, the models can become too complex. The intent of these *complex causal models* is to highlight that events cause each other and that outcomes at one point in time feed back and effect the causal forces that brought them about.

A last format in theory is a *system of propositions.* Here, relationships among statements are stated formally and linked together. For example, a simple propositions system might be this:

1. As the size of a population increases, so does the degree of differentiation of this population.
2. As a population becomes more differentiated, the basis of integration changes to one relying on mutual interdependence and centralized power.
3. As this new basis of integration increases, the population can grow further.

This series of propositions states the connections among size, differentiation, and bases of integration, seeing none of these processes as more fundamental than another. As such, this is a simple prepositional format. Formats can become more precise as well as more complex. One way to increase precision is to organize propositions in a hierarchy in which a few propositions stand at the top of the hierarchy, with the others being derivations to even more specific content. For instance, plain geometry is a hierarchal system of axioms, theorems, and corollaries in which one moves from the highest order propositions—the axioms—to theorems and corollaries. Sociological theories can be of this structure, with a general statement at the top of a hierarchy and others "following" from this most general statement. For example, we might propose the axiom "people from similar backgrounds are more likely to associate than are those from dissimilar backgrounds" and then try to explain with this axiom-like statement why ethnic groups form and sustain themselves. To do so, we would begin with our quasi-axiom about backgrounds and associations, then we would work our way through a series of derivations to the phenomenon of interest—ethnic group endogamy. Relatively few theories in sociology are truly axiomatic in a strict sense of this format, but many are formal hierarchies of propositions that resemble in form, if not precision, the structure of what is often termed *deductive theorizing* in which one makes systematic deductions from a few first principles down to more specific phenomena of interests. In our review of Herbert Spencer, for instance, we will see that his entire philosophical scheme began with a few "first principles" on evolution in all domains of the universe. Then, in trying to explain evolution in the social

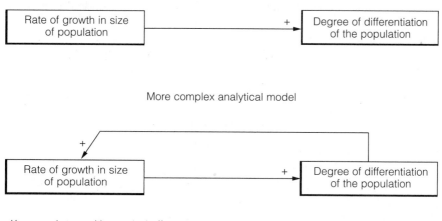

Simple, one-way causal model

More complex analytical model

Key: + = has positive causal effect on

FIGURE 3.2 Comparing Simple and Complex Causal Models

universe of human societies, he developed more specific principles that followed from the general ones.

After reviewing the theories as stated by the theorists themselves, we will conclude with a third chapter on each theorist that translates their discursive statements into more precise and formal theoretical formats. We have moved this analysis into a separate chapter so that readers who would see this kind of exercise as wandering too far from the theorists' own words and intentions can simply skip the chapter if they choose. But our goal in this third chapter is to show that these theorists stated concepts and their relations with sufficient precision that a more general and formal theory can be produced from their works. When necessary, we will first try to make concepts more abstract so that we remove the theory from a particular historical time and place. So, for example, rather than talk about the bourgeoisie and proletarians in Marx's original text, we will become more abstract and refer to superordinates and subordinates in a system of inequality. Much is lost in making these concepts more abstract, but much is also gained: The concepts can now be applied to a much longer range of history and to more diverse contexts. Then, we will convert the theories into both causal and proposition formats, just to see what doing so reveals about the power of theories. Some theories will look better in a causal format, others in propositions, and a few in both, but the point of the exercise is to show the underlying theory in its most powerful light.

Many would argue, with justification, that converting discursive text, even when arguments are closely reasoned, into the alien format of science distorts the theories. True enough, but the point of theory is to endure. Without making the concepts more abstract and without teasing out the relations specified in statements in a more precise format, the theories can look dated because they were developed in a particular historical period. The reason that sociologists read and re-read these figures of sociology's first one hundred years is to explore the power of these early theorists' ideas. We will simply be trying to make this power more explicit, and in so doing, the theories can be restated in ways that increase the likelihood that they will be part of sociology's theoretical tradition for the next one hundred years.

THE EMERGENCE OF SOCIOLOGICAL THEORY

In closing, let us emphasize that the title of this book describes its contents. Our goal is to analyze the *emergence* of sociological *theory.* Our concern is the theories themselves and, to a lesser degree, the context in which they were produced. We will certainly offer, as we already have for Comte, a sense of the figures and schools of thought that influenced each theorist, but this book is not a history of sociology. Rather, this is a history of sociology's theoretical development from roughly 1830, when Comte gave the discipline its name and called for a theoretically informed positive science, to around 1930 when the first generation of great masters had passed away. What will endure from these figures is their theories. Other concerns, such as their biographies, the now obscure figures who influenced their thinking, and the issues of the time are no longer important to a science of sociology, although they are very important to a history of sociology. We offer much history, but we offer much more theory. For in the end, it is the theories that will endure.

4

✳

The Origin and Context of Herbert Spencer's Thought

BIOGRAPHICAL INFLUENCES ON SPENCERIAN SOCIOLOGY

Herbert Spencer was born in Derby, England, in 1820. Until the age of thirteen, he was tutored by his father at home. He subsequently moved to his uncle's home in Bath, where his private education continued.[1] Except for a few months of formal education, Spencer never really attended school outside his family. Still, he received a very solid education in mathematics and science from his father and uncle, and this technical education, in the end, encouraged him to view himself as a philosopher and to propose a grand project for uniting ethics, natural science, and social science. This great project was termed *Synthetic Philosophy,* an indication that Spencer's work moved far beyond the disciplinary border of sociology. Only rather late in his career (between 1873 and 1896) did he turn his attention to sociology. He thought big in a time when the intellectual world in general, and academia in particular, was specializing and compartmentalizing.

This breadth and scope of inquiry probably accounts for the popularity of Spencer's work in the second half of the nineteenth century. He raised questions that intrigued both the lay public and scholars in particular specialties. Many of his works first appeared in serial form as installments in popular science magazines, and only later were they bound together in book volumes. His ideas remained popular; indeed, one hundred thousand copies of his books were sold before the turn of the century, an astoundingly high figure for that time and place. Even more amazing, however, is that they are not mere popularizations of ideas but, rather, academic works. Spencer's books could hardly be considered light reading, but apparently their vision and scope captured readers' imaginations. Anyone who reads Spencer today cannot help but be impressed by the power of his ideas and perhaps even their arrogance, for who now would proclaim it possible to unite all the sciences and questions of ethics under one set of general principles?

If Spencer had received a formal education and advanced degrees from established universities, as his father had, his thinking would probably have been more focused and restrained. By today's standards, elite universities of the last century offered very broad training in letters and science, but even by that yardstick, he would have been compelled by tradition and established genres to recognize that, after all, one does not undertake to explain the entire universe with a few general principles. Formal education has a tendency to limit horizons and force concentration on narrow topics, but because he avoided the halls of academia in his youth and throughout his career, he was not bound by its rules of scholarship.

In a quiet way, Spencer's work flouts the rules of academia. He never read very much; rather, he picked the brains of distinguished scholars. Instead of burying himself in the library, he frequented London clubs and was friends with the most eminent scientific and literary figures of his time. From them, no doubt, he learned much by listening carefully and asking probing questions.[2] One suspects that Spencer was a kind of intellectual sponge, absorbing ideas on contact. How else could a man write detailed works on ethics, physics, biology, psychology, sociology, and anthropology while maintaining a constant flow of pointed and popular social commentary? Moreover, unlike academics who used students to do much of their legwork, Spencer employed professional academics. His research assistants tended to be Ph.D.s who either needed the money or found the assigned tasks interesting. It seems likely, of course, that an uncredentialed private scholar employing credentialed academics represented somewhat of an affront to the academic establishment, although he managed to maintain cordial relations with many important academics.

Despite Spencer's enormous popularity with the literate lay public, however, he was an inordinately private individual. Indeed, he was rather neurotic and odd. He hardly ever gave public lectures; he spent a good part of the day in bed, either writing or complaining about real and imagined ailments; he remained a lifetime bachelor who, at best, had only one great love affair (and even here the nature of the relationship is not clear); he lived in rather sparse and puritan circumstances despite his inherited wealth and substantial royalties; and when he got older, his somewhat dour disposition became punctuated with considerable bitterness as his ideas came under increasing attack and then passed into obscurity. Yet many of those who knew him, and even the nurses who cared for him during his last years of failing health, emphasized that he was still a thoughtful and engaging individual.

It is perhaps not so surprising, then, that Spencer is an enigma to us. He was a lone and private scholar in a time when scholarship was becoming an increasing monopoly of academia, and despite his popularity, he never revealed a public presence and persona. Finally, he was a global thinker in a time of increasing specialization. The result, we can speculate, was that as his scholarly ideas were criticized by specialized academics and as his political commentary became less fashionable, he had few students and adherents to carry his case. He was too neurotic to defend himself publicly, although he did make a celebrated and trumpeted tour of the United States in the early 1900s to espouse his moral philosophy (which became an embarrassment to those who recognized the importance of his scholarly ideas). As a consequence, Spencerian philosophy and sociology disappeared very rapidly after he died. There were no students and academic colleagues to carry forward his grand synthesis, with the consequence that one of the first important theoretical works of the modern era opens with the question "Who now reads Spencer?" Today, very few academics read Spencer. As we hope to demonstrate in the next chapter, this marks a great intellectual tragedy, for we now have a stereotypical and largely inaccurate view of him, which keeps us from fully appreciating the powerful quality of his ideas. A contemporary view of him might read as follows:

> Herbert Spencer, the first self-conscious English sociologist, advocated a perspective that supported the dominant political ideology of free trade and enterprise. He naively assumed that "society was like an organism" and developed a sociology that saw each institution as having its "function" in the "body social," thereby propagating a conservative ideology and legitimating the status quo. What is even worse, Spencer coined the phrase *survival of the fittest* to describe the normal state of relations within and between societies, thus making it seem right that the elite of a society should possess privilege and that some societies should conquer others.

There are elements of truth in a surface portrayal such as this, but there is also a great distortion, as we will come to see in this chapter and the next. Spencer was indeed an ideologue, but no more so than many other sociologists of the last century—or today, for that matter. Although his ideas were progressive and radical for their time, they are now considered right wing and conservative. But in contrast with many others, such as Karl Marx and Émile Durkheim, Spencer did not include social and political ideology very often in his scientific work; rather, Spencer's ideology was packaged in separate volumes on ethics. Furthermore, as this contemporary view emphasizes, Spencer did make analogies to organic forms, but these are far more sophisticated than is typically recognized. (He had written a large, two-volume work on biology before embarking on sociology.) It is also true that he developed functional analysis, but it is not as naive or simplistic as many contend. Actually, Spencerian functionalism is highly sophisticated and avoids many of the pitfalls of contemporary functionalism. And of course, Spencer did coin the phrase *survival of the fittest,* which, we suspect, was to be his biggest mistake. But we should emphasize that in doing so, he came very close to postulating the principle of natural selection ten years before Charles Darwin published his thesis; in fact, Darwin acknowledges Spencer in the preface of *On the Origin of Species.*

But we should not be carried away with a defense of the much-maligned and enigmatic Spencer. The power of his ideas will, we believe, speak for themselves. Let us now return to tracing his biography and its impact on the development of his thought. Then we will be in a better position to appreciate his theoretical ideas.

Because Spencer had not received a formal education, he felt himself unqualified to attend college. In 1837, therefore, he sought to use his mathematical and scientific training as an engineer during the construction of the London and Birmingham Railway. The practical application of Spencer's training in mathematics had an enormous effect on his later thinking, for he was always attuned to the consequences of structural stress on the dynamics of the physical and social universe, and he expressed these consequences as equations (although the relationships were usually stated verbally). The impact of these four years as an engineer could not be foreseen in the next turns in his intellectual life.

In 1841, when the railroad was completed, Spencer returned to his birthplace in Derby. Over the next few years, he wrote several articles for the radical (for his time) press and numerous letters to the editor of a dissenting newspaper, *The Nonconformist.* In these works he argued for limiting the power of government, and although these ideas are often defined as "conservative" today, they were seen as "liberal" and "radical" in the last century. After several years as a kind of fringe figure in radical politics and journalism, he secured a permanent position as a subeditor for the London *Economist* in 1848. The appointment marked a turning point in his life, and from that time on, his intellectual career accelerated. In 1851, he published *Social Statics,*[3] a work that has hurt Spencer's reputation and has been largely responsible for our present-day view of him as a Social Darwinist, a libertarian, and perhaps a right-wing ideologue. In this work he championed the cause of laissez-faire—free trade, open markets, and non-intervention by government. He asserted that individuals had the right to do as they pleased, as long as they allowed others to do the same. Despite its negative impact on our retrospective view of Spencer, however, the book was well received and opened doors into the broader intellectual community, although it remained a burden as he began to write less ideological and more scholarly works.

In 1853, the uncle who had tutored Spencer in science and mathematics died and left Spencer a substantial inheritance. This inheritance allowed him to quit his job as an editor and assume the life of a private scholar full time. Despite his emotional problems—depression, insomnia, and reclusiveness—he was enormously productive as a private scholar. His collected works span volumes and run into thousands of pages. Moreover, with his more ideological tract out of his system, at least until the end of his life, he used the hard-nosed skills of an engineer and scientist to write a series of brilliant works. In 1854, he published *Principles of Psychology,* which was used as a text at Harvard and Cambridge. In 1862, he published *First Principles,* which marked the beginnings of his grand Synthetic Philosophy. In this book he sought to unify ethics and science under one set of elementary principles. Clearly the young Spencer, who had felt himself unqualified for college, was gaining confidence. Between 1864 and 1867, he published the several volumes of his *Principles of Biology,* in which he sought to apply the abstract "first principles" of the universe to the dynamics of the organic realm. In 1873, he began to think about the super-organic—that is, the social organization of organic forms. In particular, he initiated an analysis of how human organization, as the most obvious type of super-organic organization, could illustrate the plausibility of his first principles. He opened this movement into the domain of sociology with a methodological treatise on the problems of humans studying themselves; in so doing, he emphasized that laws of human organization could be discovered and used in the same way as in the physical and biological sciences. In 1874, the first serialized installments of his *The Principles of Sociology* appeared, and for the next twenty years, he devoted himself to sociology and to articulating the basic laws of human organization. The last portions of *The Principles of Sociology* were published in 1896. At the same time that he was preparing these last parts of his sociology, Spencer was publishing *The Principles of Ethics,* which restated the then-liberal, but now-conservative, social philosophy of laissez-faire. Because this philosophy was published in separate volumes from the scholarly work in sociology, it intrudes less than might otherwise have been the case.

Thus, although Spencer's work in sociology spans only a twenty-year period in a much longer and comprehensive intellectual career, his sociology reflects other scholarly and political concerns. In turn, these other concerns are the product of the general intellectual milieu of nineteenth-century England as well as of specific scholars in this milieu. To understand Spencerian sociology, then, we should note some of the other forces influencing his thinking.

THE POLITICAL ECONOMY
OF NINETEENTH-CENTURY ENGLAND

In contrast with France, where decades of political turmoil had created an overconcern for collective unity, England remained comparatively tranquil. As the first society to industrialize, England enjoyed considerable prosperity under early capitalism. Open markets and competition appeared to be an avenue for increased productivity and prosperity. It is not surprising, therefore, that social thought in England was dominated by ideological beliefs in the efficiency and moral correctness of free and unbridled competition not only in the marketplace but in other realms as well.[4]

Spencer advocated a laissez-faire doctrine in his philosophic works. Individuals should be allowed to pursue their interests and to seek happiness as long as they do not infringe on others' rights to do so. Government should be restrained and should not regulate the pursuits of individuals. Much like Adam Smith, Spencer assumed a kind of "invisible hand of order" as emerging to maintain a society of self-seeking individuals. Most of Spencer's early essays and his first book, *Social Statics,* represent adaptations of laissez-faire economics. But later, his social and economic philosophy was supplemented by the more scientific analyses contained in his biological works.

THE SCIENTIFIC MILIEU
OF SPENCER'S ENGLAND

Spencer's early training with his father and uncle was primarily in mathematics and science. More important, his informal contacts as a freelance intellectual were with such eminent scientists as Thomas Henry Huxley, Joseph Dalton Hooker, John Tyndall, and even Darwin. Indeed, Spencer read less than he listened, for he clearly acquired an enormous breadth of knowledge by talking with the foremost scientists of his time. Biographers have frequently commented on the lack of books in his library, especially for a scholar who wrote with such insight in several different disciplines. Despite his reliance on informal contacts with fellow scientists, however, several key works in biology and physics appear to have had considerable impact on his thought.

Influences from Biology

In 1864, Spencer wrote the first volume of his *Principles of Biology,* which at the time represented one of the most advanced treatises on biological knowledge.[5] Later, as we will see, he sought to apply the laws of biology to "super-organic bodies,"[6] revealing the extent to which biological knowledge influenced his more purely sociological formulations. He credited three sources for some of the critical insights that he later applied to social phenomena: Thomas Malthus (1766–1834), Karl Ernst Von Baer (1792–1876), and Charles Darwin (1809–1882).

Malthus Spencer was prfoundly influenced by Malthus's *Essay on Population.* (Malthus, of course, was not a biologist, but his work had an influence in this sphere and hence is discussed in this section.) In this work, Malthus emphasized that the geometric growth of population would create conditions favorable to conflict, starvation, pestilence, disease, and death. Indeed, he argued that populations grew until "checked" by the "four horsemen": war, pestilence, famine, and disease.

Spencer reached a much less pessimistic conclusion than Malthus, for the competition and struggle that ensues from population growth would, Spencer believed, lead to the "survival of the fittest" and, hence, to the elevation of society and "the races." Such a vision corresponded, of course, to Spencer's laissez-faire bias and allowed him to view free and open competition not just as good economic policy but also as a fundamental "law of the organic universe."[7] In addition to these ideological uses of Malthus's ideas, the notion of competition and struggle became central to Spencer's more formal sociology. Spencer saw evolution of societies as the result of territorial and political conflicts, and he was one of the first sociologists to understand fully the significance of war and conflict on the internal patterns of social organization in a society.

Von Baer Spencer was also influenced by William Harvey's embryological studies as well as by Henri Milne-Edward's work, which had borrowed the phrase "the physiological division of labor" from social thought. Indeed, as Spencer so ably emphasized, biologists had often borrowed from social discourse terms that he was merely borrowing back and applying in a more refined manner to the "super-organic realm." Yet he gave a German, Von Baer, the credit for recognizing that biological forms develop from undifferentiated, embryologic forms to highly differentiated structures revealing a physiological division of labor.

Von Baer's principles allowed Spencer to organize his ideas on biological, psychological, and social evolution. Spencer came to emphasize that evolution is a process of development from an incoherent, undifferentiated, and homogeneous mass to a differentiated and coherent pattern in which the functions of structures are well coordinated.[8] Conversely, dissolution involves movement from a coherent and differentiated state to a more homogeneous and incoherent mass. Thus, Spencer came to view the major focus of sociology as the study of the conditions under which social differentiation and de-differentiation occur.

Darwin The relationship between Darwin and Spencer is reciprocal in that Spencer's early ideas about development exerted considerable influence on Darwin's formulation of the theory of evolution,[9] although Darwin's notion of "natural selection" was apparently formulated independently of Spencer's emphasis on competition and struggle. Only after *On the Origin of Species* was in press did Darwin recognize the affinity between the concepts of *survival of the fittest* and *natural selection*. Conversely, his explicit formulation of the theory of evolution was to reinforce, and give legitimacy to, Spencer's view of social evolution as the result of competition among populations, with the most organizationally "fit" conquering the less fit and, hence, increasing the level and complexity of social organization. Moreover, Darwin's ideas encouraged Spencer to view differences among the "races" and societies of the world as the result of "speciation" of isolated populations, each of which adapted to varying environmental conditions. Spencer's continuous emphasis on environmental conditions, both ecological and societal, as shaping the structure of society is the result, no doubt, of Darwin's formulations.

The theory of evolution also offered Spencer a respected intellectual tool for justifying his laissez-faire political beliefs. For both organic and super-organic bodies, he argued, it is necessary to let competition and struggle operate free of governmental regulation. To protect some segments of a population is to preserve the "less fit" and thus reduce the overall "quality of civilization."[10]

From biology, then, Spencer took three essential elements: (1) the notion that many critical attributes of both individuals and society emerge from competition among individuals or collective populations; (2) the view that social evolution involves movement from undifferentiated to differentiated structures marked by interrelated functions; and (3) the recognition that differences among both individuals and social systems are the result of their adapting to varying environmental conditions. He supplemented these broad biological insights with several discoveries in the physical sciences to forge the "first principles" of his general Synthetic Philosophy.

Influences from the Physical Sciences

From informal education within his family and from contacts with the most eminent scientists of his time, Spencer acquired considerable training in astronomy, geology, physics, and chemistry. In reading his many works, it is impossible not to be impressed by his knowledge of wide varieties of physical phenomena and their laws of operation. His Synthetic Philosophy thus reflected his debt to the physical sciences, particularly for (1) the general mode of his analysis and (2) the specific principles of his philosophy:

1. All of Spencer's work is indebted to the post-Newtonian view of science—that is, the emphasis on universal laws that could explain the operation of phenomena in the world. Indeed, Spencer went beyond Newton and argued that there were laws transcending all phenomena, both physical and organic. In other words, laws of the universe or cosmos can be discovered and used to explain, at least in general terms, physical, organic, and super-organic (social) events. Spencer emphasized that each domain of reality—astronomical, geological, physical, chemical, biological, psychological, and sociological—revealed its own unique laws that pertained to the properties and forces of its delimited domain. He also believed that at the most abstract level, however, a few fundamental, or first, principles cut across all domains of reality.

2. In seeking these first principles, Spencer relied heavily on the physics of his time. He incorporated into his Synthetic Philosophy notions of force, the indestructibility of matter, the persistence of motion, and other principles that were emerging in physics. We will discuss these later when examining his scheme in depth, but we should emphasize that much of the inspiration for his grand scheme came from the promise of post-Newtonian physics.

Thus, Spencer's Synthetic Philosophy emerged from a synthesis of ideas and principles being developed in physics and biology. Yet the precise way in which he used these ideas in his sociological work was greatly influenced by his exposure to Auguste Comte's vision of a positive philosophy (see the previous chapter). Before we can fully appreciate Spencer's philosophy, therefore, we need to review his somewhat ambivalent and defensive reaction to Comte's work.

SPENCER'S SYNTHETIC PHILOSOPHY
AND THE SOCIOLOGY OF COMTE

In 1864, Spencer published an article titled "Reasons for Dissenting from the Philosophy of M. Comte," in which he sought to list the points of agreement and disagreement with the great French thinker.[11] Spencer emphasized that he disagreed with Comte about the following issues: (1) that societies pass through three stages, (2) that causality is less important than relations of affinity in building social theory, (3) that government can use the laws of sociology to reconstruct society, (4) that the sciences have developed in a particular order, and (5) that psychology is merely a subdiscipline of biology.

Spencer also noted a number of points in which he was in agreement with Comte, but he stressed that many other scholars besides Comte had similarly advocated (1) that knowledge comes from experiences or observed facts and (2) that there are invariable laws in the universe. Most revealing are the few passages where Spencer explicitly acknowledged an intellectual debt to Comte. Spencer accepted Comte's term, *sociology,* for the science of super-organic bodies, and, most important, he gave Comte begrudging credit for reintroducing the organismic analogy back into social thought. Spencer stressed, however, that Plato and Thomas Hobbes had made similar analogies and that much of his organismic thinking had been influenced by Von Baer.

Yet one gets the impression that Spencer was working too hard at dissociating his ideas from Comte's. That his most intimate intellectual companions, George Elliot and George Lewes, were well versed in Comte's philosophy argues for considerable intellectual influence of Comte's work on Spencer's initial sociological inquiries. True, Spencer would never accept Comte's collectivism, but he extended two critical ideas clearly evident in Comte's work: (1) social systems reveal many properties of organization in common with biological organisms, and thus a few principles of social organization can be initially borrowed (and altered somewhat) from biology; and (2) when viewed as a "body social," a social system can be analyzed by the contribution of its various organs to the maintenance of the social whole. There can be little doubt, then, that Spencer was stimulated by Comte's analogizing and implicit functionalism. But as Spencer incorporated these ideas, they were altered by his absorption of key insights from the physical and biological sciences.

WHY READ SPENCER?

When compared with the intellectual influences on other scholars whom we will analyze in later chapters, those on Spencer are less clear. He did not attend a university, so his mentors cannot be traced there. Nor did he ever hold an academic position, thereby avoiding compartmentalization in a department or particular school of thought. As a freelance intellectual, he borrowed at will and was never constrained by the intellectual fads and foibles that sweep through academia. The unrestrained scope of his scheme makes it fascinating, and perhaps this same feature makes his work less appealing to present-day scholars, who tend to work within narrow intellectual traditions.

Yet, as we will explore in depth in the next chapter, Spencer offered many important insights into the structure and dynamics of social systems. Although he presented these insights in the vocabulary of the physics and biology of his time, they still have considerable relevance for sociological theorizing. As we approach the analysis of Spencer's basic works, therefore, we should be prepared to appreciate not only the scope of his ideas but also the profound insights that he achieved into the nature of social systems.

NOTES

1. Jonathan H. Turner, *Herbert Spencer: A Renewed Appreciation* (Beverly Hills, CA: Sage, 1985), chap. 1; and *Herbert Spencer, an Autobiography* (London: Watts, 1926).

2. For example, see Hugh Elliot, *Herbert Spencer* (New York: Holt, Rinehart & Winston, 1917), and David Duncan, *Life and Letters of Herbert Spencer* (London: Methuen, 1908). In his *Masters of Sociological Thought* (New York: Harcourt Brace Jovanovich, 1977), Lewis Coser best summarizes Spencer's relationship with his contemporaries by noting that from informal conversations, Spencer was supplied "with scientific facts he used so greedily as building blocks for his theories. Spencer absorbed his science to a large extent as if through osmosis, through critical discussions and interchanges with his scientific friends and associates" (p. 110).

3. For complete references to this and other works by Spencer, see foot-

notes later in this chapter and in the next chapter, where these works are discussed.

4. The major legitimating work in this context was Adam Smith's *An Inquiry into the Nature and Causes of the Wealth of Nations* (London: Cadell and Davies, 1805; originally published in 1776).

5. In *Principles of Biology* (New York: Appleton-Century-Crofts, 1864–1867), Spencer formulated some original laws of biology that still stand today. For example, his formulation of the relationship among growth, size, and structure are still axiomatic in biology. Yet few biologists are aware that Spencer, the engineer turned scientist, formulated the law that among regularly shaped bodies, surface area increases as the square of the linear dimensions, and volume increases as the cube of these dimensions—hence requiring new structural arrangements to support and nourish larger bodies.

6. This was Spencer's phrase for describing patterns of social organization.

7. See, for example, his *Autobiography;* also see the long footnote in *First Principles* (New York: A. L. Burt, 1880; originally published in 1860).

8. This idea can be found in its early form in one of Spencer's early essays, "Progress: Its Law and Cause," first published in 1857 (*Westminster Review,* April 1857). Also, see Spencer's article "The Developmental Hypothesis," *The New Leader* (1852).

9. Indeed, as noted earlier, Darwin explicitly acknowledges Spencer's work in the introduction to *On the Origin of Species* (London: Murry, 1890; originally published in 1859). Moreover, at one point in his life, Darwin was moved to remark that Spencer was "a dozen times his intellectual superior." For more lines of influence, see *Life and Letters of Charles Darwin* (New York: Appleton-Century-Crofts, 1896).

10. It is not hard to see how these ideas were to be transformed into what became known as social Darwinism in America. A more accurate term would have been social Spencerianism. See Richard Hofstadter, *Social Darwinism in American Thought* (Boston: Beacon, 1955).

11. The article is conveniently reprinted in Herbert Spencer, *Reasons for Dissenting from the Philosophy of M. Comte and Other Essays* (Berkeley, CA: Glendessary, 1968). The article was written in a somewhat defensive manner in an effort to distinguish Spencer's first book, *Social Statics* (New York: Appleton-Century-Crofts, 1888; originally published in 1850), from Comte's use of these terms. Spencer appears to have "protested too much," perhaps seeking to hide some of his debt to the positive philosophy of Comte.

5

※

The Sociology
of Herbert Spencer

Herbert Spencer saw himself as a philosopher rather than as a sociologist. His grand scheme was termed Synthetic Philosophy, and it was to encompass all realms of the universe: physical, psychological, biological, sociological, and ethical. The inclusion of the ethical component makes this philosophy problematic because ideological statements do occasionally slip into Spencer's sociology. Spencer's philosophy was a grand, cosmic scheme, but when he turned to sociology, he made many precise statements and introduced a copious amount of empirical data to illustrate his theoretical ideas. Spencer was, at best, a mediocre philosopher, but he was a very accomplished sociologist, even though he took up sociology rather late in his career. We will begin with the moral philosophy, just to get it out of the way, and then we will turn to his important sociological contributions.[1]

SPENCER'S MORAL PHILOSOPHY: SOCIAL STATICS AND PRINCIPLES OF ETHICS

In his later years, Spencer often complained that his first major work, *Social Statics*,[2] had received too much attention. He saw this book as an early and flawed attempt to delineate his moral philosophy and, hence, as not representative of his more mature thought. Yet, the basic premise of the work is repeated in one of his last books, *Principles of Ethics*.[3] Despite his protests, there

is considerable continuity in his moral arguments, although we should emphasize again that his more scientific statements can and should be separated from
these ethical arguments.

Because Spencer's moral arguments did not change dramatically, we will
concentrate on *Social Statics.* The basic argument of *Social Statics* can be stated as
follows: Human happiness can be achieved only when individuals can satisfy
their needs and desires without infringing on the rights of others to do the same.
As Spencer emphasized,

> Each member of the race . . . must not only be endowed with faculties
> enabling him to receive the highest enjoyment in the act of living, but
> must be so constituted that he may obtain full satisfaction for every
> desire, without diminishing the power of others to obtain like satisfac-
> tion: nay, to fulfill the purpose perfectly, must derive pleasure from seeing
> pleasure in others.[4]

In this early work, as well as in *Principles of Ethics,* Spencer saw this view as
the basic law of ethics and morality. He felt that this law was an extension of
laws in the natural world, and much of his search for scientific laws represented
an effort to develop a scientific justification for his moral position. Indeed, he
emphasized that the social universe, like the physical and biological realms,
revealed invariant laws. But he turned this insight into an interesting moral
dictum: Once these laws are discovered, humans should obey them and cease
trying to construct, through political legislation, social forms that violate these
laws. In this way he was able to base his laissez-faire political ideas on what he
saw as a sound scientific position: The laws of social organization can no more
be violated than can those of the physical universe, and to seek to do so will
simply create, in the long run, more severe problems.[5] In contrast with Comte,
then, who saw the discovery of laws as the tools for social engineering,
Spencer took the opposite tack and argued that once the laws are ascertained,
people should "implicitly obey them!"[6] For Spencer, the great ethical axiom,
"derived" from the laws of nature, is that humans should be as free from
external regulation as is possible. Indeed the bulk of *Social Statics* seeks to
show how his moral law and the laws of laissez-faire capitalism converge and,
implicitly, how they reflect biological laws of unfettered competition and
struggle among species. The titles of some of the chapters best communicate
Spencer's argument: "The Rights of Life and Personal Liberty," "The Right
to the Use of the Earth," "The Right of Property," "The Rights of
Exchange," "The Rights of Women,"[7] "The Right to Ignore the State," "The
Limit of State-Duty," and so forth.

In seeking to join the laws of ethics, political economy, and biology, Spencer
initiated modes of analysis that became prominent parts of his sociology. First,
he sought to discover invariant laws and principles of social organization.
Second, he began to engage in organismic analogizing, drawing comparisons
between the structure of individual organisms and that of societies:

> Thus do we find, not only that the analogy between a society and a liv-
> ing creature is borne out to a degree quite unsuspected by those who

commonly draw it, but also, that the same definition of life applies to both. This union of many men into one community—this increasingly mutual dependence of units which were originally independent—this gradual segregation of citizens into separate bodies, with reciprocally subservient functions—this formation of a whole, consisting of numerous essential parts—this growth of an organism, of which one portion cannot be injured without the rest feeling it—may all be generalized under the law of individuation. The development of society, as well as the development of man and the development of life generally, may be described as a tendency to individuate—*to become a thing.* And rightly interpreted, the manifold forms of progress going on around us, are uniformly significant of this tendency.[8]

Spencer's organismic analogizing often goes to extremes in *Social Statics*—extremes that he avoided in his later works. For example, he at one point argued that "so completely . . . is a society organized upon the same system as an individual being, that we may almost say that there is something more than an analogy between them."[9]

Third, *Social Statics* also reveals the beginnings of Spencer's functionalism. He viewed societies, like individuals, as having survival needs with specialized organs emerging and persisting to meet these needs. And he defined "social health" by how well these needs are being met by various specialized "social organs."

Fourth, Spencer's later emphasis on war and conflict among societies as a critical force in their development can also be observed. While decrying war as destructive, he argued that it allows the more organized "races" to conquer the "less organized and inferior races"—thereby increasing the level and complexity of social organization. This argument was dramatically tempered in his later, scientific works, with the result that he was one of the first social thinkers to see the importance of conflict in the evolution of human societies.[10]

In sum, then, *Social Statics* and *Principles of Ethics* are greatly flawed works, representing Spencer's moral ramblings. We have examined these works first because they are often used to condemn his more scholarly efforts. Although some of the major scientific points can be seen in these moral works, and although his scientific works are sprinkled with his extreme moral position, there is, nonetheless, a distinct difference in style, tone, and insight between his ethical and scientific efforts. Thus, we would conclude that the worth of Spencer's thought is to be found in the more scientific treatises, relegating his ethics to deserved obscurity. We will therefore devote the balance of this chapter to understanding his sociological perspective.[11]

SPENCER'S FIRST PRINCIPLES

In the 1860s, Spencer began to issue his general Synthetic Philosophy by subscription. The goal of this philosophy was to treat the great divisions of the universe—inorganic matter, life, mind, and society—as subject to understanding by scientific principles. The initial statement in this rather encompassing philo-sophical scheme was *First Principles,* published in 1862.[12] In this book Spencer delineated the "cardinal" or "first principles" of the universe. Drawing from the biology and physics of his time, he felt that he had perceived, at the most abstract level, certain common principles that apply to all realms of the universe. Indeed, it must have been an exciting vision to feel that one had unlocked the myster-ies of the physical, organic, and super-organic (societal) universe.

The principles themselves are probably not worth reviewing in detail; rather, the imagery they communicate is important. For Spencer, evolution is the master process of the universe, and it revolves around movement from sim-ple to complex forms of structure. As matter is aggregated—whether this mat-ter be cells of an organism, elements of a moral philosophy, or human beings—the force that brings this matter together is retained, causing the larger mass to differentiate into varying components, which then become integrated into a more complex whole. This complex whole must sustain itself in an envi-ronment, and as long as the forces that have aggregated, differentiated, and integrated the "matter" are sustained, the system remains coherent in the envi-ronment. Over time, however, these forces dissipate, with the result that the basis for integration is weakened, thereby making the system vulnerable to forces in the environment. At certain times, these environment forces can revi-talize a system, giving it new life to aggregate, differentiate and integrate, whereas at other times, these forces simply overwhelm the weakened basis of integration and destroy the system. Thus, evolution is a dual process of build-ing up more complex structures through integration and dissolution of these structures when the force driving them is weakened.

This is all rather vague, of course, but it gives us a metaphorical vision of how Spencer viewed evolution. Evolution revolves around the process of aggregating matter—in the case of society, populations of human beings and the structures that organize people—and the subsequent differentiation and integration of this population. The forces that aggregate this matter—forces such as immigration, new productive forms, use of power, patterns of con-quest, and all those phenomena that have the capacity to bring humans together—are retained, and as a consequence, they also become the forces that differentiate and integrate the matter. For example, if war and conquest have been the basis for aggregation of two populations, the coercive and organiza-tional power causing their aggregation is also the force that will drive the pat-tern of differentiation and integration of the conquered and their conquerors. When this force is spent or proves ineffective in integrating the new society, the society becomes vulnerable to environmental forces, such as war-making from another society.

This image of evolution helps explain the issues that most concerned Spencer when he finally turned to sociology in the 1870s. His view of evo-lution as the aggregation, differentiation, integration, and disintegration of matter pushed him to conceptualize societal dynamics as revolving around increases in the size of the population (the "aggregation" component), the differentiation of the population along several prominent axes, the bases for integrating this differentiated population, and the potential disintegration of the population in its environment. Evolution is thus analysis of societal movement from simple or homogeneous forms to differentiated or heterogeneous forms as well as the mechanisms for integrating these forms in their environments. This is all we need to take from Spencer's *First Principles.*

Spencer moved considerably beyond this general metaphor of evolution, however, because he proposed many specific propositions and guidelines for a science of society. Ultimately, his contribution to sociological theorizing does not reside in his abstract formulas on cosmic evolution but, rather, in his specific analyses of societal social systems—what he called *super-organic* phenomena. This contribution can be found in two distinct works, *The Study of Sociology,* which was published in serial form in popular magazines in 1872, and the more scholarly *The Principles of Sociology,* which was published in several volumes between 1874 and 1896. The former work is primarily a methodological statement on the problems of sociology, whereas the latter is a substantive work that seeks to develop abstract principles of evolution and dissolution and, at the same time, to describe the complex interplay among the institutions of society.

SPENCER'S *THE STUDY OF SOCIOLOGY*

The Study of Sociology[13] was originally published as a series of articles in *Contemporary Review* in England and *Popular Science Monthly* in America. This book represents Spencer's effort to popularize sociology and to address "various considerations which seemed needful by way of introduction to the *Principles of Sociology,* presently to be written."[14] Most of *The Study of Sociology* is a discussion of the methodological problems confronting the science of sociology. At the same time, and in less well-developed form, there are a number of substantive insights that later formed the core of his *Principles of Sociology.* We will first examine Spencer's methodological discussion and then his more theoretical analysis, even though this division does not correspond to the order of his presentation.

The Methodological Problems Confronting Sociology

The opening paragraph of Chapter 4 sets the tone of Spencer's analysis:

> From the intrinsic natures of its facts, from our natures as observers of its facts, and from the peculiar relation in which we stand toward the facts to be observed, there arise impediments in the way of Sociology greater than those of any other science.[15]

He went on to emphasize that the basic sources of bias stem from the inadequacy of measuring instruments in the social sciences and from the nature of scientists who, by virtue of being members of society, observe the data from a particular vantage point. In a series of insightful chapters—far superior to any statement by any other sociologist of the nineteenth century—Spencer outlined in more detail what he termed *objective* and *subjective* difficulties.

Under objective difficulties, Spencer analyzed the problems associated with the "uncertainty of our data." The first problem encountered revolves around the difficulty of measuring the "subjective states" of actors and correspondingly, of investigators' suspending their own subjective orientation when examining that of others. A second problem concerns allowing public passions, moods, and fads to determine what is investigated by sociologists, because it is all too easy to let the popular and immediately relevant obscure from vision more fundamental questions. A third methodological problem involves the "cherished hypothesis," which an investigator can be driven to pursue while neglecting more significant problems. A fourth issue concerns the problem of personal and organizational interests influencing what is seen as scientifically important. Large-scale governmental bureaucracies, and individuals in them, tend to seek and interpret data in ways that support their interests. A fifth problem is related to the second, in that investigators often allow the most visible phenomena to occupy their attention, creating a bias in the collection of data toward the most readily accessible (not necessarily the most important) phenomena. A sixth problem stems from the fact that any observer occupies a position in society and hence will tend to see the world in terms of the dictates of that position. And seventh, depending on the time in the ongoing social process when observations are made, varying results can be induced— thereby signaling that "social change cannot be judged . . . by inspecting any small portion of it."[16]

Spencer's discussion is timely even today, and his advice for mitigating these objective difficulties is also relevant: Social science must rely on multiple sources of data, collected at different times in varying places by different investigators. Coupled with efforts by investigators to recognize their bias, their interests, and their positions in society as well as their commitment to theoretically important (rather than popular) problems, these difficulties can be further mitigated. Yet many subjective difficulties will persist.

There are, Spencer argued, two classes of subjective difficulty: intellectual and emotional. Under intellectual difficulties, Spencer returned to the first of the objective difficulties: How are investigators to put themselves into the subjective world of those whom they observe? How can we avoid representing another's "thoughts and feelings in terms of our own"?[17] For if investigators cannot suspend their own emotional states to understand those of others under investigation, the data of social science will always be biased. Another subjective intellectual problem concerns the depth of analysis, for the more one investigates a phenomenon in detail, the more complicated are its elements and their causal connections. Thus, how far should investigators go before they are to be satisfied with their analysis of a particular phenomenon? At what point are the basic causal connections uncovered? Turning to emotional subjective difficulties, Spencer argued that the emotional state of an investigator can directly influence estimations of probability, importance, and relevance of events.

After reviewing these difficulties and emphasizing that the distinction between subjective and objective is somewhat arbitrary, Spencer devoted separate chapters to "educational bias," "bias of patriotism," "class bias," "political bias," and "theological bias." Thus, more than any other sociologist of the nineteenth century, he had a clear recognition of the many methodological problems confronting the science of society.

Spencer felt that the problems of bias could be mitigated not only by attention to one's interests, emotions, station in life, and other subjective and emotional sources of difficulty but also by the development of "mental discipline." He believed that by studying the procedures of the more exact sciences, sociologists could learn to approach their subjects in a disciplined and objective way. In a series of enlightening passages,[18] he argued that by studying the purely abstract sciences, such as logic and mathematics, one could become sensitized to "the necessity of relation"—that is, that phenomena are connected and reveal affinities. By examining the "abstract-concrete sciences," such as physics and chemistry, one is alerted to causality and to the complexity of causal connections. By examining the "concrete sciences," such as geology and astronomy, one becomes alerted to the "products" of causal forces and the operation of lawlike relations. For it is always necessary, Spencer stressed, to view the context within which processes occur. Thus, by approaching problems with the proper mental discipline—with a sense of relation, causality, and context—one can overcome many methodological difficulties.

The Theoretical Argument

The opening chapters of *The Study of Sociology* present a forceful argument against those who would maintain that the social realm is not like the physical and biological realms. On the contrary, Spencer argued, all spheres of the universe are subject to laws. Every time people express political opinions about what legislators should do, they are admitting implicitly that there are regularities, which can be understood, in human behavior and organization.

Given the existence of discoverable laws, Spencer stressed, the goal of sociology must be to uncover the principles of morphology (structure) and physiology (process) of all organic forms, including the super-organic (society). But, he cautioned, we must not devote our energies to analyzing the historically unique, peculiar, or transitory. Rather, sociology must look for the universal and enduring properties of social organization.[19] Moreover, sociologists should not become overly concerned with predicting future events because unanticipated empirical conditions will always influence the weights of variables and, hence, the outcomes of events. Much more important is discovering the basic relations among phenomena and the fundamental causal forces of them.

In the early and late chapters of *The Study of Sociology,* Spencer sought to delineate, in sketchy form, some principles common to organic bodies. In so doing, he foreshadowed the more extensive analysis in *Principles of Sociology.* He acknowledged Auguste Comte's influence in viewing biology and sociology as parallel sciences of organic forms and in recognizing that understanding of the principles of biology is a prerequisite for discovering the principles of sociology.[20] As Spencer emphasized in all of his sociological works, certain principles of structure and function are common in all organic bodies.

Spencer even hinted at some of these principles, on which he was to elaborate in the volumes of *Principles of Sociology*. One principle is that increases in the size of both biological and social aggregates create pressures for differentiation of functions. Another principle is that such differentiation results in the creation of distinctive regulatory, operative, and distributive processes. That is, as organic systems differentiate, it becomes necessary for some units to regulate and control action, for others to produce what is necessary for system maintenance, or for still others to distribute necessary substances among the parts. A third principle is that differentiation initially involves separation of regulative centers from productive centers, and only with the increases in size and further differentiation do distinctive distributing centers emerge.

Such principles are supplemented by one of the first functional orientations in sociology. In numerous places, Spencer stressed that to uncover the principles of social organization, it is necessary to examine the social whole, to determine its needs for survival, and to assess various structures by how they meet these needs. Although this functionalism always remained somewhat implicit and subordinate to his search for the principles of organization among super-organic bodies, it influenced subsequent thinkers, particularly Émile Durkheim.

In sum, then, *The Study of Sociology* is a preliminary work to Spencer's *Principles of Sociology*. It analyzes in detail the methodological problems confronting sociology; it offers guidelines for eradicating biases and for developing the proper "scientific discipline"; it hints at the utility of functional analysis; and, most important, it begins to sketch out what Spencer thought to be the fundamental principles of social organization. During the two decades after the publication of *The Study of Sociology*, he sought to use the basic principles enunciated in his *First Principles* as axioms for deriving the more specific principles of super-organic bodies.

A NOTE ON SPENCER'S
DESCRIPTIVE SOCIOLOGY

Using his inheritance and royalties, Spencer commissioned a series of volumes to describe the characteristics of different societies.[21] These volumes were, in his vision, to contain no theory or supposition; rather, they were to constitute the "raw data" from which theoretical inductions could be made or by which deductions from abstract theory could be tested. These descriptions became the data source for Spencer's sociological work, particularly his *Principles of Sociology*. As he noted in the "Provisional Preface" of Volume 1 of *Descriptive Sociology*:

> In preparation for *The Principles of Sociology*, requiring as bases of induction large accumulations of data, fitly arranged comparison, I . . . commenced by proxy the collection and organization of facts presented by societies of different types, past and present . . . the facts collected and

arranged for easy reference and convenient study of their relations, being so presented, apart from hypotheses, as to aid all students of social science in testing such conclusions as they have drawn and in drawing others.[22]

Spencer's intent was to use common categories for classifying "sociological facts" on different types of societies. In this way, he hoped that sociology would have a sound database for developing the laws of super-organic bodies. In light of the data available to Spencer, the volumes of *Descriptive Sociology* are remarkably detailed. What is more, the categories for describing different societies are still useful. Although these categories differ slightly from volume to volume, primarily because the complexity of societies varies so much, there is an effort to maintain a consistent series of categories for classifying and arranging sociological facts. Volume 1, *The English,* illustrates Spencer's approach.

First, facts are recorded for general classes of sociological variables. Thus, for *The English,* "facts" are recorded on the following:

1. inorganic environment
   a. general features
   b. geological features
   c. climate
2. organic environment
   a. vegetable
   b. animal
3. sociological environment
   a. past history
   b. past societies from which present system formed
   c. present neighbors
4. characteristics of people
   a. physical
   b. emotional
   c. intellectual

It will be recalled that this initial basis of classification is consistent with Spencer's opening chapters in *Principles of Sociology.* (See his section on "Critical Variables.")

Second, most of *The English* is devoted to a description of the historical development of British society, from its earliest origins to Spencer's time, divided into the following topic headings:

division of labor

regulation of labor

domestic laws—marital

domestic laws—filial

political laws—criminal, civil, and industrial

general government

local government

military

ecclesiastical

professional

accessory institutions

funeral rites

laws of intercourse

habits and customs

aesthetic sentiments

moral sentiments

religious ideas and superstitions

knowledge

language

distribution

exchange

production

arts

agriculture, rearing, and so forth

land—works

habitations

food

clothing

weapons

implements

aesthetic products

supplementary materials

Third, for some volumes, such as *The English,* more detailed descriptions under these headings are represented in tabular form. *The English,* for example, opens with a series of large and detailed tables, organized under the general headings "regulative" and "operative" as well as "structural" and "functional." The tables begin with the initial formation of the English peoples around A.D. 78 and document through a series of brief statements, organized around basic topics (see previous list), to around 1850. By reading across the tables at any given period, the reader can find a profile of the English for that period. By reading down the columns of the table, the reader can note the patterns of change of this society.

The large, oversize volumes of *Descriptive Sociology* make fascinating reading. They are, without doubt, among the most comprehensive and detailed descriptions of human societies ever constructed, certainly surpassing those of

Weber or any other comparative social scientist of the late nineteenth and early twentieth centuries. Although the descriptions are flawed by the sources of data (historical accounts and travelers' published reports), Spencer's methodology is sound, and because he employed professional scholars to compile the data, they are as detailed as they could be at the time. Had the volumes of *Descriptive Sociology* not lapsed into obscurity and had they been updated with more accurate accounts, modern social science would, we believe, have a much firmer database for comparative sociological analysis and for theoretical activity.

SPENCER'S PRINCIPLES OF SOCIOLOGY

The *Principles of Sociology* is a massive work—more than two thousand pages.[23] It is filled with rich descriptive detail from *Descriptive Sociology*, but the book's importance resides in the theory that Spencer developed as the successive installments of this work were released between 1874 and 1896. In *The Principles of Sociology*, Spencer defined sociology as the study of *super-organic* phenomena—that is, of relations among organisms. Thus, sociology could study nonhuman societies, such as ants and other insects, but the paramount super-organic phenomenon is human society. Spencer employed an evolutionary model in analyzing human societies because, over the long haul of history, societies had become ever-more complex. They had followed the basic principles, articulated in *First Principles*, of evolutionary movement from small, homogeneous masses to more complex and differentiated masses. Thus, for Spencer, evolution is the process of increasing differentiation of human populations as they grow in size.

This movement from small and homogeneous to large and complex social forms is always influenced, Spencer argued, by several important factors. One is the nature of the people involved, another is the effects of environmental conditions, and a third is what he termed *derived factors* involving the new environments created by the evolution of society. This last factor is the most important because the larger and more complex societies become, the more their culture and structure shapes the environment to which people and groups must adapt. Of particular importance are the effects of (1) size and density of a population and (2) the relations of societies with their neighbors. As the size and density of a population increases, it becomes more structurally differentiated, with the result that individuals live and adapt to highly diverse social and cultural environments. As societies get larger, they begin to have contact with their neighbors, and this contact can range from cordial relations of economic exchange to warfare and conquest. As we will see, these two derived factors are related in Spencer's scheme because the nature of internal differentiation of a population is very much influenced by the degree to which it is engaged in war with its neighbors.

The Super-Organic and the Organismic Analogy

Part 2 of Volume 1 of *Principles of Sociology* contains virtually all the theoretical statements of Spencerian sociology. Employing the organismic analogy—that is, comparing organic (bodily) and super-organic (societal) organization—Spencer developed a perspective for analyzing the structure, function, and transformation of societal phenomena. Too often commentators have criticized him for his use of the organismic analogy, but in fairness to him, we should emphasize that he generally employed the analogy cautiously. The basic point of the analogy is that because both organic and super-organic systems reveal organization among component parts, they should reveal certain common principles of organization. As Spencer stressed,

> Between society and anything else, the only conceivable resemblance must be due to *parallelism of principle in the arrangement of components.*[24]

It is not surprising that Spencer, as one who saw in his *First Principles* a unity in evolutionary processes among realms of the entire universe and as one who had enumerated the principles of biology, should begin his analysis of the super-organic by trying to show certain parallels between principles of societal and bodily organization.

Spencer began his analogizing by discussing the similarities and differences between organic and super-organic systems. Among important similarities, he delineated the following:

1. Both society and organisms can be distinguished from inorganic matter, for both grow and develop.

2. In both society and organisms, an increase in size means an increase in complexity and differentiation.

3. In both, a progressive differentiation in structure is accompanied by a differentiation in function.

4. In both, parts of the whole are interdependent, with a change in one part affecting other parts.

5. In both, each part of the whole is also a microsociety or organism in and of itself.

6. And in both organisms and societies, the life of the whole can be destroyed, but the parts will live on for a while.[25]

Among the critical differences between a society and an organism, Spencer emphasized the following:

1. The degree of connectedness of the parts is vastly different in organic and super-organic bodies. There is close proximity and physical contact of parts in organic bodies, whereas in super-organic systems there is dispersion and only occasional physical contact of elements.

2. The nature of communication among elements is vastly different in organic and super-organic systems. In organic bodies, communication occurs as molecular waves passing through channels of varying degrees of coherence, whereas among humans communication occurs by virtue of the capacity to use language to communicate ideas and feelings.

3. In organic and super-organic systems, there are great differences in the respective consciousness of units. In organic bodies, only some elements in only some species reveal the capacity for conscious deliberations, whereas in human societies all individual units exhibit the capacity for conscious thought.

The Analysis of Super-Organic Dynamics

If Spencer had only made these analogies, there would be little reason to examine his work. The analogies represent only a sensitizing framework, but the real heart of Spencerian sociology is his portrayal of the dynamic properties of super-organic systems. We begin by examining his general model of system growth, differentiation, and integration; then we will see how he applied this model to societal processes.

System Growth, Differentiation, and Integration As Spencer had indicated in *First Principles,* evolution involves movement from a homogeneous state to a more differentiated state. Spencer stressed that certain common patterns of movement from undifferentiated states can be observed.

First, growth in an organism and in society involves development from initially small units to larger ones.

Second, both individual organisms and societies reveal wide variability in the size and level of differentiation.

Third, growth in both organic and super-organic bodies occurs through compounding and recompounding; that is, smaller units are initially aggregated to form larger units (compounding), and then these larger units join other like units (recompounding) to form an even larger whole. In this way organic and super-organic systems become larger and more structurally differentiated. Hence, growth in size is always accompanied by structural differentiation of those units that have been compounded. For example, small clusters of cells in a bodily organism or in a small, primitive society initially join other cells or small societies (thus becoming compounded); then these larger units join other units (thus being recompounded) and form still larger and more differentiated organisms or societies; and so on for both organic and super-organic growth.

Fourth, growth and structural differentiation must be accompanied by integration. Thus, organic and societal bodies must reveal structural integration at each stage of compounding. Without such integration, recompounding is not possible. For instance, if two societies are joined, they must be integrated before they can, as a unit, become compounded with yet another society. In the processes of compounding, growth, differentiation, and integration,

Spencer saw parallel mechanisms of integration in organisms and societies. For both organic and super-organic systems, integration is achieved increasingly through the dual processes of (1) centralization of regulating functions and (2) mutual dependence of unlike parts. In organisms, for example, as the nervous system and the functions of the brain become increasingly centralized, the organs become ever-more interdependent, whereas in super-organic systems, as political processes become more and more centralized, institutions become increasingly dependent on one another.

Fifth, integration of matter through mutual dependence and centralization of control increase the "coherence" of the system and its adaptive capacity in a given environment. Such increased adaptive capacity often creates conditions favoring further growth, differentiation, and integration, although Spencer emphasized that dissolution often occurred when a system overextended itself by growing beyond its capacity to integrate new units.

These general considerations, which were initially outlined in Spencer's 1862 *First Principles,* offer a model of structuring in social systems. In this model the basic processes are (1) forces causing growth in system size, (2) the differentiation of units, (3) the processes whereby differentiated units become integrated, and (4) the creation of a "coherent heterogeneity," which increases the level of adaptation to the environment.

Thus, for Spencer, institutionalization is a process of growth in size, differentiation, integration, and adaptation. With integration and increased adaptation, a new system is institutionalized and capable of further growth. For example, a society that grows as the result of conquering another will tend to differentiate along lines of conqueror and conquered. It will centralize authority, it will create relations of interdependence, and, as a result, it will become more adapted to its environment. The result of this integration and adaptation is an increased capacity to conquer more societies, setting into motion another wave of growth, differentiation, integration, and adaptation. Similarly, a non-societal social system such as a corporation can begin growth through mergers or expenditures of capital, but it soon must differentiate functions and then integrate them through a combination of mutual dependence of parts and centralization of authority. If such integration is successful, it has increased the adaptive capacity of the system, which can grow if some capital surplus is available.

Conversely, to the extent that integration is incomplete, dissolution of the system is likely. Thus, social systems grow, differentiate, integrate, and achieve some level of adaptation to the environment, but at some point, the units cannot become integrated, setting the system into a phase of dissolution. Figure 5.1 illustrates this process.

Thus, Spencer did not see growth, differentiation, and integration as inevitable. Rather, as differentiation increases, problems of integrating the larger social "mass" generate pressures to find solutions to these problems. For example, if the roles people fill are poorly coordinated, if crime and deviance are high, if commitments to a society's values are weak, if people have no place to work, and if many other disintegrative pressures prevail, then people will seek solutions. These disintegrative tendencies are a kind of "selection

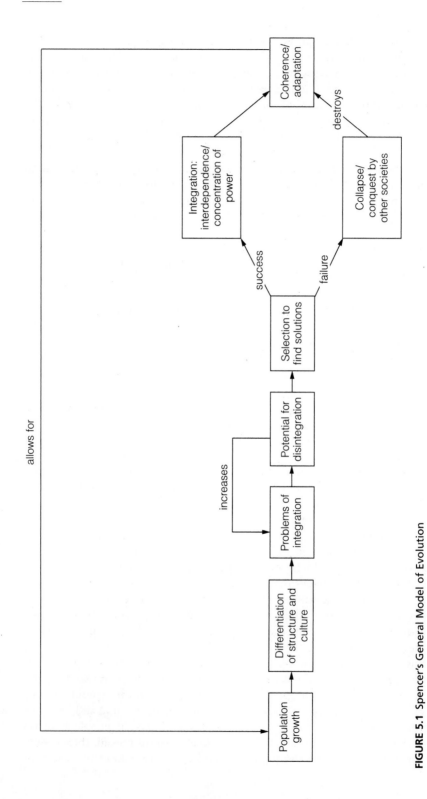

FIGURE 5.1 Spencer's General Model of Evolution

pressure" because, as problems of integration mount, members of a population see the problems and attempt to do something about them. If members find ways to develop relations of mutual interdependence and to regulate their actions with centralized authority, they can stave off these pressures for disintegration and prevent dissolution. Many societies, Spencer argued, had failed to respond adequately to pressures for integration, and as a result, they had collapsed or, more likely, had been conquered by a more integrated and powerful population. Indeed, Spencer argued that war has been an important force in human evolution because the more integrated and organized society will generally win wars against less integrated societies. As the conquered are integrated into the social structure and culture of their conquerors, the size and scale of society increases, and so, even as some societies dissolve or are conquered, the scale of societies had been slowly growing. Spencer's famous phrase "survival of the fittest" was partly intended to communicate this geopolitical dimension of societal evolution.

Geopolitical Dynamics Spencer's model of geopolitics is rather sophisticated for his time. As noted earlier, he argued that one of the most important forces increasing the size and scale of societies was war. Throughout *Principles of Sociology,* a theory of geopolitics is developed—a theory that, surprisingly, is ignored by contemporary sociology.

In this theory, Spencer posited that when power becomes centralized around its coercive base, leaders often use the mobilization of coercive power to repress conflicts within the society and, equally often, to conquer their neighbors. The reverse is also true: When leaders must deal with internal conflicts or external threats from other societies, they will centralize power to mobilize resources to deal with these sources of threat. So, for example, if there is class or ethnic conflict within a society, coercive power will be used to repress it, or if a neighboring society is seen as dangerous, political leaders will centralize coercive power to meet this perceived threat. Indeed, leaders will often use real or imagined internal and external threats as a way to legitimate their grabbing more power; once this power is consolidated, it can be used to centralize power even further.

The result is that once this cycle of threat and centralized power is initiated, it becomes self-fulfilling, for several reasons. First, when power is concentrated, it is used to usurp the wealth and resources of a population, within the result that inequality increases. Those with power simply tax or take resources from others to finance war making and supplement their privilege. And, as inequality increases, the sense of internal threat also escalates because those who have had their resources taken are generally hostile and pose a threat to elites who must then concentrate even more power to deal with this escalated threat, thereby increasing inequality and raising new threats. Over the long run, Spencer felt, this escalating cycle would potentially cause the disintegration of a society, or make it vulnerable to conquest by other societies. Second, when power is concentrated and used to make war against other societies, resources must be extracted to pay for this military effort, thus potentially causing escalated inequality and internal threat, which would compound problems of making war. As long as a society is successful in adventurism, the resentments of those who must pay for it often remain muted, but when external war making does not go well, the resentments from those who have had their resources taken will increase and pose internal threats to leaders, forcing them to mobilize more coercive power, if they can. Third, when power is concentrated to make war, and such efforts at conquest are successful, it then becomes necessary to control those who have been conquered. Needs to manage a restive and resentful population push political leaders to concentrate more power, thus extracting ever-more resources for social control. As resources are channeled to social control, inequality increases, thereby escalating internal threats, which require even more usurpation of resources to maintain social control.

For Spencer, then, concentrating power was a double-edged sword. It allowed one population to conquer another and to increase the size, scale, and complexity of human societies, but it also increased inequalities and internal threats that, unless the cycle of concentrating ever-more power could be broken, would cause the disintegration of the new, larger, and more complex society. This is why, Spencer argued, that military adventurism in the industrial era was ill advised; it drained a population's resources toward coercive and control activities and away from innovation and investment in domestic production. In essence, Spencer was arguing against the creation of what we would call today the *military-industrial complex*. Moreover, Spencer felt that once power was concentrated around the coercive base (military and police) of power, decision making by leaders in government was biased toward the use of coercion rather than alternatives, such as negotiation, compromise, use of incentives, and other alternatives to repression and tight control. For example, if Spencer had seen the rise of the Soviet Union through most of the twentieth century, his theory of geopolitics might have led him to predict its collapse in the 1990s.

Spencer's theory of geopolitics is woven throughout the pages of *Principles of Sociology,* and it is part of a much more general theory of evolution of societies moving from simple to more complex forms. Spencer conceptualized these movements as a series of prominent stages.

Stages of Societal Evolution Spencer argued that increases in the size of a social aggregate necessitate the elaboration of its structure. Such increases in size are the result of high birth rates, migrations, and joining populations together through conquest and assimilation. Although Spencer visualized much growth as the result of compounding and recompounding—that is, successive joining together of previously separate social systems through treaties, conquest, expropriation, and other means—he also employed the concept of compounding in another sense: to denote successive stages of internal growth and differentiation of social systems.

Spencer employed the terms *primary, secondary,* and *tertiary compounding,* by which he meant that a society had undergone a qualitative shift in the level of differentiation from a simpler to a more complex form.[26] These stages of compounding marked a new level of differentiation among and within what Spencer saw as the three main axes of differentiation in social systems: (1) the *regulatory,* in which structures, mobilizing and using power manage relations with the external environment, while engaging in internal coordination of a society's members; (2) the *operative,* in which structures meet system needs for production of goods and commodities and for reproduction of system members and their culture; and (3) the *distributive,* in which structures move materials, people, and information. In simple societies, these three great axes of differentiation are only incipient, being collapsed together, but as societies grow and compound, distinctive structures emerge for each of these axes. The subsequent course of evolution then occurs with further differentiation between and within these axes.

Primary compounding occurs when the simplest structures become somewhat more complex. At first, only a differentiation of regulatory and operative processes is evident. For example, the sexual division of labor between males and females might move to one where some males have more authority than do females (regulatory functions), while females began to shoulder a greater burden in gathering food and in socializing the young (operative functions). Thus, the first big shift in the level of differentiation is along the regulatory and operative axes; only with further growth and differentiation of the population does a distinctive set of structures devoted to distribution of resources, people, and information emerge. Secondary compounding occurs, Spencer argued, when the structures involved in regulatory, operative, and distributive functions undergo further differentiation. For example, internal administrative structures might become distinguished from warfare roles in the regulative system; varieties of domestic activities, with specialized persons or groups involved in these separate activities, might become evident; or distinguishable persons or groups involved in external trade and internal commerce might become differentiated. Tertiary compounding occurs when these secondary structures undergo further internal differentiation, so that one can observe distinct structures involved in varieties of regulatory, operative, and distributive processes.

Figure 5.2 represents these dynamics diagrammatically as a model. This model outlines the "stages" of societal evolution in three respects. First, Spencer saw five basic stages: (1) simple without head or leadership, (2) simple with head or leadership, (3) compound, (4) doubly compound, and (5) trebly compound. Second, he visualized each stage as being denoted by (1) a given degree of differentiation *among* regulatory, operative, and distributive processes and (2) a level of differentiation *within* each process. Third, he suggested how the nature of regulation, operation, and distribution changes with each stage of compounding (as denoted by the descriptive labels in each box in Figure 5.2).

Contained within Spencer's view of the stages of evolution is a mode of functional analysis. By viewing social structures with reference to regulatory, operative, and distributive processes, Spencer implicitly argued that these three processes represent basic "functional needs" of all organic and super-organic

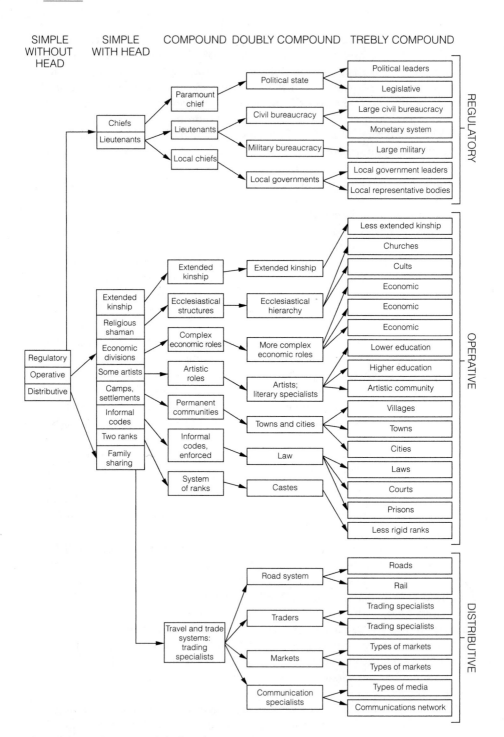

SIMPLE WITHOUT HEAD | SIMPLE WITH HEAD | COMPOUND | DOUBLY COMPOUND | TREBLY COMPOUND

REGULATORY

Chiefs / Lieutenants

Paramount chief → Political state → Political leaders / Legislative

Lieutenants → Civil bureaucracy → Large civil bureaucracy / Monetary system

→ Military bureaucracy → Large military

Local chiefs → Local governments → Local government leaders / Local representative bodies

Regulatory / Operative / Distributive

OPERATIVE

Extended kinship / Religious shaman / Economic divisions / Some artists / Camps, settlements / Informal codes / Two ranks / Family sharing

Extended kinship → Extended kinship → Less extended kinship / Churches / Cults / Economic

Ecclesiastical structures → Ecclesiastical hierarchy → Economic / Economic

Complex economic roles → More complex economic roles → Lower education / Higher education

Artistic roles → Artists; literary specialists → Artistic community

Permanent communities → Towns and cities → Villages / Towns / Cities

Informal codes, enforced → Law → Laws / Courts / Prisons

System of ranks → Castes → Less rigid ranks

DISTRIBUTIVE

Travel and trade systems: trading specialists

Road system → Roads / Rail

Traders → Trading specialists / Trading specialists

Markets → Types of markets / Types of markets

Communication specialists → Types of media / Communications network

FIGURE 5.2 Spencer's Model of Evolution

systems. Thus, a particular structure is to be assessed by its contribution to one or more of these three basic needs. But Spencer's functionalism is even more detailed, for he argued in several places that all social structures had their own internal regulatory, operative, or distributive needs, regardless of which of the three functions they fulfilled for the larger social whole in which they were located.[27] For example, the family might be viewed as an operative structure for the society as a whole, but it also reveals its own division of labor along regulatory, operative, and distributive functions.

Sequences of Differentiation Spencer devoted most of his attention to analyzing the regulatory system because he was primarily a theorist of power.[28] His discussion revolves around delineating those conditions under which the regulatory system (1) becomes differentiated from operative and distributive processes and (2) becomes internally differentiated. We consider Spencer a political theorist because of this emphasis on the regulating system—that is, the center of power in society.

If we translate differentiation between regulatory and operative functions into more modern terminology, then the first phase of differentiation is between the emergence of a political system and specialized structures involved in (a) production or the conversion of resources into usable commodities and (b) reproduction or the regeneration of people as well as their culture. Most of Spencer's sociology is devoted to the regulatory system, especially the cause and consequences of centralized power on operative and distributive processes. In general, Spencer posited the following conditions as increasing the concentration and centralization of power:

1. When productive processes become complex, they require some kind of external authority to coordinate activity to ensure that exchanges proceed smoothly, to maintain contractual obligations, to prevent fraud and corruption, and to ensure that necessary productive activities are conducted. These pressures for external authority lead to the mobilization of power. Once this capacity to regulate the economy exists, the level of production can expand further, creating new pressures for expanded use of power to coordinate more complex levels of economic activity.

2. When there are internal threats, typically arising from conflicts over inequalities, centers of power will mobilize to control the conflict. Ironically, the use of power to control conflict often increases inequality because those with power begin to usurp resources for themselves. As a result, as more power is concentrated, further inequality and conflict will ensue in a cycle of conflict, use of power to control and usurp, increased inequality, and escalated potential for conflict.

3. When there are external threats from other societies arising from economic competition or military confrontations, centers of power will mobilize coercive force to deal with such threats. Consequently, they will also set off the dynamics described under (2) above because when

power is mobilized to deal with threats, it is also used to enhance the well being of elites, thereby increasing inequality and the potential for conflict. Moreover, as noted for Spencer's theory of geopolitics, when a political system is mobilized for conflict with other societies, it will generally pursue war as the first option (rather than diplomacy), with the result that if it wins a war, this very success creates new internal threats, as specified in (2) above, revolving around the inequalities between conquerors and conquered.

As both regulatory and operative processes develop, Spencer argued, pressures for transportation, communication, and exchange among larger and more differentiated units increase. The result of these pressures is that new structures emerge as part of a general expansion of distributive functions. Spencer devoted considerable attention to the historical events causing increases in transportation, roads, markets, and communication processes, and by themselves, these descriptions make for fascinating reading. At the most general level, he concluded:

> The truth we have to carry with us is that the distributing system in the social organism, as in the individual organism, has its development determined by the necessities of transfer among inter-dependent parts. Lying between the two original systems, which carry on respectively the outer dealings with surrounding existences, and the inner dealings with materials required for sustentation [*sic*] its structure becomes adapted to the requirements of this carrying function between the two great systems as wholes, and between the sub-divisions of each.[29]

As the regulatory and operative systems expand, thereby causing the elaboration of the distributive system, this third great system differentiates in ways that facilitate increases in (1) the speed with which material and information circulate and (2) the varieties of materials and information that are distributed. As the capacities for rapid and varied distribution increase, regulatory and operative processes can develop further; as the latter expand and differentiate, new pressures for rapid and varied distribution are created. Moreover, in a series of insightful remarks, Spencer noted that this positive feedback cycle involved an increase in the ratio of information to materials distributed in complex, differentiating systems.[30]

In sum, then, Spencer's view of structural elaboration emphasizes the processes of structural growth and differentiation through the joining of separate systems and through internal increases in size. As an evolutionist, Spencer took the long-range view of social development as growth, differentiation, integration, and increased adaptive capacity; then, with this new level as a base, further growth, differentiation, integration, and adaptive capacity would be possible. His view of structural elaboration is thus highly sophisticated, and though flawed in many ways, it is the equal of any other nineteenth-century social theory.

System Dialectics and Phases As we have emphasized, Spencer saw war as an important causal force in human societies. War pushes a society to develop centralized regulatory structures to expand and coordinate internal operative and distributive processes. Yet war can have an ironic effect on a society: Once these operative and distributive processes are expanded under conditions of external conflict, they increasingly exert pressures for less militaristic activity and for less authoritarian centralization. For example, a nation at war will initially centralize along authoritarian lines to mobilize resources, but as such mobilization expands the scope of operative and distributive processes, those engaged in operation and distribution develop autonomy and begin to press for greater freedom from centralized control. In this way, Spencer was able to visualize war as an important force in societal development but, at the same time, as an impediment to development if concentrated power is used to concentrate even more power. And in an enlightening chapter on "social metamorphoses,"[31] he argued that the dynamic force underlying the overall evolution of the super-organic from homogeneous to heterogeneous states was the successive movement of societies in and out of "militant" (politically centralized and authoritarian) and "industrial" (less centralized) phases. This cyclical dynamic is presented in Figure 5.3, which views these phases somewhat more abstractly than in Spencer's portrayal.

Figure 5.3 presents one of the most interesting (and often ignored) arguments in Spencerian sociology. For Spencer, there is always a dialectical undercurrent during societal evolution (and dissolution) revolving around the relationship between regulatory and operative processes. On the one hand, each of these initial axes of differentiation encourages the growth and development of the other in a positive feedback cycle, and on the other hand, there is an inherent tension and dialectic between the two. For example, war expands regulatory functions; increased regulatory capacity allows for more extensive coordination of operative processes; greater operative capacity encourages expanded war efforts and, hence, expansion of the regulatory system. But at some point in this cycle, development of internal operative structures primarily for war making becomes counterproductive, limiting the scope and diversity of development in operative processes. Indeed, Spencer argued that too much political control of production and reproduction causes economic stagnation and, in the reproductive sphere, arouses resentments. Over time, and under growing pressures from the internal sector, as mobilization against tight control increases, the warlike profile of the regulatory system is reduced. Thus, as resentments against too much power arise, it is not inevitable that political elites will continue to concentrate power to manage such threats, as we examined earlier in Spencer's theory of geopolitics. Spencer saw an alternative: Growing resentments leads political leaders to make concessions and to recognize that they must release some of their control. Spencer never specifies the conditions under which leaders will give up power; he simply assumed that it had been an important dynamic in the evolution of human societies from simple to complex forms. When power is released, operative

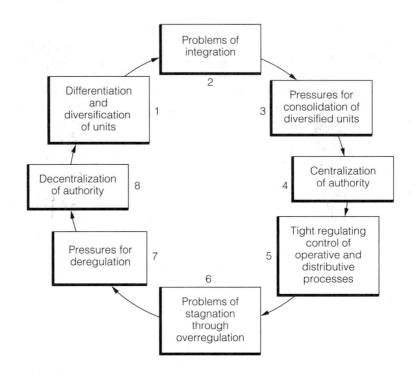

FIGURE 5.3 Phases of Institutionalization

structures expand and differentiate in many directions, but over time, these structures become too divergent, poorly coordinated, and unregulated. A war can provide, Spencer believed, the needed stimulus for greater regulation and coordination of these expanded and diversified operative processes, thus setting the cycle into motion once again. Alternatively, problems of coordination become so acute that government must step in to restore order.

Such had been the case throughout evolutionary history, Spencer thought. Curiously, he also seemed to argue that modern, industrial capitalism made the need for war and extensive regulation by a central state obsolete. No longer would it be necessary, in Spencer's capitalistic utopia, for centralized government, operating under the pressures of war, to seek extensive regulation of operative and distributive processes. These processes were, in his vision, now sufficiently developed and capable of growth, expansion, and integration without massive doses of governmental intervention. Here, Spencer's ideology clearly distorts his perceptions because advanced capitalism requires the exercise of control by government, yet, the analysis of the dialectic between militant and industrial societies allowed him to see how concentrated power could be lessened without disintegration.

Classifying Social Systems Spencer also used these models of societal evolution (Figure 5.2) and system phases (Figure 5.3) as a basis for classifying societies. His most famous typology (Table 5.1) is of what he termed *militant* and *industrial* societies—a typology, that has frequently been misunderstood by commentators. Too often it is viewed as representing a unilinear course of evolutionary movement from traditional and militant to modern and industrial societal forms. Although Spencer often addressed the evolution of societies from a primitive to a modern profile, he did not rely heavily on the militant-industrial typology in describing types or stages of evolutionary change. Rather, as is emphasized in Figure 5.3, the militant-industrial distinction is primarily directed at capturing the difference between highly centralized authority systems where regulatory processes dominate and less centralized systems where operative processes prevail.[32] The term *industrial* does not refer to industrial production in the sense of modern factories and markets but, instead, to a reduction in centralized power and to the vitality and diversity of operative processes. Both the simplest and most modern societies can be either militant or industrial; Spencer hoped that modern industrial capitalism would be industrial rather than militaristic. As we noted in the last section, however, Spencer saw societies as cycling in and out of centralized and decentralized phases. The typology is meant to capture this dynamic.

The distinction between militant and industrial societies emphasizes that during the course of social growth, differentiation, integration, and adaptive upgrading,[33] societies move in and out of militant (dominance of regulatory) and industrial (operative) phases. Militant phases consolidate the diversified operative structures of industrial phases. The causes of either a militant or industrial profile for a system at any given time are varied, but Spencer saw as critical (1) the degree of external threat from other systems and (2) the need to integrate dissimilar populations and cultures. The greater the threat to a system from external systems or the more diverse the system's population (an internal threat), the more likely it is to reveal a militant profile. Once external and internal threats have been mitigated through conquest, treaties, assimilation, and other processes, however, pressures for movement to an industrial profile increase. Such is the basic dynamic underlying broad evolutionary trends from a homogeneous to a heterogeneous state of social organization.

Spencer's other typology, which has received considerably less attention than the militant-industrial distinction, addresses the major stages in the evolution of societies. Whereas the militant-industrial typology seeks to capture the cyclical dynamics underlying evolutionary movement, Spencer also attempts to describe the distinctive stages of long-term societal development, as was modeled earlier in Figure 5.2. This typology revolves around describing the pattern and direction of societal differentiation. As such, it is concerned with the processes of compounding. As was evident in Figure 5.2, Spencer marked distinctive stages of societal growth and differentiation: simple (with and without leadership), compound, doubly compound, and trebly compound.

In Table 5.2 we have taken Spencer's narrative and organized it in a somewhat more formal way. But the listing of characteristics for simple (both those with leaders and those without), compound, doubly compound, and trebly compound societies for regulatory, operative, and distributive as well as for

Table 5.1 Spencer's Typology of Militant and Industrial Societies

Basic System Processes	Militant	Industrial
1. Regulatory processes		
a. Societal goals	Defense and war	Internal productivity and provision of services
b. Political organization	Centralized, authoritarian	Less centralized; less direct authority over system units
2. Operative processes		
a. Individuals	High degrees of control by state; high levels of stratification	Freedom from extensive controls by state; less stratification
b. Social structures	Coordinated to meet politically established goals of war and defense	Coordinated to facilitate each structure's expansion and growth
3. Distributive processes		
a. Flow of materials	From organizations to state; from state to individuals and other social units	From organizations to other units and individuals
b. Flow of information	From state to individuals	Both individuals to state and state to individuals

demographic (population characteristics) dimensions captures the essence of Spencer's intent. Several points need to be emphasized. First, although certain aspects of Spencer's description are flawed, his summary of the distinctive stages of societal is equal, or superior, to any that have been delineated recently by anthropologists and sociologists.[34] Second, this description is far superior to any developed by other anthropologists and sociologists of Spencer's time.

Spencer sought to communicate what we can term *structural explanations* with this typology. The basic intent of this mode of explanation is to view certain types of structures as tending to coexist. As Spencer concluded,

> The inductions arrived at . . . show that in social phenomena there is a general order of co-existence and sequence; and therefore social phenomena form the subject-matter of a science reducible, in some measure at least, to the deductive form.[35]

Thus, by reading down the columns of Table 5.2, we can see that certain structures are likely to coexist within a system. And by reading across the table, the patterns of change in structures with each increment of societal differentiation can be observed. Moreover, as Spencer stressed, such patterns of social evolution conformed to the general law of evolution enunciated in *First Principles.*

> The many facts contemplated unite in proving that social evolution forms a part of evolution at large. Like evolving aggregates in general, societies show *integration,* both by simple increase of mass and by coalescence and re-coalescence of masses. The change from *homogeneity* to *heterogeneity* is multitudinously exemplified; up from the simple tribe, alike in all its

parts, to the civilized nation, full of structural and functional unlikenesses. With progressing integration and heterogeneity goes increasing *coherence.* We see the wandering group dispersing, dividing, held together by no bonds; the tribe with parts made more coherent by subordination to a dominant man; the cluster of tribes united in a political plexus under a chief with sub-chiefs; and so on up to the civilized nation, consolidated enough to hold together for a thousand years or more. Simultaneously comes increasing *definiteness.* Social organization is at first vague; advance brings settled arrangements which grow slowly more precise; customs pass into laws which, while gaining fixity, also become more specific in their applications to varieties of actions; and all institutions, at first con-fusedly intermingled, slowly separate, at the same time that each within itself marks off more distinctly its component structures. Thus in all respects is fulfilled the formula of evolution. There is progress towards greater size, coherence, multiformity, and definiteness.[36]

In sum, then, Spencer provided two basic typologies for classifying societal systems. One typology—the militant-industrial distinction—emphasizes the cyclical phases of all societies at any stage of evolution. The second typology is less well known but probably more important. It delineates the structural fea-tures and demographic profile of societies at different stages of evolution. Embedded in this typology is a series of statements on what structures tend to cluster together during societal growth and differentiation. This typology is, in many ways, the implicit guide for Spencer's structural and functional analysis of basic societal institutions, which comprises Parts 3 through 7 in Volumes 1 and 2 of *Principles of Sociology.* We should, therefore, close our review of *Principles of Sociology* by briefly noting some of the more interesting generalizations that emerge from Spencer's description of basic human institutions.

THE ANALYSIS
OF SOCIETAL INSTITUTIONS

Fully two-thirds of *Principles of Sociology* is devoted to an evolutionary description and explanation of basic human institutions.[37] For Spencer, insti-tutions are enduring patterns of social organization that (1) meet fundamen-tal functional needs or requisites of human organization and (2) control the activities of individuals and groups in society. Spencer employed a "social selection" argument in his review of institutional dynamics. The most basic institutions emerge and persist because they provide a population with adap-tive advantages in a given environment, both natural and social. That is, those patterns of organization that facilitate the survival of a population in the nat-ural environment and in the milieu of other societies will be retained, or "selected"; as a consequence, these patterns will become institutionalized in the structure of a society. Because certain problems of survival always confront

Table 5.2 Spencer's Stages of Evolution

System Dimensions	Simple Society	
	Headless	*Headed*
1. Regulatory system	Temporary leaders who emerge in response to particular problems	Permanent chief and various lieutenants
2. Operative system		
a. Economic structure	Hunting and gathering	Pastoral; simple agriculture
b. Religious structure	Individualized religious worship	Beginnings of religious specialists: shaman
c. Family structure	Simple; sexual division of labor	Large, complex; sexual and political division of labor
d. Artistic-literary forms	Little art; no literature	Some art; no literature
e. Law and customs	Informal codes of conduct	Informal codes of conduct
f. Community structure	Small bands of wandering families	Small, settled groupings of families
g. Stratification	None	Chief and followers
3. Distributive system		
a. Materials	Sharing within family and band	Intra- and interfamilial exchange and sharing
b. Information	Oral, personal	Oral, personal
4. Demographic profile		
a. Size	Small	Large
b. Mobility	Mobility within territory	Less mobility; frequently tied to territory

Compound Society	Doubly Compound	Trebly Compound (Never Formally Listed)
Hierarchy of chiefs, with paramount chief, local chiefs, and varieties of lieutenants	Elaboration of bureaucratized political state; differentiation between domestic and military administration	Modern political state
Agricultural; general and local division of labor	Agricultural; extensive division of labor	Industrial capitalism
Established ecclesiastical arrangements	Ecclesiastical hierarchy; rigid rituals and religious observance	Religious diversity in separate church structures
Large, complex; numerous sexual, age, and political divisions	Large, complex; numerous sexual, age, and political divisions	Small, simple; decreased in sexual division of labor
Artists	Artists; literary specialists; scholars	Many artistic literary specialists; scholars
Informal codes; enforced by political elites and community members	Written law and codes	Elaborate legal codes; civil and criminal
Village; permanent buildings	Large towns; permanent structures	Cities, towns, and hamlets
Five or six clear ranks	Castes; rigid divisions	Classes; less rigid
Travel and trade between villages	Roads among towns; considerable travel and exchange; traders and other specialists	Roads, rail, and other non-manual transportation; many specialists
Oral, personal; at times, mediated by elites or travelers	Oral and written; edicts; oracles; teachers and other communications specialists	Oral and written; formal media structures for edicts; many communications specialists
Larger; joining of several simple societies	Large	Large
Less mobility; tied to territory; movement among villages of a defined territory	Settled; much travel among towns	Settled; growing urban concentrations; much travel; movement from rural to urban centers

the organization of people, it is inevitable that among surviving populations a number of common institutions would be evident for all enduring societies—for example, kinship, ceremony, politics, religion, and economy. Spencer discusses more than these five institutions, but our review will emphasize only these, because they provide some of the more interesting insights in Spencerian sociology.

Domestic Institutions and Kinship

Spencer argued that kinship emerged to meet the most basic need of all species: reproduction.[38] Because a population must regulate its own reproduction before it can survive for long, kinship was one of the first human institutions. This regulation of reproduction involves the control of sexual activity, the development of more permanent bonds between men and women, and the provision of a safe context for rearing children.

Spencer's discussion of kinship was extremely sophisticated for his time. After making the previous functional arguments, he embarked on an evolutionary analysis of varying types of kinship systems. Although flawed in some respects, his approach was nonetheless insightful and anticipated similar arguments by twentieth-century anthropologists. Some of the more interesting generalizations emerging from his analysis are the following:

1. In the absence of alternative ways of organizing a population, kinship processes will become the principal mechanism of social integration.

2. The greater the size of a population without alternative ways of organizing activity, the more elaborate will be a kinship system, and the more it will reveal explicit rules of descent, marriage, endogamy, and exogamy.

3. Those societies that engage in perpetual conflict will tend to create patrilineal descent systems and patriarchic authority; as a consequence, they will reveal less equality between the sexes and will be more likely to define and treat women as property.[39]

Ceremonial Institutions

Spencer recognized that human relations were structured by symbols and rituals.[40] Indeed, he tended to argue that other institutions—kinship, government, and religion—were founded on a "preinstitutional" basis revolving around interpersonal ceremonies, such as the use of (1) particular forms of address, (2) titles, (3) ritualized exchanges of greetings, (4) demeanors, (5) patterns of deference, (6) badges of honor, (7) fashion and dress, and (8) other means for ordering interactions among individuals. Thus, as people interact, they "present themselves" through their demeanor, fashion, forms of talk, badges, titles, and rituals, and in so doing they expect certain responses from others. Interaction is thereby mediated by symbols and ceremonies that structure how individuals are to behave toward one another. Without this control of relations through symbols and ceremonies, larger institutional structures could not be sustained.

Spencer was particularly interested in the effects of inequality on ceremonial processes, especially inequalities created by centralization of power (as is the case in the militant societies depicted in Table 5.1). These interesting generalizations emerge from his more detailed analysis:

1. The greater the degree of political centralization that exists in a society, the greater the level of inequality will be and, hence, the greater the concern for symbols and ceremonials demarking differences in rank among individuals will be.

2. The greater the concern over differences in rank, (1) the more likely people in different ranks are to possess distinctive objects and titles to mark their respective ranks, and (2) the more likely interactions between people in different ranks are to be ritualized by standardized forms of address and stereotypical patterns of deference and demeanor.

3. Conversely, the less the degree of political centralization and the less the level of inequality, the less people are concerned about the symbols and ceremonies that demark rank and regulate interaction.[41]

Political Institutions

In his analysis of political processes in society, Spencer also developed a perspective for examining social class structures.[42] In his view, problems of internal conflict resulting from unbridled self-interest and the existence of hostility with other societies have been the prime causal forces behind the emergence and elaboration of government. Although governments reveal considerable variability, they all evidence certain common features: (1) paramount leaders, (2) clusters of subleaders and administrators, (3) large masses of followers who subordinate some of their interests to the dictates of leaders, and (4) legitimating beliefs and values that give leaders "the right" to regulate others. Spencer argued that once governmental structures exist, they are self-perpetuating and will expand unless they collapse internally for lack of legitimacy or are conquered from without. In particular, war and threats of war centralize government around the use of force to conquer additional territories and internally regulate operative processes, with the result that governmental structures expand. Moreover, the expansion of government and its centralization create or exacerbate class divisions in a society because those with resources can use them to mobilize power and political decisions that further enhance their hold on valued resources. Thus, Spencer developed a very robust political sociology, and although a listing of only a few generalizations cannot do justice to the sophistication of his approach, some of his more interesting conclusions are the following:

1. The larger the number of people and internal transactions among individuals in a society, the greater will be the size and degree of internal differentiation of government.

2. The greater the actual or potential level of conflict with other societies and within a society, the greater will be the degree of centralization of power in government.

3. The greater the centralization of power, the more visible class divisions will be; and the more these divisions create potential or actual internal conflict.

Religious Institutions

Spencer's analysis emphasized that all religions shared certain common elements: (1) beliefs about supernatural beings and forces, (2) organized groupings of individuals who share these beliefs, and (3) ritual activities directed toward those beings and forces presumed to have the capacity to influence worldly affairs.[43] Religions emerge in all societies, he argued, because they increase the survival of a population by (1) reinforcing values and beliefs through the sanctioning power of the supernatural and (2) strengthening existing social structural arrangements, especially those revolving around power and inequality, by making them seem to be extensions of the supernatural will.

Spencer provided an interesting scenario on the evolution of religion from primitive notions of "ancestor spirits" to the highly bureaucratized monotheistic religions that currently dominate the world. He saw the evolution and structural patterns of religion as intimately connected to political processes, leading him to propose the following generalizations:

1. The greater the level of war and conquest by a society, the greater are the problems of consolidating diverse religious beliefs, thereby forcing the expansion of the religious class of priests to reconcile these diverse religions and create polytheistic religions.

2. The greater the political centralization and the greater the level of class inequalities in a society, the more likely is the priestly class to create a coherent pantheon of ranked deities.

3. The more government relies on the priestly class to provide legitimation through a complex system of religious beliefs and symbols, the more this class extracts wealth and privilege from political leaders, thereby consolidating their distinctive class position and creating an elaborate bureaucratic structure for organizing religious activity.

4. The more centralized a government is and the more it relies on religious legitimation by a privileged and bureaucratized class of priests, the greater is the likelihood of a religious revolt and the creation of a simplified and monotheistic religion.

Economic Institutions

For Spencer, the long-term evolution of economic institutions revolves around (1) increases in technology or knowledge about how to manipulate the natural environment, (2) expansion of the production and distribution of goods and services, (3) accumulation of capital or the tools of production, and (4) changes in the organization of labor.[44] In turn, these related processes are the result of efforts to achieve greater levels of adaptation to the environment and to meet constantly escalating human needs. That is, as one level of economic adaptation is created, people's needs for new products and services escalate and generate pressures for economic reorganization. Thus, as new technologies, modes of production, mechanisms of distribution, forms of capital, and means for organizing labor around productive processes are developed, a more effective level of adaptation to the natural environment is achieved; as this increased adaptive capacity is established, people begin to desire more. As a result, economic production becomes less and less tied to problems of survival in the natural environment during societal evolution and increasingly the result of escalating wants and desires among the members of a society.

Spencer further argued that war decreased advances in overall economic productivity because mobilization for war distorts the economy away from domestic production toward the development of military technologies and the organization of production around military products or services. For Spencer, war depletes capital, suppresses wants and needs for consumer goods, encourages only military technologies, and mobilizes labor for wartime production (while killing off much of the productive labor force). Only during times of relative peace, then, will economic growth ensue. Such growth in the domestic economy will be particularly likely to occur when there are increases in population size. In Spencer's view, escalating population size under conditions of peace creates pressure for expanded production while increasing needs for new products and services. These and many other lines of argument in his analysis of the economy have a highly modern flavor, but unlike his approach to other institutions, he presents few abstract generalizations, so we will not attempt to conclude with any here.

This brief summary of Spencer's analysis of basic institutions does not do justice to the sophistication of his approach. As much as any scholar of his time, or of today, he saw the complex interrelationships among social structures. One reason for this sophistication in his analysis is his in-depth knowledge of diverse societies, which he acquired through the efforts of researchers hired to construct descriptions of historical and contemporary societies. Throughout his work, his ideas are illustrated by references to diverse societies. Such familiarity with many historical and contemporary societies came from his efforts to build a "descriptive sociology."

CRITICAL CONCLUSIONS

Herbert Spencer is, without doubt, the most neglected of the early sociological theorists. Comte is, of course, also neglected but unlike Spencer, he never really developed a theory. Spencer did articulate a theory that, for the most part, is ignored by contemporary sociologists. Why should this be so?

There can be no doubt that Spencer's moral philosophy stigmatized him, especially his view that government should not intervene too extensively to help the unfortunate. Such a view ran counter to the expansion of the welfare state in the twentieth century. This ideology taints Spencer's sociology, and it has clearly made scholars reluctant to give it a fair reading.

Spencer's coining of the phrase "survival of the fittest" and the use of this idea in much twentieth-century conservative philosophy, and even worse, in the eugenics movement of the last century further stigmatized his sociology. Indeed, those advocating the selective breeding of humans, or alternatively, the natural death of the "less fit" have at times made appeals to Spencer, a fact which certainly has not helped our retrospective view of him.

Spencer also was the supreme generalist at a time when academic disciplines were beginning to specialize. Spencer's sociology is a part of a much larger, almost cosmic vision of evolution in all domains of the universe. Twentieth-century sociologists were less likely to embrace such grandiose and rather vague pronouncements, and this is even more the case for the discipline today where hyper-specialization is rapidly occurring.

Spencer's emphasis on evolution as the master societal process was also to get him into trouble. By the second decade of the twentieth century, evolutionary thinking was under heavy attack, and as the supreme evolutionary thinker in the social sciences, Spencer was under constant criticism. When the evolutionary paradigm collapsed and fell into obscurity in the 1930s, so did Spencer's sociology. Even with the revival of evolutionary thinking in the 1960s in sociology, Spencer was never resurrected, except by a few dedicated scholars.

Spencer probably wrote too much. The key ideas of Spencer's sociology must be extracted from thousands of pages, and most sociologists are unwilling to read all these materials. Spencer's works are filled with far too many examples and illustrations, and his work gets sidetracked on issues that are of little interest to sociologists today. This is also true of other theorists in sociology's early pantheon, but contemporary scholars are unwilling to make the effort to wade through all of the pages.

Still, if scholars will have the patience to read through these many pages, there are many strong points in Spencer's sociology that deserve a re-hearing. First, Spencer developed a very sophisticated theory of politics in his sociology. This theory emphasizes that the concentration of power dramatically transforms all other institutional system, as can be seen by the propositions that we have listed in the text, and it sets into motion both geopolitical and dialectical dynamics. Even by today's standards, this portion of Spencer's sociology is rather sophisticated. Indeed, Spencer should be considered a political theorist as much as a functionalist or evolutionary thinker, and if this fact were recognized, perhaps sociologists would be willing to give his work another reading. Second, Spencer's views on the dynamics of differentiation are worth revisiting. The basic relationships among system size, level of differentiation, and integration through interdependence and power do represent some of sociology's most powerful laws, and although more contemporary sociologists have worked with these ideas, they seem to forget from where they come. And third, even though the use of so much data from his *Descriptive Sociology* makes reading *The Principles of Sociology* an arduous task, there is much to be learned from these materials. Few sociologists have ever documented their arguments with so much ethnographic and historical detail. In some ways, Spencer can serve as a model for how this should be done.

Yet, there are problems in Spencer's sociology. Although the ideological tracks are not as prominent, they are still in *The Principles of Sociology,* and they do indeed detract from his arguments and erode his credibility. Spencer's general arguments about evolution—"matter in motion," and the like—are vague; although contemporary general systems theory tries to employ similar concepts across system levels, the lack of broad acceptance of this modern approach tells us why Spencer's efforts at a general systems theory will never be useful. Spencer's organismic analogy, along with the functionalism that inheres in this analogy, also pose problems, especially in recent decades where functionalism as a theoretical approach now appears to be dead. To analyze structures through the needs they meet—in the case of Spencer's sociology, needs for regulation, operation, and distribution—is always problematic, because outcomes often appear to cause themselves. For example, if a structure is seen to meet a need for regulation, it is easy to imply that regulation, or the outcome, caused the structure meeting this outcome to emerge. Functional arguments often have the circular reasoning, which is why sociologists are suspicious of such arguments. There are many questionable elements in Spencer's sociology, but no more so than for other founders of the discipline. In the end, there is a prejudice against Spencer, one not founded on a careful reading because most sociologists have never read Spencer, but a prejudice that has been passed down from one generation of sociologists to the other.

Still, there is powerful theory in Spencer's sociology. In the next chapter, we will try to ignore the problematic aspects of Spencer's grand Synthetic Philosophy and, instead, extract the useful theoretical arguments. These arguments, we believe, can still inform sociology.

NOTES

1. Spencer's complete works, except for his *Descriptive Sociology* (see later analysis), are conveniently pulled together in the following collection: *The Works of Herbert Spencer,* 21 vols. (Osnabruck, Germany: Otto Zeller, 1966). However, our references will be to the separate editions of each of his individual works. Moreover, many of the dates for the works to be discussed span several years because Spencer sometimes published his works serially in several volumes (frequently after they had appeared in periodicals). Full citations will be given when discussing particular works. For a recent review of primary and secondary sources on Spencer, see Robert G. Perrin, *Herbert Spencer: A Primary and Secondary Bibliography* (New York: Garland, 1993).

2. Herbert Spencer, *Social Statics: or, the Conditions Essential to Human Happiness Specified, and the First of Them Developed* (New York: Appleton-Century-Crofts, 1888). This was originally published in 1851; the edition cited here is an offset print of the original.

3. Herbert Spencer, *Principles of Ethics* (New York: Appleton-Century-Crofts, 1892–1898). A very high-quality and inexpensive edition of this work is published by Liberty Press, Indianapolis.

4. Spencer, *Social Statics,* p. 448.

5. Ibid., pp. 54–57.

6. Ibid., p. 56.

7. Spencer's arguments here are highly modern and, when compared with Marx's, Weber's, or Durkheim's, are quite radical.

8. Spencer, *Social Statics,* p. 497.

9. Ibid., p. 490.

10. Ibid., p. 498.

11. It should be remembered that this perspective was developed between 1873 and 1896. For a more complete and detailed review of Spencer's sociology during this period, see Jonathan H. Turner, *Herbert Spencer: A Renewed Appreciation* (Beverly Hills, CA: Sage, 1985).

12. Herbert Spencer, *First Principles* (New York: A. L. Burt, 1880; originally published in 1862). The contents of this work had been anticipated in earlier essays, the most important of which are "Progress: Its Law and Cause," *Westminster Review* (April 1857), and "The Ultimate Laws of Physiology," *National Review* (October 1857); moreover, hints at these principles are sprinkled throughout the first edition of *Principles of Psychology* (New York: Appleton-Century-Crofts, 1880; originally published in 1855).

13. Herbert Spencer, *The Study of Sociology* (Boston: Routledge & Kegan Paul, 1873).

14. Ibid., p. iv.

15. Ibid., p. 72.

16. Ibid., p. 105.

17. Ibid., p. 114.

18. Ibid., pp. 314–326.

19. Ibid., pp. 58–59.

20. Ibid., p. 328.

21. The full title of the work reads *Descriptive Sociology, or Groups of Sociological Facts.* The list of volumes of *Descriptive Sociology* is as follows: vol. 1: *English* (1873); vol. 2: *Ancient Mexicans, Central Americans, Chibchans, Ancient Peruvians* (1874); vol. 3: *Types of Lowest Races, Negritto, and Malayo-Polynesian Races* (1874); vol. 4: *African Races* (1875); vol. 5: *Asiatic Races* (1876); vol. 6: *North and South American Races* (1878); vol. 7: *Hebrews and Phoenicians* (1880); vol. 8: *French* (1881); vol. 9: *Chinese* (1910); vol. 10: *Hellenic Greeks* (1928); vol. 11: *Mesopotamia* (1929); vol. 12: *African Races* (1930); and vol. 13: *Ancient Romans* (1934). A revised edition of vol. 3, edited by D. Duncan and H. Tedder, was published in 1925; a second edition of vol. 6 appeared in 1885; vol. 14 is a redoing by Emil Torday of vol. 4. In addition to these volumes, which are folio in size, two unnumbered works appeared: Ruben Long, *The Sociology of Islam,* 2 vols. (1931–1933); and John Garstang, *The Heritage of Solomon: An Historical Intro-duction to the Sociology of Ancient Palestine* (1934). For a more detailed review and analysis of these volumes, see Jonathan H. Turner and Alexandra Maryanski, "Sociology's Lost Human Relations Area Files," *Sociological Perspectives* 31 (1988), pp. 19–34.

22. *The English,* classified and arranged by Herbert Spencer, compiled and abstracted by James Collier (New York: Appleton-Century-Crofts, 1873), p. vi.

23. Herbert Spencer, *The Principles of Sociology,* 3 vols., 8 parts (New York: Appleton-Century-Crofts, 1885; originally initiated in 1874). This particular edition is the third and is printed in five separate books; subsequent references are all to this third edition. Other editions vary in volume numbering, although part numbers are consistent across various editions.

24. Spencer, *Principles of Sociology,* vol. 1, p. 448 (emphasis in original).

25. This particular listing is taken from Jonathan H. Turner, *The Structure of Sociological Theory* (Belmont, CA: Wadsworth, 1998).

26. Spencer, *Principles of Sociology,* vol. 1, pp. 479–483.

27. Spencer, *Principles of Sociology,* vol. 1, p. 477.

28. Spencer, *Principles of Sociology,* vol. 1, part 2, pp. 519–548.

29. Ibid., vol. 1, p. 518.

30. Of course, the absolute amounts of both increase, but the processing of information—credits, accounts, ideas, purchase orders, and so on—increases as a proportion of things circulated.

31. Spencer, *Principles of Sociology,* vol. 1, pp. 577–585.

32. The misinterpretation of Spencer's intent stems from his introduction of the typology at several points in *Principles of Sociology*. From its usage in his discussion of political and industrial (economic) institutions, it would be easy to see the typology as his version of the stages of evolution. But if one reads the more analytical statement in the early chapter on social types and constitutions in vol. 1, paying particular attention to the fact that this chapter precedes the one on social metamorphoses, then our interpretation is clear. Because Spencer uses another typology for describing the long-run evolutionary trends, it seems unlikely that he would duplicate this effort with yet another typology on militant-industrial societies. See, in particular, *Principles of Sociology*, vol. 1, part 2, pp. 569–580.

33. We are using Parsons' terms here because they best connote Spencer's intent. See Parsons, *Societies*, cited in note 34.

34. See, for example, Parsons, *Societies* and *The System of Modern Societies* (Englewood Cliffs, NJ: Prentice-Hall, 1971); Gerhard Lenski, Jean Lenski, and Patrick Nolan, *Human Societies* (New York: McGraw-Hill, 1991); and Morton H. Fried, *The Evolution of Political Society* (New York: Random House, 1967).

35. Spencer, *Principles of Sociology*, vol. 1, part 2, p. 597.

36. Ibid., p. 596.

37. See Turner, *Herbert Spencer* (cited in note 11), for a more detailed review of Spencer's institutional analysis.

38. Spencer, *Principles of Sociology*, vol. 1, part 3, pp. 603–757. See also Leonard Beeghley, "Spencer's Analysis of the Evolution of the Family and the Status of Women: Some Neglected Considerations," *Sociological Perspectives* (formerly *Pacific Sociological Review*) 26 (August 1983), pp. 299–313.

39. See Turner, *Herbert Spencer* (cited in note 11), p. 115.

40. Spencer, *Principles of Sociology*, vol. 2, part 4, pp. 3–216.

41. See Turner, *Herbert Spencer*, p. 122.

42. Spencer, *Principles of Sociology*, vol. 2, part 5, pp. 229–643.

43. Ibid., vol. 2, part 6, pp. 3–159. We should note how close this view of religious functions is to that to be developed by Durkheim.

44. Ibid., vol. 2, part 8, pp. 327–608.

6

✳

Spencer's
Theoretical Legacy

The power of a theory is in the models and principles that it can generate. The theories of the early founders were, for the most part, stated discursively as texts, whereas in science theoretical ideas are best stated in
more precise formats, such as complex causal models or elementary principles.
In this chapter, and its counterpart for other theorists, our goal is to convert
the most interesting ideas of theorists into more formal formats. This kind of
exercise is not everyone's cup of tea, of course, but it is one way to appreciate
the contribution of the founding generation to contemporary sociology. In
Spencer's case, the core theoretical models and principles are readily extracted
from discursive text. Moreover, for all the vagueness of Spencer's cosmic view
of evolution, it does provide a consistent metaphor that gives his theory continuity, unity, and coherence.

SPENCER'S UNDERLYING CAUSAL MODEL

Spencer was a functional theorist who often argued that structures exist to
meet the basic requisites for regulation, operation, and distribution. All such
functional arguments obscure causality because they imply that needs for regulation, operation, and distribution somehow "cause" the structures that meet
these needs to come into existence. But how does this occur? How can the
outcome bring about the cause of this outcome? All functional theories

encounter this problem; they assume that needs will magically bring about the necessary structures to meet these needs. As a result, the causal sequence of events is vague, if not somewhat mysterious. If this is the case, however, how do we draw a causal model of the processes involved?

There is a solution to this problem. Functional arguments need to be recast into selection statements. That is, we should view problems of regulation, operation, and distribution not so much as needs of the social system that must be fulfilled, but as selection pressures placed on a population. A *selection pressure* is a problem of adaptation that people, as well as the structures organizing them, confront. For example, if there is not enough food to support a growing population, this crisis represents a selection pressure: find a way to feed more people, or many people will soon be dead. Or, if growth of the population creates problems of coordinating and controlling people's activities, this too is a selection pressure that forces individuals and collective units to search for solutions. Under these kinds of selection pressures, people and collective actors like groups and organizations try to find solutions. If by luck, invention, borrowing, or some other act they find a way to respond to the selection pressure, the population can remain viable in its environment. Thus, if a larger population cannot be coordinated or fed, these selection pressures send actors scrambling to find a way to increase the level of regulation or operation, to use Spencer's terms; and if they find a solution, then the population will maintain itself in its environment.

Thus, functional statements almost always contain an implicit appeal to selection processes. Many sciences use functional statements; for example, medical biologists often make functional arguments such as the "function of the heart is to distribute oxygen to parts of the body." By itself, this kind of argument obscures the causal processes whereby the heart evolved, but there is an implied selection argument in this statement: In the distant evolutionary past, natural selection worked to give larger organisms a pump to circulate blood to parts of the body, thereby increasing the chances that such organisms could survive in their environment. Similarly, all functional arguments in sociology can invoke the same selectionist argument. Although these kinds of selectionist statements do not fully resolve the problem of vagueness in the causal chain, they do avoid the problem of viewing the outcome as the cause of the processes producing this outcome.

To illustrate what selectionist causal arguments look like, Figure 6.1 translates Spencer's statements about the relationship between population size and differentiation along the regulatory, operative, and distributive axes into a complex causal model. As the size of the population grows, problems of maintaining and regulating the members of the population increase; these problems can be considered escalating *logistical loads* or new burdens on a population that require a solution. These escalating logistical loads put selection pressures on the population to find new ways to manage the escalating burden. In Spencer's analysis of population growth and size, he argued that as logistical loads mount, selection pressures are placed on the population to consolidate and use power to coordinate more people, to expand production and reproduction so that

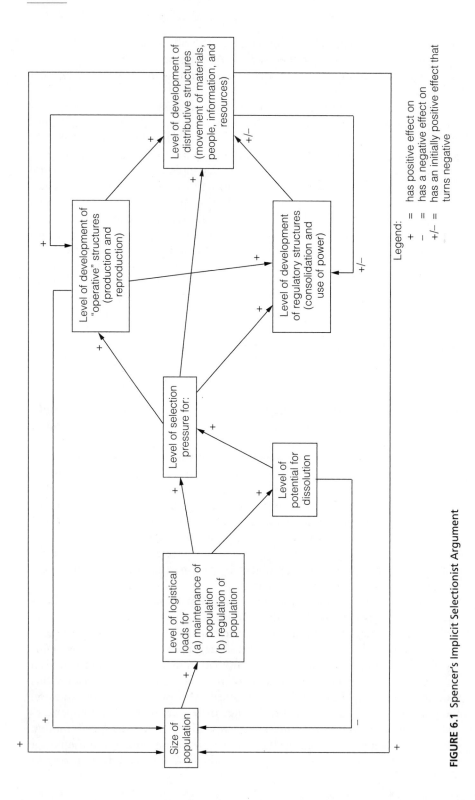

FIGURE 6.1 Spencer's Implicit Selectionist Argument

members of the population can be sustained and regenerated, and to build new systems for distributing resources, people, and information. The positive signs on the arrows in Figure 6.1 indicate that increases in one causal force increase the values for others, whereas the negative signs indicate just the opposite. Therefore, growth of the population increases (via the positive signed arrow in the model) logistical loads, which, in turn, increase selection pressures that raise the potential for dissolution or, alternatively, cause actors to develop new regulatory, operative, and distributive structures.

An important aspect of complex causal models is reverse causal effects that are represented by the arrows flowing from right to left in the model in Figure 6.1. As emphasized in Chapter 3, complex causal models typically address the recursive nature of social reality because outcomes of one causal sequence generally feed back and affect the very forces that brought them about. Thus, in the model in Figure 6.1, as the potential for dissolution increases, this potential exerts a negative effect (symbolized by the negatively signed arrow) on the size of the population. In contrast, as new kinds of regulatory, operative, and distributive structures are generated, these structures have a positive effect on population size, allowing it to grow larger. As Spencer emphasized, with differentiation of new regulatory, operative, and distributive structures comes a greater capacity to support a larger "social mass."

The model also reveals some mixed signs or arrows, such as the +/− sign of the direct causal path from development of regulatory structures to distributive as well as the reverse causal path back to regulatory from distributive processes. This sign denotes a positively curvilinear causal relation between these forces; that is, the causal relation is initially positive with development of regulatory structures causing the expansion of distributive processes (through building roads, ports, and other infrastructures as well as the coinage of money and enforcement of contracts in markets), but too much use of regulatory power turns the relationship negative, decreasing the dynamism of markets. Many processes in the social world are of this nature, with forces operating in one causal direction for a while, only to reverse the causal effect as the values for this force reach high levels. Spencer recognized that a certain amount of regulatory power had to be exerted on markets; and moreover, distribution requires government financing and building of infrastructures for moving people, information, and resources around. Consequently, the effects of power use on distribution are positive, until government begins to overregulate and stagnate market forces.

Figure 6.2 illustrates the basic elements of all complex causal models by portraying Spencer's entire scheme as a complex causal model. Figure 6.2 displays all of the elements of Spencer's theory in one causal model. The model starts, as Spencer intended, with those forces increasing the size of the population. At first, a growing population will simply create more structures of the same kind, or as the model highlights, the rate of "segmentation" will increase. For example, as a population of hunter-gatherers grows, it may simply create more bands of nuclear families as its first response to the increase in the logistical loads that come with growth. Segmentation can, for a time, manage the

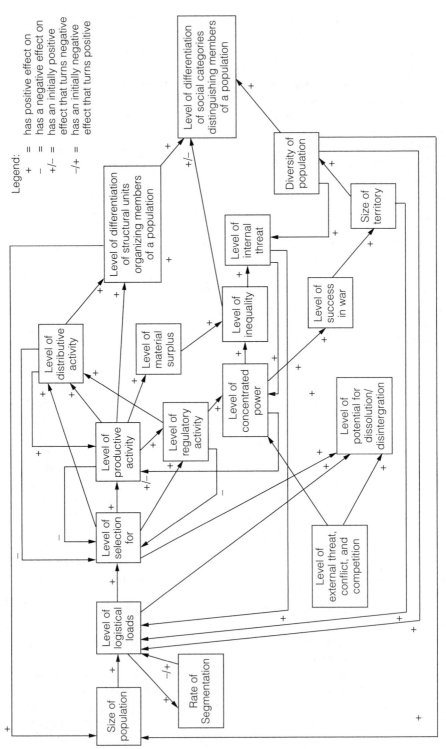

FIGURE 6.2 Spencer's General Theory of Societal Dynamics

increasing logistical loads of growing size, but as the arrow indicates, this is a curvilinear effect. As is denoted by the $-/+$ sign of the arrow from segmentation to logistical loads, segmentation initially increases logistical loads (the negative sides of the curvilinear relationship), but eventually, segmentation is insufficient to manage the escalation loads of growth. Thus, further segmentation begins to increase loads (the positive side of the curvilinear relationship). The larger size of the population and the declining effectiveness of segmentation increase logistical loads and selection pressures on the population to find new ways to increase production to support the population; to coordinate, control, and regulate the population; and to distribute resources, information, and people about the society. Segmentation alone will no longer work; as logistical loads and selection pressures mount, the population must differentiate into new kinds of structures, or face the possibility of dissolution (as denoted by the arrow from logistical loads to dissolution/disintegration). If these selection pressures are not met with an effective response, the likelihood of dissolution and disintegration of the population increases. If, however, differentiation of new kinds of regulatory, operative, and distributive structures can meet these selection pressures by managing the escalated logistical loads, then the potential for societal disintegration is reduced (as emphasized by the negatively signed reverse causal arrows from regulatory, operative, and distributive activity to selection pressures which, when reduced, will travel along positively signed arrows to lower the potential for dissolution). As the overall level of structural differentiation around new productive, reproductive, regulatory, and distributive structures increases, this exerts a positive effect on population size (as denoted by the positive reverse causal arrow at the top of the model back to population size from structural differentiation). With increased differentiation, then, the society can now support a larger "social mass," to use Spencer's terminology. These portions of the model thus summarize Spencer's core evolutionary argument about size and differentiation without all the baggage of evolutionary thinking. Spencer's model can be applied to any system— from a society to an organization, for example—because he argues that there is a fundamental set of causal relations among population size and structural differentiation.

We can now turn to theory of power of dynamics that Spencer developed within this general theory of growth and differentiation. As regulatory processes expand, the level of concentrated power increases. Such an increase in mobilization of power, however, depends on expansion of production to generate the material surplus that can be used to finance the mobilization of power (as denoted by the positive signs on the arrows from productive activity to material surplus to concentrated power). It takes resources, after all, to support police, military, and administrative personnel, and as Spencer emphasized, without the expansion of production, the necessary material surplus cannot be generated to consolidate and concentrate power. Power will become even more concentrated, especially around the capacity for coercion and around administrative structures monitoring conformity, under conditions of threat. Threats can be external, as is the case when societies engage in war or when they compete economically, or these threats can be internal, especially as inequality increases. External threat and conflict increase mobilization of coercive power, as indicated by the positively signed arrow from threat, conflict, and competition to concentrated power. External threat also increases the potential for dissolution and disintegration, as is emphasized by the positively signed arrow from external threat to dissolution. Turning to internal threats, inequality increases with concentrated power, as elites tax and usurp surplus for their own use and consumption. As inequality increases, people become resentful, which forces political elites to concentrate power even more in a cycle that can ratchet itself up to the point of causing dissolution (as indicated by the reverse causal arrow from internal threat to logistical loads, back down to dissolution).

Whether from real sources of external threat, or from manufactured external threats that political elites use to deflect attention away from inequalities that could generate internal threats, governments often go to war. Success in war often increases the size as well as the cultural diversity of the population and territory to be controlled. These changes—expanded territories to govern, increased size and diversity of the consolidated populations—all increase the potential for dissolution because logistical loads are compounded: A larger population, per se, raises logistical loads (as denoted by the causal chain from size of territory to size of population to logistical loads). A larger territory presents entirely new logistical loads revolving around how to maintain control (as denoted by the arrow from size of territory to logistical loads). Finally, an increasingly diverse population, especially one that has been conquered and is resentful, poses logistical problems and internal threats (as indicated by the arrows from diversity of population to internal threat and logistical loads). As these logistical loads mount, the potential for dissolution also increases, as do the selection pressures to differentiate new regulatory, operative, and distributive structures. If new structures are differentiated and prove effective, then the larger population can be more integrated and remain viable in its environment. If, however, these structures cannot be generated or prove ineffective, then disintegration is more likely.

As we can see, then, Spencer's ideas can be translated without any real distortion into a complex causal model. This model, which is more than one hundred years old, can still provide new leads for sociological theory. And when converted into propositions or theoretical principles, the model offers many interesting hypotheses, if not laws of human organization.

SPENCER'S THEORETICAL PRINCIPLES

Ultimately, it is desirable to express a theory as a series of propositions or principles. Spencer's major work is titled *The Principles of Sociology,* and it should not be surprising that the principles are easily extracted. Propositions emphasize key causal relations, and although we lose a sense of the robust direct and reverse causal effects among the forces in the theory, we gain parsimony with propositions. Indeed, as is evident, the model in Figure 6.2 is complex—perhaps too complex. Translating key causal arguments into a smaller set of principles helps reduce this complexity.

1. The larger a population, and the greater its rate of growth, the more intense are logistical loads and the greater are the selection pressures for differentiation and elaboration of new productive, reproductive, regulatory, and distributive structures.

This law is the core of Spencer's theory; all else in his theoretical scheme follows from this statement of the relationship between population size and growth, on one side, and differentiation along the regulatory, operative (productive and reproductive), and distributive axes on the other size. This is one of the basic laws of macro sociology.

2. The development of new technologies and the differentiation of new structures for gathering resources and converting them into usable commodities (production) will increase with

   a. The size and rate of growth of a population.
   b. The degree of access to natural resources, whether a society's own or those of another society.
   c. The expansion of distributive capacities, especially those revolving around markets.
   d. The consolidation of centers of power that are engaged in low-to-moderate levels of regulation of productive structures and that are able to avoid prolonged wars with neighboring societies.

This principle on production dynamics emphasizes that population growth will expand production as new ways to sustain the increased numbers of people in a society are sought. Moreover, access to natural resources is also important in two ways. First, the level of available resources is critical; societies without many natural resources will, in general, be less productive than will those with these resources. Second, societies with the power, technology, or distributive facilities to gather resources from afar can expand production even with relatively few natural resources of their own. Thus, a society's distributive capacities within its own borders, or with other societies, is an important mediating factor in production. Moreover, Spencer felt that once markets develop to a threshold where they are free and open, encouraging competition and productive innovations, they would increasingly be the engine that drives production. Well-developed and open markets, coupled with the infrastructure to move information, people (labor), and resources rapidly, will dramatically increase demand for new kinds of productive outputs. In addition, this level of development in distributive structures will increase the volume, velocity, and scope of distribution in ways that create pressure to expand production. In Spencer's eye, markets would no longer be a place where productive outputs are placed for distribution; rather, they would be the force that dictated how production would be conducted. Power is also very important in production, for Spencer consistently argued that low-to-moderate regulation of distributive and productive processes was conducive to the development of new technologies and productive structures. Too much centralization of power—what he termed "militant societies"—would destroy incentives for new technologies and expansion of the distributive infrastructure as well as high-volume, high-velocity, and far-reaching markets. War is, he felt, the biggest inhibitor of production because it consumes productive capital, removes incentives to technological development outside of war-making machines, and inhibits investments in domestic production. Internal threats can similarly consume resources better devoted to the domestic economy, but never on the scale of war making against other societies.

3. The differentiation of new reproductive structures for restocking the population, for socializing the population into a common culture, and for sustaining existing members of the population and the structures that organize them increases with:

    a. The size and rate of growth of the population.
    b. The level of differentiation of social structures.
    c. The level of differentiation of culture.

Even though Spencer never directly addressed the question of reproduction, in the contemporary sense of this term, *The Principles of Sociology* is filled with analyses of reproductive structures. He tended to collapse these structures under what he termed "operative" or "sustaining" processes, so it is necessary to tease out his analysis. For Spencer, differentiation of reproductive structures follows from the differentiation of social structures and culture. The more diversity in roles to be played in different kinds of structures, and the more diversity in the cultural symbols of a society, the greater are the logistical loads for reproduction, and hence, the more diverse and numerous will be distinctive kinds of reproductive structures.

4. The differentiation of centers of regulatory power will increase with

    a. The size and rate of growth of a population.
    b. The material surplus from production.
    c. The volume, velocity, and scope of exchange.
    d. The level of inequality and internal threat.
    e. The level of external threat from the environment, both biophysical and sociocultural.
    f. The size of territories to be regulated.

Spencer argued that a larger population required new mechanisms for controlling and coordinating activities. Such structures can emerge, however, only with an increase in production to generate the material surplus to finance coercive and administrative structures engaged in regulation. Threats, whether internal, environmental (such as a natural disaster, famine, or decline in resources) or sociocultural (threats from other societies) will also increase the level of differentiation of regulatory power, although these threats will tend to centralize power (see next proposition). Distributive structures also increase the differentiation of power because at least some of the infrastructures for markets must be financed and built by government, and although Spencer encouraged a "hands-off" policy by government toward production and markets, he recognized that distinctive power structures had to evolve to regulate many market aspects, such as coining money, enforcing contracts, and regulating crime and fraud. Thus, expanding distribution also will force the differentiation of new kinds of regulatory structures. Finally, Spencer recognized that extended territories pose special logistical loads for regulation, and as the amount of territory to be controlled increases, many distinctive levels and centers of power across the expanded territory will emerge, thus increasing the level of differentiation of government.

5. Power will increasingly be consolidated and centralized with

   a. Threat from any source, whether from biophysical or sociocultural environments, stratification and inequalities, or cultural and ethnic diversity.
   b. Failure of existing regulatory structures to coordinate and control productive and distributive processes.

A good portion of Spencer's sociology concerns the degree of centralization of power, its causes, and its consequences. Threat will always lead to the mobilization of power, especially its coercive base as well as its administrative base for giving orders and monitoring conformity to directives. Also, when existing levels of power prove ineffective in coordinating and controlling activities, pressures will mount for increased centralization of power.

6. Power will be increasingly decentralized with

   a. High rates of political mobilization against overregulation, supported by liberal ideologies and large numbers of actors.
   b. Increasing stagnation in productive and distributive structures.

The argument is that concentrated power, per se, will increasingly be resented, leading to mobilizations against such regulation. But, as Spencer's law on the centralization of power underscores, such mobilizations may be perceived as a threat, leading the political system to concentrate and centralize power even more. Spencer appears to see stagnation in production and markets as a key condition that explains when pressures for decentralizing power will be successful. When the economy is stagnant, the resources for tight regulation become ever-more scarce, and Spencer appears to have assumed that political leaders will recognize the necessity to encourage innovation and investment in productive activities, especially when the population is mobilized to demand less regulation. This might have been simple wishful thinking on Spencer's part, but principles 3 through 6 on power give us a sense of the forces involved in the dynamics of consolidation and centralization/decentralization of power. Clearly much more work needs to be done on specifying the conditions under which power is consolidated, centralized, or decentralized, but for his time, Spencer's analysis is rather sophisticated.

7. The probability that a population will disintegrate and cease to be a coherent and distinctive system in its environment increases as logistical loads mount and selection pressures cannot be addressed. Logistical loads and selection pressures will increase with

   a. The size and rate of growth of a population.
   b. The degree of cultural diversity in a population.
   c. The size of the territory occupied by a population.
   d. The level of inequality among members of a population, especially if strata are highly correlated with ethnic cultural diversity.
   e. The rate and intensity of conflict with other societies.
   f. The level of environmental crises, whether from natural disasters or depletion of resources.

From the publication of *First Principles,* Spencer emphasized that dissolution was part of the process of evolution. True, the long-term trend had been for increasing complexity of societal forms, but this trend historically had been punctuated by collapse and conquests of populations. Thus, within *The Principles of Sociology* are several consistent lines of argument about the conditions under which populations disintegrate. Population size and rate of growth do not automatically lead to differentiation of effective regulatory, operative, or distributive structures. Population diversity, especially cultural and ethnic diversity, can push people away from each other or cause internal conflicts among diverse members of a population. Large territories are difficult to govern, and they pose constant problems of how to distribute information and resources. Populations spread out in space will develop cultural diversity, thus aggravating the logistical loads associated with this phenomenon. Inequalities stemming from too much concentrated power and usurpation of resources will eventually exceed government's capacity for coercive control. War and conflict with other societies always drain resources, stagnate production, and aggravate inequalities, to say nothing of the possibility of losing a war and being conquered. As Spencer always emphasized, populations must adapt to the natural environment, and moreover, populations can change their environments, often depleting resources and degrading the environment's ability to support productive activities. In addition to environmental depletion and degradation, episodic disasters can destroy a population.

In conclusion, Spencer clearly took seriously the emphasis on "principles" in his great book. In addition to the empirical generalizations that he offered for each institution, summarized in the previous chapter, Spencer's sociology offers many interesting leads for further theoretical or empirical inquiry. It is a great tragedy that Spencer is so infrequently read by contemporary sociologists because even this cursory review of the core model and principles of his sociology demonstrates the power of his theory.

These principles are, unfortunately, surrounded by text; they are not so much buried in text because they are quite easy to spot once one is looking for them, but they are embedded in so much verbiage that one must be patient. *Principles of Sociology* is not much longer than the collected works of Karl Marx, Max Weber, and Émile Durkheim, and so length alone cannot explain the reluctance of contemporary sociologists to read Spencer. There is clearly a bias and prejudice against his conservative moral philosophy (which was liberal in Spencer's time), against his big theoretical scheme, against his emphasis on evolution, and against his functionalism. It is perhaps understandable why Spencer is not read given these modern sensibilities about political ideologies, grand theory, evolution, and functionalism, but this does not take away from basic unfairness of relegating Spencer to a minor place in sociology's early pantheon. Comte's place is understandable because he did not develop explanatory theory, but Spencer's is indeed a great tragedy because he did develop explanatory theory.

7

✳

The Origin and Context
of Karl Marx's Thought

BIOGRAPHICAL INFLUENCES
ON MARX'S THOUGHT

Karl Marx, theorist and revolutionary, was born to Heinrich and Henrietta Marx on May 5, 1818, in the city of Trier. Located in the Rhineland, Trier was (and is) the commercial center of the Moselle wine-growing area of Germany. Descended from a long line of rabbis on both sides of the family, the young Marx lived in a stable bourgeois (or middle-class) household. His father, a lawyer and lover of ideas, converted to Lutheranism in 1817 to protect his position. Although Jewish by heritage, the elder Marx appears to have had little interest in organized religion, being attracted to the deism characteristic of the Enlightenment. The young Marx was apparently close to his father and learned of Voltaire, Rousseau, and other writers on individualism and human progress from him.

As Marx grew up, he was also influenced by an upper-class Prussian, Ludwig von Wesphalen, whose daughter Jenny he eventually married. Despite status differences between the two families, von Wesphalen took a liking to Marx, encouraging him to read and introducing him to works of the great German writers of the time, Johann Goethe and Friedrich Schiller, as well as to the classical Greek philosophers.

This intellectual background paved the way for his subsequent study of the philosophy of G. W. F. Hegel and the political economy of Adam Smith, leading eventually to a theoretical critique of the capitalist social order. Just as important, however, these aspects of his background made Marx peculiar among nineteenth-century revolutionaries, for he was neither thwarted nor persecuted as a young man. Thus, although he was arrogant, vain, and vindictive toward enemies, Marx was also positive and self-confident throughout his adult life.[1]

Hegel and the Young Hegelians

After graduating from the Trier gymnasium (or high school), the seventeen-year-old Marx enrolled at the University of Bonn in 1835. After a year, however, he left for the more cosmopolitan and sophisticated University of Berlin. Here he encountered Hegel's idealism. The great philosopher, who died only a few years before, dominated intellectual life in Germany at that time. Marx also met youthful academic interpreters of Hegel, who called themselves Young Hegelians. They constituted Marx's first contact with people who did not blindly accept the dominant values and norms of German society.

The Young Hegelians, including such forgotten men as Max Stirner, Bruno Bauer, David Strauss, and Ludwig Feuerbach, saw themselves as radicals. And they were, in fact, irreligious and liberal. They questioned the established order in Prussia (where Berlin was located). Marx noted their influence on him in a now-famous letter to his father. "There are moments in one's life," he wrote, "which are like frontier posts marking the completion of a period but at the same time clearly indicating a new direction." After studying Hegel's idealism, he continued, "I arrived at the point of seeking the idea in reality itself."[2] The last phrase is important, for Marx was asserting that he had rejected Hegel's idealism in favor of studying "reality itself," as defined by the Young Hegelians. In effect, he began to question the status quo. He thus began the long process of transforming philosophy into social science.

This transition, however, occurred in a very despotic social context. During most of the nineteenth century, Prussia was perhaps the most repressive nation in Europe, with organized religion supporting the state's activities. Those who questioned the established order, religious or political, were treated as subversive. Hence, over time the Young Hegelians saw their writings censored and found themselves dismissed from faculty positions.

Nonetheless, the young Marx prepared himself for a life in academia. In addition to studying philosophy, he wrote hundreds of poems, a novel, a play modeled after a Greek tragedy, and much more. In 1841, Marx received a doctorate based on a thesis titled "The Difference between the Democritean and Epicurean Philosophy of Nature."[3] Unfortunately, his academic patrons had been dismissed from their posts and were unable to obtain a position for him. Marx was thus left without career prospects.

Lacking alternatives, Marx tried journalism, becoming a writer for—and eventually editor of—a liberal newspaper, the *Rheinisch Zeitung* (or *Rhineland News*). In this role, he battled the Prussian censors constantly, writing articles on the poverty of the Moselle valley winegrowers, the harsh legal treatment received by peasants who stole timber to heat their homes in winter, and the repressiveness of various European governments. Within six months, the Prussian authorities suppressed the paper and Marx was out of work, a situation that recurred frequently during his life. In the aftermath, he turned again to studying Hegel. The result was "A Contribution to the Critique of Hegel's *Philosophy of Right.*"[4] Although unpublished at the time, this essay constitutes Marx's decisive break with Hegel's idealism, particularly its religious and philosophical justification of the political status quo in Germany.

Paris and Brussels

Marx, now married to Jenny von Wesphalen, moved to Paris in 1843; he was twenty-five years old. Paris was the intellectual center of Europe at that time, and the years Marx spent there allowed him to meet many radicals and revolutionaries: the Russian Mikhail Bakunin, the poet Heinrich Heine, and the tailor Wilhelm Weitling, among others. In addition, Marx encountered the emerging discipline of political economy during this period, reading Adam Smith, David Ricardo, Pierre Proudhon, and many more. Perhaps most important, however, in September 1844, Marx met the man who became his lifelong friend and partner: Friedrich Engels (1820–1895). The son of a wealthy German industrialist, Engels wrote the first great urban ethnography, *The Condition of the Working Class in England in 1844,* along with an essay, "Outlines of a Critique of Political Economy," during this same period.[5] These works helped Marx to see the new urban working class, the proletarians, as real human beings with practical problems made worse by the systematic exploitation characteristic of capitalism at that time. One result was that Marx now rejected the ideas of the Young Hegelians as politically timid. In fact, the first product of his collaboration with Engels, a pompous and nearly unreadable tome titled *The Holy Family,* consisted of a diatribe against the Young Hegelians.[6] As we will discuss later, of all the Young Hegelians, only Feuerbach had a long-term impact on Marx's works. Another, more significant result was that Marx wrote a series of notebooks, the now famous *Economic and Philosophic Manuscripts,* in which he set forth his initial interpretation of capitalism as inherently exploitive and alienating.[7] In 1845, Marx moved to Brussels after being forced to leave Paris by the French government.

Shortly after arriving in Brussels, Marx and Engels wrote *The German Ideology,* a more effective work, which they intended as a final settling of accounts with the Young Hegelians. According to Marx and Engels, the German philosophers were less concerned with "reality itself" than with ideas about reality; they had not, in other words, really rejected Hegel. Although we will describe the theoretical implications of *The German Ideology* in the next chapter, Marx and Engels used the opportunity to poke fun at Stirner, Bauer, and the others, as in the following example:

> Once upon a time an honest fellow had the idea that men were drowned in water only because they were possessed with the idea of gravity. If they were to knock this idea out of their heads, say, by stating it to be a

superstition, a religious idea, they would be sublimely safe against any danger from water. His whole life long he fought against the illusion of gravity, of whose harmful results all statistics brought him new and manifold evidence. This honest fellow was the prototype of the German revolutionary philosophers of our day.[8]

In contrast with the Young Hegelians, Marx wanted to understand the practical problems people face. He also saw himself as a true revolutionary, dedicated to the overthrow of capitalist society—violently if necessary. Thus, he and Engels joined with other European émigrés and radicals in a variety of revolutionary organizations: the League of the Just, the German Workers' Educational Association, and the Communist League. Both Marx and Engels were dominating personalities, determined to lead working-class people toward a revolutionary reorganization of society. Here is a prophetic description of Marx by Paul Annenkov, a Russian who knew him during these years:

> He was most remarkable in his appearance. He had a shock of deep black hair and hairy hands and his coat was buttoned wrong; but he looked like a man with the right and the power to demand respect, no matter how he appeared before you and no matter what he did. His movements were clumsy but confident and self-reliant, his ways defied the usual conventions in human relations, but they were dignified and somewhat disdainful; his sharp metallic voice was wonderfully adapted to the radical judgments that he passed on persons and things. He always spoke in imperative words that would brook no contradiction and were made all the sharper by the almost painful impression of the tone which ran through everything he said. This tone expressed the firm conviction of his mission to dominate men's minds and prescribe them their laws. Before me stood the embodiment of a democratic dictator such as one might imagine in a daydream.[9]

In 1847, Marx and Engels decided to compose a statement of revolutionary principles under the aegis of the Communist League. Accordingly, Engels wrote an initial draft in catechism form titled "Principles of Communism" and sent it to Marx.[10] During the early days of 1848, Marx completely rewrote the draft and, although the final version incorporated many of Engels's ideas, the document printed in February of that year was strikingly different and original: *The Communist Manifesto.*[11] Although it had little immediate impact, the publication of the *Manifesto* occurred during great political ferment in Europe. Many observers, not all of them radicals, believed that some form of communist revolution was inevitable in West European societies. Later that year, revolts broke out all over the continent. In Paris, for example, workers held the city against the onslaught of the French army for six weeks. Ultimately, however, the workers and peasants were defeated throughout Europe, often after bloody battles. In 1849, Marx returned to Paris, still (like many others) believing that a communist insurrection was imminent. Subsequently, under pressure from the French government, he left for London, where he lived the remainder of his life.

The London Years

Now thirty years old, Marx withdrew from public life altogether for about fifteen years, concentrating instead on devising his theoretical analysis of capitalism. He studied and wrote copiously, producing notebook after notebook of observations about the nature of capitalist societies and criticism of economics as then practiced. These materials, almost all unpublished at the time, eventually appeared as *The Grundrisse* (or *Notebooks*), *The Theory of Surplus Value,* and *A Contribution to the Critique of Political Economy.*[12] Finally, Marx's greatest book appeared in 1867, when he was forty-nine years old: *Capital,* Volume 1.[13]

Although he intended to produce a multivolume work, only Volume 1 appeared at the time, and it usually stands alone as a theoretical analysis of capitalism. Although Engels subsequently edited and published the second and third volumes, Engels observed that the first "is in a great measure a whole in itself and has for more than twenty years ranked as an independent work."[14] As we will explain in the next chapter, *Capital* is more than a narrow work of economics; it is, rather, a theoretical analysis of capitalist social systems.

The tremendous quantity of work, however, did not bring in much money. Although Marx's income was adequate, neither he nor Jenny could manage money very well, with the result that the family lived in constant financial peril through most of these years. During much of this period, Marx served as European correspondent for the *New York Daily Tribune,* and the income from these articles constituted his main source of financial support. In addition, Engels, who benefited from an inheritance, periodically sent Marx money or ghostwrote articles for the *Tribune.* Apart from their economic circumstances and the death of two children in infancy, however, Marx and his family appear to have enjoyed a settled and happy life during these years. Only after the death of his mother in 1863 and the receipt of a bequest from a socialist, Wilhelm Wolff, did Marx's financial worries decline.

Although aloof from public life during the years in London, Marx, like many other radicals, still believed that economic crises would produce some form of workers' revolt. In 1864, the International Working Man's Association was formed in London. Composed of working people from most European nations, the organization proposed to destroy the capitalist system and substitute some form of collective control of the society. Abandoning his long reticence, Marx joined the group and, characteristically, quickly became its dominating force. Apart from ongoing work on *Capital,* all his energies were devoted to the International (as it was called). One side benefit, perhaps intended, was that *Capital* received considerable publicity. Unlike Marx's previous works, which had been generally ignored, *Capital* was widely read and quickly translated into French, Russian, English, and Italian—with Marx supervising these efforts. Aside from this activity, he immersed himself in political life, attempting to show how theory and revolution could be combined in practice.

In 1871, the long-awaited workers' revolt occurred in the aftermath of the Franco-Prussian War. As in 1848, however, the proletarians were suppressed, again with much loss of life. At this time, Marx produced his last great political pamphlet, *The Civil War in France,* in which he defended Paris workers protesting the government.[15] Soon afterward, the International split apart and ceased to exist. This was Marx's last effective political role.

In the years after 1870, Marx finally achieved a comfortable lifestyle. Engels, very wealthy by this time, gave him a bequest, and Marx settled into the life of a Victorian gentleman—albeit a radical one. A famous man, revered by socialists and revolutionaries around the world, Marx was sought out for advice by those who would defend the rights of working people. But he wrote far less and without much creativity. It was as if relative prosperity had robbed him of his anger, the source of his insight.

Jenny's death in 1881 deprived Marx of his lifelong companion. His oldest daughter, also named Jenny, died in January 1883, and on March 14 of that year, Marx died in his armchair. He was sixty-five years old.

Karl Marx's analysis of capitalism represents one of the most striking and original achievements in the history of social thought. As we will show in Chapter 8, he constructed a theoretical analysis that sought to account for the origins of capitalism, its historical stability, and its eventual demise. In the process, he combined social theory and revolutionary action in a way that has never been duplicated. That his work is shortsighted in some respects and misbegotten in others does not detract from its evocativeness. Like all scholars, however, he benefited from the legacy of concepts and ideas that had been advanced by others.

Marx was a voracious reader, and his writings are filled with detailed analyses of the philosophers and political economists of the day. In the remainder of this chapter, we sketch the ways in which he was influenced by Hegel, Feuerbach and the other Young Hegelians, Adam Smith and the other capitalist political economists, and, of course, Engels.

G. W. F. HEGEL AND KARL MARX

The origin of Marx's sociological theory lies in his youthful reaction to the writings of Georg Wilhelm Friedrich Hegel (1770–1831). In four main books, *The Phenomenology of Mind* (1807), *The Science of Logic* (1816), *The Encyclopedia of Philosophy* (1817), and *The Philosophy of Right* (1821), Hegel developed one of the most original, complex, and obscure philosophical doctrines ever devised.[16] Marx transformed Hegel's philosophy into an empirically based social science, albeit a peculiar one, decisively rejecting Hegel's idealism while retaining his reliance on dialectical analysis and applying it to the material world. To appreciate Hegel's influence on Marx, we need to briefly discuss idealist philosophy and Marx's major criticisms of it. Only then will the continuity and discontinuity between the two men's ideas become clear.

Hegel's Idealism

In Hegel's writing, idealism is a complex philosophical doctrine that can only be superficially sketched here. Its essence consists of the denial that things in the finite world—such as trees, houses, people, or any other physical object—are ultimately real. In Hegel's words, idealism "consists in nothing else than in recognizing that the finite has no veritable being."[17] For Hegel, true reality is embodied in that which is discovered through reason. In thus emphasizing the importance of thought, he followed a philosophical tradition that originated with Plato. From this point of view, the objects perceived by the senses are not real: They are merely the phenomenal appearance of a ultimate reality of ideas. Only "logical objects," or concepts, constitute ultimate reality. As Hegel wrote, "it is *only* in thought that [an] object is truly in and for itself; in intuition or ordinary perception it is only an appearance."[18] Hegel continued by asserting that if only concepts are real, then the ultimate concept is God, and Hegel's philosophy is essentially an attempt at proving the existence of God through the application of reason. According to Hegel, previous philosophers had seen only finite things as real and had relegated the infinite (or God) to the "mere 'ideal.'" He argued that this separation was artificial and could not show how God existed and acted through people because it involves a logical impossibility: The infinite, which is absolute and cannot perish, is kept separate from finite things, which must inevitably perish, and is placed in an abstract and mentally conceived "beyond." If this latter were true, Hegel argued, God could not have come to earth in the form of Jesus, and the bread and wine of the Last Supper were merely bread and wine.

Hegel argued that there was an inherent dialectical relationship between God (the infinite) and people (the finite). The essence of the dialectic is contradiction: Each concept implies its opposite, or in Hegel's terms, each concept implies its negation. Thus, after proposing that "the finite has no veritable being," Hegel immediately said, "the finite is ideal"; that is, its essence lies in that which contradicts it: the infinite, God. In this way the finite world of flesh and blood is annihilated (at least in thought), and the "infinite can pass over from the beyond to the here and now—that is, become flesh and take on earthly attire," as Jesus did a long time ago.[19] Hence, although this phrase states the issue too simply, Hegel believed that human history could be considered the autobiography of God because history only "exists" through its negation by the infinite and the latter's manifestations in this world. As in Christianity, even as the finite world of things is destroyed, it is saved. In Hegel's words, "the finite has vanished in the infinite and what *is*, is only the *infinite*," or everlasting life.[20] One implication of this analysis is a belief in the reality of transubstantiation (that the bread and wine become the body and blood of Jesus). Another implication, which is also characteristic of some forms of Christianity, is a relatively passive acceptance of the political status quo. For example, Hegel said, "All that is real is rational; and all that is rational is real."[21] Statements like this were taken by many as a sanctification of the Prussian state, with its despotism, police government, star-chamber proceedings, and censorship. Hence, the Prussian government glorified Hegel's philosophy for its own purposes and, when he died, gave him a state funeral.

Marx's Rejection of Hegel's Idealism

Marx reacted strongly against Hegel's idealism, criticizing it in a number of ways. First, and most important, he completely rejected Hegel's assertion that finite or empirical phenomena are not ultimately real. All his other criticisms follow from this basic point. Marx believed that when empirical phenomena were understood only as thoughts, people's more significant practical problems were ignored. Neither material objects nor relationships can be changed by merely thinking about them. The puerile quality of Hegel's point is evident, Marx suggested, in a simple example: If people are alienated such that they have no control over their lives or the material things produced by their labor, they cannot end their alienation by changing their perception of reality (or by praying, for that matter).[22] Rather, people must change the social structure in which they live; that is, they must make a revolution in this world rather than wait for the next world. Marx believed that life in this world posed a variety of very practical problems that people could solve only in hardheaded ways, and that human reason was of little use unless it was applied to the problems that exist in the finite world.

Second, according to Marx, Hegel's emphasis on the ultimate reality of thought led him to misperceive some of the essential characteristics of human beings. For example, Marx contended that although Hegel correctly "grasps labor as the essence of man," "the only labor which [he] knows and recognizes is abstractly mental labor."[23] Yet, people have physical needs, Marx noted, such as those for food, clothing, and shelter, which can be satisfied only by productive activity in the finite world. Hence, for Marx, the most significant labor is productive activity rather than mental activity. Similarly, Marx said that Hegel's belief in the unreality of finite things had led him to a position in which people were regarded as nonobjective, spiritual beings. But Marx asserted that people were "natural beings"; that is, they have physical needs that can be satisfied only in this world:

> As a natural, corporeal, sensuous, objective being [a person] is a suffering, conditioned and limited creature, like animals and plants. That is to say, the objects of his instincts exist outside him, as objects independent of him; yet these objects are objects that he needs—essential objects, indispensable to the manifestation and confirmation of his essential powers. To say that man is a corporeal, living, real, sensuous, objective being full of natural vigor is to say that he has real, sensuous objects as the objects of his being or of his life, or that he can only express his life in real, sensuous objects.[24]

Marx's third criticism was also an outgrowth of the first, in that he rejected the religious motif that pervades Hegel's work. As noted earlier, Hegel denied reality to the finite world to prove the existence of God, albeit a Christian

God. Nevertheless, Marx believed that when "reason" is applied to such impractical problems, people are prevented from recognizing that they are exploited and that they have an interest in changing the status quo in this world. For Marx, the next world is a religious fantasy not worth worrying about. Thus, he was particularly vitriolic, yet strangely poetic, in his denunciation of the religious implications of Hegel's philosophy:

> Religion is the sigh of the oppressed creature, the sentiment of a heartless world, and the soul of soulless conditions. It is the opium of the people. The abolition of religion as the illusory happiness of men, is a demand for their real happiness. The call to abandon their illusions about their conditions is a call to abandon a condition which requires illusions. The criticism of religion is, therefore, the embryonic criticism of this vale of tears of which religion is the halo.[25]

Marx believed that one of the main functions of religion was to blind people to their true situations and interests. Religion does this by emphasizing that compensation for misery and exploitation on earth will come in the next world.

Marx's fourth criticism of Hegel was that idealism was politically conservative rather than revolutionary. Idealism creates the illusion of a community of people rather than the reality of a society riddled with opposing interests. This illusion results partly from Hegel's assertion that the state, a practical and physical entity, emerges from the Spirit, or thought. In this way, Hegel imbued the state with a sacred quality. As Marx noted, Hegel "does not say 'with the will of the monarch lies the final decision' but 'the final decision of the will is—the monarch.'"[26] When the state is sacred, history can be seen as part of an overall divine plan that is not only reasonable but necessary. For this reason Marx interpreted Hegel's philosophy as politically conservative.

Marx's Acceptance of Hegel's Dialectical Method

Despite his complete rejection of idealism, Marx saw a significant tool in Hegel's use of the dialectic. In Hegel's hands, however, the entire analysis is couched in terms of a mystical theology. Thus, as Marx noted in *Capital,* Hegel's dialectic "is standing on its head. It must be turned right side up again, if you would discover the rational kernel within the mystical shell."[27] As we will show in Chapter 8, the process by which Marx turned the dialectic right side up involved its application to the finite world where people make history by producing their sustenance from the environment. Rather than being concerned with the existence of God, Marx emphasized that the focus must be on concrete societies (seen as social systems) and with actual people who have conflicting interests.

The significance of turning Hegel "right side up" is that for Marx, no product of human thought or action can be final; there can be no absolute truth that, when discovered, need only be memorized. From this point of view, science can only increase knowledge; it cannot discover absolute knowledge. Moreover, there can be no end to human history, at least in the sense of attaining an unchanging utopia, a perfect society. Such social structures can exist only in the imagination. Rather, every society is only a transitory state in an endless course of human development. This development occurs as conflict is systematically generated from people's opposing interests. Although each stage of history is necessary, and hence justified by the conditions in which it originated, progress occurs as the old society inevitably loses its reason for being. In Marx's work, the dialectical method means that nothing can be final or absolute or sacred: Everything is transitory and conflict is everywhere.

LUDWIG FEUERBACH AND KARL MARX

The Young Hegelians also affected Marx's sociology. The most important influence among them was unquestionably Ludwig Feuerbach. In this section, we outline some of the Young Hegelians' ideas and then suggest more specifically how Feuerbach's ideas altered the direction of Marx's thought.

The Young Hegelians and Marx's Thought

Like Hegel, the Young Hegelians tried to understand the nature of reality and the relationship between religious beliefs and reality. However, because religion legitimated oppressive political conditions, the Young Hegelians rejected the political conservatism that seemed inherent to Hegel's thought. They reacted in this way because during most of the nineteenth century Prussia was an extremely repressive nation, with religion serving as one of the chief pillars of the repressive state. The Young Hegelians believed that the church's emphasis on the sanctity of tradition, authority, and the renunciation of worldly pleasures helped to prop up an oppressive governmental apparatus. But because political agitation was not possible (without being arrested or expatriated), they sought to criticize the state indirectly by investigating the sacred texts, doctrines, and practices of Christianity.

For example, in 1835 David Strauss published *The Life of Jesus Critically Examined,* in which he tried to show that the Gospels were not accurate historical narratives.[28] This book prompted great controversy because it was thought that if the life of Jesus as portrayed by the Gospels was not to be believed, then the authority of the church would be undermined. Shortly thereafter, Bruno Bauer published a series of articles in which he denied the historical existence of Jesus altogether and tried to explain the Gospels as works of pure fiction.[29] By debunking the nature and logic of Christian tenets (and hence the church) in this way, the Young Hegelians hoped also to impugn the authority of the state. The Prussian government recognized the seditious implications of these works, however, and as a result, the Young Hegelians suffered varying degrees of surveillance, political harassment, and dismissal from their university posts.

Nonetheless, despite their political stance, all these men were still Hegelian in orientation, and this eventually led to Marx's split with them. For example, in *The Ego and His Own: The Case of the Individual Against Authority* (1844), Max Stirner argued that nothing was objective outside the individual.[30] According to Stirner, social institutions, such as the church, are oppressive to the individual's spirit. Like a true Hegelian, Stirner then asserted that reality was not based on people's sense perceptions. As Hegel claimed, reality is created by the imagination and will of each person and, as a corollary, there is no objective reality apart from the ego. Thus, according to Stirner, individuals should avoid participating in the society as much as possible, and in this way they can also avoid being oppressed by authority. With this argument, he anticipated the development of anarchist thought some years later. Marx, however, believed that Stirner's position was politically futile because social institutions must be controlled rather than ignored.

As will be seen in our discussion of *The German Ideology* in Chapter 8, Marx believed that the Young Hegelians were intellectual mountebanks, and he wrote hundreds of pages of vituperation against them. For example, he and Engels made fun of Stirner, Bauer, and others by calling them "The Holy Family" and referring to them as "Saint Max" and "Saint Bruno." More generally, Marx developed four main criticisms of the Young Hegelians, all of which can be seen as variations on his criticisms of Hegel. First, their writings treated the development of theology independently of the actual activities of the church and other social institutions that were pervaded by theological ideas. Such an emphasis ignored the fact that the development of ideas never proceeds apart from human practices. Second, the Young Hegelians were essentially idealists, in that the origin of religious as well as other kinds of thought was to be found in the Spirit. But for Marx, religion and all other ideas emerge from people's actual social relationships and in their need to survive. As he would emphasize some years later in the *Communist Manifesto,* people's ideas, worldviews, and political interests depend on their positions in society.[31] Third, the Young Hegelians' writings were fatalistic in that the historical process was seen as automatic and inexorable, either because it was directed by the Spirit or because it was directed by individuals (such as the Prussian king) who were somehow seen as connected with the Spirit. For Marx, although history has direction and continuity, it is shaped by human action. Fourth, and most fundamental, the Young Hegelians foolishly believed that by changing ideas they could change human behavior. Therefore, they fought a war against the state, using words as the primary weapons. Wars must be fought with guns, Marx believed, and those who do not recognize this elementary fact are very unrealistic.

Marx made one exception to his indictment of the Young Hegelians, however. The only member of the group Marx did not vilify, even though the two men disagreed, was Feuerbach.

Feuerbach and Marx's Thought

Like the other Young Hegelians, Feuerbach was also interested in the religious implications of Hegel's philosophy, but unlike the others, he fundamentally altered the direction of Marx's thought. This alteration occurred in Marx's critique of Hegel and in the development of Marx's peculiar but highly effective version of social theory.

In his book *The Essence of Christianity* (1841), Feuerbach undercut both Hegel and the Young Hegelians by arguing that religious beliefs arose from people's unconscious deification of themselves.[32] According to Feuerbach, human beings have taken all that they believe is good in themselves and simply projected these characteristics onto God. He showed how the "mysteries" of Christianity—the Creation, the suffering God, the Holy Trinity, the Immaculate Conception, the Resurrection, and the like—all represented human ideals. Thus, he argued that theology was simply a mythical vision of human aspirations and that "what man praises and approves, that is God to him; what he blames [and] condemns is the nondivine."[33] The true essence of religion, Feuerbach believed, is to be found in anthropology, not theology, for "religion is man's earliest . . . form of self-knowledge."[34]

This analysis reveals Feuerbach to have been the most original of the Young Hegelians. Whereas most of them were content to analyze and critique Christian theology, Feuerbach decisively rejected any analysis that treated theology as existing independently of empirical activities. Moreover, although many Young Hegelians still accepted the idea that God necessarily directed human affairs, Feuerbach argued that an abstract and amorphous Spirit could not be the guiding force in history because people were simply worshipping projections of their own characteristics and desires. Finally, although the other Young Hegelians continued to be mired in idealism, Feuerbach was a materialist in the sense that he believed that people's consciousness of the world was the product of their brains and, hence, of physical matter. To Marx and others, this position seemed clearsighted after the obfuscations and puerile logic of Hegel, Strauss, Bauer, and Stirner.

Feuerbach's argument had yet another consequence for Marx. In Feuerbach's work, Marx found the key to criticizing Hegel and, ultimately, to developing a social theory designed to promote revolutionary action. Marx realized that Feuerbach's analysis of religion as an expression of human desires could be generalized to people's relationships to other social institutions (especially the state) and to any situation in which human beings were ruled by their own creations. Thus, following Feuerbach, Marx reversed Hegel's argument, which asserted that the state emerged from the spirit, by arguing that the modern state emerged from capitalist social relationships (which he called "civil society"). This argument has important implications, for if the state is the product of human action, it can be changed by human action. Marx's mature social theory follows from this fundamental insight.

ADAM SMITH AND KARL MARX

By the late eighteenth century, England had already become a relatively indus-trialized and commercial nation. As such, it constituted the first fully capitalist society, with the result that scholars attempted to account for the origins of capitalism, its nature, and its future development. Such men as Adam Smith, David Ricardo, and many others developed a new mode of analysis, called political economy, and sought to understand the characteristics of industrial capitalism. After being introduced to the study of political economy by Engels and others, Marx began to deal with the topics characteristic of the new dis-cipline. For example, in *The Economic and Philosophical Manuscripts,* he analyzed (among other things) the origin of the value of commodities, the origin of profit, the role of land in a capitalist economy, and the accumulation of capi-tal. However, his most detailed analyses and criticisms did not occur until the 1850s in his notebooks (subsequently published as the *Grundrisse*) and *A Contribution to a Critique of Political Economy.* From these efforts Marx's great work, *Capital,* eventually emerged.

Political Economy and Marx's Thought

Marx's detailed analyses of various political economists are less important today than are his more general criticisms of their works. In his opinion, the literature in political economy displayed two fundamental defects. First, capitalist social relations were assumed to reflect "irrefutable natural laws of society." Because of this emphasis, basic types of social relations, such as exchange, exploitation, and alienation, were all assumed (at least by implication) to be historically immutable. Second, the political economists analyzed each part of society sep-arately, as if it had no connection with anything else.[35] For example, even such strictly economic categories as production, exchange, distribution, and con-sumption were generally treated as if they were separate and unconnected phe-nomena. But Marx had learned from Hegel and Feuerbach that history moves in a dialectical pattern. As a result, Marx saw capitalism as a historically unique pattern of social relationships that would inevitably be supplanted in the future. Thus, he set himself the task of developing a scientific analysis of capitalist soci-ety that could account for both its development and eventual demise.

Although Marx regarded most political economists as simply bourgeois ideologues defending the status quo, he believed that Adam Smith and David Ricardo were the two most objective and insightful observers of the econom-ics of capitalism. In the course of analyzing their work, Marx achieved many of his fundamental insights into the dynamics of capitalism. For illustrative pur-poses, we focus here on Smith's work.

Adam Smith's Influence

Adam Smith was a moral philosopher as well as a political economist. In his first book, *The Theory of Moral Sentiments,* originally published in 1759, he argued that there was a natural order to the world, including both its physical and social aspects, that had been created by God and carefully balanced to benefit all species.[36] Hence, he emphasized the beneficent qualities of the natural order and the general inadequacy of human institutions that tried to change or alter this order. His subsequent book, *An Inquiry into the Nature and Causes of the Wealth of Nations,* published in 1776, represented his attempt at applying the principles of naturalism to the problems of political economy.[37]

The *Wealth of Nations* focuses on three main issues. First, Smith wanted to discover the "laws of the market" holding society together. In dealing with this issue, he hoped to show both how commodities acquired value and why this value included profit for the capitalist. Second, he wanted to understand the laws of evolution characteristic of capitalist society. Third, like most work in political economy (at least according to Marx), *The Wealth of Nations* is a thoroughgoing defense of capitalist society, a defense Marx found inadequate.

Laws of the Market Smith's attempt at showing how the economic laws of the market hold society together begins with the assertion that people act out of self-interest when they produce commodities for other members of the society to purchase. For "it is not from the benevolence of the butcher, brewer, or the baker, that we expect our dinner but from their regard to their own interest. We address ourselves, not to their humanity, but to their self-love, and never talk to them of our own necessities but of their advantages."[38] And the advantage that accrued to the butchers, bakers, and other capitalists is profit. Indeed, in Smith's view, the exchange of commodities for profit becomes a fundamental characteristic of human society whenever the division of labor and private property develop beyond a certain point. Marx believed, however, that Smith had been guilty of trying to make patterns of interaction that were characteristic of capitalist social relationships valid for all times and places. As an alternative, Marx envisioned a modern society without exchange relationships because he felt that they were inherently exploitive. Nonetheless, for Smith the origin of value and profit resided in the process of commodity exchange, and by distinguishing between the "use value" and the "exchange value" of commodities, he achieved an insight that later guided Marx's thought.

Smith went on to formulate a version of the labor theory of value in which the amount of labor time going into a product was the source of its value, a thesis Marx embraced some ninety years later. But if labor is the source of value, Smith could not account for the origin of profit because those who profit generally contribute very little labor to the creation of the product. They merely invest money and reap a return on it. Thus, although *The Wealth of Nations* displays much vacillation and confusion, Smith ultimately dropped the labor theory of value and simply argued that profit was added on to the costs of production by the capitalist. As we will see in the next chapter, Marx was able to adopt the labor theory of value and still account for the origin of profit by distinguishing between the workers' labor and their labor power (or capacity to work).

By arguing that profit was merely part of the cost of production, Smith created a potential problem: The "natural price" of a commodity is difficult to determine because nothing prevents capitalists from constantly and arbitrarily raising prices. His solution was to argue that competition prevented avaricious persons from pushing prices too high. Those capitalists who try to raise prices unduly (and Smith was very aware that they constantly try to do just that) will inevitably find other enterprising persons underselling them, thereby forcing prices back down. Similarly, capitalists who attempt to keep wages too low will find that they have no workers because others offer the workers higher wages. In this way, then, both profits and wages are more or less automatically regulated—as if by an "invisible hand." Paradoxically, people's selfish motives promote social harmony through the natural operation of the market, even though that goal is not their objective. In Smith's words, "by directing that industry in such a manner as its produce may be of the greatest value, he intends only his own gain . . . he is in this, as in many other cases, led by an invisible hand to promote an end which was no part of his intention."[39]

The final step in Smith's analysis was to argue that these laws of the market also ensured that the proper quantities of products were produced. For example, if the public prefers to own coats rather than tables, a greater number of the former will be produced because the profit involved in making tables will fall, and capitalists (and workers) will turn to the manufacture of coats. Thus, natural mechanisms inherent to the market govern the allocation of resources in the society and, hence, the production of goods. Once again, this process occurs because of people acting in their own self-interests.

Laws of Evolution During the latter portion of the eighteenth century and well into the nineteenth, many political economists speculated that as capitalism advanced, the rate of profit on investment would fall. Smith had a rather optimistic view of the process of history, however, and he did not believe this calamity would occur. To Smith, in addition to being self-regulating, society seemed to be improving because of the operation of two relatively simple laws of evolution. The first can be called the "law of capital accumulation." Smith saw that capitalists continuously tried to accumulate their savings or profits, invest them, accumulate even more savings or profits, and invest them again. The impact is to increase both production and employment. Thus, from Smith's point of view, selfish motives can be seen once again to redound to the public good, because the expansion of production and employment helps everyone in some way. (Smith did not worry about whether savings would be invested; that became a problem for later economists.)

Some observers argued, however, that if accumulation was to continue and production to expand, more and more workers were required. When the supply of workers is exhausted, then profits will fall, and hence the rate of accumulation will also fall—just as many feared. Smith dealt with this problem by formulating a "law of population," his second law of evolution. This hypothesis asserts that when wages are high, the number of workers will increase; when

wages are low, the number of workers will decrease. Smith meant this statement literally, as people living and dying, not their periodic ventures into or
out of the labor market. Mortality rates, especially among children, were
extraordinarily high in those days; it was common for a woman to have a
dozen or more children and have only one or two survive. Yet it was still possible for a higher standard of living to affect decisively people's ability to feed,
clothe, and protect their children. As a result, Smith argued, higher wages
would allow greater numbers of children to survive and become workers
themselves. Lower wages, of course, would have the reverse effect. Thus, Smith
believed that the advance of capitalism would be accompanied by an increase
in population and that this increase would, in turn, allow capital accumulation
to continue. Therefore, according to Smith, the rate of profit will not fall and
a capitalist society will constantly improve itself, all because of the natural
forces, unencumbered by rules and regulations established by the state.
Although he recognized that an expanding population would always act to
deflate wages, as long as capital accumulation continued, wages had to remain
above the level of subsistence. Of course, Smith's argument, in Marx's view,
assumes that capitalist social relations are somehow irrefutable "natural laws" of
the social universe.

The Defense of Capitalism As enunciated by Smith, the logical implication of *The Wealth of Nations* is fairly simple: Leave the market alone. From
Smith's point of view, this stricture meant that the natural regulation of the
market would occur as consumers' purchasing practices forced businesses to
cater to their needs. Such a process could only happen, he believed, if business
was not protected by the government and did not form monopolies. Hence,
he opposed all efforts to protect business advantage. His analysis, however,
quickly became an ideological justification for preventing government regulation in some important areas. Moreover, because any act of government could
be seen as interfering with the natural operation of the market, *The Wealth of
Nations* was used to oppose humanitarian legislation designed to protect workers from the many abuses already apparent in Smith's time.

For Marx, this result showed the inherent weakness in classical political
economy, for there is nothing natural about the operation of the market or any
other social relationship. The market could not be left alone without the use
of governmental power, which was impossible because the capitalists, like all
ruling classes, controlled the government. Hence, Marx argued that theory
must take into account the interconnections among the parts of society, with
special attention to how political power is used to justify and enforce exploitive
social relationships. The capitalists, like all ruling classes, also controlled the dissemination of ideas, which suggested to Marx why they were able to use *The
Wealth of Nations* for their own ideological purposes. In his theory, Marx
emphasized the importance (and the difficulty) of stimulating an awareness in
the working classes of their true interests.

FRIEDRICH ENGELS AND KARL MARX

Friedrich Engels and Karl Marx were friends and collaborators for more than forty years. When possible, they saw each other every day; at other times, they corresponded about every other day. Despite Marx's sometimes-churlish temperament, the two men never broke off their relationship. Most commentators see Engels's role in the development of Marx's theory as secondary, and with regard to their joint works, especially *The German Ideology* and *The Communist Manifesto,* this appears to be an accurate assessment. Yet two of Engels's own writings, "Outlines of a Critique of Political Economy" (1844) and his much-neglected classic, *The Condition of the Working Class in England* (1845), fundamentally influenced the development of Marx's thought at a time when he was still searching for a way of understanding and changing the world.

Engels's Critique of Political Economy

Engels's short and angry essay, "Outlines of a Critique of Political Economy," appeared in the same journal as did Marx's critique of Hegel's philosophy. In this work, which is characterized by the excessively acerbic prose of a young man, Engels indicted both the science of political economy and the existence of private property. He began by noting caustically that political economy ought to be called "private economy" because it existed only to defend the private control of the means of production. He continued (although in a quite disorganized way) by stridently attacking the institution of private property. According to Engels, a modern industrial society based on the private ownership of property is inevitably inhumane, inefficient, and alienating. In the process of his attack, Engels also suggested (albeit vaguely) that, despite these faults, capitalism was historically necessary for a communist society to emerge in the future.

From Engels's point of view, capitalism is inhumane for two reasons. First, people do not and cannot trust one another. When private property exists in an industrial context, Engels wrote, trade and competition are the center of life. And because everyone seeks to buy cheap and sell dear, people must distrust and try to exploit one another. In Engels's words, "trade is legalized fraud."[40] The second reason capitalism is inhumane is that competition generates an increased division of labor, one of the major manifestations of which is the factory system. As we will see, Engels regarded factory work and the urban lifestyle accompanying it as one of the most inhumane and exploitive forms of social organizations in history. The factory system was becoming more pervasive in the 1840s, however, and as a result, capitalist society appeared to be dividing into two groups: those who owned the means of production and those who did not.

Engels argued that capitalism was inefficient because those who dominated it could neither understand nor control the recurrent and steadily worsening economic crises that afflicted every nation. Capitalist society is, therefore, beset by a curious paradox: Although its productive power is incredibly great, overproduction periodically results in misery and starvation for the masses. "The economist has never been able to explain this mad situation," Engels wrote.[41] Moreover, he believed that people living under capitalism were inevitably alienated because they had no sense of community. In the competitive environment characteristic of capitalism, each person's interests are always opposed to every other person's. As a result, "private property isolates everyone in his own crude solitariness," with the consequence that people's lives have little meaning and carry no intrinsic rewards.[42] Underlying this entire argument, however, is Engels's belief that the rise of capitalism is historically necessary to make a communist society possible, for only now are people "placed in a position from which we can go beyond the economics of private property" and end the "unnatural" separation of individuals from one another and from their work.[43]

Before 1843, the still youthful Marx was relatively unfamiliar with political economy. Engels's essay, as much as any other event, introduced Marx to the topic and made him recognize its importance in developing a theory of society. Thus, after reading the essay, he began an intensive study of political economy that lasted for more than twenty years. Ultimately, he indicted the science of political economy for essentially the same reason as had Engels: It defended capitalist society. In addition, all the main ideas that we will describe subsequently appeared in a more sophisticated fashion in Marx's theory.

Engels's Analysis of the Working Class

To continue his business training at the textile mills in which his father was part owner, Engels left his native Germany for Manchester, England, in 1842. At that time, Manchester was the greatest industrial city in the most industrialized nation in the world; to many observers it was the epitome of the new kind of society forming as a result of the rise of capitalism and the industrial revolution. Engels spent two years in Manchester, leading something of a double life because he not only learned the textile business but also gathered the materials for his book. During this period, nearly all his leisure time was spent walking through Manchester and the surrounding towns, talking to and drinking with working-class people and reading the many governmental reports and other descriptions of living conditions in Manchester. The result was the first urban ethnography—and a damning indictment of the English ruling class.

Engels's analysis of *The Condition of the Working Class* in England can be divided into three parts. First, he sketched an idyllic rural society that existed before industrialization and briefly suggested the factors that had destroyed that society. Second, he described the conditions of working-class life in Manchester. Third, he indicted the attitudes of the bourgeoisie toward the proletariat and concluded that a violent revolution was inevitable.

Peasant Life Before Industrialization   Like many other observers, Engels saw that the Industrial Revolution was utterly transforming Western society. Like other observers, he believed that feudal society had been better for people in many ways. Consequently, he described the feudal past in an idyllic manner.

Although he has been justifiably criticized for idealizing the past, it is not altogether clear how (in the middle of the nineteenth century) he could have obtained a sound or accurate portrayal of feudal society. Thus, *The Condition of the Working Class* begins with a description of simple, God-fearing peasants who lived in a stable and patriarchal society where "children grew up in idyllic simplicity and in happy intimacy with their playmates." Engels saw feudal life as "comfortable and peaceful" and believed that most peasants generally had a higher standard of living in the past than did factory workers in 1844:

> They were not forced to work excessive hours; they themselves fixed the length of their working day and still earned enough for their needs. They had time for healthy work in their gardens or smallholdings and such labor was in itself a recreation. They could also join their neighbors in various sports such as bowls and football and this too kept them in good physical condition. Most of them were strong, well-built people, whose physique was virtually equal to that of neighboring agricultural workers. Children grew up in the open air of the countryside, and if they were old enough to help their parents work, this was only an occasional employment and there was no question of an eight- or twelve-hour day.[44]

At the same time, Engels argued, these peasants were "spiritually dead" because they were ignorant, concerned only with their "petty private interests," and contented with their "plantlike existence."[45]

Although this depiction of life before industrialization is clearly not accurate, it does identify some themes that Engels used in his indictment of capitalist society: People are forced to work excessive hours, they are in chronic ill health, and child labor is pervasive. In addition, his sketch of feudal life also implied the historical inevitability of a communist revolution; according to Engels, industrialization not only shattered forever this idyllic lifestyle, it also made people aware of their subordination, exploitation, and alienation. As we will see, Marx and Engels believed that this recognition was the first necessary step to a communist revolution. Thus, Engels's portrayal of the atrocities characteristic of urban life in the 1840s should be seen in light of his optimistic vision of the historical development of a revolutionary proletariat capable of seizing the world for itself. All Marx's subsequent work was imbued with this vision, which he and Engels shared and tried to actualize in the political arena.

Having described peasant life before industrialization, Engels noted the four interrelated factors that went into making the modern working class he observed in Manchester. First, the use of water-and-steam power in the productive process meant that, for the first time in human history, muscle power was not the primary motive force in producing goods. Second, the massive introduction of modern machinery into the productive process signaled not only that machines rather than people set the pace of work but also that more goods were being produced than ever before. Third, the intensification of the division of labor meant that the number of tasks in the productive process increased while the requirements for each task were simplified. Fourth, the tendency in modern society for concentration of both work and ownership caused not only the rise of the factory system but also a division of society into owners and producers. According to Engels, and he was not alone, these factors were the "great levers" of the Industrial Revolution that had been used to "heave the world out of joint."[46] In *Capital,* Marx took these same ideas and placed them in a theoretical context that, in his mind, allowed him to demonstrate why a proletarian revolution was inevitable.

Working-Class Life in Manchester. The world Engels saw was indeed out of joint. His description began with a portrayal of the neighborhoods in which working-class people were forced to live. Manchester had grown from a town of twenty-four thousand people in 1773 to a metropolitan area of more than four hundred thousand in 1840. Throughout this period, it had no effective city government, little police protection, and no sewer system. Engels observed that middle-class people and the owners of the factories and mills lived apart and provided themselves with city services, police protection, and sewage disposal. In contrast, the working classes were forced to live with pigs in the slums available to them. When Engels said that human beings lived with pigs (and, unavoidably, like pigs), he meant it literally.

Because there were no modern sewage facilities in Manchester, people had to use public privies. In some parts of the city, more than two hundred people used a single receptacle. In a city without government, there were few provisions for cleaning the streets or removing debris. Engels described the result in some detail; for example, in one courtyard, "right at the entrance where the covered passage ends, is a privy without a door. This privy is so dirty that the inhabitants can only enter or leave the court by wading through puddles of stale urine and excrement."[47] Thus, in *The Condition of the Working Class,* Engels portrayed a situation in which thousands of men, women, and children were living amid their own bodily wastes. If it can be imagined, the situation was even worse for those thousands of people living in cellars, below the waterline. As Steven Marcus has observed, "that substance [their bodily waste] was also a virtual objectification of their social condition, their place in society: that was what they were."[48]

Engels continued by describing the neighborhoods where pigs and people lived together:

> Heaps of refuse, offal and sickening filth are everywhere interspread with pools of stagnant liquid. The atmosphere is polluted by the stench and is darkened by the smoke of a dozen factory chimneys. A horde of ragged women and children swarm about the streets and they are just as dirty as the pigs which wallow happily on the heaps of garbage and in the pools of filth. In short, the horrid little slum affords as hateful and repulsive a spectacle as the worst courts to be found on the banks of the Irk [river]. The inhabitants live in dilapidated cottages, the windows of which are broken and patched with oilskin. The doors and the door posts are broken and rotten. The creatures who inhabit these dwellings and even their dark, wet cellars, and who live confined amidst all this filth and foul

air—which cannot be dissipated because of the surrounding lofty buildings—must surely have sunk to the lowest level of humanity.[49]

It is not hard to conclude, as many did, that a society in which people have gone back to living like animals has something terribly, deeply wrong with it. For many observers, however, Manchester epitomized a new and better world, an industrial world.

The Bourgeoisie, the Proletariat, and Revolution. Engels concluded *The Condition of the Working Class* by describing the attitudes of the bourgeois toward the proletarians. In his prose the bourgeoisie are portrayed as debased people who know nothing except greed and see all human ties as having a "cash nexus." He used the following vignette to illustrate these traits:

> One day I walked with one of these middle-class gentlemen into Manchester. I spoke to him about the disgraceful unhealthy slums and drew his attention to the disgusting condition of that part of the town in which the factory workers lived. I declared I had never seen so badly built a town in my life. He listened patiently and at the corner of the street at which we parted company he remarked: "And yet there is a great deal of money made here. Good morning, Sir."[50]

Yet the proletarians were sometimes capable of responding to their condition in life. Although much self-destructive behavior always occurs among oppressed people (as in the use of drugs, alcohol, and the like), Engels noted that Manchester was "the mainspring of all working-class movements" in England[51] And he described the long history of working-class efforts at organizing in opposition to the factory owners, for only by acting together rather than competing with one another could they effectively oppose the capitalists. More generally, however, he argued that the proletarians' true interest was in establishing a noncompetitive society, which meant the abolition of the private ownership of the means of production (although this last point was not made explicitly):

> Every day it becomes clearer to the workers how they are affected by competition. They appreciate even more clearly than the middle classes that it is competition among the capitalists that leads to those commercial crises which cause such dire suffering among the workers. Trade unionists realize that commercial crises must be abolished, and they will soon discover *how* to do it.[52]

The Condition of the Working Class ends with Engels's prophecy of a violent proletarian revolution. Although Engels believed this "revolution must come," he had not shown why; his work accounted for neither how capitalist society functioned nor why it would inevitably be destroyed. He had not, in short, developed a theory to explain what he had observed. But at a time when Marx was searching for the underlying dynamics of society, Engels demonstrated the significance of the proletariat. Furthermore, he recognized (although the point was not made very clearly) that the evils of capitalism were a necessary prelude to a communist revolution. Yet Marx, rather than Engels, developed a set of theoretical concepts and propositions that purported to show why a revolution would occur in capitalist societies. In developing these theoretical arguments, Marx contributed to the emergence of sociological theory.

NOTES

1. Isaiah Berlin, *Karl Marx: His Life and Environment* (New York: Oxford University Press, 1963), p. 33.

2. Karl Marx, "Discovering Hegel" (Marx's letter to his father), in *The Marx-Engels Reader,* ed. Robert C. Tucker (New York: Norton, 1978), pp. 7–9.

3. Karl Marx, "The Difference between the Democritean and Epicurean Philosophy of Nature," in *Activity in Marx's Philosophy,* ed. Norman D. Livergood (The Hague: Martinus-Nijhoff, 1967), pp. 57–109.

4. Karl Marx, "A Contribution to the Critique of Hegel's *Philosophy of Right,"*
in Tucker, ed., *Marx-Engels Reader,* pp. 16–26, 53–66.

5. Friedrich Engels, *The Condition of the Working Class in England* (Stanford, CA: Stanford University Press, 1968). The current translation omits the year 1844 from the title. Friedrich Engels, "Outlines of a Critique of Political Economy," in *The Economic and Philosophic Manuscripts,* ed. Karl Marx (New York: International, 1964), pp. 197–228.

6. Karl Marx and Friedrich Engels, *The Holy Family* (Moscow: Foreign Languages Publishing House, 1956).

7. Marx, *Economic and Philosophic Manuscripts.*

8. Karl Marx and Friedrich Engels, *The German Ideology* (New York: International, 1947), p. 3.

9. Quoted in David McClellen, *Karl Marx: His Life and Thought* (New York: Harper & Row, 1973), p. 452.

10. Friedrich Engels, "Principles of Communism," in *The Birth of the Communist Manifesto,* ed. Dirk Struik (New York: International, 1971), pp. 169–192.

11. The edition we are using is reprinted in Struik, ed., *Birth of the Communist Manifesto,* pp. 85–126.

12. Karl Marx, *The Grundrisse* (New York: Random House, 1973), *The Theory of Surplus Value* (Moscow: Foreign Languages Publishing House, 1963), and *A Contribution to the Critique of Political Economy* (New York: International, 1970). Only the last was published in Marx's lifetime, in 1859.

13. Karl Marx, *Capital,* vol. 1 (New York: International, 1967).

14. Friedrich Engels, "Preface to the First English Edition," in Marx, *Capital,* vol. 1, p. 5.

15. Karl Marx, "The Civil War in France," in Karl Marx and Friedrich Engels, *Selected Works,* vol. 2 (Moscow: Progress, 1969), pp. 178–244.

16. G. W. F. Hegel, *The Phenomenology of Mind* (New York: Macmillan, 1961), *The Science of Logic* (London: Allen & Unwin, 1969), *The Encyclopedia of Philosophy* (New York: Philosophical Library, 1959), and *The Philosophy of Right* (Oxford: Clarendon, 1942).

17. Hegel, *Science of Logic,* p. 154.

18. Ibid., p. 585 (emphasis in original).

19. Lucio Colletti, *Marxism and Hegel* (Atlantic Highlands, NJ: Humanities, 1973), p. 12. This is a good Marxist source. One of the best non-Marxist commentaries is John N. Findlay,

Hegel: A Re-Examination (London: Allen & Unwin, 1958).

20. Hegel, *Science of Logic,* p. 138 (emphasis in original).

21. Quoted in Friedrich Engels, "Ludwig Feuerbach and the End of Classical German Philosophy," in *Marx and Engels, Selected Works,* vol. 3, p. 337.

22. Marx, *Economic and Philosophical Manuscripts,* p. 175.

23. Ibid., p. 177.

24. Ibid., p. 181.

25. Marx, "Contribution to the Critique of Hegel's *Philosophy of Right,*" p. 54.

26. Quoted in Sidney Hook, *From Hegel to Marx* (Ann Arbor: University of Michigan Press, 1962), p. 23.

27. Marx, *Capital,* p. 20.

28. David Strauss, *The Life of Jesus Critically Examined* (London: Swan Sonneschein, 1902).

29. On Bauer, see Hook, *From Hegel to Marx.*

30. Max Stirner, *The Ego and His Own: The Case of the Individual Against Authority* (New York: Libertarian Book Club, 1963).

31. Karl Marx and Friedrich Engels, "The Communist Manifesto," pp. 85–125 in *The Birth of the Communist Manifesto,* ed. Dirk Struik (New York: International, 1971).

32. Ludwig Feuerbach, *The Essence of Christianity* (New York: Harper & Row, 1957).

33. Quoted in Hook, *From Hegel to Marx,* p. 246.

34. Feuerbach, *Essence of Christianity,* p. 13.

35. Marx, "Introduction," in *Contribution to the Critique of Political Economy,* pp. 188–217.

36. Adam Smith, *The Theory of Moral Sentiments* (Oxford: Clarendon, 1976).

37. Adam Smith, *An Inquiry into the Nature and Causes of the Wealth of Nations* (Oxford: Clarendon, 1976).

38. Ibid., pp. 26–27.

39. Ibid., p. 73.

40. Engels, "Outlines of a Critique of Political Economy," p. 202.

41. Ibid., p. 217.

42. Ibid., p. 213.

43. Ibid., pp. 199, 212.

44. Engels, *Condition of the Working Class,* p. 10.

45. Ibid., pp. 11–12.

46. Ibid., pp. 27–29.

47. Ibid., p. 58.

48. Steven Marcus, *Engels, Manchester, and the Working Class* (New York: Vintage, 1975), pp. 184–185.

49. Engels, *Condition of the Working Class,* p. 71.

50. Ibid., p. 312.

51. Ibid., p. 50.

52. Ibid., p. 249 (emphasis in original).

8

✳

The Sociology
of Karl Marx

Industrialization and capitalism destroyed feudal social relationships that had existed for a millennium, but in Karl Marx's eyes, these changes had produced a paradoxical result. Industrialization and capitalism meant that sustenance and amenities could be available for everyone, and yet, only those who owned capital (income-producing assets) actually benefited. These capitalists exploited the masses, who lived in great misery and depravity. To remedy this situation, Marx tried to stimulate people to reorganize social arrangements so that everyone's needs could be met. He argued that such a change was inevitable, the only question being when it would occur. Throughout his life, he served as a participant, organizer, and leader of revolutionary groups dedicated to ending the exploitation of the masses.

Of all the classical sociologists, Marx was unique in that he acted as revolutionary and social scientist, a combination that constitutes the greatest weakness in his sociology. His orientation can be summarized in the following way: As a revolutionary, he sought to overthrow the existing order and substitute collective control of society by the people so that, in a cooperative context, they could be free to develop their potential as human beings. As a social scientist, he tried to show that such collective control was historically inevitable. According to Marx, history has a direction that can be observed. This direction, he and Engels wrote in *The Communist Manifesto,* will lead inevitably to a communist society in which "the free development of each is the condition for the free development of all."[1] In such a context, Marx believed, the few will no longer exploit the many.

THE GERMAN IDEOLOGY

The German Ideology was completed in 1846, when Marx was twenty-eight years old and Engels was twenty-six. Much of the rather lengthy book is given over to heavy-handed and satirical polemics against various Young Hegelians. The publisher declined to accept the manuscript at the time, perhaps for political reasons, because Marx was already well known as a radical and had been expelled from both Germany and France, or perhaps because of the arcane writing style. In any case, Marx later recalled, the manuscript was "abandoned to the gnawing criticism of the mice . . . since we had achieved our main purpose—self-clarification."[2]

Marx opened *The German Ideology* with a bitter attack on the Young Hegelians, whom he described at one point as engaging in "theoretical bubble blowing."[3] For the Young Hegelians, Marx observed, great conflicts and revolutions take place only in the realm of thought because no buildings are destroyed and no one is injured or dies. Thus, despite their excessive verbiage, Marx believed, Young Hegelians merely criticized the essentially religious nature of Hegel's work and substituted their own negative religious canons. "It is an interesting event we are dealing with," he said caustically, "the putrescence of the absolute spirit."[4] In the process of debunking the Young Hegelians' writings, however, Marx developed an understanding of social theory, a description of the characteristics of all societies, and a theoretical methodology for understanding those characteristics.

The Nature of Social Theory

As an alternative to the "idealistic humbug" of the Young Hegelians, Marx argued that theoretical analyses should be empirically based. Social theory, he said, should be grounded on the "existence of living human individuals" who must survive, often in a relatively hostile environment.[5] This orientation is necessary because human beings are unlike other animals in that they manipulate the environment to satisfy needs. They "begin to produce their means of subsistence, a step which is conditioned by their physical [that is, social] organization."[6] This idea implies that people are "conscious"—that is, self-reflective. Thus, human beings are also unlike other animals in that they can look at themselves and their environment and then act rationally in their own interests. This means that consciousness arises from experience, an argument directly opposed to Hegel's idealism, in which notions of morality, religion, and all other forms of awareness are considered to exist independently of human beings. Put in modern language, Marx was asserting that people produced their ideas about the world in light of the social structures in which they live and the experiences they have in these structures. Further, as social structures change, the content of people's ideas (their consciousness) changes as well. In breaking with the idealists in this way, however, Marx did not imply a simple-minded materialist orientation. He did not see the human mind as a passive receptacle; rather, it is active, both responding to and changing the material world.

According to Marx, then, social theory should focus on how people influence and are influenced by their material conditions: for example, their degree of hunger, degree of protection from the environment, opportunity to enjoy the amenities of life, and ability to realize their creative potential. This emphasis constitutes a fundamental epistemological break with idealism. In effect, Marx stood Hegel "right side up" by transforming philosophy into an empirical social science.

The Characteristics of All Societies

Based on this vision of social theory, Marx emphasized that theoretical analyses should be oriented to what he called "the real process of production"—that is, the most essential characteristics that all societies have in common. These characteristics (Marx called them *moments*) do not refer to evolutionary stages of development but, rather, to social conditions that have "existed simultaneously since the dawn of history and the first men, and . . . still assert themselves in history today."[7] Marx's language is significant. He used a phrase that appears to have narrow, economic connotations—"the real process of production"—to refer to a more general sociological issue. Such phrasing occurs frequently in his writings.

The first characteristic of all societies is that human beings, unlike other animal species, produce sustenance from the environment to live and thereby "make history." Marx noted that human "life involves before anything else eating and drinking, a habitation, clothing, and many other [material] things."[8] Such needs are satisfied by employing technology to manipulate the environment in some socially organized manner. For Marx, this clearly implied that social theory had to deal with more than just ideas. It had to be grounded in "the existence of living human individuals," who have material needs that must be satisfied through production. From this angle of vision, the task of social theory is to explain how people "produce their means of subsistence."

The second characteristic of all societies is that people create new needs over time. Need creation occurs because production (or work) always involves the use of tools or instruments of various sorts, and these tools are periodically improved, yielding more and better consumer goods. Thus, Marx said the processes of production and consumption always feed back on each other in a cumulative fashion, so that as one set of needs is satisfied, new ones emerge.[9] See Figure 8.1 for a graphic representation.

This close connection between production and consumption led Marx to assert that the two were "identical" to or "simultaneously" each other because it is not possible to consider one apart from the other. Thus, the process of need creation, as indicated by the changing modes of production and consumption, implies that social theory must deal with historical change, its direction, and its source. He believed that human history displayed an evolutionary pattern from less complex to more complex social structures and that the origin of change was internal to each society.

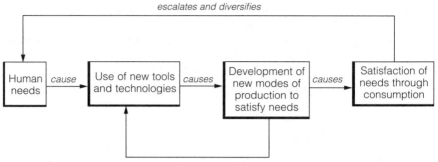

As this cycle is repeated, societies evolve from simple to ever-more complex forms, with various "epochs" of history evident with transformations in tools and technologies, modes of production, and patterns of consumption. Societies "have a history" because escalating human needs drive the development of new tools and technologies generating new modes of production and new patterns of consumption.

FIGURE 8.1 Mark's View of Human Needs, Production, and History

The process of need creation involves the desire not only for improved food, clothing, and shelter but also for the various amenities of life. Marx observed that in the production and consumption of goods beyond the minimum necessary for survival—what are called amenities—people became "civilized" in the sense that they could distinguish their uniquely human characteristics from those of other species. Thus, in *The Economic and Philosophic Manuscripts* (written in 1844), he described productive work as serving a dual purpose: (1) to satisfy physical needs and (2) to express uniquely human creativity. According to Marx, this duality is why other animals work only to satisfy an "immediate physical need, whilst man produces even when he is free from physical need and only truly produces in freedom therefrom."[10] Unfortunately, Marx believed, most people are prevented from expressing their human potential through work because the exploitation and alienation inherent in the division of labor prevent it.

The third characteristic of all societies is that production is based on a division of labor, which in Marx's writings always implies a hierarchical stratification structure, with its attendant exploitation and alienation. The division of labor means the tasks that must be done in every society—placating the gods, deciding priorities, producing goods, raising children, and so forth—are divided among members of the society. But Marx observed that in all societies the basis for this division was private ownership of land or capital, which he called the *means of production*. Private ownership of the means of production produces a stratification system composed of the dominant group, the owners, and the remaining classes arrayed below them in varying degrees of exploitation and alienation. Non-owners are exploited and alienated because they cannot control either the work they do or the products produced. For example, capitalists, not employees, organize a production line to produce consumer goods, and capitalists, not employees, own the finished products. But because employees, whom Marx called *proletarians,* need these products to survive, they are forced to return their wages to the capitalists, who use the money to make more consumer goods and enrich themselves further. In this context, alienation takes the form of a fantastic reversal in which people feel themselves to be truly free only in their animal-like functions—such as eating, drinking, and fornicating—whereas in their peculiarly human tasks, such as work, they do not feel human because they control neither the process nor the result. On this basis, Marx concluded, in capitalism "what is animal becomes human and what is human becomes animal."[11] Thus, paradoxically, the division of labor means that proletarians continually re-create that which enslaves them: control of capital by the few.

In some form or another, Marx argued, exploitation and alienation occur in all societies' characterized by private ownership of the means of production. That is, in all societies, members of the subordinate classes are forced to continuously exchange their labor power for sustenance and amenities so they can keep on producing goods to benefit the members of the dominant class. For Marx, this situation implied that social theory had to focus on who benefits from existing social arrangements by systematically describing the structure of stratification that accompanies private ownership of the means of production. This situation also implied for Marx that only collective ownership could eliminate these problems.

The fourth characteristic of all societies is that ideas and values emerge from the division of labor. Put differently, ideas and values result from people's practical efforts at obtaining sustenance, creating needs, and working together. As a result, ideologies usually justify the status quo. "Ideologies" are systematic views of the way the world ought to be, as embodied in religious doctrines and political values. Thus, Marx argued, religious and political beliefs in capitalist societies state that individuals have a right to own land or capital; they have a right to use the means of production for their own rather than for the collectivity's benefit. It is perverse, he noted, for everyone to accept these values even though only a few people, such as landowners and capitalists, can exercise this right.

Marx believed the values (or *ideologies,* to use his word) characteristic of a society are the tools of the dominant class because they mislead the populace about their true interests. This is why he described religion as "the opium of the masses."[12] He reasoned that religious belief functioned to blind people so they could not recognize their exploitation and their real political interests. Religion does this by emphasizing that salvation, compensation for misery and alienation on earth, will come in the next world. In effect, religious beliefs justify social inequality. For Marx, the fact that ideas and values emerge from the division of labor implies that social theory must focus on both the structural sources of dominant ideas and the extent to which such beliefs influence people.

Marx's Theoretical Methodology

The exposition in *The German Ideology* is an early example of Marx's dialectical materialism. Although he did not use this phrase, it expresses the discontinuity and continuity between Hegel and Marx. Marx rejected Hegel by grounding social theory in the real world, where people must satisfy their physical and psychological needs. The term *materialism* denotes this. Having rejected the substance of Hegel's idealism, however, Marx continued to use the Hegelian method of analysis. The term *dialectical* denotes this. In Marx's hands, *dialectical materialism* transforms historical analysis.

Dialectical materialism has four characteristics. First, society is a social structure, or *system*. Marx did not use this modern term, but it means that societies can be seen as having interrelated parts, such as classes, social institutions, cultural values, and so forth. These parts form an integrated whole. Thus, the observer's angle is very important when viewing a society. In tracing the connections among the parts of the stratification system, for example, it can be seen that from one angle a specific label can be applied (for example, bourgeoisie), whereas from another angle an opposing label can be applied (for instance, proletariat). But there is an inherent connection between the two classes, which is why Marx noted in *The Communist Manifesto* that it was tautologous to speak of wage labor and capital, for one cannot exist without the other. Similarly, this is why he described production and consumption as "identical," or as occurring "simultaneously." He meant that they were parts of a coherent structure, or system, and that there was an inherent connection between them. Furthermore, the process of production and consumption (which today would be called the economy) is connected to stratification. More generally, class relations are reflected in all arenas of social behavior: the economy, kinship, illness and medical treatment, crime, religion, education, and government. Although Marx emphasized the primacy of economic factors, especially ownership of the means of production, his work is not narrowly economic; it is, rather, an analysis of how social structures function and change.

Second, social change is inherent in all societies as people make history by satisfying their ever-increasing needs. For Marx the most fundamental source of change comes from within societies rather than from outside them. The force behind these internally generated changes is the *contradiction* inherent in the system. Not only are all the parts of society connected, they also contain their own inherent contradictions, which will cause their opposites to develop. For example, as will be described in the next section, Marx argued that feudalism contained within itself the social relations that eventually became capitalism. Similarly, in the *Manifesto* and *Capital,* Marx contended that capitalism contained within itself the social relations that would inevitably engender a new form of society: communism.

Third, social change evolves in a recognizable direction. For example, just as a flower is inherent in the nature of a seed, so the historical development of a more complex social structure, such as capitalism, is inherent in the nature of a less complex one, such as feudalism. The direction of history is from less complex to more complex social structures, which is suggested by the pattern of need creation depicted earlier. As Robert Nisbet comments, Marx was a child of the Enlightenment, and he believed in the inevitability of human progress.[13] He had a vision of evolutionary development toward a utopian end point. For Marx, this end point was a communist society.

Fourth, freely acting people decisively shape the direction of history given the predictable patterns of opposition and class conflict that develop from the contradictions in society. As with all Marx's concepts, his use of the term *class* is sometimes confusing. The key to understanding this concept lies in the idea of opposition, for he always saw classes as opposed to one another. It should be remembered, however, that this opposition occurs within a stratification structure; classes are opposed but still connected.[14]

Thus, regardless of their number or composition, the members of different classes are enemies because they have opposing interests. This is not a result of choice, but of location within the stratification structure. For example, if the position of an aggregate of people makes obtaining food and shelter a constant problem and if these people cannot control their own activities or express their human potential, they are clearly in a subordinate position in relationship to others. In their alienation, they have an interest in changing the status quo, whether they are aware of it or not. On the other hand, if the position of an aggregate of people is such that their basic needs are satiated, if they can control their daily activities, and if they can devote themselves to realizing their human potential, such people have an interest in preserving the status quo. Marx believed that these opposing interests could not be reconciled.

Hence, given a knowledge of the division of labor in capitalism, the differing interests and opportunities of the proletarians and capitalists are predictable, as is the generation of class conflict. The latter, however, is a matter of choice. History does not act, people do. From this point of view, Marx's theoretical task was to identify the social conditions under which people will recognize their class interests, unite, and produce a communist revolution. As will become clear later, Marx believed that he had achieved this goal. The important point to remember is that his theoretical methodology combines determinism, or direction, with human freedom: A communist revolution is a predictable historical event ushered in by freely acting people who recognize and act in their own interests.

Dialectical materialism can thus be summarized in the following way: Within any society, a way of producing things exists, both for what is produced and the social organization of production. Marx called this aspect of society the *productive forces*.[15] In all societies, the productive forces are established and maintained through a division of labor. Those few who own the means of production make up the dominant class, which benefits from the status quo. The masses make up the subordinate class (or classes). They are exploited and alienated because they have little control over their lives, and hence they have an interest in change. Over time, new ways of producing things are devised, whether based on advances in technology, changes in the way production is organized, or both. Such new forces of production better satisfy old needs and stimulate new ones. They are in the hands of a new class, and they exist in opposition to current property relationships and forms of interaction. Over the long run, the tension between these opposing classes erupts into revolutionary conflict, and a new dominant class emerges.

Marx's methodology is unique because it is a logically closed theoretical system that cannot be refuted. This separates Marxist and non-Marxist social scientists today. Among non-Marxists, theories are evaluated by observations, which means they can be disproved. The goal is to develop abstract statements that summarize patterns of social organization. From this point of view, the social sciences resemble the natural sciences in orientation.

Among Marxists, however, theories are evaluated by what they lead people to do (or not do), which means they cannot be disproved. Because the goal is to assess where a society is along an evolutionary continuum, theories are constantly adjusted with changing political conditions.[16] The end point of this continuum is a communist society, a communal social organization in which there is collective control of the means of production (in today's societies, this is capital) so that people, acting cooperatively, can be free. In such a social context, Marx argued, exploitation and alienation will not exist because the division of labor will not be based on private ownership of property. From this point of view, the social sciences are radically different from the natural sciences.

The German Ideology constitutes the first presentation of Marx's theory. It is, however, incomplete. It does not, for example, raise one of the most crucial issues: How are the oppressed proletarians to become aware of their true interests and seize control of the society for the benefit of all? This and other problems of revolutionary action are dealt with in *The Communist Manifesto.*

THE COMMUNIST MANIFESTO

In 1847, Marx and Engels joined the Communist League, which they soon dominated. Under their influence, the League's goal became the overthrow of bourgeois society and the establishment of a new social order without classes and private property. To this end, Marx and Engels decided to compose a manifesto that would publicly state the Communist League's doctrines. The result constitutes one of the greatest political pamphlets ever written.

The *Manifesto* opens with a menacing phrase that immediately reveals its revolutionary intent: "A specter is haunting Europe—the specter of Communism. All the Powers of old Europe have entered into a holy alliance to exorcise this specter." In a political context where opposition parties of all political orientations were called communist, Marx wrote, it was time for the communists themselves to "meet this nursery tale of the specter of Communism with a Manifesto of the party itself."[17] The remainder of the *Manifesto* is organized into four sections, which are summarized as follows.

Bourgeoisie and Proletarians

Marx presented his theoretical and political position early in the text when he emphasized, "the history of all hitherto existing society is the history of class struggles." He continued by observing that in every era "oppressor and oppressed stood in constant opposition to one another [and] carried on an uninterrupted, now hidden, now open fight, a fight that each time ended either in a revolutionary reconstitution of society at large or in the common ruin of the contending classes."[18] Put differently, Marx believed that in every social order those who own the means of production always oppress those who do not. Thus, in his view, bourgeois society merely substituted a new form of oppression and, hence, struggle in place of the old feudal form. Marx argued, however, that bourgeois society was distinctive in that it had simplified class antagonisms, because the "society as a whole is splitting up more and more into two great hostile camps, into two great classes directly facing each other: Bourgeoisie and Proletariat."[19] Because one class owns the means of production and the other does not, the two have absolutely opposing interests: the bourgeoisie in maintaining the status quo and the proletariat in a complete reorganization of society so that production can benefit the collectivity as a whole. This situation reflected a long historical process. As in *The German Ideology,* the analysis in the *Manifesto* is an example of Marx's dialectical materialism.

Historically, Marx argued, capitalism emerged inexorably from feudalism. "From the serfs of the Middle Ages sprang the chartered burghers of the earliest towns. From these burgesses the first elements of the bourgeoisie [capitalists] were developed."[20] Such changes were not historical accidents, Marx said, but the inevitable result of people acting in their own interests. The rise of trade and exchange, stimulated by the European discovery of the Americas, constituted new and powerful productive forces, which faced a feudal nobility that had exhausted itself by constant warfare. Further, as they were increasingly exposed to other cultures, the members of the nobility wanted new amenities, and so they enclosed the land to raise cash crops using new methods of production. It should be recalled that production and consumption reciprocally affect each other—they are part of a social system—and they are tied to the nature of the class structure. As this historical process occurred, the serfs were forced off the land and into the cities, where they had to find work.

During this same period, a merchant class arose. At first the nascent capitalists existed to serve the needs of the nobility by facilitating trade and exchange. Over time, however, capital became the dominant productive force. This process occurred as new sources of energy (such as steam) were discovered, as machines were invented and used to speed up the production process, and as the former serfs were pressed into service in new industries as wage laborers. The result, Marx noted, was that in place of feudal retainers and patriarchal ties, there was "left no other nexus between man and man than naked self-interest, than callous 'cash payment.'"[21]

The *Manifesto* summarizes the situation in the following way:

The feudal system of industry, under which industrial production was monopolized by closed guilds, now no longer sufficed for the growing wants of the new markets. The manufacturing system took its place; the guild masters were pushed on one side by the manufacturing middle class; division of labor between the different corporate guilds vanished in the face of division of labor in each single workshop.

Meantime, the markets kept ever growing, the demand ever rising. Even manufacture no longer sufficed. Thereupon steam and machinery revolutionized industrial production. The place of manufacture was taken by the giant, modern industry, the place of the industrial middle class by industrial millionaires, the leaders of whole industrial armies, the modern bourgeois. . . .

We see then: The means of production and of exchange, on whose foundation the bourgeoisie built itself up, were generated in feudal society. At a certain stage in the development of these means of production and of exchange, the conditions under which feudal society produced and exchanged, the feudal organization of agriculture and manufacturing industry, in one word, feudal relations of property, became no longer compatible with the already developed productive forces; they became so many fetters. They had to be burst asunder, they were burst asunder.[22]

Thus, the rise of capitalism meant that the forces of production were revolutionized, and therefore, the class structure changed as well. Marx said that although these developments had been the result of freely acting people pursuing their self-interests, they had also been predictable—indeed, inevitable—historical events. Furthermore, because of the rise of capitalism, the class structure became simplified. Now there existed a new oppressed class, the proletarians, who had to sell their labor to survive. Because these people could no longer produce goods at home for their own consumption, they constituted a vast exploited and alienated work force that was constantly increasing in size. Opposed to the proletarians was a new oppressor class, the bourgeoisie (or capitalists), as a few former artisans and petty burghers became entrepreneurs and eventually grew wealthy. These people owned the new productive forces on which the proletarians depended.

Marx then described the truly revolutionary nature of the capitalist mode of production. As a result of the Industrial Revolution, the bourgeoisie "has accomplished wonders far surpassing Egyptian pyramids, Roman aqueducts, and gothic cathedrals; it has conducted expeditions that put into the shade all former Exoduses of nations and crusades."[23] For the bourgeoisie to exist, Marx predicted, it must constantly develop new instruments of production and thereby create new needs that can be filled by manufactured products. As this process occurs, the bourgeoisie also seizes political power in each country, so that "the executive of the modern state is but a committee for managing the common affairs of the whole bourgeoisie."[24]

Having described the great historical changes accompanying the rise of capitalism, Marx then made two of his most famous predictions concerning the ultimate demise of the capitalist system. First, capitalism is inherently unstable. Periods of economic growth and high employment are followed by economic decline and unemployment. For Marx, these cycles—what today we call the "business cycle"—are endemic to capitalism. Capitalists and proletarians cannot escape them because eventually too many goods are produced relative to the demand for them, causing production to be cut back, and thereby forcing capitalists to lay off labor. Once this process begins, it accelerates as those who have been laid off can no longer afford to purchase goods, pushing capitalists to terminate the employment of even more workers in an escalating cycle that can lead to an economic depression. Capitalists try to avoid this cycle in many ways. For example, they may destroy older products and sell only new ones; they may try to eliminate their competitors and thus exploit their markets more efficiently; and they may seek new markets. Try as they might, however, they cannot escape the inherent tendency of capitalist economies to experience recessions and depressions. As proletarians' lives are made more miserable by these circumstances, they begin to sense that their interests do not reside with capitalists, leading Marx to make his second great prediction.

Marx's second prediction was that "the modern working class, the proletarians" would become increasingly impoverished and alienated under capitalism. Because they could no longer be self-supporting, the proletarians had become "a class of laborers who live only so long as they find work, and who find work only so long as their labor increases capital."[25] Thus, in a context characterized by the extensive use of machinery owned by others, proletarians have no control over their daily lives or the products of their activities. Each person becomes, in effect, a necessary but low-priced appendage to a machine. In this situation, Marx said, even women and children are thrown into the maelstrom. Thus, under capitalism, human beings are simply instruments of labor whose only worth is the cost of keeping them minimally fed, clothed, and housed. Confronted with their own misery, Marx predicted, the proletarians will ultimately become class conscious and overthrow the entire system, especially as they live through cycles of recession and depression where their lives are made ever-more miserable.

The rise of the proletariat as a class proceeds with great difficulty, however, primarily because individual proletarians are forced to compete among themselves. For example, some are allowed to work in the capitalists' factories, and others are not. Within the factories, a few are allowed to work at somewhat better-paying or easier jobs, but most labor at lower-paying and more difficult tasks. After work, proletarians with too little money still compete with one another for the inadequate food, clothing, and shelter that is available. Under these competitive conditions, it is difficult to create class consciousness. Marx showed, however, that as the bourgeoisie introduce improvements in education, force the proletarians to become better educated (to work the machines), and drag the proletarians into the political arena, the proletarians' ability to recognize the source of their exploitation increases. But this process is slow and difficult; when workers did revolt, they usually directed their attacks against the instruments of production rather than the capitalists. When they did organize, the proletarians were often co-opted into serving the interests of the bourgeoisie.[26] With the development of large-scale industry, however, the proletariat constantly increases in size. Like many other observers of nineteenth-century society, Marx predicted that the number of working-class people would continually increase as elements of the lower-middle class—artisans, shopkeepers, and peasants—were gradually absorbed into it. Furthermore, he believed that even those in professions such as medicine, law, science, and art would increasingly become wage laborers. He thought that all the skills of the past were being swept aside by modern industry, creating but two great classes.

The revolutionary development of the proletariat would, Marx argued, be aided by the fact that it was becoming increasingly urban, and hence its members were better able to communicate with one another. Further, they were becoming better educated and politically sophisticated, partly because the bourgeoisie constantly dragged them into the political arena. Although the proletarians' efforts at organizing against the bourgeoisie were often hindered, Marx believed that they were destined to destroy capitalism because the factors mentioned here would stimulate the development of their class consciousness.

Proletarians and Communists

As Marx expressed it, the major goal of the communists could be simply stated: the abolition of private property. After all, he noted, under capitalism nine-tenths of the population has no property anyway. As might be imagined, the bourgeoisie were especially critical of this position. But Marx felt that just as the French Revolution had abolished feudal forms of private property in favor of bourgeois forms, so the communist revolution would abolish bourgeois control over capital—without substituting a new form of private ownership. Marx emphasized, however, that the abolition of the personal property of the petty artisan or the small peasant was not at issue. Rather the communists wished to abolish bourgeois "capital, i.e., that kind of property which exploits wage labor and which cannot increase except upon condition of begetting a new supply of wage labor for fresh exploitation."[27]

To change this situation, the proletarians periodically organized and rebelled during the nineteenth century. Indeed, shortly after publication of the *Manifesto*, revolts occurred throughout Europe. Even though such efforts were always smashed, Marx believed that the proletariat was destined to rise again, "stronger, firmer, mightier," ready for the final battle.

Marx viewed this process as an inevitable evolutionary development. In the *Manifesto*, Marx emphasized that "the theoretical conclusions of the Communists . . . express, in general terms, actual relations springing from an existing class struggle, from an historical movement going on under our very eyes."[28] According to Marx, just like the feudal nobility before it, "the Bourgeoisie [has] forged the weapons that bring death to itself." This process occurred because the productive forces of capitalism make it possible for all people to satisfy their needs and realize their human potential. For this possibility to occur, Marx contended, productive forces must be freed from private ownership and allowed to operate for the common good. Furthermore, the bourgeoisie has also "called into existence the men who are to wield those weapons—the modern working class—the proletarians." Marx believed the working classes in all societies would, in their exploitation and alienation, eventually bring about a worldwide communist revolution.

Although Marx did not say much about the future, he knew that the transition to communism would be difficult, probably violent. This is because the communists aimed at destroying the core of the capitalist system: private ownership of the means of production. To achieve this goal, Marx believed that the means of production had to be "a collective product" controlled by the "united action of all members of the society." Such cooperative arrangements are not possible in bourgeois society, with its emphasis on "free" competition and its apotheosis of private property. Collective control of the society, Marx thought, is only possible under communism, where capital can be used as a means to widen, to enrich, to promote the existence of the laborer." This drastic change required a revolution.

The first step in a working-class revolution, Marx argued, would be for the proletariat to seize control of the state. Once attaining political supremacy, the working class would then wrest "all capital from the bourgeoisie," "centralize all instruments of production in the hands of the state," and "increase the total of productive forces as rapidly as possible."[29] Furthermore, the following measures would also be taken in most countries:

1. Abolition of private ownership of land
2. A heavy progressive income tax
3. Abolition of all rights of inheritance
4. Confiscation of the property of emigrants and rebels
5. Centralization of credit and banking in the hands of the state
6. Centralization of communication and transportation in the hands of the state
7. State ownership of factories and all other instruments of production
8. Equal liability of all to labor
9. Combination of agricultural and manufacturing industries to abolish the distinction between town and country
10. Free public education for all children and the abolition of child labor

Marx understood perfectly that these measures could only be implemented arbitrarily, and he forecast a period of temporary communist despotism in which the Communist party acted in the interests of the proletariat as a whole. In an essay written many years after the *Manifesto,* Marx labeled this transition period the "revolutionary dictatorship of the proletariat."[30] Ultimately, however, his apocalyptic vision of the transition to communism

was one in which people would become free, self-governing, and cooperative instead of alienated and competitive. They would no longer be mutilated
by a division of labor over which they had no control. "The public power
will lose its political character," Marx wrote. "In place of the old bourgeois
society with its classes and class antagonisms, we shall have an association in
which the free development of each is the condition for the free development of all."[31] It is a splendid vision; unfortunately, it is not that of the sorcerer, but of the sorcerer's apprentice.

Socialist and Communist Literature

In the third section of the *Manifesto,* Marx attacked the political literature of the
day. He recognized that in all periods of turmoil and change, some inevitably
desire to return to times past or to invent fantastic utopias as the way to solve
humankind's ills. He believed that such dreams were, at best, a waste of time and,
at worst, a vicious plot on the part of reactionaries. Thus, this section of the
Manifesto is a brief critique of socialist literature as it then existed. He classified
this literature as (1) reactionary socialism (including here feudal socialism, petty-
bourgeois socialism, and German "true" socialism); (2) conservative, or bourgeois, socialism; and (3) critical-utopian socialism.

Reactionary Socialism Because the bourgeoisie had supplanted the feudal
nobility as the ruling class in society, the remaining representatives of the aristocracy attempted revenge by trying to persuade the proletarians that life had
been better under their rule. Marx characterized this literature as "half lamen-
tation, half lampoon; half echo of the past, half menace of the future" and said
that their efforts were misbegotten primarily because the mode of exploitation
was different in an industrial context and a return to the past was not possible.
    Petit bourgeois socialism is also ahistorical and reactionary. Although its
adherents have dissected capitalist society with great acuity, they also have little to offer but a ridiculous return to the past: a situation in which corporate
guilds exist in manufacturing and patriarchal relations dominate agriculture.
Because they manage to be both reactionary and utopian, which is difficult,
this form of socialism always ends "in a miserable fit of the blues." Marx had
previously criticized German, or "true," socialism in *The German Ideology.* In
the *Manifesto* he merely emphasized again (with typically acerbic prose) that
the Germans had written "philosophical nonsense" about the "interest of
human nature, of Man in General, who belongs to no class, has no reality, who
exists only in the misty realm of philosophical fantasy."[32]

Conservative, or Bourgeois, Socialism In Marx's estimation bourgeois
socialists wanted to ameliorate the miserable conditions characteristic of proletarian life without abolishing the system itself. Today he might call such persons
liberals. In any case, Marx believed that this goal was impossible to achieve, for
what Proudhon and others did not understand was that the bourgeoisie could
not exist without the proletariat and all the abuses inflicted on it.

Critical–Utopian Socialism Utopian socialists had many critical insights into the nature of society, but Marx believed that their efforts were historically premature because the full development of the proletariat had not yet occurred, and so they were unable to see the material conditions necessary for its emancipation. As a result, they tried to construct a new society independent of the flux of history. For the utopian socialists, the proletarians were merely the most suffering section of society rather than a revolutionary class destined to abolish the existence of all classes.

Communist and Other Opposition Parties

In the final section of the *Manifesto,* Marx described the relationship between the Communist party, representing the most advanced segment of the working class, and other opposition parties of the time. Basically, in every nation the communists were supportive of all efforts to oppose the existing order of things, for Marx believed that the process of opposition would eventually "instill into the working class the clearest possible recognition of the hostile antagonism between the bourgeoisie and the proletariat."[33] In this regard, communists would always emphasize the practical and theoretical importance of private property as the means of exploitation in capitalist society.

Marx, the revolutionary, concluded the *Manifesto* with a final thundering assault on the bourgeoisie:

> The communists disdain to conceal their views and aims. They openly
> declare that their ends can be attained only by the forcible overthrow of
> all existing social conditions. Let the ruling classes tremble at a
> Communist revolution. The proletarians have nothing to lose but their
> chains. They have a world to win. WORKING MEN OF ALL COUN-
> TRIES, UNITE![34]

Marx's View of Capitalism in Historical Context

Reading *The Communist Manifesto* makes it clear that Marx saw human societies as having developed through a series of historical stages, each characterized by its unique class divisions and exploitations. His vision is summarized in Table 8.1.[35]

Marx believed that humans originally lived in hunting and gathering societies in which everyone worked at the same tasks to subsist. Private property did not exist. Nor did a division of labor. Hence, there were no classes and no exploitation based on class. These societies, in short, were communist, with all members contributing according to their abilities and taking according to their needs. But this primitive communism collapsed, in his rendering of history, as social organization changed.

The first system of exploitation was slavery, in which rank and position were determined by ownership of other human beings. In slave societies, the interests of owners and slaves were obviously opposed. Slaves had an interest in minimizing daily work demands, improving their living conditions, providing

Table 8.1   Marx's View of the Stages of History

Stage	Oppressing Class	Oppressed Class
Primitive communism	No classes	
Slavery	Slave owners	Slaves
Feudalism	Landowners	Serfs
Capitalism	Bourgeoisie	Proletariat
Socialism	State managers	Workers
Communism	No classes	

mechanisms by which they could work their way out of bondage, and preventing the inheritability of slave status (so their children would be born free). Slave owners had an interest in maximizing daily work (productivity), minimizing expenditures for food and other maintenance costs, making it difficult for slaves to escape bondage, and ensuring the inheritability of slave status. These conflicts of interest grew more difficult to control as the number of slaves increased and owners competed with one another in ways that increased the plight of the slaves—for example, by demanding more work while reducing food rations. The resulting conflict, in Marx's interpretation, led to a revolution in which slaves rose up and abolished the mechanism of their exploitation: the system of slavery.

Slavery was followed by feudalism, in which landless serfs and landowners represented the two great classes. Again, they had opposing interests. Those who owned the land wanted to increase productivity and, over time, to generate more cash income. Serfs were obliged to work the land under the presumption that they would share in a portion of its bounty. Their interest was to retain as much control over their crops as possible. In countries such as England, feudalism declined because landowners cleared the countryside of peasants to make room for products that would generate cash. For example, sheep were raised not for meat but as a source of raw material for the nascent wool industry. Sheep generated more profit, enabling landowners to purchase valued goods and amenities.

As described in the *Manifesto,* the feudal epoch gave way to capitalism. The name signifies that capital rather than land became the source of exploitation. The two great classes, of course, are the proletariat and the bourgeoisie (capitalists). Capitalists hire proletarians only if they generate profit, which is why capitalists are often described as leeches in Marx's writings. He believed that capitalism would grow like a giant octopus, spreading its tentacles over the entire globe, until nearly all human activity became debased because it was a commodity subject to purchase.

Marx argued that as the contradictions inherent in capitalism grew, it would collapse and be replaced by socialism. He described this stage as a transitory "dictatorship of the proletariat" in which the Communist party would seize control of the state in the name of the working class and expropriate private property (capital). Eventually, he believed, communism would emerge, a classless society in which all would give according to their ability and take according to their needs. The circle would be complete.

This depiction of the stages of history is superficial and, indeed, quite wrong.[36] Remember, however, that Marx did not have access to the data available to modern historians. But his vision does reveal Marx's view of history as successive systems of exploitation in which change emerges from within a society as people with competing interests attempt to satisfy their expanding needs. Thus, it reflects the use of dialectical materialism as a historical method. Moreover, despite its empirical flaws, it is possible to construct a model of stratification and conflict that remains useful.

Marx's Model of Stratification and Class Conflict

Modern readers often have two contrasting reactions when studying *The Communist Manifesto,* neither of which is very clearly articulated. On the one hand, it is easy to see how aspects of Marx's analysis can be applied to societies today. After all, exploitation does occur, and people in different classes do have opposing interests. On the other hand, Marx's political orientation seems both naive and threatening. It appears naive because a truly cooperative industrial society is hard to imagine. It appears threatening because subsequent history shows that a totalitarian government (like that in the former Soviet Union) seems to follow from any application of his ideas. Both reactions reflect Marx's peculiar combination of revolution and theory, which constitutes the greatest weakness in his writings. Nonetheless, it is possible to extrapolate a useful model of social stratification and class conflict from *The Communist Manifesto.*

Before doing so, however, we must recognize that any discussion of Marx's legacy demands a political confession: We are not Marxists. Thus, in what follows, the analysis implies nothing about the inevitability of a communist revolution or the transformation of society. Rather, it implies a concern with those ideas in Marx's writings that can still serve sociological theory.

Figure 8.2 displays a model of stratification and class conflict taken from the *Manifesto.* It illustrates some key variables to look for in studying social stratification and conflict, and it implies a modern sociological orientation. Marx asserted that in a stable social structure, goods are produced to satisfy the material needs of people, a process necessitating a division of labor and justified in terms of dominant values. This situation is depicted in the first box in Figure 8.2.

Many past observers have construed Marx's emphasis on productive activity to be a form of economic determinism. But this is too narrow a reading. Marx's point is not that economic activity determines behavior in other areas but, rather, that all social action is conditioned by, and reciprocally related to, the type of productive activity that exists. For example, family life is likely to be different in a hunting-and-gathering society than in an industrial one, as are the forms of government, education, religious beliefs, law, cultural values, and so on. These variations occur, in part, because the way people obtain food, clothing, and shelter differs. Alternatively, however, in two societies at

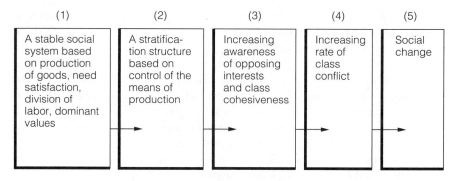

FIGURE 8.2 Marx's Model of the Generation of Stratification, Class Conflict, and Change

the same level of economic development, the organization of economic activity is likely to vary, because of differences in religious beliefs, law, family life, and so on.[37]

The recognition of such variation implies an essential sociological orientation: The range of options available to people is shaped by the nature of the society, its way of producing goods, its division of labor, and its cultural values. This orientation is fundamental to sociology today. Some writers like to begin with economic issues, others focus on some aspect of the division of labor (such as the family or criminal justice), and still others start by looking at how values circumscribe behavior. In every case, however, sociologists emphasize that society is a social system with interrelated parts and that social facts circumscribe behavior.

Marx argued—and he is probably correct—that a structure of stratification emerges in all societies based, at least in part, on control of the means of production. This fact, which is depicted in the second box in Figure 8.2, means the upper class also has the capacity to influence the distribution of resources because it dominates the state. Thus, those who benefit because they control the means of production have an interest in maintaining the status quo, in maintaining the current distribution of resources, and this interest is pervasive across all institutional arenas. For example, classes in the United States today have different sources of income, they have different political resources, they are treated differently in the criminal justice system, they provide for their children differently, they worship at different churches, and so forth.[38]

In assessing what modern sociologists can learn from Marx, the use of the word "control" rather than "ownership" in box 2 in Figure 8.2 is an important change because control over the means of production can occur in ways that he did not realize. For example, in capitalist societies the basis of social stratification is private ownership of property, whereas in communist societies the basis of social stratification is Communist party control of property. In effect, the Communist party is a new kind of dominant class ushered in by the revolution.[39] In both cases the group controlling the means of production exploits those who do not, while acting to justify its benefits by dominating the state and promulgating its values among the masses that legitimize its exploitation.

When he looked at social arrangements, Marx always asked a simple question, one that modern sociologists also ask: Who benefits? For example, the long empirical sections of *Capital* (to be examined shortly) are designed to show how attempts at lengthening the working day and increasing productivity also increased the exploitation of the working class to benefit the capitalists. Marx, however, also applied this question to non-obvious relationships. For example, his analysis of the "fetishism of commodities" in the early part of *Capital* shows how people's social relationships are altered by the reification (or worship) of machines and products that commonly occurs in capitalist societies, again to the benefit of capitalists. In effect, Marx teaches modern observers that an emphasis on who is benefiting from social arrangements and public policies can always improve an analysis. For example, macroeconomic decisions that emphasize keeping inflation low and unemployment high benefit the middle class and rich in American society at the expense of working people. In every arena—at home, at work, in court, at church, in the doctor's office, and so forth—it is useful to ascertain who is benefiting from current social arrangements.

The second box in Figure 8.2 is important in another way as well. As emphasized in the *Manifesto,* Marx divided modern capitalist societies into two great classes: bourgeoisie and proletariat. Although he recognized that this basic distinction was too simplistic for detailed analyses, his purpose was to highlight the most fundamental division within these nations. Whenever he chose, Marx would depict the opposed interests and experiences of various segments of society, such as bankers, the "lower middle classes," or the *lumpenproletariat* (the very poor).

Boxes 3, 4, and 5 in Figure 8.2 outline the process of class conflict and social change. Under certain conditions, members of subordinate classes become aware that their interests oppose those of the dominant class. In such a context, Marx taught, class conflict ensues, and social change occurs.

In Marx's work, of course, this process is linked to assumptions about the direction of history and the inevitability of a communist revolution. But this need not be the case. Members of a class can become aware of their true interests and be willing to act politically without seeking a revolutionary transformation of society. This process occurs because, although classes might be opposed to one another in any ongoing social structure, they are also tied to one another in a variety of ways. As Reinhard Bendix argues, citizenship, nationalism, religion, ethnicity, language, and many other factors bind aggregates of people together despite class divisions.[40] Furthermore, to the extent that a subordinate class participates effectively in a political system, as when it obtains some class-related goals, it then acquires an interest in maintaining that system and its place within it. In the United States, at least, most mass movements composed of politically disenfranchised people have sought to get into the system rather than overthrow it. The labor movement, various racial and ethnic movements, and the feminist movement are all examples of this tendency. Thus, although the middle class and rich dominate the political process in the United States, subordinate classes also have resources that can influence public policy. This militates against a revolutionary transformation of U.S. society.

The emphasis on class conflict that pervades Marx's writings implies what sociologists today call a *structural* approach—that is, a focus on how rates of behavior among aggregates of people are influenced by their location in the society. Their differing locations dictate that classes have opposing interests. Moreover, Marx usually avoided looking at individual action because it is influenced by different variables. Rather, he wanted to know how the set of opportunities (or range of options) that people had influenced rates of behavior. For example, his analysis of the conditions under which proletarians transform themselves into a revolutionary class does not deal with the decision-making processes or cost-benefit calculations of individuals; rather, it shows that urbanity, education, political sophistication, and other factors are the social conditions that will produce class consciousness among the proletarians. Sociology at its best deals with structural variables. Although his work is misbegotten in many ways, Marx was a pioneer in this regard.

CAPITAL

In *The German Ideology,* Marx attacked the Young Hegelians because they had avoided an empirical examination of social life. In *Capital,* he demonstrated the intent of this criticism by analyzing capitalist society. Using England (and copious amounts of British government data) as his primary example, he sought to show that the most important characteristic of the capitalist mode of production was the constant drive to accumulate capital using exploited and alienated labor. As a result of the need to accumulate capital, Marx argued, the processes of production are incessantly revolutionized, and over the long run, the instability and degradation of people characteristic of capitalist society will lead to its complete transformation. Thus, in contrast with the *Manifesto,* which is a call to arms, *Capital* is a scholarly attempt to show why such a transformation of capitalist society will inevitably occur. As such, *Capital* is much more than a narrow work of economics; it is an analysis of capitalist social structure and its inevitable transformation.

The Labor Theory of Value

Marx sketched the labor theory of value in the opening chapter of *Capital.* Although he approached this issue from what appears to be a strictly economic vantage point—the nature and value of commodities—his discussion turns out to have considerably broader implications. A *commodity* is "an object outside of us, a thing that by its properties satisfies human wants of some sort or another."[41] For his purposes, both the origin of people's wants and the manner in which commodities satisfy them are irrelevant. The more important problem is what makes a commodity valuable. The answer provides the key to Marx's analysis of capitalist society.

Two different sources of value are inherent to all commodities. One resides in their *use value*—that is, in the fact that they are produced to be consumed. For example, people use paper to write on, autos for transportation, and so forth. Clearly, some things that have value, such as air and water, are not produced but are there for the taking (at least they were in the nineteenth century). Marx, however, was primarily interested in manufactured items. Commodities having use value are qualitatively different from one another; for example, a coat cannot be compared with a table. As a result, the amount of labor required to produce them is irrelevant.

Another source of value is the *exchange value* of commodities, which provides a basis for comparing the labor time required to produce them. Essentially, then, Marx's labor theory of value states that the value of commodities is determined by the labor time necessary to produce them. He phrased the labor theory of value in the following way:

> That which determines the magnitude of the value of any article is the amount of labour socially necessary, or the labour-time socially necessary for its production. Each individual commodity, in this connection, is to be considered as an average sample of its class. Commodities, therefore, in which equal quantities of labour are embodied, or which can be produced in the same time, have the same value. The value of one commodity is to the value of any other, as the labour-time necessary for the production of the one is to that necessary for the production of the other. As values, all commodities are only definite masses of congealed labour-time.[42]

Marx supplemented the labor theory of value in five ways. First, different kinds of *useful labor* are not comparable. For example, the tasks involved in producing a coat are qualitatively different from those involved in producing linen. All that is comparable is the expenditure of human labor power in the form of brains, nerves, and muscles. Thus, the magnitude of exchange value is determined by the quantity of labor as indicated by its duration in hours, days, or weeks. Marx called this quantity *simple average labor.*

Second, although different skills exist among workers, Marx recognized that "skilled labour counts only as simple labour intensified, or rather, as multiplied simple labour."[43] Thus, to simplify the analysis, he assumed that all labor was unskilled. In practice, he asserted, people make a similar assumption in their everyday lives.

Third, the value of a commodity differs according to the technology available. With mechanization, the labor time necessary to produce a piece of cloth is greatly reduced (and so, by the way, is the value of the cloth—at least according to Marx). During the initial stages of his analysis, Marx wished to hold technology constant. Therefore, he asserted that the value of a commodity is determined by the labor time socially necessary to produce an article under the normal conditions of production existing at the time.

Fourth—and this point will become very important later on—under capitalism, labor itself is a commodity with exchange value, just like linen and coats. Thus, "the value of labor power is determined as in the case of every other commodity, by the labour time necessary for the production, and consequently, the reproduction, of this special article."[44]

Fifth, an important implication of the labor theory of value is the development of what Marx called the *fetishism of commodities* whereby people come to believe that commodities possess human-like attributes and that exploitation as well as alienation arise from relations with machines, as a kind of commodity, rather than from those who own the machines. In capitalist society, the fetishism of commodities manifests itself in two different ways: (1) Machines (as a reified form of capital and a commodity) are seen as exploiting workers, which is something only other people can do. Thus, products that were designed and built by people and can be used or discarded at will come to be seen not only as having human attributes but even as being independent participants in human social relationships. (2) When machines are seen to exploit workers, the social ties among people are hidden, so that their ability to understand or alter the way they live is impaired. In this context, Marx wrote, "there is a definite social relation between men, that assumes, in their eyes, the fantastic form of a relation between things."[45]

In later chapters of *Capital*, Marx illustrated what he meant by the fetishism of commodities by showing that machines rather than laborers set the pace and style of work and by showing that machines rather than their owners "needed" the night work of laborers so they could be in continuous operation. Hidden behind machines stand their owners, capitalists, who are the real villains in this exploitive and alienating relationship.

Capitalists have little interest in the use value of the commodities produced by human labor. Rather, it is exchange value that interests them. Marx writes, "the restless never-ending process of profit-making alone is what [the capitalist] aims at."[46] His term for profit was *surplus value*.

Surplus Value

Because Marx believed that the source of all value was labor, he had to show how laborers create surplus value for capitalists. He did this by distinguishing between "labor" and "labor power." *Labor* is the work people actually do when they are employed by capitalists, whereas *labor power* is the capacity to work that the capitalist purchases from the worker. As Marx put it, "by labour-power or capacity for labour is to be understood the aggregate of those mental and physical capabilities existing in a human being, which he exercises whenever he produces a use-value of any description."[47] Labor power is a commodity just like any other, and it is all the workers have to sell. Marx noted that the laborer, "instead of being in the position to sell commodities in which his labour is incorporated, [is] obliged to offer for sale as a commodity that very labour-power, which exists only in his living self."[48] Furthermore, in a capitalist society the proletarians can sell their labor power only to capitalists, who own the means of production. The two meet, presumably on an equal basis, one to sell labor power and the other to buy it. In reality, Marx saw labor as always at a disadvantage in this exchange.

The value, or selling price, of labor power is "determined, as in the case of any other commodity, by the labour-time necessary for the production, and consequently also the reproduction, of this special article."[49] Thus, labor power is, at least for the capitalist, a mass of congealed labor time—as represented by the cost of food, clothing, shelter, and all the other things necessary to keep the workers returning to the marketplace with their peculiar commodity. Because workers must also reproduce new generations of workers, the cost of maintaining entire families must be included. Having discovered that labor power is the source of surplus value, Marx wanted to calculate its rate. To do so, he distinguished between absolute and relative surplus value.

Absolute surplus value occurs when capitalists lengthen the working day to increase laborers' productivity. This issue became a matter of conflict throughout the nineteenth century. Hence, Marx spent a considerable amount of space documenting the way in which the early capitalists had forced laborers to work as many hours as possible each day.[50] The data that he presented are significant for two reasons. First, despite their anecdotal quality (by today's standards), they are clearly correct: Capitalists sought to extend the working day and keep the proletarians in an utterly depraved condition. In general, Marx thought that the effort to lengthen the number of hours laborers worked was inherent to capitalism and that proletarians would always be helpless to resist. Second, these remarkable pages of *Capital* probably constitute the first systematic use of historical and governmental data in social scientific research. Marx took great satisfaction in using information supplied by the British government to indict capitalism.

Relative surplus value occurs when capitalists increase laborers' productivity by enabling them to produce more in the same amount of time. This result can be achieved in two ways, he said. One is to alter the organization of work—for example, by placing workers together in factories. Another, more prevalent as capitalism advances, is to apply advanced technology to the productive process. By using machines, laborers can produce more goods (boots, pens, computers, or anything else) in less time. This means that capitalists can undersell their competitors and still make a profit. Because the reorganization of the workplace and the use of machines were methods of exploiting laborers, they were also the locus of much conflict during the nineteenth century. For such changes meant that proletarians had to work either harder or in a more dehumanizing environment. As in his analysis of absolute surplus value, Marx spent much time documenting the capitalists' efforts to increase relative surplus value.[51] By using historical and governmental data, he again showed how productivity had increased steadily through greater exploitation of proletarians.

This analysis of the sources of surplus value provided Marx with a precise definition of exploitation. In his words, "the rate of surplus value is therefore an exact expression of the degree of exploitation of laborer-power by capital, or of the laborer by the capitalist."[52] In effect, surplus value is value created by workers but skimmed off by capitalists just as beekeepers take a (large) fraction of the honey from the bees who make it.

More broadly, exploitation is not simply a form of economic injustice, although it originates the labor theory of value. The social classes that result from the acquisition of surplus value by one segment of society are also precisely defined. That class accruing surplus value, administering the government, passing laws, and regulating morals is the *bourgeoisie,* and that class being exploited is the *proletariat.*

By discovering the advantages of increasing productivity, Marx thought he had uncovered the hidden dynamic of capitalism that would lead inexorably to increasing exploitation of the proletarians, more frequent industrial crises, and, ultimately, the overthrow of the capitalist system itself. His rationale was that the capitalists' increased profits were short-lived, because others immediately copied any innovation, and thus the extra surplus value generated by rising productivity disappeared "so soon as the new method of productivity has become general, and has consequently caused the difference between the individual value of the cheapened commodity and its social value to vanish."[53] The long-term result, Marx predicted, would be the sort of chaos originally described in *The Communist Manifesto.*

The Demise of Capitalism

Marx's description of surplus value was a systematic attempt at showing the dynamics of capitalist exploitation. His next task was to reveal the reasons why, despite its enormous productivity, capitalism contained the seeds of its own destruction. He proceeded in two steps.

The first deals with what he called *simple reproduction.* It occurs as workers continuously produce products that become translated into surplus value for capitalists and wages for themselves. Proletarians use their wages in ways that perpetuate the capitalist system. Because capitalists own the means of production and the commodities produced with them, proletarians must give their wages back to the capitalists as they purchase the necessities of life. The capitalists, of course, use that money to make still more money for themselves. In addition, after minimally satisfying their needs, workers return to the marketplace ready to sell their labor power and prepared once again to augment capital by creating surplus value. Over time, then, capitalist society is continuously renewed, because proletarians produce not only commodities, not only their own wages, and not only surplus value but also capitalist social relations: exploited and alienated workers on one side and capitalists on the other.

The second step focuses on what Marx called the *conversion of surplus value into capital.* Today, we refer to the reinvestment of capital. Thus, after consuming a small part of the surplus value they obtain from proletarians, capitalists reinvest the remainder to make even more money. As Marx observed, "the circle in which simple reproduction moves, alters its form and . . . changes into a spiral."[54] The result is a contradiction so great that the demise of capitalism and its transformation into "a higher form of society" becomes inevitable.

On this basis, Marx made three now-famous predictions. The first was that proletarians would be forever separated from owning or controlling private property, even their own labor. Workers would always be at a disadvantage in labor markets, and as a result, they would sell their labor power and, thereby, give capitalists surplus value. Without this surplus value, the proletariat would never own or control private property. They would have just enough, perhaps, to survive and reproduce the next generation of exploited labor. Yet, paradoxically, the laborers have not been defrauded—at least according to capitalist rules of the game—for as we saw earlier, the capitalists merely pay laborers for the value of their commodity, labor power. Moreover, because proletarians have only labor power to sell, they have little choice but to participate according to the capitalists' rules.

Marx's second prediction was that proletarians would become increasingly impoverished and that an industrial reserve army of poor people would be created. This outcome would increasingly occur as capitalists used ever more machines in the factories to make labor more productive and lower the price of goods; as a result, fewer laborers would be needed, and their labor power could be purchased at a lower price. Thus, Marx predicted not only that proletarians would continuously reproduce their relations with the capitalists— that is, selling their labor and making profits for capitalists—but also that they would produce the means by which they were rendered a superfluous population forced to work anywhere, anytime, for any wages. Under these extreme conditions, Marx believed, proletarians will become a self-conscious revolutionary class.

Marx's third prediction was that the *rate of profit* would fall and bring on industrial crises of ever-greater severity. As capitalists compete with each other, the price of commodities would have to fall to the point where it was not possible to make a profit. Even more efficient organization of work or the adoption of new technologies reducing costs for capitalists could not, in the end, keep the profits from falling. As Marx emphasized, each new innovation was soon copied by competitors, thereby eroding any pricing advantages. Yet, the cutthroat competition would continue, forcing capitalists to lower prices relative to costs. Eventually, an insurmountable crisis would begin to emerge: capitalists would have ever-more problems generating profits. And, this situation would be aggravated by the fact that capitalists had laid off workers as they adopted new technologies, with the result that the ability of workers to buy commodities at any price was now diminished. As an outcome, capitalism would fall. The contradictions built into its very nature, Marx felt, would increasingly disrupt the operation of the capitalist system, while at the same time making the proletariat more aware of their interests in overthrowing the bourgeoisie. Indeed, capitalism is locked into several self-destructive cycles described in Table 8.2. Thus, according to Marx, the logic of capitalist development will produce the conditions necessary for its overthrow: an industrial base along with an impoverished and class-conscious proletariat. Ultimately, these dispossessed people will usher in a classless society in which production occurs for the common good.

Table 8.2 Marx's Views on Why Capitalism Would Collapse

1. Capitalists must exploit labor—that is, extract surplus value from labor power—to make profits. This exploitation cannot be hidden from workers, especially as capitalists continuously increase the rate of exploitation.

2. Capitalists must compete with each other, forcing them to lower prices and to find new ways to reduce costs to maintain a profit. As they seek to find new ways to lower costs, capitalists gain only a short-term advantage until competitors copy cost-cutting efforts, but they increase the longer-term likelihood that the proletariat will become aware of their interests.

   a. As capitalists build larger factories to take advantage of the cost benefits that come with "economies of scale," they amass workers so that they can better communicate their grieves with each other and form a more effective revolutionary force.

   b. As capitalists adopt new technologies to reduce reliance on labor, they increase unemployment, which makes labor more hostile to capitalists but also reduces the demand for the commodities produced by capitalists.

   c. As capitalists copy each others' innovations, a new round of price competition occurs, eventually creating a declining rate of profit that begins to destroy capitalist enterprises.

3. Capitalism will always overproduce commodities relative to demand, causing recessions and depressions that make workers even more aware of their misery and of who is to blame.

Capitalism in Historical Context

Marx's analysis of capitalism presupposed that it was an ongoing social system. Thus, in the final pages of *Capital,* he once again sketched the origins of capitalism, which he now called the process of *primitive accumulation.* We should recall that capitalist social relations occur only under quite specific circumstances; that is, the owners of money (the means of production) who desire to increase their holdings confront free laborers who have no way of obtaining sustenance other than by selling their labor power. Thus, to understand the origins of capitalist social relations, Marx had to account for the rise of both the proletariat and the bourgeoisie. Typically, he opted for a structural explanation.

According to Marx, the modern proletariat arose because self-supporting peasants were driven from the land (and from the guilds) and transformed into rootless and dependent urban dwellers. This process began in England during the fifteenth and sixteenth centuries and then spread throughout Western Europe. Using England as his example, Marx argued that this process had begun with the clearing of the old estates by breaking up feudal retainers, robbing peasants of the use of common lands, and abolishing their rights of land tenure under circumstances he described as "reckless terrorism." In addition, Marx argued, one of the major effects of the Protestant Reformation was "the spoilation of the church's property" by its conversion into private property—

illegally, of course. Finally, the widespread theft of state land and its conversion into privately owned property ensured that nowhere in England could peasants continue to live as they had during medieval times. In all these cases (although this analysis is clearly too simplistic) the methods used were far from idyllic, but they were effective, and they resulted in the rise of capitalist agriculture capable of supplying the needs of a "free" proletariat. Further, given that they had nowhere to go, thousands of displaced peasants became beggars, robbers, and vagabonds. Hence, throughout Western Europe beginning in the sixteenth century, there was "bloody legislation against vagabondage" with severe sanctions against those who would not work for the nascent capitalists who were then emerging.

Marx believed the emergence of the capitalist farmer and the industrial capitalist occurred concomitantly with the rise of the modern proletariat. Beginning in the fifteenth century, those who owned or controlled land typically had guarantees of long tenure, could employ newly "freed" workers at very low wages, and benefited from a rise in the price of farm products. In addition, they were able to increase farm production, despite the smaller number of people working the land, through the use of improved methods and equipment, which increased cooperation among workers in the farming process and concentrated land ownership in fewer hands. Thus, primitive accumulation of capital could occur.

Marx believed that industrial capitalism had developed as the result of a variety of interrelated events. First, he emphasized, usury and commerce existed throughout antiquity—despite laws against such activity—and laid a basis for the primitive accumulation of capital to occur. Second, the exploration and exploitation of the New World brought great wealth into the hands of just a few people. In this regard, Marx pointed especially to the discovery of gold and silver, along with the existence of native populations that could be exploited. Finally, he noted the emergence of a system of public credit and its expansion into an international credit system. On this basis, he claimed, capitalism emerged in Western Europe.

CRITICAL CONCLUSIONS

How Can Marx Be Refuted?

Marx believed that he had discovered the pattern of history and that a communist revolution and the destruction of capitalism were inevitable historical events. In drawing this conclusion, he was a political activist, arguing for the destruction of one type of society and advocating the creation of a new type. Although Marx did not see himself as a scientist in the same manner as did Spencer, he nonetheless posited "laws" of capitalism and based his predictions about the fate of capitalism on these laws. If his predictions do not come true, are the laws invalid? Or, were the predictions wishful thinking, and can they be separated from the laws? Modern-day Marxists are generally willing to concede problems in the predictions but are usually committed to the laws, or at least the basic arguments about the self-destructive nature of capitalism.

Indeed, Marxists often argue that dialectical materialism is not designed to make predictions at all, that it is merely a useful guide to reality.[55] They further point to Marx's and Engels' many assertions about the importance of studying actual historical events. Nonetheless, the attempt to show the inherently contradictory character of capitalist society leads inexorably to an interpretation of the historical pattern and to predictions about the future. Such analyses are based on a leap of faith: that a communist revolution and the destruction of capitalism are inevitable. These tenets constitute the core of Marxist thought.

The results, however, are peculiar. Such an orientation can explain anything.[56] This means, of course, that it explains nothing because the theory cannot be disproved because contrary evidence is simply disregarded. For example, if a communist revolution has not yet occurred in a capitalist society, such as the United States, Marxists commonly argue that all the inherent contradictions have not yet worked themselves out. After all, they can demonstrate that most workers are exploited and alienated because they have little control over their work. Thus, Marxists continue to maintain that working people in America will eventually become aware of their true interests and act politically to overthrow the entire social order. Similarly, if the state becomes oppressive after a communist revolution and if a new dominant class emerges as in the Soviet Union, Marxists argued that until recently, the postrevolutionary society remains in a period of transition that can last for centuries.[57] After a while, they assert, the state will "wither away," and a true communist society will evolve, one in which "the free development of each is a condition for the free development of all." In both cases, Marx's theoretical methodology allows uncomfortable observations to be explained away by positing the need for further political action that will verify the "scientific" prediction. The demise of the Soviet Union suggests how flawed this theoretical strategy is.

Although brilliantly conceived, Marx's orientation is not scientific because theories must ultimately be subjected to critical tests. That is, a situation must be devised that is capable of refuting the theory. Dialectical materialism cannot be tested in this way and therefore cannot be scientific. Thus, a "scientific political doctrine" is a contradiction in terms, for political action can be justified only through values, not science. Marx constructed a political doctrine that proved to be of enormous historical significance but, by itself, is not scientific.

Substantive Contradictions

Marx had a utopian vision of a classless society within which people acted cooperatively for the common good and, in the process, realized their human potential. Paradoxically, he believed that this goal could be achieved through the centralization of political power in the hands of the state. This belief is why we described him previously as a sorcerer's apprentice. The image is that of a leader without wisdom who inadvertently releases the power of the nether world on the earth. Put bluntly, Marx's vision of the transition from capitalism to communism invites the establishment of a regime in which the individual is subordinate to the state and there is strict control over all aspects of life; it invites, in other words, modern totalitarianism.

To understand why Marx proceeded in this way, we need to appreciate the dilemma he faced. As a revolutionary, he sought to overthrow a brutal and exploitive society in favor of a humane and just community. It is worth remembering that Engels' description of the living conditions of the working class was horribly accurate, and many nineteenth-century observers saw the situation as becoming steadily worse. Thus, as Marx saw it, the problem was to get from a competitive society to a communal one, which would free individuals to realize their potential as human beings.

So Marx made a series of proposals that are worth restating: the abolition of private ownership of land, confiscation of the property of emigrants and rebels, centralization of credit by the state, centralization of communication and transportation by the state, ownership of factories by the state, and several others. These measures imply a belief that unrestrained political power can be redemptive, that the way to freedom is through totalitarian control. As Marx put it, the transition to communism would require a temporary dictatorship of the proletariat. But experience has shown that this strategy can only mean total rule by the Communist party, which justifies its exploitation of the masses by invoking the common good. Now the political issue is not whether the ends justify the means. It is, rather, whether the means can produce the ends; that is, can power, unfettered by accountability, produce freedom for individuals? The answer is no. There is no evidence that totalitarianism can produce freedom. Despite its grandiose vision, Marx's writings had perverse political consequences.

Where Prophecy Fails

Marx's predictions or prophecies go wrong not only because he failed to recognize that power, once given, does not "wither away," but also because he assumed that it would be the proletariat who would rise up and overthrow capitalism. Yet, the great communist revolutions in Russia and China were really outcomes longer-term civil wars, and the key actors were not the proletariat but peasants. The state did not wither away in either case, and ironically, only with the rise of capitalism in these countries, particularly China, can some hope for a less totalitarian regime be found. Clearly, Marx's grand predictions contain several significant miscalculations.

First, Marx saw the value of commodities as inhering in the labor power necessary to produce them (less other costs like machines and marketing). This assumption is perhaps his most fundamental mistake, and despite efforts by contemporary Marxists to stay with this idea because it offers a measure of exploitation, it is fundamentally flawed. Value inheres in what one is willing to pay, or must pay under constrained conditions, for something in a market, and though Marxists would decry this as imposing capitalist categories, it is nonetheless true. People can still be exploited when they are paid little and forced to work under terrible conditions, but we need not invoke the value theory of labor as our measure of the degree of exploitation. Rather, we invoke other values, such as fair wages, basic standards of living, and good working conditions to assess exploitation. The notion of exploitation is thus evaluative; it cannot be an objective construct, as Marx sought to make it. We can still see capitalist profits and wealth as coming from exploitation, if we choose, but the value theory labor is an ideology disguised as science. Like Adam Smith, who had worked with the idea, we should abandon the notion of a labor theory of value because it is not useful.

Second, Marx miscalculated the extent to which capitalists and proletarians were inexorably on a collision course. The early capitalism that Marx observed did indeed seem to be locked into a self-destructive system, but Marx simply assumed that these crises of recession/depression and labor discontent were unresolvable within the framework of capitalism. Part of the reason for this miscalculation was that Marx overestimated the extent to which the state is simply a tool of the bourgeoisie, whereas in fact, capitalism is associated with the rise of political democracies in which all citizens have some say, despite the fact that the rich certainly have more influence than the poor. Another part of the reason for Marx's miscalculations is that persistent crises force the state to seek agreements between capitalists and labor over wages and working conditions, and these crises have pulled the state into regulating markets and capitalists as recessions threatened political stability. Marx assumed that such flexibility by the state and capitalists could not exist. In fact, during early capitalism, this flexibility is not so evident, but over time, masses of urban workers have been able to gain political power.

Third, Marx incorrectly assumed that workers were, and always would be, powerless in labor markets. Although this was certainly true in early capitalism as rural peasants migrated to cities (and indeed is still true for rural migrants to cities), workers were able to gain political power to force political interventions in labor markets. This gain in power was partly the result of labor unions that were far more successful than Marx could have envisioned at the time he was writing. Moreover, there is not always a perpetual shortage of labor or reserve labor force that can be drawn upon when existing workers demand higher wages. Labor shortages do emerge, and under these conditions, the proletariat is in a better bargaining position. Furthermore, as investments in technology and facilities mount, the costs of unused capital investments arising from prolonged labor disputes have often forced capitalists to bargain with workers rather than leave big machines idle.

Fourth, Marx did not anticipate the rise of the middle classes. Indeed, he made the opposite prediction: Most people would be pushed into the proletariat. History shows, however, that as the economy expands, especially as new technologies drive the expansion, the proportion of skilled, white-collar workers grows and eventually comes to constitute the majority of workers. These more skilled workers are in a much better bargaining position over wages and working conditions than were early industrial workers.

Fifth, Marx did not recognize the importance of government as a large employer. He tended to see government as the tool of the bourgeoisie, but as government intervenes in all spheres of society—from schools to economic regulation—a significant proportion of the labor force comes to work for government. As a result, it is difficult to typify the interests of government workers as part of conflicts between an exploited proletariat and capitalists.

Sixth, late in his writings, Marx began to see some of the implications of joint stock companies, but he could not have predicted the revolution ushered in by the issuing of stocks in markets. Ownership was to become more diffused, with many workers having a stake in capitalism as they acquired stock. Moreover, not only was ownership diffused, but also it was separated from management such that owners would not directly manage companies and, hence, relations with labor. Under these conditions, management would increasingly be interested in rationalizing relations with labor to keep production going.

Thus, many specific forces in capitalism mitigated against Marx's predictions. Some of these forces he could not be expected to have anticipated, however, he might have seen the effects of others if he had not been so committed ideologically to overthrowing capitalism. It is always hazardous to make predictions based on an historical trend, as Marx did, because specific historical events can change the trajectory of a prediction. Marx had confidence in his prophecies because his entire intellectual scheme forced these predictions, but he never questioned some of the assumptions on which this scheme was based; when some proved questionable, the entire system collapsed.

Is Marx Still Relevant?

So, why should sociologists still read Marx? After all, have not most of his predictions about revolution and the spread of communism failed to materialize? The communist revolution by the proletariat never really occurred, the class structure of capitalist system did not polarize, but instead, became ever more complex, the state did not "wither away" in communist countries (indeed, just the opposite occurred), and the world has become more capitalist rather than communist. Some contemporary sociologists continue to hold out, arguing that as the capitalism goes completely global, the contradictions in the system will finally emerge and usher in the communist revolution. Others have sustained an interest in exploitation and the value theory of labor, reworking these ideas to fit more contemporary conditions. Yet, it must be said that much of the Marxian system of thinking, especially its more ideologically-loaded portrayals of the future has not held up, at least to date.

Still Marx anticipated and framed many of the issues that occupy discussion of the economy and society today. Marx saw, more than any other scholar of the last century has, that the economy is the driving force of society, and he predicted that capitalism would spread or, in today's vocabulary, become a global force. He understood that economic power and political power are highly cor-related and that those with power could disproportionately influence the for-mation of ideologies and the other elements of culture. He explained the incredible wealth-generating capacities of free markets, but this dynamism is tempered by the inequalities, exploitation, and alienation generated by such a system as well as by the inherent tendency of the system to cycle in and out of ever-deeper recessions and depressions. Moreover, he even anticipated the power of big capitalism to standardize activities, to impoverish small businesses and artisans, and to destroy old cultures in the relentless drive to make produc-tion more efficient and to penetrate all markets. Thus, Marx had a very good sense for many of the outcomes of capitalism, once unleashed. Why, then, did his more specific predictions about the revolution of the proletariat go wrong? The answer must reside in Marx's ideological fervor. Marx was blinded by his convictions and, hence, could not see that the state, bourgeoisie, and workers could change the capitalist system in ways that make it more benign.

Even with his biases, Marx's works are still worth rereading for the insights that they contain about the dynamics of capitalism. And there is another rea-son to study Marx, even if he would not agree with the rationale for doing so: Marx did articulate some powerful theoretical principles, or laws, about the dynamics of human organization. These can be seen when the substantive con-tent of his analysis—that is, the dynamics of capitalist modes of production and the coming communist revolution—are made more abstract, rising above the particulars of any historical epoch, time, or place. When Marx's ideas are con-verted into a general theory, their enduring power is more evident; perhaps more important, we can see where his predictions went wrong. Thus, even though Marx saw his "laws" as historically bound, we argue that they can be stated more abstractly and made theoretically useful. For those interested in this exercise, we offer the next chapter on Marx's theoretical principles.

NOTES

1. Karl Marx and Friedrich Engels, "The Communist Manifesto," in *The Birth of the Communist Manifesto,* ed. Dirk Struik (New York: International, 1971), p. 112.

2. Karl Marx, "Preface," *A Contribution to the Critique of Political Economy* (New York: International, 1970), p. 22.

3. Karl Marx and Friedrich Engels, *The German Ideology* (New York: Inter-national, 1947), p. 3. Only part 1 of the text is translated, and it is gener-ally assumed that Engels' contribution to this portion of the book was mini-mal. This is mainly because the text appears to be an elaboration of Marx's

"Theses on Feuerbach," which he outlined for himself in 1845; see pp. 43–45 in *The Marx-Engels Reader,* ed. Robert C. Tucker (New York: Norton, 1978). In addition, Engels stated repeatedly that Marx had already developed his conception of history before their collaboration. Therefore, in what follows we will generally refer only to Marx.

4. Marx and Engels, *German Ideology,* p. 3.

5. Ibid., p. 7.

6. Ibid., p. 8.

7. Ibid., p. 18.

8. Ibid., p. 16.

9. Marx, "Introduction," *Contribution to the Critique of Political Economy,* pp. 188–217.

10. Karl Marx, *The Economic and Philosophical Manuscripts* (New York: International, 1964), p. 111.

11. Ibid., p. 111.

12. Karl Marx, "A Contribution to the Critique of Hegel's Philosophy of Right," in Tucker, ed., *Marx Engels Reader,* pp. 16–26, 53–66.

13. Robert A. Nisbet, *Social Change and History* (New York: Oxford University Press, 1968).

14. Bertell Ollman, "Marx's Use of 'Class,'" *American Journal of Sociology* 73 (March 1968), pp. 573–580.

15. Sometimes Marx uses the phrase *forces of production* narrowly, so that it refers only to the instruments used in the productive process. Sometimes, however, he uses the phrase so that it refers to both the instruments used in production and the *social organization* that accompanies their use. By social organization is meant not only the organization of work (as in factories) but also family life, law, politics, and all other institutions. This tactic occurs with many of Marx's key concepts. See Bertell Ollman, *Alienation: Marx's Conception of Man in Capitalist Society* (New York: Oxford University Press, 1976).

16. See Richard Appelbaum, "Marx's Theory of the Falling Rate of Profit: Towards a Dialectical Analysis of Structural Change," *American Sociological Review* 43 (February 1978), pp. 73–92.

17. Marx and Engels, "Communist Manifesto," p. 87.

18. Ibid., p. 88.

19. Ibid., p. 89.

20. Ibid., p. 90.

21. Ibid., p. 91.

22. Ibid., pp. 90, 94.

23. Ibid., p. 92.

24. Ibid., p. 91.

25. Ibid., p. 96.

26. See Karl Marx, "The Civil War in France," in Karl Marx and Friedrich Engels, *Selected Works* (Moscow: Progress, 1969), pp. 178–244. Marx shows here how the proletarians actively participated in subjecting other classes to the rule of the bourgeoisie.

27. Marx and Engels, *Communist Manifesto,* p. 104.

28. Ibid., pp. 103–104.

29. Ibid., p. 111.

30. Karl Marx, "Critique of the Gotha Program," in Marx and Engels, *Selected Works,* pp. 9–11.

31. Marx and Engels, *Communist Manifesto,* p. 112.

32. Ibid., p. 117.

33. Ibid., p. 125.

34. Ibid., p. 125.

35. Marx, "Preface," A Contribution to the Critique of Political Economy, p. 22.

36. See Fernand Braudel, *Civilization and Capitalism, 15th–18th Centuries,* vol. 1 (New York: Harper & Row, 1981), and Immanuel Wallerstein, *The Modern World System,* vols. 1 and 2 (New York: Academic, 1980).

37. All of these factors constitute what Marx called the forces of production. See footnote 15.

38. See Leonard Beeghley, *The Structure of Stratification in the United States,* 2nd ed. (Boston: Allyn & Bacon, 1996).

39. See Milovan Djilas, *The New Class* (New York: Praeger, 1965), and *Rise and Fall* (New York: Harcourt Brace Jovanovich, 1985). See also Michael Voslensky, *Nomenklatura: The Soviet Ruling Class* (Garden City, NY: Doubleday, 1986).

40. Reinhard Bendix, "Inequality and Social Structure: A Comparison of Marx and Weber," *American Sociological Review* 39 (April 1974), pp. 149–161.

41. Karl Marx, *Capital: A Critical Analysis of Capitalist Production,* vol. 1 (New York: International, 1967). The original spelling is retained in all quotations.

42. Ibid., pp. 39–40.

43. Ibid., p. 44.
44. Ibid., p. 170.
45. Ibid., p. 72.
46. Ibid., p. 149.
47. Ibid., p. 167.
48. Ibid., pp. 168–169.
49. Ibid., p. 170.
50. Ibid., pp. 231–312.
51. Ibid., pp. 336–507.

52. Ibid., p. 218.
53. Ibid., p. 319.
54. Ibid., p. 581.
55. Ollman, *Alienation*.
56. Karl Popper, *Conjectures and Refutations* (New York: Harper & Row, 1968), pp. 33–37.
57. Harry Braverman, *Labor and Monopoly Capital* (New York: Monthly Review Press, 1974).

9

✳

Marx's
Theoretical Legacy

Karl Marx considered himself a general theorist of history, portraying the evolution of society as a series of inevitable conflicts between superordinates and subordinates in the system of stratification. Although he sought to develop theoretical principles for each great epoch of human evolution—that is, primitive communism, slavery, feudalism, capitalism, and communism—a more general theory that transcends any historical period can be found in his work. In this chapter, we first model this theory and, then, translate it into a series of theoretical principles. While Marx himself, and certainly most Marxist scholars today, would abhor this effort, we believe that Marx's ultimate contribution to sociology resides in this general theory. Moreover, by presenting Marx's theory in more formal terms, we can perhaps see where his predictions about the collapse of capitalism and the rise of communism go wrong. We might add that, at a time when the impact of Marxism as a political force is receding, the conversion of Marx's ideas to a more general theory provides one way to save the powerful theoretical legacy that Marx left to sociology.

MARX'S UNDERLYING CAUSAL MODEL

The General Model of History and Evolution

In Figure 9.1, we outline the general causal model that can be gleaned from Marx's discursive text. In presenting Marx's ideas in this form, we have made his concepts more abstract, converted them into variables, and outlined the direct

159

160

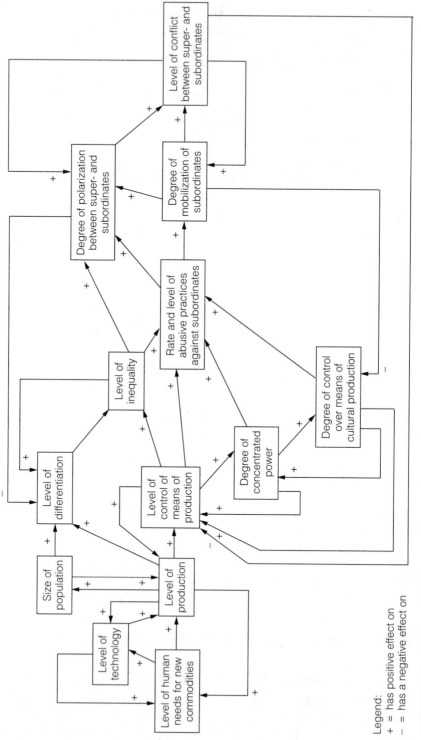

FIGURE 9.1 Marx's View of Social Organization

Legend:
+ = has positive effect on
− = has a negative effect on

and reverse causal connections among the forces denoted by the concepts listed in each box. We have had to make many inferences, but nonetheless, the model represents Marx's argument once it is taken out of historical context.

Starting at the far left of the model, Marx saw a series of fundamental relationships among human needs, production, and technology. Human needs ultimately drive production and the invention of new technologies used in production. As Marx emphasized, the reverse causal processes are critical to understanding why societies have a history. As one level of needs is satisfied, new needs arise and push technological innovation and production even further. These positive connections among needs, production, and technology are all represented by positive signs that signal Marx's view that this escalating cycle is, ultimately, one of the key engines to history.

Marx also recognized, like Spencer and most nineteenth century theorists, that increasing production allowed the population to grow and differentiate in a positive cycle. But this demographic process, found mostly in *The German Ideology,* is not the core of Marx's theory. Rather, Marx's theory of society centers on the relationships among control of the means of production, the instruments of power, and the production of culture. Those who own and control the means of production also are able to concentrate power among themselves and to use symbols, especially ideologies, to legitimate their power and their ownership. As the model indicates, the arrows among these variables are all positive, emphasizing that control of one of these critical forces allows control of the others. Those who owned the means of production could also control the societal *superstructure,* as Marx called their ability to hold power and use culture to blind subordinates to their real interests.

This control of production (as well as power and symbols) increases the level of inequality in a society, as is indicated by the positively signed arrow from production to inequality. Moreover, those who control production, power, and symbols are free to engage in practices that disrupt the routines and lives of subordinates, making them ever-more miserable. At times, economic, political, and cultural elites are driven to do so by the nature of the means of production, as was the case for capitalists locked into cutthroat competition and a falling rate of profit. The end result is that these practices make subordinates more aware of their interests, leading them to mobilize ideologically, thereby attempting to break the control of economic and political elites on the production of cultural symbols (as is denoted by the negatively signed reverse causal arrow from mobilization by subordinates to control of the means of cultural production). Because of abusive practices and mobilization, Marx saw society as polarizing into two potentially warring camps of superordinates and subordinates. As polarization occurred, it would, presumably, lower the level of differentiation, as indicated by the negatively signed arrow from polarization to differentiation. As mobilization proceeds, Marx felt that the growing tension between subordinates and superordinates would move beyond ideology to political organization. With political organization, subordinates would take control of the means of production from elites (as indicated by the reverse negative arrow from conflict to control of the means of production); as a result, the old elites' control of power and symbolic production would be broken.

Except for his communist utopia in which the temporary "dictatorship of the proletariat" would somehow "wither away," a new elite group would own the means of production, setting into the motion the dialectic once again. The model in Figure 9.1 cannot capture this argument; it only takes Marx's model for one historical epoch. The model challenges Marx's assumption that the very dynamics that had driven history would be obviated with communism. Marx assumed that capitalist productivity had increased the capacity to support all members of society to such a degree that private ownership of the means of production was no longer necessary, and indeed contradictory, to the social organization of production. With collective ownership of the means of production, somehow the superstructure—power and symbols—would not be used to forge a new basis of inequality.

When we model Marx's argument, however, there is no basis for this assumption; it is just a wish on his part. Even if the means of production is collectively owned, or more realistically owned by the state, an entirely new basis for inequality is set into motion: control by the state of the means of economic and cultural production. The great contradiction in Marx's ideology, therefore, is that the state does not "wither away," but becomes a new basis for setting into motion the dynamics outlined in Figure 9.1. Indeed, when ideology is stripped away, we can see an abstract model of how societal organization systematically generates conflict. The key to this conflict is that elites engage in abusive practices that lead to mobilization of subordinates. This has never happened on a revolutionary scale in a capitalist society, and so, we need to understand why Marx appears to have been so wrong on this issue.

Part of the answer resides in the assumption that control of production would be concentrated, which is generally true in capitalists economies but not to the degree that Marx prophesized (primarily because of stock markets that diffuse ownership and often separate ownership from management). Thus, in practice, capitalism does not meet the key variable in Marx's model: increasing production does not always lead to high degrees of concentration of ownership. If this is so, then power and symbol production are not as concentrated as Marx would have predicted. Indeed, ironically, when the state controls the means of production (as was the case of the old Soviet Union) we see high degrees of concentrated power in industrial systems, and as a result, the state also controls the means of cultural production. But translating Marx's ideas into an abstract causal model shows that he was fundamentally correct: When control of the means of production is concentrated, as is the case in feudalism and industrial communism, power is used to regulate cultural production which, in turn, leads to high levels of inequality and abusive practices by elites.

Another part of the answer about why Marx's scheme goes wrong resides in his prediction that those who control the means of production will be driven to escalate their abusive practices because they have no choice or because they have the power to do so. Marx underestimated the extent to which those who do not own the means of production could mobilize power and symbols in nonrevolutionary ways, forcing centers of power and owners/managers of production to make concessions.

Still another part of the answer can be seen in the causal chain from production to differentiation to inequality to polarization. Marx clearly underestimated the extent to which increasing levels of production differentiates a population in general and the class structure in particular. Escalating production increases the number of classes rather than pushing people into two warring camps. Marx appears to have argued that abusive practices of elites would polarize the population, thereby reducing differentiation and creating a simple two-class system; obviously, very high levels of production do just the opposite, differentiating the population to such a degree that class boundaries become more numerous and less distinct. And if such is the case, the polarizing dynamics so essential to Marx's theory never emerge. As a result, revolutionary conflict does not occur, at least not the way Marx predicted. Indeed, it is in less productive economies where elites do control the means of production and the societal superstructure that one sees fewer classes, abusive practices, and potential polarization of the population.

Thus, the general model presented in Figure 9.1 exposes both the strengths and weaknesses of Marx's thinking. The model is, in broad contours, correct on the key variables and relationships among them, except some of the signs are wrong. If we were to correct the model to reflect what has actually happened historically, we would need only to change some of the signs on a few key arrows. We would change the signs from differentiation to inequality to negative, or perhaps to a curvilinear relationship in which initial differentiation increases inequality but at high levels reduces inequality (the sign would thus be $+/-$ rather than positive). We might also change the signs from production to control of the means of production from positive to more positively curvilinear; that is, increasing production does indeed lead to increased control of the means of production (and the societal superstructure as well), but at very high levels of production, this control is not so great, especially in capitalist systems where ownership is more diffused (although corporate power is still very strong, but hardly absolute). With these changes, the model holds up rather well. Marx had the basic elements right, but his vision was blinded by his ideology, which emerged in reaction to the abuses of early capitalism.

The Model of Conflict

In *The Communist Manifesto,* Marx and Engels outline a theory of conflict that we have only glossed over in Figure 9.1. Figure 9.2 presents this conflict model in a more robust form; although this model is powerful, it makes some fundamental mistakes in the signs on the causal arrows much like the more general one. These mistakes account for why Marx's predictions about the revolutionary overthrow of capitalism go wrong. Before pointing out these mistakes, however, let us review the model as Marx might have drawn it.

The model begins on the left with Marx's view that control of production translates into control of the coercive/administrative apparatus of the state

164

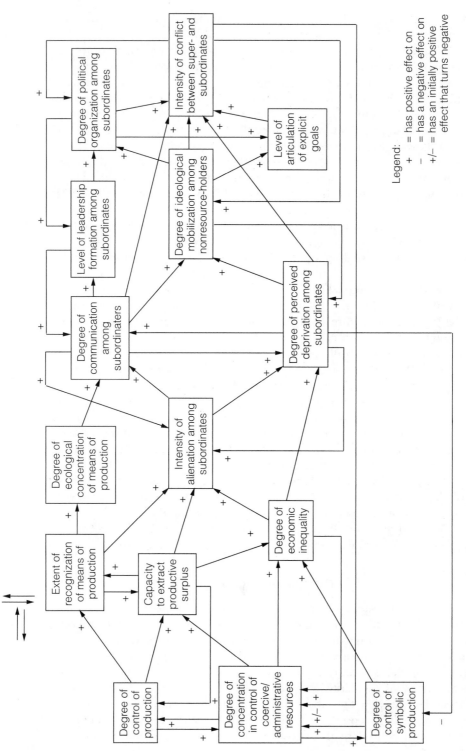

FIGURE 9.2 Marx's Model of Conflict

which, in turn, gives economic elites power over symbolic resources. This concentrated control of production, power, and symbols increases the degree of inequality, as is indicated by the positive arrows. Those who own and control the economy extract productive surplus for their privilege, thus increasing economic inequality, and they use their surplus to buy power and to manipulate the symbolic means of production. Thus, Marx argues, the overall level of inequality will increase under these conditions.

Whether they were driven to do so by the nature of production or by their effort to secure more valued resources, those who control the means of production are able to reorganize the productive processes to suit their needs and interests. Coupled with resentments by subordinates over inequalities, this reorganization sets into motion the dynamics that Marx felt would lead subordinates to become aware of their interests in changing the system, and eventually, to their political and ideological mobilization to pursue revolutionary conflict.

One causal path goes across the top of the figure. When reorganization by elites concentrates labor, the latter can better communicate their grievances and nurture leaders who can articulate ideologies, with these together leading to political organization among nonresource holders or subordinates in the system of stratification. As subordinates become a class "for itself" because of political organization, they will pursue violent and intense conflict with superordinates. The positive signs on these causal paths emphasize that Marx saw these dynamics as inevitable.

Other causal paths similarly emphasize the inevitability of the causal forces unleashed. Inequality, per se, but more fundamentally, the reorganization of the means of production so that labor has no say in what they produce, how they produce, and to whom they sell products all increase alienation among subordinates. Marx argues that the intensity of alienation leads people to communicate their grievances which, at the same time, raises their sense of deprivation. This sense of deprivation increases as communication of grievances occurs, and vice versa. As people communicate their grievances and sense of deprivation, they become more ideologically mobilized, especially as leaders emerge to articulate and codify a revolutionary ideology. As is evident with all the positive signs connecting the causal paths, these forces are mutually reinforcing, escalating, and driving the process of conflict to more intense levels. As political leaders articulate ideologies and as political organization ensues, the goals of subordinates become more clearly articulated (as Marx tried to do in *The Communist Manifesto*). All together, ideology, clear articulation of goals, and political organization lead to intense and violent conflict against superordinates. A class moves from one "of itself" to one "for itself," where interests, ideologies, goals, and organization drive subordinates to pursue conflict. In Marx's version, all the direct causal arrows, flowing from left to right, are positive, making the revolutionary conflict inevitable. The only negative signs are on reverse causal arrows, emphasizing how deprivations and conflict loosen the elites' grip in symbolic, coercive, and administrative power.

Where did Marx go wrong in this model? The answer is that many of the relationships that he posited are not purely positive. Critical causal paths leading to conflict between superordinates and subordinates are, in fact curvilinear. That is, the variables feeding into conflict (the last box on the right of the model) initially increase the potential for violent conflict, but as these variables reach high levels, they begin to lower the potential for violent conflict. Let us review these key causal connections.

First, the relationship between political organization of subordinates and intensity of conflict is curvilinear. Initial political organization increases violent conflict, but high degrees reduce conflict. The union movement in America is a good example; violence was greatest at the early stages but declined as labor became ever-more organized. With organization, goals are more explicit, and leaders are more likely to bargain and compromise rather than risk the high costs of violent conflict. Second, ideology also has a curvilinear relationship to conflict. Initial ideological mobilization raises emotions but does not generally specify in detail goals beyond general pronouncements; under these conditions of diffused emotional arousal, violence is more likely. But as the ideology becomes ever-more codified, goals and ends are more clearly articulated, and compromise is more likely, particularly when subordinates are well organized politically.

Thus, Marx had the variables right, but he made some key miscalculations on the signs for several of the critical forces feeding into the last variable in the model: intensity and violence of conflict. If we were to redraw the model, it would look much the same expect for the variables from degree of political organization to conflict, ideological mobilization to conflict, and level of articulation of goals to conflict; all of these would now be curvilinear, increasing the potential for violent conflict for a while and then decreasing this potential. Marx was living through the early stages of the conflict process, where increasing conflict between superordinates and subordinates was evident, but as the forces driving this conflict reached high levels—that is, high degrees of political organization, well-codified ideologies, and clearly articulated goals—they began to reduce the tendency for violence. At least, this is what the empirical record shows.

MARX'S THEORETICAL PRINCIPLES

Marx's ideas are rather easily converted into theoretical principles. In essence, such principles translate the causal models developed earlier into a series of abstract statements that, as we have suggested for the models, can be altered slightly to produce a more powerful theory.

Principles of Social Organization

The German Ideology, The Communist Manifesto, and *Capital* all contain, to varying degrees, a set of fundamental ideas about the nature of social organization. Despite the varied contexts in which Marx presents these ideas, they imply a question that virtually all social theorists ask themselves: How is it that patterns of social organization emerge, persist, and change over time? Or, as Marx phrased it in *The German Ideology:* How is it possible for human societies to have a history? His answer to this question can be seen in a number of interrelated principles explained in the section to follow. Let us begin with five background propositions that summarize how Marx conceptualized the forces producing inequality.

1. Production increases the level of technology, and vice versa.
2. The level of differentiation in a society increases with productivity, and vice versa.
3. The more productivity and differentiation reciprocally affect each other, the greater the rate of population growth, and vice versa.
4. The larger a population is and the greater is the level of differentiation, the more concentrated is control of the means of production and the greater is the level of inequality.
5. The greater the control of the means of production and the more concentrated power, the more economic and political elites can control the production and dissemination of symbols, especially ideologies justifying inequalities.

These principles emphasize Marx's conviction that the level of technology, productivity, social differentiation, population size, inequality, concentration of power, and use of ideologies to justify inequality are fundamentally related. We have phrased Marx's ideas more neutrally than he did so that the insights they offer into the nature of social organization can be divorced from his polemics.

Marx argued that members of social systems use technology to facilitate productivity and that increased productivity encourages social differentiation and population growth. As noted for his model, this portion of Marx's view of social organization does not go beyond Herbert Spencer's and others' analyses of these processes. The unique part of Marx's view is his analysis of how integration in differentiated systems is achieved. Marx argued that inequality, concentration of power, and manipulation of idea systems by those in power are the means by which differentiated social systems are maintained. As he also emphasized, however, this basis of organization contains the seeds for its transformation because inequality inexorably sets into motion forces for conflict and change. Marx's ultimate contribution to sociological theory resides in the principles generating conflict in systems of inequality.

Principles of Inequality and Change in Social Systems

Marx focused on how inequality generated in capitalist social systems leads to class conflict and the revolutionary transformation of society. Our goal, however, is to liberate Marxian principles from their historical embeddedness and to state them at a sufficiently high level of abstraction so they have relevance beyond the nineteenth century. For Marx, the key question is how a system of inequality in which those with power use beliefs and norms to perpetuate this situation inexorably generates the potential for conflict and change. He visualized the genesis of conflict and change by recognizing that those holding different shares of a system's resources have differing and conflicting interests. Those with resources have an interest in preserving their privileges, whereas those with few resources have an interest in taking them from the privileged. Power is the most critical resource because it can be used to secure other resources, so the unequal distribution of power signals the greatest conflict of interest. Marx's ideas can be expressed in the following propositions:

6. The greater the level of inequality, the greater the conflict of interest between dominant and subordinate segments of a population, with this conflict of interest increasing as

   a. Those with power use their power to consolidate control over other material and symbolic resources.
   b. Those with power limit upward mobility and access to resources by subordinates.

These propositions follow from Marx's observation that a conflict of interest is inherent in inequality and that those in advantageous positions will seek to increase their privileges at the expense of those in lower-ranked positions.

Marx's next theoretical task is thus quite straightforward: to document the conditions under which awareness of interests causes subordinates to question the legitimacy of inequality in resource distribution. These conditions are summarized in the following propositions:

7. The more subordinate segments become aware of their interests, the more they question the legitimacy of the unequal distribution of scarce resources, with awareness of subordinates' interests increasing as

   a. Social changes produced by dominant segments disrupt existing relations among subordinates.
   b. Practices of dominant segments create alienation among subordinates.
   c. Subordinates communicate their grievances to each other, which increases with
      1. The degree of spatial concentration among subordinates.
      2. The level of education and rates of literacy among subordinates.
   d. Subordinate segments develop unifying systems of ideologies that increase with
      1. The capacity of subordinates to recruit or generate ideological spokespersons.
      2. The inability of dominant groups to regulate the socialization process and communication networks among subordinates.

Marx's great insight is that the concentration of power causes those with power to act in ways increasing subordinates' awareness of their interests. As those with power seek to consolidate their position, they often disrupt the routines of subordinates, while creating alienation. Or, as those with power seek or organize subordinates to increase their productivity (and hence the resources of those with power), they create conditions favoring communication among subordinates and awareness of their common interests. Although Marx visualized these processes as occurring at the societal level, as the bourgeoisie used their power to exploit the proletariat in the interest of greater productivity and profit, his insights are, we believe, applicable to a broader range of social units and to any historical period. What principles 7-a, 7-b, 7-c, and 7-d underscore is the inherent tension built into the unequal distribution of resources, especially power, and the tendency of concentrated power to be used in ways that create sources of counter-power.

For these counter sources of power to become effective, however, Marx thought an awareness of common interests must be translated into organization among subordinates. In capitalist systems, he saw this process as one in which the proletariat went from a class "of itself" to one organized "for itself." Marx's insight applies, we would argue, to more than social classes in societal social systems. The process of organization among subordinates who are aware of their interests is both transhistorical and applicable to any social unit revealing inequalities in power. The following proposition specifies some conditions that Marx believed important in translating an awareness of interests and a questioning of legitimacy into organizational forms designed to pursue conflict:

8. The more subordinates are aware of their interests, the greater is their questioning of the legitimacy of the distribution of scarce resources; and the more they question legitimacy, the more likely are they to organize and initiate conflict against superordinates, especially as

   a. Deprivations of subordinates move from an absolute to a relative basis.
   b. Superordinates fail to become aware of their common interests.
   c. Subordinates develop leadership structures.

In these propositions, Marx summarized some of the conditions leading to those forms of organization among subordinates that, in turn, will result in overt conflict. The first key question in addressing this issue is why an awareness of conflicting interests and a questioning of legitimacy of the system would lead to organization and the initiation of conflict. Seemingly, awareness would have to be accompanied by intense emotions if people are to run the risks of opposing those holding power. Presumably, Marx's proposition on alienation would indicate one source of emotional arousal because for Marx, alienation goes against human beings' basic needs. Further, ideological spokespersons would, as Marx's own career and works testify, arouse emotions through their prose and polemics. The key variable in the Marxian scheme is "relative deprivation." The emotions aroused by alienation and ideological spokespersons are necessary but insufficient conditions for taking the risks of organizing and initiating conflict against those with power. Only when these conditions are accompanied by rapidly escalating perceptions of deprivation by subordinates is the level of emotional arousal sufficient to prompt organization and open conflict with superordinates. Such organization, however, is not likely to be successful unless dominant groups fail to organize around their interests and unless leaders among the subordinates can emerge to mobilize and channel emotional energies.

Thus, although Marx assumed that conflict is inevitable, his theory of its origins was elaborate, setting down a series of necessary and sufficient conditions for the occurrence of conflict. Marx's great contribution to a theory of conflict resides in these propositions, for his subsequent propositions appear to be simple translations of his dialectical assumptions into statements of covariance, without the careful documentation of the necessary and sufficient conditions that would cause these conflict processes to occur.

Marx also sought to account for the degree of violence in the conflict between organized subordinates and superordinates. The key variable here is polarization, a concept denoting the increasing divisions within a system.

9. The greater is the unity among subordinates around common beliefs and the more developed is their political leadership structure, the greater is the polarization between superordinates and subordinates.

10. The more polarized superordinates and subordinates, the more violent their ensuing conflict becomes.

In contrast with his previous principles, propositions 9 and 10 do not specify the conditions under which polarization will occur, nor do they indicate when polarized groups will engage in violent conflict. Marx assumed that such would be the case as the dialectic mechanically unfolds. Presumably, highly organized subordinates in a state of emotional arousal will engage in violent conflict. As a cursory review of actual historical events underscores, such a state often results in just the opposite: less violent conflicts with a considerable degree of negotiation and compromise. This fact points to the Marxian scheme's failure to specify the conditions under which polarization first occurs and then leads to violent conflict. It is not just coincidental that, at this point in his scheme, Marx's predictions about class revolutions in capitalistic societies begin to go wrong. Thus, the Marxian legacy points rather dramatically to a needed area of theoretical and empirical research: Under what condition is conflict likely to be violent? More specifically, under what conditions is conflict involving highly organized and mobilized subordinates likely to be violent and under what conditions are less combative forms of conflict likely to occur?

The final proposition in the Marxian inventory also appears to follow more from a philosophical commitment to the dialectic than from carefully reasoned conclusions:

11. The more violent is the conflict, the greater are the structural changes and the greater is the redistribution of scarce resources.

This proposition reveals Marx's faith in the success of the revolution as well as his assertion that these successful revolutionaries would establish new sets of superordinate–subordinate relations of power. As such, the proposition is ideology rephrased in the language of theory, especially because no conditional statements are offered about just when violent conflict leads to change and redistribution and just when it does not. Had Marx not assumed conflicts to become polarized and violent, then he would have paid more attention to the degrees of violence and nonviolence in the conflict process, and this interest in turn, would have alerted him to the variable outcomes of conflict for social systems.

CORRECTING MARX'S THEORY

As the causal models in Figures 9.1 and 9.2 suggest, Marx's theoretical principles need to be reformulated, in several subtle but important ways. First, the relationship among technology, productivity, and societal differentiation is positive, and it generates an ever-more complex class system. Thus, in contrast to Marx's position, stratification does not inevitably polarize into two basic classes; rather, the greater is the productivity, the more complex will be the class system.

Second, the relationship between ownership of the means of production, on the one hand, and the concentration of power and symbolic production, on the other, is probably curvilinear. That is, as production increased through the agrarian era, the relationships posited by Marx hold essentially true, but as production reached very high levels under capitalism, (1) ownership became more dispersed, (2) ownership and management often became separated, and (3) economic elites' ability to control the societal superstructure declined. Today in advanced capitalism, the less-propertied classes can also exert both ideological and political pressure and, thereby, limit the efforts of those managing the economy from being too abusive. This empirical fact, as it has become evident with advanced capitalism, changes the whole equation about the relationship between ownership of the means of production, control of the societal superstructure, "immiseration" of the working class, polarization of society, and violent revolution. The relationships posited by Marx are more typical of agrarian-feudal societies of peasants and elites than of capitalistic societies revealing many diverse social classes. Ironically, the relationships among control of the economy, state, and ideological apparatus are even more typical of state ownership in communist societies than of capitalistic ones.

Third, the model of conflict needs to be adjusted to account for the curvilinear relationship between the intensity and violence of conflict, on one side, and the level of ideological mobilization, the clarity of goals and ends, and the degree of political organization, on the other. Initial ideological and political organization is more likely to generate violent or, at least, intense conflict than is the case when codification of ideologies, clear statements of goals, and well-structured political organizations have emerged. When Marx was writing, he was living through the early phases of mobilizations by labor, but he did not live long enough to see that the intensity and violence of conflict declines as subordinates become well-organized and successful in conflict with capitalists.

Finally, Marx did not see that the capitalists could be flexible because he assumed that they were caught in a downward spiral of intense competition leading to a declining rate of profit that forced them to engage in the very activities that would facilitate mobilization of workers. He was essentially correct in much of his analysis, but he clearly miscalculated the extent to which (1) the proletariat could also exert both ideological and political pressure on the state to respond to their interests, (2) the state could manage the points of tension between capitalists and labor, while mitigating the business cycle, and (3) the capitalists could avoid locking themselves into ruinous competition that forced them to be abusive and pushed the economy into ever-deeper recessions.

Even with these miscalculations, Marx was correct in his assessment of the key forces organizing societies in general and producing conflicts between sub- and superordinates in particular. For this reason, we believe, Marx remains an important theorist today, especially as the more ideologically loaded portions of Marxism recede in relevance.

10

✳

The Origin and Context of Max Weber's Thought

BIOGRAPHICAL INFLUENCES
ON WEBER'S THOUGHT

Max Weber, the first of seven children, was born to Max and Helene Weber on April 21, 1864, in the city of Erfurt in Thuringia. Thuringia was located in Prussia, the most powerful of the German states at that time. Weber was descended from Protestants on both sides of his family. His father's ancestors were Lutheran refugees from Austria, and his mother's forebears were Huguenot emigrants from France. As we will see, Weber's Protestantism weighed heavily on him, serving as a source of torment and eventually as motivation for one of the greatest sociological analyses ever written, *The Protestant Ethic and the Spirit of Capitalism*.[1]

THE EARLY YEARS

Weber's father, a lawyer and judge in Erfurt, became a politician in Berlin, where the family moved in 1869. In Berlin, the elder Weber began his political career as a city councilor and subsequently served as a member of the Landtag (Regional Assembly) and the Reichstag (Imperial Parliament). In this context, the Weber family entertained a wide assortment of distinguished people. For example, the historians Theodor Mommsen and Wilhelm Dilthey lived nearby and frequently visited the Weber household.[2] This background allowed the young Weber to meet the leading politicians and scholars of the day, listen to and participate in their discussions, and become aware of the issues facing the nation.

By all accounts, Weber's father enjoyed the freewheeling lifestyle of a German politician, with its emphasis on material success, its lack of religiosity, and its rough-and-tumble world of gossip, deals, and accommodation. He was a hedonist, a man who enjoyed bourgeois living to the fullest. Within the family, however, the senior Weber ruled absolutely. He did not tolerate young people holding opinions different from his own and felt compelled, as a patriarch, to control his wife's behavior in myriad ways. The elder Weber was nonetheless devoted to his children, active in supervising their education and taking them on outings in the countryside. During his youth, the young Weber was close to his father, an orientation that would later change.

Weber's mother was altogether different from her husband. Helene Weber, a shy and sensitive woman, was religiously devout. When she was sixteen, an older friend of the family sexually attacked her, and one result of this episode was that she came to hate sexuality. Marianne Weber, Max Weber's wife, reports "the physical aspect of marriage was to her not a source of joy but a heavy sacrifice and also a sin that was justified only by the procreation of children. Because of this, in her youthful happiness she often longed for old age to free her from that 'duty.'" A loving and affectionate mother, Helene Weber nonetheless adhered to strict Calvinist standards of hard work, ascetic behavior, and personal morality, which she tried to instill in her children. "She was never satisfied with herself and always felt inadequate before God," with the result that her life was marked by great inward struggle.[3] Moreover, because they were so mismatched, Weber's parents became permanently estranged very early in their marriage, a conflict that affected Weber throughout his life. Marianne Weber claims plausibly that he believed not only that he had to choose between his parents but also that this choice would be decisive for his own personality development. This "choice" became a source of emotional agony that Weber lived with throughout his life. Indeed, it can be argued that his sociological writings are an attempt at working through the inner conflicts he experienced.[4]

Weber was a sickly child. He contracted a serious disease, possibly meningitis, at age two, and the experience left him smaller and less physically capable than other children. Nonetheless, he was intellectually precocious. His youthful letters, many of which survive, are filled with reflections on the classical Greek and Roman writers as well as on the philosophers Johann Goethe, Benedict de Spinoza, and Immanuel Kant. Weber's conversations at home also ensured that he became politically sophisticated at a very young age, a characteristic that apparently made him a discipline problem in school, where he thought the level of instruction too low and the ignorance of his classmates appalling. More generally, the problem of the nature and use of authority preoccupied him throughout his life, both personally and intellectually.

In 1882, Weber graduated from the gymnasium (high school) and enrolled at the University of Heidelberg. Like his father, he chose the law as a field of study and professional training. In addition, however, he also stud-ied economics, history, philosophy, and theology. Sociology was not offered at that time. Weber became active in his father's fraternity, joining in the rit-ual dueling and drinking bouts characteristic of German university life in those days. The large amount of beer consumed and the hedonistic lifestyle transformed the frail youth into a rather heavy-set young man, complete with fencing scars on his face.

In 1883, Weber served an obligatory year of military service and, during that period, came under the influence of an aunt, Ida Baumgarten (his mother's sister), and an uncle, the historian Herman Baumgarten. It proved to be a turning point in Weber's life. Stronger and more forceful than her sister, Ida Baumgarten led a simple and ascetic religious life. In so doing, she helped Weber understand and appreciate his mother's Christian piety. As a result, although this way of phrasing the issue is somewhat simplified, he began to identify with his mother rather than with his father.

Before the Breakdown

Weber returned in the following year to Berlin, where he enrolled at the University of Berlin and lived at home. He remained there for seven years, financially dependent on a father he increasingly disliked and condemned, while completing his apprenticeship in law. Like Marx, Weber was a person with encyclopedic learning. While working for several years as a full-time unpaid legal apprentice, he also completed a Ph.D. dissertation titled "The History of Trading Companies in the Middle Ages" and a postdoctoral thesis titled "Roman Agrarian History," which qualified him to teach at the univer-sity level.[5] He also joined the Evangelical Social Union, a Protestant political group reacting against the excesses of industrialization in Germany, and the Social Political Union, an academic organization committed to doing research on social problems. Under the aegis of the latter, he investigated the conditions of rural peasants. The result, a nine-hundred-page book titled *The Situation of Farm Workers in Germany East of the Elbe River,* established his reputation as a young scholar.[6] To produce three books while working full time as a junior barrister, Weber "repressed everything," living an ascetic life strictly regulated by the clock.[7] These characteristics of his own life assumed intellectual signif-icance in his subsequent work, *The Protestant Ethic and the Spirit of Capitalism.*

Although convinced he was not a "true scholar," Weber nonetheless decided to pursue a combined academic and legal career. Thus, in 1892 he accepted an instructor's position at the University of Berlin. During this same period he courted and married his cousin, Marianne Schnitger, whose loving biography of her husband remains the standard source about his life.

Weber had a passion for work: "Hardly was one [task] complete when his restless intellect took hold of a new one."[8] Thus, his chronic overwork and unhealthy lifestyle became a cause of concern for both his mother and wife, who urged him to slow down. Their remonstrations, however, had little effect. In 1894, the couple moved to Freiburg, where he took a position as professor of political economy. According to Marianne Weber, the workload there "surpassed everything up to then."[9] Over the next several years, Weber maintained a punishing academic, legal, and political schedule. He was apparently regarded as an outstanding professor, a promising lawyer, and a man with a future in public service. In addition, the Webers (who were by all accounts, not just Marianne's, happily married) maintained an unconventional lifestyle for the period. Over time Marianne Weber became a student under Heinrich Rickert, a historian and social worker and a supporter of women's rights. She apparently converted Weber because he soon became "more of a feminist than she was."[10]

Against this background, Weber's long-simmering anger toward his father erupted in 1897, with disastrous consequences for all.[11] Each year Helene Weber usually spent several weeks visiting her children and their families. The elder Weber, however, always made these trips difficult, believing that he should control his wife's every activity. During the summer, father and son clashed violently over this issue and parted without reconciliation. Shortly thereafter the old man died. Soon after that, Weber, at age thirty-three and now a professor of political economy at the University of Heidelberg, suffered a complete nervous breakdown, which incapacitated him for more than five years.

It is intriguing, of course, to speculate about the causes of Weber's psychic break.[12] Although there exists little doubt that the fight with, and subsequent death of, his father constituted the precipitating incident, the more general issues contributing to Weber's psychological trauma were unresolved difficulties of identification with his parents and inner conflicts over their contradictory values. Further, it makes sense to assert that his chronic overwork served as both a symptom of his underlying stress and an additional cause of the breakdown. In any case, and it should be recognized that although these remarks are very superficial, Weber spent the following years unable to work. During a period before the availability of psychotherapy, the only "cure" for those afflicted with any form of mental fatigue was rest and relaxation.

Weber did little work between 1897 and 1903. Sustained by an inheritance, he traveled widely, periodically recovering for short periods, only to collapse repeatedly. In 1900, Weber was retired by the University of Heidelberg. He did not teach again for nearly two decades.

The Transition to Sociology

Beginning in 1903 Weber found himself able to write again. He first produced a rather laborious work criticizing the German historical economists Wilhelm Roscher and Karl Knies.[13] Shortly afterward, he wrote an important methodological essay, "'Objectivity' in Social Science and Public Policy," in which he analyzed the place of values in the emerging social scientific disciplines.[14] This piece was followed in 1904 and 1905 by the seminal book for which Weber is primarily remembered, *The Protestant Ethic and the Spirit of Capitalism,* in which he outlined the historical significance of Protestantism for the development of capitalist cultural values. These last two works mark the beginning of Weber's self-conscious identification as a sociologist.

The transition to sociology is important because Weber's writings before his breakdown were composed from the point of view of historical economics. Thus, as a professor of political economy, he infused his empirical analyses with a definite value standard. For example, in his study of farm workers he argued that agrarian policy in Germany should be determined by the interests of the state, an angle of vision that disapproved of Polish workers being brought into the eastern portions of the nation because their presence undermined the German claim of sovereignty over the area. As he began to recover the ability to work, however, he studied Rickert, Dilthey, and others and saw more clearly the necessity for separating social scientific research and values. His subsequent writings emphasized that the two spheres—research and values—had to be kept as separate as possible. Thus, although social scientific knowledge can inform political decisions, such actions can be justified only in terms of values.

Between 1906 and 1914, Weber continued research and writing, now confining himself to the role of private scholar. He studied religion, the origin of cities, and social scientific methodology, producing a series of books and essays. Among them are the methodological *Critique of Stammler* (1907), *The Sociology of Religion* (1912), *The Religion of China* (1913), *The Religion of India* (which appeared in 1916–1917), and *Ancient Judaism* (which appeared in 1917).[15]

In addition to scholarly work, Weber also participated in the social life of German intellectuals.[16] Max and Marianne's home served as a meeting place for distinguished persons in many fields. The sociologists Georg Simmel and Robert Michels, the historian Heinrich Rickert, and the philosopher Karl Jaspers were among the many scholars who regularly took part in wide-ranging discussions of politics and social science. In 1910, Weber helped found the German Sociological Association, serving as its secretary for several years. In this context, he continued to press his views on the nature of sociology, especially the importance of objectivity in social research.

With the outbreak of World War I, Weber, a passionate German nationalist, became a hospital administrator in the Heidelberg area. Over time, however, he began to oppose the German conduct of the war, advocating limited aims and prophesying defeat if unrestricted submarine warfare brought the United States into the conflict. Few people paid any attention.

In 1918, Weber accepted an academic position at the University of Vienna and offered a course for the first time in twenty years. In the following year he taught at the University of Munich, giving two of his most famous addresses: "Science as a Vocation" and "Politics as a Vocation."[17] During this period, he began reworking the material from the prewar years, writing what became Part 1 of his *Economy and Society*.[18] He also gave a series of lectures that were posthumously published under the title *General Economic History*.[19]

In the twilight of his life, Weber evidently found some release from the traumas of the past. Although he had little time for relaxation, Marianne Weber says his capacity for work became steadier and his sleep more regular. During the summer of 1920, Max Weber developed pneumonia. He died on June 14.

Although Weber stands as one of the greatest classical sociologists, the exact lines of his theoretical contributions are sometimes difficult to ascertain. Part of the reason for this lack of precision is the breadth of his work. Indeed some have argued that no sociologist before or since has displayed his intellectual range or sophistication. He analyzed the historical significance of the Protestant Reformation, the characteristics of Indian and Chinese social structure and religion, the genesis of modern legal systems, the nature of modern bureaucracies, the types of political domination, the origin of the city in the West, and many other topics. Because of this concern with a variety of substantive topics, his influence on the development of modern sociological theory remains unclear. The scope of his empirical studies suggests that he was not primarily interested in the development of abstract laws of human behavior and organization. Nonetheless, we will show that although his theoretical goals were limited, his works have contributed enormously to sociological theory.

The origins of Weber's sociology lie in his reaction to intellectual trends in Germany at the turn of the century. In the remainder of this chapter, we will focus on four of the most important influences on his thought. First, and perhaps most significantly, Weber rejected both Marx and Marxism as too simplistic and inherently nonscientific. Second, Weber rejected as nonproductive the long-standing debate about the nature of the social sciences that dominated late nineteenth-century German thought. This debate, called *Methodenstreit* (or methodological controversy), involved two competing schools of thought whose attitudes toward the practice of social science were quite at odds: the neoclassical or theoretical economists and the historical economists. The third major influence on Weber's thought was Wilhelm Dilthey, who emphasized the importance of understanding the subjective meanings that people attach to their behavior. In a somewhat altered form, this idea became one of the cornerstones of Weber's thought. Fourth, Weber took many of the methodological precepts developed by Heinrich Rickert for the study of history and altered them to facilitate his own brand of sociology

KARL MARX AND MAX WEBER

Despite the fact that Marx is rarely cited in Weber's works, Weber carried on a "silent dialogue" with the dead revolutionary. Hans Gerth and C. Wright Mills have even argued that Weber's writings should be seen as an effort at "rounding out," or supplementing, Marx's interpretation of the rise and fall of capitalist society.[20] Although this point of view ignores the fundamental differences between Marx and Weber, many modern scholars have agreed with Gerth and Mills. Hence, it is worth noting some points of similarity between the two men before emphasizing their essential differences.

First, in *The Protestant Ethic and the Spirit of Capitalism,* Weber showed the relationship between the cultural values associated with the Protestant Reformation and the rise of the culture of capitalism in the West, but he did not deny the importance of the material factors that Marx had previously identified. Apart from the transformative impact of Puritanism, Marx and Weber generally agreed on the structural factors involved in the rise of modern society. Second, both men can be seen as "systems theorists" in the sense that their conceptual schemes represent an attempt at mapping the connections among the situational and environmental contexts in which people act. Third, both scholars recognized the extent to which individuals' freedom of action was limited in capitalist societies, although they did so in somewhat different ways.[21] In Marx's work, people are alienated because they do not control the means of production, whereas in Weber's work individuals often find themselves in an "iron cage" constructed by increasingly omnipresent and "rationalized" bureaucracies. Finally, despite the constraints just noted, both Marx and Weber observed the importance of human decision making in shaping history. For Marx, who was always a hopeful utopian and revolutionary, action will usher in a new era of freedom for all people; for Weber, who was less hopeful about the future, individuals have a wider range of choices in modern societies than was possible in the traditional communities of the past.

Despite these areas of similarity, Weber's work was different from Marx's in origin, purpose, and style. Marx combined revolution and theory to explain what he saw as the pattern of history. Weber helped establish an academically based sociology committed to the objective observation and understanding of historical processes, which he regarded as inherently unpredictable. These differences in orientation cannot be reconciled without obliterating the distinctiveness of each man's work. Hence, rather than "rounding out" Marx, Weber tried to refute Marxist thought as it existed at the turn of the century. For example, in the *Protestant Ethic,* Weber went out of his way to note that his findings flatly contradicted those postulated by "historical materialism," and he wondered at the naiveté of those Marxists who espoused such doctrines.[22] More generally, Weber disagreed with Marx and the Marxists (the two are not the same) on three interrelated and fundamental topics: (1) the nature of science, (2) the inevitability of history, and (3) economic determinism.

The Nature of Science

As we saw in Chapter 8, Marx combined science and revolution in such a way that theories were verified by action, by what they led people to do (or not do) based on their material interests. Weber, on the other hand, saw science as the search for truth and argued that knowledge was verified by observation. In making observations, he said, research must be "value-free" in the sense that concepts are clearly defined, agreed-upon rules of evidence are followed, and logical inferences are made. Only in this way, he argued, can there be an objective science of sociology.[23]

Although recognizing that Marxists were often motivated by moral outrage at the conditions under which most people were forced to live, Weber asserted that ethical positions were not scientifically demonstrable, no matter how laudable they might be. Further, by combining science and revolution to justify their view of the future, Weber asserted, Marxists inevitably confuse "what is" and "what ought to be," with the result that their ethical motives are undermined.[24] Such confusion should be eliminated as much as possible, Weber insisted, by making social science objective through an exclusive emphasis on "what is." Nonetheless, Weber recognized that social scientists' values inevitably intrude into social inquiry because they influence the topics considered important for research. This fact, Weber contended, does not preclude the possibility that the process of research can and should be objective. Thus, Weber believed that science could not tell people how to live or how to organize themselves but could provide them with the sort of information necessary to make such decisions. On this basis, he sought to understand the origin and characteristics of modern societies by developing a set of concepts that could be used in understanding social action.

The Inevitability of History

Marx posited the existence of historical laws of development, with the result that he saw feudalism as leading inevitably to capitalism and the latter leading inexorably to a more humane communist society. Against this position, Weber argued that there were no laws of historical development and that capitalism had arisen in the West as a result of a series of historical accidents.

As will be shown in Chapter 11, Weber's sociology is oriented to understanding how modern Western societies could have arisen when and where they did. Essentially, he argued that a number of historical processes had occurred together and resulted in the rise of modern capitalism in the West. Among these were the following: industrialization, the rise of a free labor force, the development of logical accounting methods, the rise of free markets, the development of modern forms of law, the increasing use of paper instruments of ownership (such as stock certificates), and the rise of what Weber called the *spirit of capitalism*.[25] As will be seen, he believed the last factor to be the most significant. Further, he argued that none of these phenomena could have been predicted in advance; rather, they were all dependent on chance. From his point of view, societies are always perpetually balanced between the opposing forces of determinism and chance, for the course of history is often altered by unforeseen political struggles, wars, ecological calamities, or the charisma of single individuals.

Economic Determinism

By the beginning of the twentieth century, many Marxists were arguing that certain economic arrangements, especially the private ownership of the means of production, inevitably caused specific political forms as well as other social structures to develop. Although this crude form of economic determinism distorts Marx's analysis and eliminates its subtlety, it had the advantage of allowing quick and easy (not to mention nasty) assessments of modern capitalist societies. Weber attempted to refute this rather congealed form of Marx's analysis in two different ways. First, in the *Protestant Ethic*, Weber showed the importance of religious ideas in shaping the behavior of the Puritans and, by extrapolation, all Western people. Second, in *Economy and Society,* he outlined the extent to which systems of domination are maintained because they are viewed as legitimate by citizens, a commitment that generally overwhelms the class divisions that always exist. In Weber's words "it is one of the delusions rooted in the modern overestimation of the 'economic factor' . . . to believe that national solidarity cannot survive the tensions of antagonistic economic interests, or even to assume that political solidarity is *merely* a reflection of the economic substructure."[26]

THE *METHODENSTREIT* AND MAX WEBER

The methodological controversy that dominated German academic life in the latter half of the nineteenth century can only be understood in light of two interrelated factors. First, in Germany there tended to be a rather rigid division between the natural sciences and the cultural disciplines so that only natural phenomena—such as those studied in physics, chemistry, biology, and the like—were seen as readily amenable to theoretical (that is, scientific) analysis. Based on the philosophy of Kant (and Georg Hegel, but to a much lesser extent), it was believed that the social and cultural realms, the world of the "spirit," was beyond analysis in scientific terms. Hence, studies of natural and social phenomena developed in much different directions in Germany.[27]

Second, after the early work of Adam Smith and David Ricardo, non-Marxist economic theory became stagnant, with the result that economists had great difficulty trying to explain the workings of actual industrial economies as they existed in the nineteenth century. There were two main ways of dealing with the problem. One was to develop better theory, and the other was to eschew science altogether and to concentrate on depicting the historical development of particular economic systems. The members of the historical school of economics chose the latter course, a position that fit comfortably with the dominant German intellectual tradition. Nonetheless, a number of scholars (although they were a minority in German academic circles) chose to develop non-Marxist economic theory. For the most part, these theoretical economists were non-Germans who came from a positivistic background roughly similar to Émile Durkheim's.

The major figures in the German historical school were individuals who are generally not remembered today, largely because their writings have not proven to be of enduring significance. Wilhelm Roscher, Bruno Hildebrand, and Karl Knies, all contemporaries of Marx, are generally credited with founding the movement during the middle portion of the nineteenth century. Later such men as Lujo Brentano and Gustav Schmoller added to and modified this perspective. Although there were differences in the way each of these scholars approached the study of economics and, by extrapolation, social science in general, they shared a number of basic criticisms of theoretical economics as well as a relatively common methodological approach to their subject.

On the other side of the conflict were the members of the theoretical school, many of whom remain well-known figures in the history of economic thought, largely because their writings furthered the development of the discipline as a science. Among these scholars are Léon Walras, W. S. Jevons, Eugen Böhm-Bawerk, and Karl Menger. Menger is by far the most important because he discovered the theory of marginal utility, an idea that went a long way toward solving the theoretical dilemmas that had plagued economics throughout the latter half of the nineteenth century. Because the *Methodenstreit* is primarily remembered for the acrimonious and often vicious debate between Menger and Schmoller that occurred during the 1870s, we will refer to them as the representative of each school of thought.

Methodological Issues
Dividing Historical and Theoretical

The historical school and the theoretical school disagreed about four fundamental issues, all stemming from the divergence between economic theory and economic reality.[28] The first involved the relative importance of deduction and induction. Schmoller and the historical economists charged that the theoreticians' use of deductive methods was faulty, chiefly because their theories could not explain reality. Hence, the historical economists emphasized as an alternative the importance of observing and describing people's concrete patterns of action (often down to the smallest details), and they spent many years compiling such data. Unlike some of the other historians, for whom description quickly became an end in itself, Schmoller asserted that the long-run result of this descriptive work would be the discovery of economic laws by inductive methods. He believed that the resulting propositions would better describe reality because they would take the complexity of people's actual behavior into account. Alternatively, Menger and the theoretical economists charged (correctly, as it turned out) that the historians were so immersed in data that no laws would ever result. Further, Menger said, the more realistic response to the inadequacies of economic theory is to develop better theories, which was precisely what he and others were doing at that time.

The second issue dividing the two schools had to do with the universality versus the relativity of findings. Schmoller and the historical economists asserted that the theoreticians' emphasis on the universal applicability of economic laws was absurd. Rather, from the historians' point of view, their empirical research had shown that economic development occurs in evolutionary stages unique to each society, which implies that it is possible to understand a society's present stage of economic advancement only by ascertaining previous stages. Menger and the theoreticians responded by observing that theory, whether in the social sciences or the natural sciences, is oriented toward that which is common rather than toward that which is unique. Hence, economic theories can (at least in principle) explain certain aspects of human behavior that are common to all societies, but, admittedly, not every element of social action can be explained theoretically. On this basis, Menger argued that there was a place for both theory and history in economics and the other social sciences.

The third issue of debate in the *Methodenstreit* had to do with the degree of rationality versus nonrationality in human behavior. Schmoller and the historical economists believed that the theoretical economists' view of economic man as rational and motivated only by narrow self-interest was unrealistic. They went on to assert that there was a unity to all of social life, in the sense that people act out of a multiplicity of motives, which are not always rational. Thus, to obtain a comprehensive view of social reality, historical research often went far beyond the narrow confines of economic action, dealing with the interrelationships among economic, political, legal, religious, and other social phenomena. Although Schmoller was correct, Menger simply replied that economic theory dealt with only one side of human behavior (that is, people's attempts at material need satisfaction) and that the other social sciences must focus on other aspects of social action. Over the long run, Menger believed, the result will be a comprehensive understanding of human behavior.

Finally, the fourth issue separating the two schools had to do with economics as an ethical discipline versus economics as a science. Schmoller and other members of the historical school unquestionably saw economics as an ethical discipline that could help solve many of the problems facing German society, with the result that their scholarly writings often had an avowedly political intent. This attitude was partly a consequence of the long-standing German division between the natural sciences and the cultural disciplines and partly a consequence of the fact that Schmoller and many of the others held important university and governmental positions. In opposition, Menger charged that Schmoller's political value judgments were hopelessly confused with his scholarly analyses, to the detriment of both. In science, Menger said, the two must be kept separate.

Weber's Response to the *Methodenstreit*

In economics, the *Methodenstreit* eventually dissipated, although more by the force of theoretical developments than the rhetoric of the participants. On several occasions, however, Weber appears to have used the arguments raised in the controversy as a baseline from which to develop his own methodological orientation.[29]

In regard to the first issue, the relative importance of inductive and deductive methods, Weber tried to bridge the gap between the two schools to create a historically based social science. With the historical economists, he argued that if the social sciences imitated the natural sciences by seeking to discover general laws of social behavior, then not very much useful knowledge would be produced. His reasoning was that any social science oriented toward the development of timelessly valid laws would, of necessity, emphasize those patterns of action that were common from one society to another, with the result that ideographic events would inevitably be omitted from consideration. Unique phenomena, such as the Protestant Reformation, are often the most significant factors influencing the development of a society. Hence, a science seeking to understand the structure of social action must necessarily focus on precisely those factors not amenable to lawlike formulations. Put differently, Weber argued that the social sciences had to make use of historical materials. Nonetheless, with the theoretical economists he asserted that the development of abstract concepts was absolutely necessary to guide empirical research. As will be seen in the next chapter, his goal was an objective (that is, scientific) comprehension of modern Western society, and for that reason he needed to develop a set of clear and precise concepts, which he called *ideal types,* that could be used in understanding historical processes.

Weber's response to the second issue dividing the two schools follows from the first. That is, a historically based social science cannot be universally applicable; rather, findings are always relative to a particular culture and society. One implication of this point of view is that although Weber tried to understand the origins of modern Western society, his findings might not have any relevance for the process of modernization in the Third World today because those societies are operating in a rather different historical context. It should be emphasized, however, that Weber strongly disagreed with the historical economists' evolutionary interpretations. Rather, he believed that economic development did not occur in evolutionary stages because unpredictable events, such as wars, ecological changes, charismatic leaders, and myriads of other phenomena, alter the course of history.

The third issue in the *Methodenstreit* became essential to Weber's sociology, for the protagonists inadvertently identified one of the fundamental characteristics of modern Western society: the tension between rational and nonrational action. Thus, Menger's argument that rational economic behavior needs to be conceptually distinguished from other modes of action seemed reasonable to Weber because he had observed that action in the marketplace was characterized by an emphasis on logic and knowledge, which was often absent in other arenas. At the same time, however, Schmoller's emphasis on the unity of social life and people's multiplicity of motives, some of which are based on values other than logic, also seemed reasonable, Hence, Weber tried to conceptually summarize the "types of social action" to systematically distinguish modern Western societies from the traditional ones that had preceded them and to show the wider range of behavioral choices available to occidental people.[30]

Weber's reaction to the fourth issue in the methodological controversy was similar to his response to Marx and the Marxists; that is, he asserted that Menger was absolutely correct: The social sciences must be value-free. Although Schmoller and the other historical economists were generally political liberals with whom Weber was in sympathy, he believed that there could be no scientific justification for any ethical or political point of view. Rather, he argued that objective scientific analyses could provide people with the knowledge necessary to make intelligent ethical decisions based on their values.

WILHELM DILTHEY AND MAX WEBER

The origin of Weber's response to the *Methodenstreit* can be found in the works of Wilhelm Dilthey and Heinrich Rickert. Essentially, Weber built his sociology with the methodological tools they provided, although he went beyond each of them in a number of fundamental ways. Neither Rickert nor Dilthey is very well known in the English-speaking world, primarily because the problems they addressed were peculiar to the German intellectual scene during the late nineteenth century. Given the traditional idealist separation of the worlds of nature and human activity, the establishment of the social sciences as sciences was an extremely vexing problem.

Dilthey's Methodology of the Social Sciences

Essentially, Dilthey argued that although both the sphere of human behavior and the sphere of nature could be studied scientifically, it had to be recognized that they were different subjects and produced different kinds of knowledge. He then went on to explore some of the implications of this argument.[31]

The logic of Dilthey's analysis can be seen in three steps. First, and most obvious, the two sciences have different subject matters. The natural sciences are oriented toward the explanation of physical or natural events, whereas the social sciences are oriented toward the explanation of human action. Second, and as a result of the first, researchers in each field obtain quite different forms of knowledge. In the natural sciences, knowledge is external in the sense that physical phenomena are affected by one another in ways that can be seen and explained by timelessly valid laws. In the social sciences, however, knowledge is of necessity internal in the sense that each person has an "inner nature" that must be comprehended in some way to explain events. Third, as a result, researchers in the two spheres must have altogether different orientations to their subject. In the natural sciences, it is enough to observe events and relationships. For example, an object falling through space can be explained by the force of gravity, and this explanation is true regardless of the cultural background of different researchers who concern themselves with this topic. In the social sciences, however, scholars must go beyond mere observation and seek to understand (*verstehen*) each person's "inner nature" to explain events and relationships. Further, the explanations offered might vary depending on the cultural background of the different researchers.

For Dilthey, then, the means by which observers obtain an understanding of each person's "inner nature" is the key to the scientific knowledge of human action. In this light, he tried to classify the various fields devoted to the study of social behavior by their typical mode of analysis. The first type of analysis consists of descriptions of reality, of events that have occurred; this is the field of history. Unlike Rickert, Dilthey does not appear to have been very concerned with whether historical descriptions were accurate or objective. He seems to have assumed this possibility. The second way of discussing human action consists of value judgments made in light of historical events; this is the field of ethics or politics. The third way of dealing with social behavior consists of formulating abstractions from history; this is the field of social sciences. This last mode of analysis is the most important for understanding action, Dilthey asserted, because abstract concepts provide the tools necessary for comprehending behavior. He was unable to face the implications of this insight, however, for he went on to argue that the systematic development of abstract concepts would not be of much long-term use in understanding people's "inner nature." He opted instead for the necessity of relying on intuition (what he called the "fantasy of the artist") in comprehending social action. Such intuitive understanding occurs when, in some unexplainable and imperfect way, observers reexperience in their own consciousness the experiences of others. The result of this emphasis on the manner in which one mind becomes aware of another was that Dilthey's point of view led ultimately to a dead end. In the long run, as Weber realized, an excessive reliance on the researcher's subjective impressions cannot lead to an objective social science.

Weber's Response to Dilthey's Work

From Weber's point of view, Dilthey's methodological orientation was useful in three ways.[32] First, Dilthey was correct in noting that the social sciences could obtain a quite different form of knowledge than the natural sciences. Second, social scientific statements are different from and, Weber added, must be kept separate from value judgments of any sort. Third, the key to social scientific knowledge is to understand (*verstehen*) the subjective meanings people attach to their actions.

Weber believed that the major problem in Dilthey's work lay in his emphasis on understanding each person's "inner nature," as if an objective social science could be founded on some sort of mystical and intuitive reexperiencing of others' desires and thoughts. Hence, Weber developed a different way of emphasizing the importance of *verstehen,* one that proved to be a great deal more successful than Dilthey's. Essentially, Weber argued, although social action can be understood only when "it is placed in an intelligible and more inclusive context of meaning," the key to such understanding resides in the development of a set of abstract concepts (ideal types) that classifies the dimensions of social action and reflects the norms appropriate in different spheres. By focusing his work in this way, Weber emphasized the importance of understanding individual behavior while he was able to assess the significance of historical events (such as the Reformation) in an objective manner.

HEINRICH RICKERT AND MAX WEBER

Like Dilthey, Rickert was concerned with the problems created by the disjunction between the world of nature and the world of human activity that had been created by idealist philosophy. The two men had somewhat different solutions

in mind, however. Dilthey addressed the problem of the dissimilar subjects char-
acterizing the natural and social sciences by emphasizing that the different forms
of knowledge in each sphere required distinct methodological orientations by
researchers. Rickert, however, had a more narrow interest; he tried to show that
history could be an objective scientific discipline because the knowledge it produced was based on a valid principle of concept selection. Like Dilthey, Rickert
was not entirely successful in his task, largely because he misperceived the nature
of science and drifted into metaphysical speculation.[33] Nonetheless, Rickert's
writings constituted an important influence on Weber's work: although much
more practical than Rickert, Weber adapted some of Rickert's methodological
principles for his own more general purposes.

Rickert on the Objectivity of History

Rickert began his attempt at demonstrating that history can be an objective
science by dealing with a number of relatively noncontroversial epistemological issues. He argued that empirical reality was infinite in space and time,
which for him meant that reality could, in principle, be divided into an infinite number of objects for study and that these objects could in turn be dissected into an unlimited number of parts. An important implication of this
view is that reality can never be completely known because there will always
be some other way of looking at it. The practical problem, then, becomes how
people can know anything at all about the world around them, and Rickert's
answer was that by formulating concepts, human beings select those aspects of
reality that are important to them. Thus, concepts are the means by which we
know the world, for without them people could not distinguish among its significant parts. Given this necessity, Rickert came to the peculiar conclusion
that the essence of science centered on the problem of concept formation.
From this point of view, a discipline can be regarded as a science if it uses a
principle of concept selection that everyone agrees produces objective knowledge. Not surprisingly, Rickert said there were two valid principles of concept
selection—those used in the natural sciences and history—and in this way he
tried to show that history was a scientific discipline.

In the natural sciences, Rickert noted, concepts are designed to identify the
common traits of the empirical objects to which they refer. This tactic allows
concepts to become increasingly abstract and, hence, fit into a theory that summarizes empirical regularities (for example, the movements of the planets and
their effects on one another through the force of gravity). The result is a set of
general concepts that can, at least in principle, be used in a single, all-embracing
law of nature. On this basis, Rickert concluded that the principle of concept
selection used in the natural sciences was valid because it succeeded in identifying regular and recurrent features of the physical environment.

In history, however, Rickert argued that scholars' interests were altogether
different, which means that the principle of concept selection must differ as
well. To chronicle the events of the past and their significance for the present,
historians must focus on their uniqueness. With this purpose in mind, historical

concepts are formulated to identify those aspects of the past that make them distinctive and different from one another (for example, "traditional society" or the "spirit of capitalism"). Thus, historians produce concepts, which Rickert called "historical individuals," that summarize a complex set of events for their historical significance (that is, their uniqueness). On this basis, Rickert concluded that the principle of concept selection used in history was valid because it allowed observers to understand how particular societies had developed their specific characteristics. This result would be impossible if the historians imitated the natural sciences and conceptualized only those aspects of the past that were common to all societies. Hence, despite these differences in concept formation, according to Rickert, history is a science.

Rickert next confronted the problem of how scholars select topics for study, and at this point his emphasis on concept formation as the essence of science trapped him in a nonproductive philosophical argument. Essentially, Rickert asserted, the researchers' choice of topics is made by "value-relevance." That is, some events are seen as worth conceptualizing based on the scientists' interpretation of what the members of a society value. This emphasis on value-relevance implies a subjective rather than objective conception of knowledge, however, because scientists are inevitably forced to rely on their own values in determining what topics are worth knowing about, or conceptualizing. Rickert tried to avoid this implication by postulating that a kind of "normal consciousness" characterized all human beings. On this basis, he argued, there are areas of concern shared by all members of every society—for example, religion, law, the state, customs, the physical world, language, literature, art, and the economy. But this postulate is inherently metaphysical (and typically idealist) because it assumes that values have an existence independent of human beings.

Ultimately, as did Dilthey's, Rickert's analysis led to a dead end, for he failed to recognize that concept formation was only one essential aspect of science and, partly as a result, found himself entangled in idealism. Nonetheless, Rickert's work provided a fundamental baseline from which Weber could establish sociology as a science.

Weber's Response to Rickert

Weber was intimately familiar with Rickert's writings, as is indicated by acknowledgments of Rickert in his early methodological essays.[34] However, these citations do not indicate the extent to which he adapted some of Rickert's ideas for his own rather different purposes. Weber's preoccupation with refuting Marxism and solving the dilemma created by the *Methodenstreit* probably allowed him to recognize what Rickert had failed to see: The essence of science involves not only a coherent conceptual scheme but also, and just as important, the use of logical and systematic procedures in the interpretation of observations. Hence, even though the social sciences have different goals because they must deal with quite different data than the natural sciences do

(Dilthey's "inner nature" of human beings), what unites the two as sciences is their procedural similarity. This insight pervades all Weber's writings and constitutes the basis for his response to Rickert.

First, Weber simply accepted Rickert's argument that reality was infinite and human beings could only know it as the concepts used to select significant aspects of the world for examination.[35] Second, unlike Rickert, Weber recognized that it did not matter why a scholar chose one topic over another for study because the only practical basis for such choice can be one's ultimate values. What matters, Weber argued, is that the research process is objective, and this goal is achieved only when the data are clearly conceptualized and systematically analyzed.[36] Third, Weber adapted Rickert's notion of "historical individuals" for his more general purposes. That is, he sought to understand the origin of modern Western society, and to do this he needed to develop a set of concepts that captured the distinctiveness of historical processes. However, rather than historical individuals, he called his concepts *ideal types,* a phrase that seemed to convey more clearly what he meant: Concepts that are logically perfect in the sense that they summarize a "conceptually pure type of rational action."[37] With these and other methodological tools, Weber was able to study modern societies in what he felt was an objective and scientific manner.

WEBER'S THEORETICAL SYNTHESIS

In adapting some of the conceptual tools provided by Dilthey and Rickert, Weber was able to forge a response to Marx and the *Methodenstreit* that constitutes a continuing legacy to sociology. As we will see in the next chapter, Weber's methodology of the social sciences began with a consideration of the overriding importance of objective sociology. No scientific analysis can include ethical values within it and be regarded as objective. The second methodological problem that Weber confronted was that of how to treat social and historical data, which he resolved by emphasizing the importance of understanding social action as ideal types. Because these concepts are formulated as rational models, they allow actual historical processes to be dealt with in an objective manner. The way in which Weber went about this task can be seen in his substantive works. Because he did not have modern means of gathering or analyzing data available to him, he was forced to construct "logical experiments" designed to show that sociological analyses could be done using scientific procedures. Thus, the next chapter depicts his demonstration of the manner in which cultural values circumscribe and direct social action in the *Protestant Ethic.* Similarly, in *Economy and Society* he provided subsequent researchers with a system of concepts that has proved to be of enormous use in understanding the nature of modern societies. Chapter 10 illustrates these aspects of his work by focusing on his analysis of stratification and domination in Western societies.

CHAPTER TEN

NOTES

1. Max Weber, *The Protestant Ethic and the Spirit of Capitalism* (New York: Scribner's, 1958). The original appeared in two parts, in 1904 and 1905.

2. Marianne Weber, *Max Weber: A Biography* (New York: Wiley, 1975), p. 39. The original was published in 1926. Unless otherwise noted, all biographical material comes from this source.

3. Ibid., pp. 21–30.

4. Ibid., p. 84. On the relationship between Weber's psychic turmoil and his scholarly work, see Randall Collins, *Max Weber: A Skeleton Key* (Beverly Hills, CA: Sage, 1986).

5. Neither of these works has been translated into English.

6. Only a fragment of this book has been translated under the title "Development Tendencies in the Situation of East Elbian Rural Laborers," in Keith Tribe (ed.), *Reading Weber* (London: Routledge, 1989), pp. 158–187. The entire work is summarized in Reinhard Bendix, *Max Weber: An Intellectual Portrait* (Garden City, NY: Doubleday, 1962), pp. 14–30.

7. Marianne Weber, *Max Weber*, p. 149.

8. Ibid., p. 195.

9. Ibid., pp. 195–201.

10. Ibid., p. 229.

11. See John Patrick Diggins, *Max Weber: Politics and the Spirit of Tragedy* (New York: Basic Books, 1996), pp. 62–63.

12. See Hans Gerth and C. Wright Mills, "Introduction: The Man and His Work" in Gerth and Mills (eds.) *From Max Weber* (New York: Oxford University Press, 1946), pp. 3–32; Arthur Mitzman, *The Iron Cage: A Historical Interpretation of Max Weber* (New York: Knopf, 1970); and Collins, *Max Weber: A Skeleton Key.*

13. Max Weber, *Roscher and Knies: The Logical Problems of Historical Economics* (New York: Free Press, 1975).

14. Max Weber, "'Objectivity' in Social Science and Social Policy," in *The Methodology of the Social Sciences* (New York: Free Press, 1949), pp. 50–112. The original appeared in 1904.

15. Max Weber, *Critique of Stammler* (New York: Free Press, 1977), *The Sociology of Religion* (Boston: Beacon, 1963), *The Religion of China* (New York: Free Press, 1951), *The Religion of India* (New York: Free Press, 1958), and *Ancient Judaism* (New York: Free Press, 1952).

16. See Diggins, *Max Weber,* pp. 110–113.

17. These essays are reprinted in Hans Gerth and C. Wright Mills (eds.), *From Max Weber,* pp. 7–158.

18. Max Weber, *Economy and Society* (New York: Bedminster, 1968).

19. Max Weber, *General Economic History* (New York: Collier, 1961).

20. Gerth and Mills, "Introduction," in *From Max Weber,* pp. 3–76.

21. Karl Löwith, *Marx and Weber* (London: Routledge, 1993).

22. Weber, *Protestant Ethic,* pp. 55, 75, 90–92, 266–277.

23. Weber, "Science as a Vocation." See note 17.

24. Guenther Roth, "[Weber's] Historical Relationship to Marxism," in *Scholarship and Partisanship: Essays on Max Weber,* eds. Reinhard Bendix and Guenther Roth (Berkeley: University of California Press, 1971), pp. 227–252.

25. Weber, *General Economic History,* pp. 207–276.

26. Weber, quoted in Roth, "[Weber's] Historical Relationship to Marxism," p. 234 (emphasis in original).

27. See Talcott Parsons, *The Structure of Social Action* (New York: Free Press, 1948), pp. 473–486.

28. The following paragraphs have benefited from Thomas Burger, *Max Weber's Theory of Concept Formation: History, Laws, and Ideal Types* (Durham, NC: Duke University Press, 1976),

pp. 140–50; Joseph Schumpeter, *Economic Doctrine and Method* (New York: Oxford University Press, 1954), pp. 152–201; and Charles Gide and Charles Rist, *A History of Economic Doctrine* (Lexington, MA: D. C. Heath, 1948), pp. 383–409.

29. For Weber's views on Menger and the theoretical economists, see his "'Objectivity'" and his "Marginal Utility Theory and the So-Called Fundamental Law of Psychophysics," *Social Science Quarterly* 56 (June 1975), pp. 48–159. For his views on the historical economists, see his *Roscher and Knies.*

30. Weber, *Economy and Society,* pp. 24–26.

31. See Wilhelm Dilthey, *Meaning and History: W. Dilthey's Thoughts on History and Society* (Winchester, MA: Allen & Unwin, 1961), and *Selected Writings* (New York: Cambridge University Press, 1976). Among secondary sources, see H. P. Rickman, *Wilhelm Dilthey: Pioneer of the Human Studies* (New York: Cambridge University Press, 1979); Rudolph A. Makkreel, *Dilthey: Philosopher of the Human Studies* (Princeton, NJ: Princeton University Press, 1992).

32. Although Weber never wrote a formal commentary on Dilthey's work, his writings suggest an easy familiarity with Dilthey's teachings. See Diggins,

Max Weber, p. 114. For Weber's analysis of *verstehen,* and its relationship to ideal types, see his *Economy and Society,* pp. 8–20.

33. Heinrich Rickert's works remain untranslated. This account draws on H. H. Bruun, *Science, Values, and Politics in Max Weber's Methodology* (Copenhagen: Muunksgaard, 1972), pp. 84–99; Burger, *Weber's Theory of Concept Formation,* pp. 3–56; and H. Stuart Hughes, *Consciousness and Society: The Reorientation of German Social Thought, 1890–1930* (New York: Vintage, 1958), pp. 190–191.

34. See Weber, "'Objectivity,'" p. 50, and *Roscher and Knies,* pp. 211–218.

35. Weber, "'Objectivity,'" pp. 78–79.

36. Weber, "Science as a Vocation" (see note 17) and "Critical Studies in the Logic of the Cultural Sciences," in *Methodology of the Social Sciences,* pp. 113–88.

37. Weber, *Economy and Society,* pp. 18–20, and "'Objectivity,'" pp. 87–112. Rickert's term *historical individuals* appears in Weber's essay on *Roscher and Knies* and (once) in the *Protestant Ethic,* p. 47. Weber appears to have adopted the term *ideal type* from George Jellinek; see Bendix and Roth, *Scholarship and Partisanship,* pp. 160–164.

11

✳

The Sociology
of Max Weber

In one of his last works, Max Weber defined the fledgling discipline of sociology in the following way:

> Sociology . . . is a science concerning itself with the interpretive understanding of social action and thereby with a causal explanation of its course and consequences. We shall speak of "action" insofar as the acting individual attaches a subjective meaning to his behavior—be it overt or covert, omission or acquiescence. Action is "social" insofar as its subjective meaning takes account of the behavior of others and is thereby oriented in its course.[1]

Weber believed that this definition would allow him to achieve two interrelated goals that, taken together, signify an altogether original approach to the study of social organization.[2] First, he wanted to understand the origin and unique characteristics of modern Western societies. Second, he wanted to construct a system of abstract concepts that would be useful in describing and, hence, understanding social action in such societies. Without a set of clear and precise concepts, Weber argued, systematic social scientific research is impossible. The result was a series of concepts designed to increase understanding of the modern world.

WEBER'S METHODOLOGY
OF THE SOCIAL SCIENCES

In 1904, Weber posed a fundamental question: "In what sense are there 'objectively valid truths' in those disciplines concerned with social and cultural phenomena?"[3] All his subsequent writings can be seen as an answer to this simple query. Indeed Weber's goal was to show that objective research was possible in those academic disciplines dealing with subjectively meaningful phenomena. The way he pursued this goal is presented here in two parts. First, his depiction of the problem of values in sociological research is shown. This was the central methodological issue for Weber; if sociology were to be a true science of society, he believed, it has to be objective. Second, he thought every science required a conceptual map, an inventory of the key concepts describing the phenomenon being studied, and he began to develop such a system of concepts, labeling them "ideal types."

The Problem of Values

During Weber's time, many observers did not think that an objective social science was plausible because it seemed impossible to separate values from the research process. So most scholars attempting to describe human behavior infused their analyses with political, religious, and other values. Marx's writings constitute an extreme example of this tactic. Weber confronted the problem of values by observing that sociological inquiry should be objective, or, to use his term, *value-free*. Having said that, however, he then suggested how values and economic interests were connected to social scientific analyses.

Value-Free Sociology Weber's use of the term *value-free* is unfortunate, because it implies that social scientists should have no values at all, plainly an impossibility. What he meant, however, was that researchers' personal values and economic interests should not affect the process of social scientific analysis. He believed that if such factors influenced the research process, the structure of social action could not be depicted objectively. This fundamental concern with attaining objective and verifiable knowledge links all the sciences, natural and social. Objective analyses are only possible, Weber argued, if sociologists use a "rational method" in which the research process is systematic; that is (1) empirical data must be categorized in terms of clearly formulated concepts, (2) proper rules of evidence must be employed, and (3) only logical inferences must be made.[4]

This methodological orientation carries with it an important implication: Sociology should not be a moral science. Thus, it is not possible to state scientifically which norms, values, or patterns of action are correct or best but, rather, it is only possible to describe them objectively. Weber believed that such descriptions would represent a considerable achievement. After all, they did not then exist. Thus, unlike many others, Weber explicitly distinguished between "what ought to be," the sphere of values, and "what is," the sphere of science, arguing that sociology should focus only on the latter. This distinction implies Weber's view of the underlying value that ought to guide social scientific inquiry: the search for truth.[5]

Another implication of Weber's argument for a value-free sociology is that the new science reflects an ongoing historical process in which magic and other forms of inherited wisdom become less acceptable as means for explaining events. Weber referred to this change as the process of *rationalization,* and it is the dominant theme in his work. Unlike Marx, who used the dialectic as a leitmotif, Weber believed that social life is becoming increasingly "rationalized" in the sense that people lead relatively methodical lives: They rely on reason buttressed by objective evidence. The rationalization of the economy—for example, by means of improved accounting, the use of technology, and other methods—produced modern capitalism. The rationalization of government—by reliance on technical training and legal procedures, for example—resulted in the rise of the modern political state. The sciences, of course, are the archetypal methodical disciplines.[6] In a "rationalized discipline," values should not affect the research process. But they remain relevant.

The Connection Between Values and Science Although Weber knew that the separation between values and science was difficult to maintain in practice, the distinction highlighted the relevance of values before and after the research process. In effect, whereas facts provide information, only values guide action. This is true for both social scientists and policymakers.

Social scientists are faced with a very practical problem: how to choose topics on which to do research. Is there, for example, a scientific way of deciding whether poverty or premarital sex is a more interesting or important topic to study? Weber's answer was simple: no. In considering this response, the phrasing in the previous section should be recalled: The research process must be objective. The choice of topics comes before the research takes place. The only basis for making such a decision is scientists' religious beliefs, economic interests, and other values, which lead some of them to each topic. But once having chosen a topic for study, according to Weber's dictum, scientists must follow an objective research process.

The situation is more complex when dealing with public policy issues. Is there, for example, a scientific way for policymakers to decide about funding for public assistance or national defense? Again, Weber's answer is simple: no. This response, however, does not mean that the social sciences are irrelevant to public policy. Sociology can describe the facts. Given a specific political goal, for example, Weber said sociologists could determine (1) the alternative strategies for achieving it, (2) the consequences of using different strategies, and (3) the consequences of attaining the goal.[7] He believed that they could perform these tasks objectively by categorizing the data as clearly formulated concepts, following proper rules of evidence, and making logical deductions. Once that is done, however, there is no scientific way of choosing public policies. Selecting one goal rather than another and one strategy rather than another ultimately depends on people's political values, their economic interests, and non-objective factors.

Having said that the research process must be objective and that the sphere of values and the sphere of science must be kept separate, Weber drew a unique conclusion. Unlike nearly all the other classical sociologists (except Marx), Weber rejected the search for general laws in favor of historical theories that provide an "interpretive understanding of social action and . . . a causal explanation of its course and consequences."[8] He took this position not for scientific reasons but for epistemological ones. He saw that any system of abstract and timeless laws must focus on events that are typical and recurrent, as in the natural sciences. Thus, a search for universal laws necessarily excludes from consideration important and unique historical events. Weber summarized his position in the following way:

> For the knowledge of historical phenomena in their concreteness, the most general laws, because they are most devoid of content are also the least valuable. The more comprehensive the validity—or scope—of a term, the more it leads us away from the richness of reality since in order to include the common elements of the largest possible number of phenomena, it must necessarily be as abstract as possible and hence devoid of content. In the [social] sciences, the knowledge of the universal or general is never valuable in itself.[9]

In effect, then, Weber was most interested in focusing on the "big empirical questions," such as why capitalism had originated in the West rather than somewhere else, and he knew that an emphasis on the development of general theories would not allow for an examination of such issues. Ideal types were his method for dealing with these issues.

Ideal Types

All science requires a conceptual map that identifies the parts of the phenomena under investigation. For example, biologists give names not only to each bone and tissue in the body but also to the processes through which information, food, and other elements are transported around. Astrophysicists, chemists, and all other scientists do the same thing for their objects of study. Yet, a conceptual map of society did not exist during Weber's time, mainly because human beings seemed so unpredictable. This unpredictability resulted, in part, from not having concepts that summarize patterns of social action. So Weber set about developing them. His goal was "a system of concepts of such universal scope as to be consistent with even the most diverse value attitudes."[10]

In pursuing this goal, however, Weber had to confront a fundamental problem: Sociology is inherently different from the natural sciences because, as noted earlier, its essential task is "the interpretive understanding of social action and thereby . . . a causal explanation of its course and consequences."[11] "To understand" is the usual translation of the German word *verstehen,* and there has been considerable controversy over the years about the theoretical and methodological implications of this term. Weber's argument, however, can be presented in a reasonably straightforward manner, for he believed that the use of ideal types could lead to an "interpretive understanding of social action" and hence to a "causal explanation" of both historical events and individual behavior.[12]

Although there has been some confusion because of his use of the word *ideal,* Weber did not intend these concepts to have a normative (or value) connotation. Rather, they were designed "to be perfect on logical grounds," most of the time by summarizing a "conceptually pure type of rational action." Weber used ideal types in different ways and for somewhat different purposes, and unfortunately, he did not make any clear or explicit distinction among them, with the result that scholars have often complained about the inconsistent way in which he used his conceptual tools.[13] Despite some areas of inconsistency, however, we can distinguish two kinds of ideal types in Weber's work. The first can be called *historical ideal types,* and the second can be termed *classificatory ideal types.* Although Weber did not use these labels, they provide a convenient way of organizing our discussion.

Historical Ideal Types Historical ideal types are reconstructions of past events, or ideas, in which some aspects are accentuated so that they are logically (or *rationally,* to use Weber's word) integrated and complete. By conceptualizing historical events in this way, it is possible to compare them systematically with the ideal type and, by observing deviations from the rational model, arrive at causal judgments. This strategy enabled Weber to place historical processes, such as the Protestant Reformation, in "an intelligible and more inclusive context of meaning" and thereby to understand their significance for the development of the modern world.

In 1904, Weber noted that the historical ideal type "has the significance of a purely ideal *limiting* concept with which the real situation or action is *compared* and surveyed for the explication of certain of its significant components."[14] On this particular point, there was great continuity in his thought, for in Part 1 of *Economy and Society* (written around 1919) he made a similar observation:

> The construction of a purely rational course of action . . . serves the sociologist as a type (ideal type) which has the merit of clear understandability and lack of ambiguity. By comparison with this it is possible to understand the ways in which actual action is influenced by irrational factors of all sorts, such as affects and errors, in that they account for the deviation from the line of conduct which would be expected on the hypothesis that the action [is] purely rational.[15]

Classificatory Ideal Types During the years between 1904 and 1920, however, Weber apparently became increasingly aware of the need for an inventory of concepts that would serve as a means for more precisely describing the fundamental social processes occurring in all societies. As a result of this awareness, Part 1 of *Economy and Society* simply enumerates a system of abstract concepts to be used in understanding the structure of social action. We have labeled these concepts *classificatory ideal types,* and they constitute the conceptual core of the discipline of sociology as Weber ultimately perceived it. Although Weber's death prevented completion of his system of concepts, his intent can be illustrated by examining his conceptualization of the types of social action.[16]

According to Weber, people's actions can be classified in four analytically distinct ways.[17] The first type of action is the *instrumentally rational,* which occurs when means and ends are systematically related to each other based on knowledge. Weber knew that the knowledge that people possess might not be accurate. Thus, both the rain dance and the timing of a stock purchase are instrumentally rational acts, from the point of view of the dancers and the buyers, even though the means-end link might be based on magical beliefs or rumors. Thus, instrumentally rational action occurs in all societies. Nonetheless, Weber said, the archetypal form of instrumentally rational action is based on objective, ideally scientific, knowledge. Action buttressed by objective knowledge is more likely to be effective. Its effectiveness is one reason why the spheres in which instrumentally rational action occurs have widened over time. Business people producing goods at the lowest possible cost, coaches eliciting outstanding athletic performance, educators teaching effectively, lawyers persuading a jury, physicians healing patients, politicians asking the General Accounting Office to study pollution, and even parents raising children according to (their notion of) "right" values all use scientific knowledge to achieve their goals. These are examples of what Weber meant by instrumentally rational behavior. Its pervasiveness in modern societies reflects the historical process of rationalization.

The second type of action is *value-rational,* which is behavior undertaken in light of one's basic values. Weber emphasized "value rational action always involves 'commands' or 'demands' which, in the actor's opinion are binding."[18] Religious people avoiding alcohol use because of their faith, parents paying for their children's braces and college education, politicians passing laws, and soldiers obeying orders are acting as a result of their values. Not to do so would be dishonorable. Thus, the essential characteristic of value-rational action is that it constitutes an end in itself. It is not a means to monetary success, increasing knowledge, or any other instrumental goal. As the examples indicate, such behavior occurs in every sphere of life. The sociological task, which Weber left incomplete, is to specify the contexts in which instrumentally rational and value-rational action occur in modern societies.

The third type of action is *traditional,* which is behavior "determined by ingrained habituation." Weber's point is that in a context where beliefs and values are second nature and patterns of action have been stable for many years, people usually respond to situations from habit. In a sense, they regulate their behavior by customs handed down across generations. In such societies, people resist altering long-established ways of living, which are often sanctified in religious terms. As a result, when confronted with new situations or choices, they often continue in the old ways. Traditional action typifies behavior in contexts where choices are (or are perceived to be) limited. According to Weber, traditional action characterizes people in preindustrial societies. This is why he used a variant of the same term, *traditional societies,* to describe them. He believed that the distinction between instrumentally rational and value-rational action was unnecessary to understand patterns of behavior in traditional societies. For example, the household usually constitutes both a productive and a consumptive unit in such contexts. This means not only that raising children and obtaining life's necessities occur in the same sphere but also that people make both family and economic decisions in the light of custom. Custom nearly always precludes the calculating, logical evaluation of means and ends based on objective data that is typical of instrumentally rational action. Similarly, custom usually precludes a decision about which values are appropriate guides, such as family loyalty or acquisitiveness, that is typical of value-rational action. Nonetheless, Weber clearly states that instances of traditional action occur in modern societies. Examples can be seen whenever people justify their behavior by saying that "this is the way we do things here. . . ." Companies sometimes fail (or lose market share) because the people running them do not adapt to new situations. Their actions have become ingrained, habitual.

The fourth type of action is *affectual,* which is behavior determined by people's emotions in a given situation. The parent slapping a child and the football player punching an opponent are examples. This type of behavior occurs, of course, in all societies. Alas, it constitutes a residual category that Weber acknowledged but did not explore in detail.

These types of action classify behavior by visualizing its four "pure forms." Although Weber knew that actual situations would not perfectly reflect these concepts, they provide a common reference point for comparison. That is, a variety of empirical cases can be systematically compared with one another and with the ideal type, in this case, the types of social action. This strategy is presented in Figure 11.1. Ideal types thus represent for Weber a quasi-experimental method. The "ideal" serves as the functional equivalent of the control group in an experiment. Variations or deviations from the ideal are seen as the result of causal forces (or a stimulus in a real laboratory experiment), and effort is then undertaken to find these causes. In this sense, Weber could achieve two goals: (1) to analytically and logically accentuate the elements of social action and (2) to discover the causes of unique variations in specific instances. For example, by noting the extent to which actual empirical cases compare with instrumental rational action, the investigator can assess the causes of conformity to, or deviation from, this ideal. In this way, the unique aspects of empirical cases can be emphasized and yet systematically and logically analyzed.

The nature of such analyses, however, remains unclear. Although Weber participated in several early survey research projects, modern statistical procedures were unavailable to describe what he called "mass phenomena." Hence, hypotheses were not seen as verifiable using such techniques. A new way of doing research had to be invented, which for Weber turned out to be ideal types.[19] The following sections provide two examples of how he envisioned the use of ideal types.

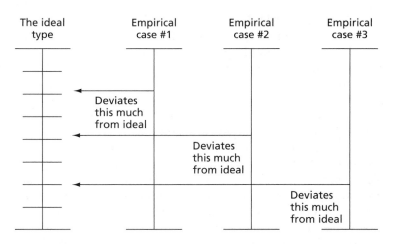

The ideal type / Empirical case #1 / Empirical case #2 / Empirical case #3

Deviates this much from ideal

Deviates this much from ideal

Deviates this much from ideal

Analytical accentuation
of generic properties
of phenomena

Actual observations
in three different
empirical contexts

By recording actual deviations from the ideal in each empirical case, the
investigator compares the cases using a common reference point. Then,
by asking what caused the deviations, the investigator can isolate the
causes of empirical events in each case and compare them.

FIGURE 11.1 The Ideal Type Methodology

WEBER'S IMAGE
OF SOCIAL ORGANIZATION

Weber's analysis of social organization is detailed and complex, and indeed, it is
often difficult to get a sense for how he visualized society as a whole. As noted
at the beginning of the chapter, Weber defined sociology as the study of *social
action,* and as we have seen in the analysis of ideal types, he felt that there are four
basic types of action: instrumental-rational, value-rational, traditional, and
affectual.[20] Thus, human behavior is guided or, in Weber's terms, "oriented," by
considerations of rationality, tradition, or affect. These types of action, however,
need not be mutually exclusive; they can be combined, although some orientations are more compatible with each other than others are. For example, affectual and value-rational are more likely to be combined than, say, are
instrumental-rational and traditional. Still, even when combined, Weber
implied that one type of action will generally dominate a social relationship.

As is typical of Weber, the nature of social relationships,[21] like the actions
forming them, is portrayed as an ideal type. There are two basic kinds of social
relationships arising out of social action: one is *communal* relationships, which are
formed by individuals' feelings for each other, with such feelings based on affectual or traditional actions; the other is *associative* relationships, which are based
on rationality, whether instrumental- or value-rational. Thus, in Weber's eye, the

two basic types of social relationships—communal and associational—are motivated by a basic split in the four types of action, with one of these splits revolving around the two types of rational action (value and instrumental) and the other around affectual and traditional orientations.

Social relationships, whether communal or associative, are generally connected to what Weber termed *legitimated orders.*[22] An "order" appears to be Weber's way of conceptualizing the larger structures that are built from social relationships. Action and social relationships almost always occur within the context of an existing legitimated order. Such orders "guarantee" that actions and social relationships will be conducted in accordance with "maxims" or rules, the violation of which will bring about negative sanctions on those failing to meet their obligations. Thus, the structure connecting the more micro processes of action and social relationships to more macro levels of reality is the legitimated order. Like so many concepts in Weber's work, there is a classification of orders into two basic types. One is organized around "subjective" guarantees that social relationships will proceed in accordance with the rules of the order, with this subjectivity arising from one of three routes: (a) affect, or "emotional surrender" to the order, (b) value rationality, or a belief in the absolute validity of the order as the most efficient means to an end, and (c) religious beliefs that salvation depends on the order. The other type of order is organized by expectations among actors for certain "external effects" that are predictable outcomes to actions undertaken; thus, because actors calculate their actions in accordance with expected outcomes, Weber implied that this kind of order is organized by instrumental rationality.

Weber then shifted to the basis of legitimation of orders—that is, routes by which actors ascribe rights to "the order" to control their conduct. Again, as is typical with Weber, there are several basic types of legitimation: (a) tradition, or the way things have always been; (b) affectual, or emotional attachments to the ways things are organized; (c) value-rational, or the "deduction" that the current order is the best possible way of organizing actions; and (d) legal, which is composed of binding agreements (entered into by considerations of instrumental-rationality) among actors or by an external authority that is considered to have the right impose and enforce agreements.

In Figure 11.2, we have diagrammed what we think is Weber's intent, although we must confess that, despite all the definitions and categories, Weber's analytical scheme is far from precise or clear. The subject matter of sociology is social action, whereby actors take cognizance of each other's behaviors. Actions are "oriented" to either affect, tradition, value-rationality, or instrumental-rationality; here, Weber implied that these orientations are cultural or part of the values, beliefs, and ideologies of a society, but they also become motivations that pushes actors to behave in certain ways. Various oriented and motivated social actions then lead to the formation of more stable social relationships that can be either communal or associative, depending on the configuration of cultural orientations and motivations involved. Communal relations are guided by affectual and traditional orientations and

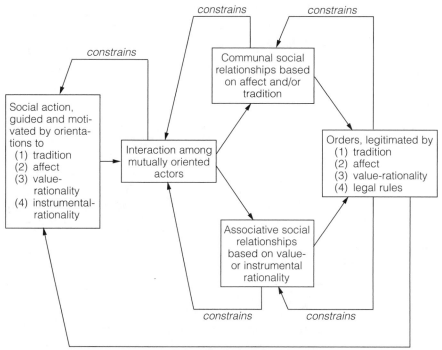

constrains   *constrains*

constrains

Social action,
guided and moti-
vated by orienta-
tions to
(1) tradition
(2) affect
(3) value-
    rationality
(4) instrumental-
    rationality

Interaction among
mutually oriented
actors

Communal social
relationships based
on affect and/or
tradition

Orders, legitimated by
(1) tradition
(2) affect
(3) value-rationality
(4) legal rules

Associative social
relationships
based on value-
or instrumental
rationality

constrains   *constrains*

Determines orientations to

FIGURE 11.2 Weber's Conception of Action, Relationships, and Orders

motivations, whereas associational relations are composed from considerations of rationality, whether instrumental or value rationality. Social relationships are typically part of an order that structures action in accordance with rules. Such orders are organized by cultural orientations emphasizing affect, value-rationality, religion, and rationality; the order's basic legitimation can be either traditional, affectual, value-rational, or legal.

At this point, Weber's view of social organization seems rather vague about how the model in Figure 11.2 leads us to the major topics of Weberian sociology. The definitional distinctions in the model were written rather late in Weber's career, after he had written much of his sociology; thus, the model does not provide clear guidelines back into the substantive topics addressed by Weber earlier in his career. Still, let us make an effort, if only to set the stage for our discussion in this chapter of Weber's sociology. Figure 11.3 begins where Figure 11.2 ends, with the formation of legitimated orders. Weber implies, but does not clearly state, that there are two basic types of orders: (1) *organizational orders* composed of structures revealing a division of labor and pursuing particular goals and (2) *stratification orders* composed of categories of individuals in a system of inequality. These are not mutually exclusive because a system of inequality is sustained by organizations, whereas an organization

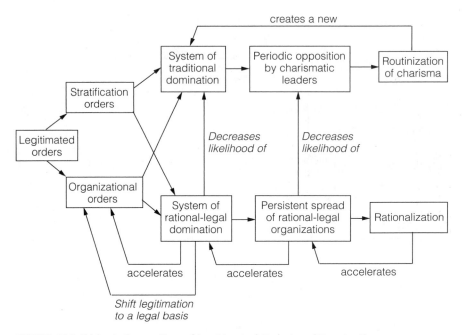

FIGURE 11.3 Weber's Conception of Legitimated Orders and Domination

can exist as the result of a particular configuration of inequality in the distribution of resources. These come together under Weber's concept of *domination,* as we will see shortly. By this term, Weber meant that some segments of a society have the authority to tell others what to do and, as a result, those with authority can control the distribution of resources. Legitimated orders, therefore, generate systems of domination.

In Weberian sociology, then, a society cannot be understood without inquiry into its patterns of domination. Weber saw the long-term trends or the movement of human society as revolving around a shift in the basis of domination. For Weber, history had been typified by periods of relative stability in patterns of domination revolving around traditional authority, punctuated by periodic emergence of charismatic leaders who had mobilized opposition movements and established new patterns of domination based on their charismatic authority that, over time, tended to turn into a new form of traditional authority. With the expansion of markets during capitalism, however, domination increasingly comes from *rational-legal authority* as personified by law and bureaucratic organizations. These organizations were, in Weber's eyes, gaining control of all legitimated orders, displacing the affectual, traditional, and even value-rational legitimation of orders with the rule of law, while orienting action in all spheres of social life to instrumental rationality. The social world was thus becoming "rationalized," as bureaucratic organizations in the state and economy became the basis for domination in society.

This is Weber's general view of the social world when he looked at historical trends and industrial capitalism as it was emerging in Germany around the turn into the twentieth century. He wanted to explain the shift in patterns of domination, and this led him to explore a variety of substantive topics—bureaucratic organizations, stratification, cities, law, religion, geopolitics, and markets—to see how these topics could help explain the shift in domination toward rational-legal authority. Alas, there is no clear theory in all of these substantive concerns, only a set of topics organized around the theme of rationalization.

WEBER'S ANALYSIS OF DOMINATION

Types of Domination

A society can be typified by its system of domination. In German, the term *herrschaft* connotes both domination and authority, and Weber probably meant this to be the case.[23] Any system of domination is ultimately built from what we termed stratification orders and organizational orders. All orders must be legitimated, so those who hold power seek to legitimate their power as "authority" in the eyes of those who are subject to this power.[24] Domination also requires organizational orders to administer and monitor conformity to directives given to subordinates. As was typical for Weber, he saw three basic types of domination—charismatic, traditional, and rational-legal—with each type relying on a different basis of legitimation and a different kind of administrative apparatus.[25]

Charismatic Domination The first type of domination is called charismatic. The term *charisma* has a religious origin and literally means "gift of grace," implying that a person is endowed with divine powers.[26] In practice, however, Weber did not restrict his use of charisma to manifestations of divinity but, rather, employed the concept to refer to those extraordinary individuals who somehow identify themselves with the central facts or problems of people's lives and who, by the force of their personalities, communicate their inspiration to others and lead them in new directions. Thus, people in other than religious roles can sometimes be considered charismatic: for example, politicians, soldiers, or artists.[27]

Weber believed that charismatic leadership emerges during times of crisis, when dominant ways of confronting the problems faced by a society seem inappropriate, outmoded, or inadequate. In such a context, charismatic domination is revolutionary. People reject the past in favor of a new direction based on the master's inspiration. As Weber put it, every charismatic leader implicitly argues that "it is written . . . but I say unto you. . . ." Thus, charismatic domination is a vehicle for social change in both traditional and rational-legal contexts, which are the other two types of domination.

The legitimacy of charismatic domination lies both in the leader's demonstration of extraordinary insight and accomplishment and in the followers' acceptance of the master. It is irrelevant, from Weber's point of view, whether a charismatic leader turns out to be a charlatan or a hero; both Hitler and Gandhi were charismatic leaders. Rather, what is important is that the masses are inspired to freely follow the master. Weber believed that charisma constituted an unstable form of authority over extended periods because its legitimacy depended on the leader's claim to special insight and accomplishment. Thus, if success eludes the leader for long and crises are not resolved satisfactorily, the masses will probably reject the charismatic figure, and his or her authority will disappear.

In charismatic domination, the leader's administrative apparatus usually consists only of a band of faithful disciples who serve the master's immediate personal and political needs. Over the long run, however, every regime led by a charismatic leader faces the "problem of routinization," which involves both finding a successor to the leader and handling the day-to-day decisions that must be made.

Weber noted that the problem of succession could be resolved in a variety of ways: for example, by the masses searching for a new charismatic leader, by the leader's designation of a successor, or by the disciples' designation of a successor. But all these methods involve political instability. For this reason, either customs or legal procedures allowing the orderly transfer of power usually develop over time.

The problem of making day-to-day decisions (that is, of governing) is usually resolved by either the development of a full-fledged administrative staff or the takeover of an already existing organization. In both cases, however, the typical result is the transformation of the relationship between charismatic leader and followers from one based on beliefs in the master's extraordinary qualities to one based on custom or law. "It is the fate of charisma," Weber wrote, "to recede before the powers of tradition or of rational association after it has entered the permanent structures of social action."[28] These new bases of legitimation represent the other two types of domination.

Traditional Domination The second type of domination is based on tradition. In Weber's words, "authority will be called traditional if legitimacy is claimed for it and believed in by virtue of the sanctity of age-old rules and powers."[29] Put differently, traditional domination is justified by the belief that it is ancient and embodies an inherent (often religiously sanctified) state of affairs that cannot be challenged by reason. For example, over time it became customary for the firstborn child of a British monarch to be the legitimate successor to the throne, and this pattern carries religious sanction.[30] In such a context, the subordinate classes obey edicts in recognition of the ruler's rightful place, out of personal loyalty, and, of course, because of their economic and political dependence. Thus, Weber's analysis suggests how traditional types of social action are generalized into a system of domination. The stratification hierarchy in such societies is usually fairly rigid because people's positions in the social system are dictated at birth by custom.

Weber distinguished between two forms of traditional authority, only one of which has an administrative apparatus. *Patriarchalism* is a type of traditional domination occurring in households and other small groups where the use of an organizational staff to enforce commands is not necessary. *Patrimonialism* is a form of traditional domination occurring in larger social structures that require an administrative apparatus to execute edicts.

In the patrimonial form of traditional domination, the administrative apparatus consists of a set of personal retainers exclusively loyal to the ruler. Weber observed that in addition to its grounding in custom, the officials' loyalty is based on either their dependence on the ruler for their positions and remuneration or their pledge of fealty to the leader, or both. As an ideal type, the essence of patrimonialism (traditional authority coupled with an administrative staff) is expressed by the following characteristics:

1. People obtain positions based on custom and loyalty to the leader.
2. Officials owe obedience to the leader issuing commands.
3. Personal and official affairs are combined.
4. Lines of authority are vague.
5. Task specialization is minimal.

In such a context, decisions are based on officials' view of what will benefit them and what the leader wants. Moreover, officials appropriate the means of production themselves or are granted them by the leader. Hence, where their jurisdiction begins and ends remains uncodified. A sheriff, for example, might both catch criminals and collect taxes (skimming off as much as possible). But how these tasks are accomplished will be idiosyncratic, subject to official whim rather than law. Thus, it should not be surprising that in *Economy and Society* Weber described traditional domination as inhibiting the development of capitalism, primarily because rules are not logically established, officials have too wide a range for personal arbitrariness, and they are not technically trained.[31] Modern capitalism requires an emphasis on logic, procedure, and knowledge. Furthermore, as will be shown, modern class structures are not possible in social systems in which statuses and roles are circumscribed by tradition.

Rational-Legal Domination The third type of domination is that based on law, what Weber called rational-legal authority. As he phrased it, "legal domination [exists] by virtue of statute. . . . The basic conception is that any legal norm can be created or changed by a procedurally correct enactment.[32] Thus, the basis for legitimacy in a system of rational-legal domination lies in procedure. People believe that laws are legitimate when they are created and enforced in what is defined as the proper manner. Similarly, people see leaders as having the right to act when they obtain positions in procedurally correct ways—for example, through election or appointment. In this context, then, Weber defined the modern state as based on the monopoly of physical coercion, a monopoly made legitimate by a system of laws binding both leaders and citizens. The rule of law, rather than of persons, reflects the process of rationalization, as described earlier. Nowhere is this more clearly observed than in a modern bureaucracy.

Weber called the administrative apparatus in a rational-legal system a *bureaucracy* and observed that it was oriented to the creation and enforcement of rules in the public interest. A bureaucracy is the archetypal example of instrumentally rational action. Although many people today condemn bureaucracies as inefficient, rigid, and incompetent, Weber argued that this mode of administration was the only means of attaining efficient, flexible, and competent regulation under a rule of law. In its logically pure form (as an ideal type), a bureaucratic administrative apparatus has different characteristics from those in traditional societies[33]:

1. People obtain positions based on knowledge and experience.
2. Obedience is owed to rules uniformly applicable to all.
3. Personal and official affairs are kept separate.
4. Lines of authority are explicit.
5. Task specialization is great.

According to Weber, bureaucratic administration in a rational-legal system is realized to the extent that staff members "succeed in eliminating from official business love, hatred, and all purely personal, irrational, and emotional elements."[34] But this is an ideal type. Weber knew that no actual bureaucracy operated in this way. People often obtain positions based on whom they know. Rules are often applied arbitrarily. Personal and official matters are often combined. Thus, the empirical task becomes assessing the degree to which a bureaucracy conforms to the ideal type (recall Figure 11.1). The issue is important because the bureaucratic ideal type reflects a fundamental value characteristic of modern societies: Political administration should be impersonal, objective, and based on knowledge, for only in this way can the rule of law be realized. Further, Weber emphasized that although this value seems commonplace today, it is historically new. It arose in the West and has become the dominant form of authority only in the last few hundred years. Finally, Weber's definition of bureaucracy points toward a fundamental arena of conflict in modern societies: Who is to make laws, and who is to administer them through their control of the bureaucracy?

Within the context of a rational-legal system of authority, political parties are the forms in which social strata struggle for power. As Weber puts it, "a political party . . . exists for the purpose of fighting for domination" to advance the economic interests or values of the group it represents, but it does so under the aegis of statutory regulation.[35] In general, the point of the struggle is to direct the bureaucracy via the creation of law, for in this way the goals of the various social strata are achieved. For example, the very rich who own income-producing property in the United States act to make sure that their economic interests are codified into law. Similarly, people in all social strata act to protect their interests and values, and the needs of those who do not participate are ignored.[36] The political process in Western societies, then, reflects basic cultural values: Economic and social success are to be achieved through competition under the rule of law, and the process is rational in the sense of being pursued in a methodical manner. Weber's distinction between classes and status groups shows in a different way how the structure of stratification in modern societies also mirrors these values.

Social Strata: Class and Status

In his analysis of the structure of stratification, Weber tried to provide observers with a conceptual map outlining the parts of the stratification system. He believed that such an inventory of concepts would allow an objective description of stratification processes in modern capitalist societies. At the core of his scheme are two ideal types: social class and social status.

Unfortunately, Weber's discussion of these two concepts is somewhat confusing. Class-oriented behavior is concerned with economic issues, especially the amount of and source of income. As such, it is an example of instrumentally rational action. Status-oriented behavior is concerned with values, mainly the honor or prestige attached to one's lifestyle. People's values are reflected in their choice of housing, friends, marriage partners, leisure activities, and other aspects of their lifestyle. As such, status-oriented behavior is an example of value-rational action. Class and status are interrelated in that lifestyle is made possible by income and income is made possible by lifestyle. Thus—and this is often misunderstood—status groups do not compete with classes. Rather, status and class are different bases for action displayed by people in each stratum, who act to protect their economic interests and values. We should note that Weber rarely used the term *strata,* which refers to a set of ranked positions. The title of this section, "Social Strata: Class and Status," is designed to emphasize our interpretation that class and status groups are coterminous.

Social Class According to Weber, a *social class* consists of those persons who have a similar ability to obtain positions in society, procure goods and services for themselves, and enjoy them via an appropriate lifestyle.[37] It should be recognized immediately that a class is defined, in part, by status considerations: the lifestyle of the stratum to which one belongs. In Weber's terminology, classes are statistical aggregates rather than groups. Behavior is class-oriented to the extent that the process by which people obtain positions, purchase goods and services, and enjoy them is characterized by an individualistic rather than a group perspective. For example, even though investors trying to make money on the stock market might have some common interests, share certain kinds of information with one another, and even join to prevent outsiders from participating, each acts individually in seeking profits or in experiencing losses. Further, in the process of seeking profits, their behavior is typically characterized by an instrumentally rational orientation; that is, action reflects a systematic calculation of means and ends based on knowledge (even if such knowledge is imperfect).

In Weber's analysis, classes are essentially economic phenomena that can exist only in a legally regulated money market where income and profit are the desired goals. In such a context, people's membership in a class can be determined objectively, based on their power to dispose of goods and services. For this reason Weber believed that one's "class situation is, in this sense, ultimately [a] market situation."[38] The most important characteristics of a money market are that in its logically pure form it is impersonal and democratic. Thus, all that should matter in the purchase of stock, groceries, housing, or any other commodity are such factors as one's cash and credit rating. Similarly, a person's class situation is also objectively determined, with the result that people can be ranked by their common economic characteristics and life chances.

A key problem, of course, involves the basis on which people possessing similar amounts of economic power are placed together in classes. Whatever sorting mechanism is used, it must be both helpful to observers and subjectively meaningful to the participants. Like Marx, Weber began by distinguishing between those who had property and those who did not. As he put it, "'property' and 'lack of property' are . . . the basic categories of all class situations."[39] This is because the possession or nonpossession of income-producing property (or capital) allows fundamentally different styles of life—and differences in lifestyle are the key to status distinctions in modern societies.

Considering first those who own income-producing property, Weber argued that they differed according to the use to which their possessions were put: "The propertied, for instance, may belong to the class of rentiers or to the class of entrepreneurs."[40]

In Weber's terminology, *rentiers* are those who live primarily off fixed incomes from investments or trust funds. For example, the large German landowners of his time were rentiers because these families had controlled much of the land for several generations and received their incomes from the peasants or tenant farmers who actually worked it. As a result of their possession of capital and values that they had acquired over time, the landowners chose to lead a less overtly acquisitive lifestyle. Weber called them rentiers because they did not work to increase their assets but simply lived off them, using their time for purposes other than earning a living. For example, they might hold public office or lead lives of idleness.

According to Weber, *entrepreneurs* are those, such as merchants, shipowners, and bankers, who own and operate businesses. Weber called them a commercial, or entrepreneurial, class because they actually work their property for the economic gain it produces, with the result that in absolute terms the members of the entrepreneurial class often have more economic power, but less social honor (or prestige), than rentiers do.

This distinction between the uses to which income-producing property is put allowed Weber to differentiate between those who work as an avocation and those who work because they want to increase their assets; that is, this distinction reflects fundamental differences in values. In most societies there exist privileged status groups, such as rentiers, the members of which "consider almost any kind of overt participation in economic acquisition as absolutely stigmatizing" despite its potential economic advantages.[41] Usually these families have possessed wealth for a long time, over several generations. Thus, Weber argued, even though economic-oriented (or class-oriented) action is individualistic and dominated by instrumentally rational action, value-rational behavior also occurs. Action at every stratum level varies along these two dimensions.

Nonetheless, even though the two classes can be distinguished, Weber asserted that the possession of capital by both rentiers and entrepreneurs provided them with great economic and political power and sharply distinguished them from those who did not own such property.[42] Both rentiers and entrepreneurs can monopolize the purchase of expensive consumer items. Both pursue monopolistic sales and pricing policies, whether legally or not. To some extent, both control opportunities for others to acquire wealth and property. Finally, both rentiers and entrepreneurs monopolize costly status privileges, such as education, that provide young people with future contacts and skills. In these terms, then, rentiers and entrepreneurs can be seen to have (roughly) similar levels of power and, partly because they are always a small proportion of the population, they often act together to protect their lifestyles. Even though they live rather differently, their source of income (ownership of capital) sets them apart from the other social classes. The distribution of property, in short, tends to prevent nonowners from competing for highly valued goods and perpetuates the structure of stratification from one generation to another. This does not mean that mobility in either direction is impossible, for such movement often occurs; it does, however, mean that mobility is difficult.

In constructing his conceptual map of the class structure, Weber next considered those who do not own income-producing property. Despite not possessing the means of production, such people are not without economically and politically important resources in modern societies, and they can be meaningfully differentiated into a number of classes. The main criteria Weber used in making class distinctions among those without property are the worth of their services and the level of their skills because both factors are important indicators of people's ability to obtain positions, purchase goods, and enjoy them. In Weber's classificatory scheme, the "middle classes" comprise those individuals who today would be called white-collar workers because the skills they sell do not involve manual labor: public officials, such as politicians and administrators; managers of businesses; members of the professions, such as doctors and lawyers; teachers and intellectuals; and specialists of various sorts, such as technicians, low-level white-collar employees, and civil servants. Because their skills are in relatively high demand in industrial societies, these people generally have more economic and political power than those who work with their hands do.[43]

According to Weber, the less-privileged, property-less classes comprise people who today would be called blue-collar workers because their skills primarily involve manual labor. Without explanation, Weber said that such people could be divided into three levels: skilled, semiskilled, and unskilled workers. He did not elaborate much on the lifestyles of those without property.

By means of these ideal types, Weber described the parts of a modern class structure in which the key factors distinguishing one class from another are the uses to which property is put by those who own it and the worth of the skills and services offered by those who do not own property. These factors combine in the marketplace to produce identifiable aggregates, or classes, the members of which have a similar ability to obtain positions, purchase goods, and enjoy them via an appropriate lifestyle.

The final topic of importance in Weber's analysis of social class is the possibility of group formation and unified political action by the propertyless classes. Like Marx, Weber said that this phenomenon was relatively rare in history because those who did not own property generally failed to recognize their common interests. As a result, action based on a similar class situation is often restricted to inchoate and relatively brief mass reactions. Nonetheless, throughout history perceived differences in life chances have periodically led to class struggles, although in most cases the point of the conflict focused on rather narrow economic issues, such as wages or prices, rather than on the nature of the political system that perpetuates their class situation.[44]

Although Weber alluded only briefly to the conditions under which the members of the property-less classes might challenge the existing political order, he identified some of the same variables that Marx had:

1. Large numbers of people must perceive themselves to be in the same class situation.

2. They must be ecologically concentrated, as in urban areas.

3. Clearly understood goals must be articulated by an intelligentsia. Here Weber suggested that people had to be shown that the causes and consequences of their class situation resulted from the structure of the political system itself.

4. The opponents must be clearly identified.

When these conditions are satisfied, Weber indicated, an organized class results. We turn now to the other basis for action displayed by people in each stratum: social status.

Social Status In Weber's work, *social status* refers to the evaluations that people make of one another, and a status group comprises those individuals who share "a specific, positive or negative, social estimation of honor."[45] As described earlier, Weber used the concepts of "status" and "status group" to distinguish the sphere of prestige evaluation, expressed by people's lifestyles, from that of monetary calculation, expressed by their economic behavior. Although the two are interrelated, the distinction emphasizes that people's actions cannot be understood in economic terms alone. Rather, their values often channel action in specific directions.

The difference between class and status can be summarized in the following way: On the one hand, because the income from a person's job provides the ability to purchase goods and enjoy them, class membership is objectively determined based on a simple monetary calculation. On the other hand, because status and honor are based on the judgments that people make about another's background, breeding, character, morals, and community standing, a person's membership in a status group is always subjectively determined. Hence, status-oriented behavior illustrates value-rational action—that is, action based on some value or values held for their own sake. Thus, rather than behaving for their economic interests, status-oriented people act as members of a group with whom they share a specific style of life and level of social honor. In Weber's words, "in contrast to classes, *Stände* (status groups) are normally groups. They are, however, of an amorphous kind."[46] That is, they are not tightly organized. For example, corporate executives dining together during the lunch hour rather than with their blue-collar subordinates are engaged in value-rational action because they are acting in terms of their values, or ideas of honor, and they are expressing their common lifestyle. In principle, prestige or honor can be attributed by virtually any quality that is both valued and shared by an aggregate of people.

It is important to remember that from Weber's point of view, status groups in any modern society are generally coterminous to the social classes identified previously. Thus, skilled blue-collar people (say, unionized construction workers) are as much a status group as corporate executives, and the differences between the two, expressed by a lack of commensality (an unwillingness to eat with one another), suggest both the defining quality of status groups and the link between class and status: Prestige or honor results from a specific style of life expected of all those who would belong to the group. Thus, despite the fact that the two concepts refer to analytically distinct phenomena, in practice classes and status groups coalesce such that social strata are formed.

With some prescience, Weber noted that individuals developed styles of living as a result of their parental background and upbringing, formal education, and occupational experiences, factors subsequently shown as fundamental to the process of status attainment.[47] On these bases people at all levels tend to associate with others whom they perceive as having similar lifestyles, and they frequently try to prevent the entry of outsiders, those seen as having different values, into the group. For despite their amorphous qualities, the members of status groups are both aware of their situation and active in maintaining it. The mechanism by which this is accomplished is based on a subjective judgment, but it is consciously used and powerful in its consequences: social discrimination. Reinhard Bendix describes the significance that Weber saw in this mechanism. Although he expresses himself in what is now considered sexist language, Bendix intends to refer to both genders:

> Status groups are rooted in family experience. Before the individual reaches maturity, he has participated in his family's claim to social prestige, its occupational subculture and educational level. Even in the absence of concerted action, families share a style of life and similar attitudes. Classes without organization achieve nothing. But families in the

same status situation need not communicate and organize in order to discriminate against those people they consider inferior.[48]

Essentially, Weber argued, status "always rests on distance and exclusiveness," in the sense that members of a status group actively express and protect their lifestyles in a number of specific ways: (1) People extend hospitality only to social equals. Thus, they tend to invite into their homes, become friends with, eat with, and socialize with others who are like themselves in that they share similar lifestyles. (2) People restrict potential marriage partners to social equals (this practice is called connubium). Thus, they tend to live in areas and send their children to school with the children of others who are like themselves, with the result that their offspring generally marry others with similar values and ways of living. (3) People practice unique social conventions and activities. Thus, they tend to join organizations, such as churches and clubs, and spend their leisure time with others who share similar beliefs and lifestyles. And (4) people try to monopolize "privileged modes of acquisition," such as their property or occupations.[49]

This last tactic is important, for those in common status positions act politically to close off social and economic opportunities to outsiders to protect their capital or occupational investments. For example, because particular skills (say, in doctoring or carpentry) acquired over time necessarily limit the possibility for acquiring other skills, competing individuals "become interested in curbing competition" and preventing the free operation of the market. So they join together and, despite continued competition among themselves, attempt to close off opportunities for outsiders by influencing the creation and administration of law. Such attempts at occupational closure are ever-recurring at all stratum levels, and they are "the source of property in land as well as of all guild [or union] monopolies."[50]

Hence, "privileged modes of acquisition" are retained, people's lifestyles are protected, and the system of stratification is maintained. It should be noted again how status and class considerations coalesce.

As a final point in the analysis of status, Weber argued that the attempt by members of status groups to discriminate against others to protect their style of life could have extreme consequences, for segregation based on status considerations can develop into castes, rather than strata, in which positions are closed by legal and religious sanctions. However, he believed that caste distinctions usually occurred only when based on underlying ethnic or racial differences, as in the United States. He emphasized, however, that patterns of ethnic segregation did not inevitably, or even normally, produce caste relations. The latter are always dependent on unique historical events.[51]

Weber's Model of the Class Structure

As described in Chapter 9, Marx posited that modern societies displayed a basic division between capitalists, those who own income-producing property, and proletarians, those who are forced to sell their labor power to survive. In his words, "society as a whole is splitting up more and more into two great hostile camps, into two great classes facing each other: Bourgeoisie and Proletarian."[52]

Marx knew that this assertion was an exaggeration. He meant that historical evolution placed these two groups at the center of a class struggle, which would inevitably produce a communist society. But when examining actual historical events, Marx often looked at specific segments of society, such as bankers or the "lower middle class" or the *lumpenproletariat* (the very poor), analyzing their different experiences and interests with great insight.[53] He did this, however, on an ad hoc basis. He did not develop a more systematic, abstract model (or map) of the class structure. Weber did. Marx's and Weber's models are compared in Table 11.1 (note that we are now using the terms *class* and *stratum* as synonyms, which has become standard in sociology). Although Weber's argument that people's behavior at each class or stratum level is based sometimes on their economic interests and sometimes on their values is exactly right, his terminology has not been adopted.

Table 11.1 shows that both Marx and Weber regarded the differences between those who own capital and those who do not as fundamental divisions in the class structure. The dashed line separating the different sections of each model is designed to suggest a semipermeable boundary that is very difficult, but not impossible, to cross.

The table also shows that Weber's map of the class structure is much more detailed than Marx's. Both those who own property and those who do not can be separated into classes (or strata) in a way that is useful to observers and subjectively meaningful to ordinary people. Thus, those capitalists who do not lead acquisitive lives are rentiers, and those who do are entrepreneurs. Sometimes the members of these two strata act together to preserve and protect their source of income, their property, but sometimes they do not, mainly because they have different values. Weber's distinction alerts observers to the need for specifying the conditions under which each occurs.

Similarly, middle-class people see themselves as different from those who are skilled workers. As a result, they tend to live in different neighborhoods, make different friends, attend different schools, and participate in different leisure activities. Current research reveals that a semipermeable boundary separates middle-class people (white-collar workers) from working-class people (blue-collar workers) at all skill levels.[54]

Although Weber could not prove this fact, because the necessary research had not been done, his model alerts observers to be sensitive to the different lifestyles displayed by members of those social strata who do not own property.

Weber's Model of Social Change

Weber's analysis of systems of domination, especially his distinction among the three types of authority, implies a model of social change. Like Marx, Weber saw the source of change as endogenous (or internal) to the society. In effect,

Table 11.1 Marx's and Weber's Models of the Class Structure

Marx's Model	Weber's Model
1. Capitalists	1. Propertied
2. Proletarians	a. Renters
	b. Entrepreneurs
	2. Nonpropertied
	a. Middle classes
	b. Skilled workers
	c. Semiskilled workers
	d. Unskilled workers

Weber posited—implicitly, to be sure—that crises associated with a disjunction among cultural values, patterns of action, and psychological orientations produce processes of social change in every society. Figure 11.4 depicts Weber's vision of the internal system dynamics that occur as societies try to resolve the dilemmas they face.

Weber argued that the struggle for power was a continuous process in every society. For example, in traditional societies rulers attempt to enlarge their areas of discretion and power at the expense of administrative officials and notables. They do so by various methods—for example, the use of military force, economic coercion, co-optation, and political alliances. In all cases, they justify their action as customs handed down over generations. Conversely, officials and notables attempt to increase their power at the expense of the rulers, and they use many of the same tactics. Although this pattern can remain stable for many years, the gradual development of structural disjunctions—as signified by incongruity among values and beliefs, patterns of action, and psychological orientations—is common. As indicated in Figure 11.4, charismatic leaders are most likely to emerge and attempt to change the course of history during these crises emerging from such malintegration. If such leaders are ineffective, the crises can continue and even escalate, with the result that leaders will eventually lose their following as the population turns to some other charismatic figure for solutions. If the charismatic leaders are effective in resolving crises, however, the process of routinization occurs as their authority is implemented over time. The inevitable result is the transformation of charismatic authority into either a traditional or rational-legal form of domination. In this context, Weber argued, the development of rational-legal authority makes a recurrence of traditional domination less likely, as was emphasized in Figure 11.2. It is more likely that social systems characterized by rational-legal forms of authority will resolve integrative crises in such a way that patterns of social action reflect either instrumental- or value-rational orientations. However, the choice between these two alternatives is not inevitable because

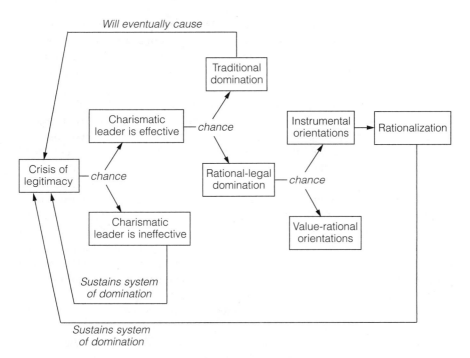

FIGURE 11.4 Weber's View of Chance and Social Change

Weber emphasized the importance of chance in human history. Hence, Figure 11.4 builds in the possibility (even if unlikely) that a society can resolve integrative problems by returning to a form of traditional authority.

Weber's Model of Stratification and Geopolitics

Weber was a sociologist of power, seeking to explain how power is used in systems of domination. In particular, Weber was interested in the development of the state as an organizational order that could be used to administer power. As we have just explored, Weber saw state power as operating to sustain a stratification order, with its built-in tendencies for opposition movements led by a charismatic leader. Weber also proposed a geopolitical theory of power, seeing both the legitimacy of political authority and the potential for the emergence of charismatic leaders as related to the relations of a society with other societies. In general, Weber argued that there is often a competition for prestige among states, with those that are successful in war and economic competition enjoying more prestige than do those that are not so successful.[55] Moreover, prestige in external relations with other states increases the legitimacy bestowed on political authority by the masses within the society. Thus, the administration of power, the degree of legitimacy given to those with power, and the potential for conflict are often tied to external, geopolitical events outside a society's borders.

Weber saw these dynamics as driven by several interrelated factors. The first was the size of the state, or the scale of the administrative structures used to exercise power. For the state to grow, production in the economy must be sufficient to create the surplus necessary to support specialized administrative personnel.

Another critical relation is between key economic actors and the state. When actors in the economy depend on the state for their right to engage in particular kinds of economic activity, as is the case with a chartered corporation or a company given a monopoly by the state, these economic actors will place pressure on the state to engage in external conflict if their interests are tied to success in the external system. For example, chartered corporations in America before the Revolutionary War gave economic actors in England a strong incentive to have the English government wage war or use coercion to protect their interests. When, however, the dependence of economic actors on the state is low but they still have interests in the external system, these economic actors are more likely, Weber believed, to exert pressures on the state to engage on co-optive strategies, such as trade agreements, rather than war. The success of the state in either war-making or deal-making in trade determines the prestige of society and its ruling elites not only vis-à-vis other states, but also in relation to the masses within a society. For example, Japan's prestige in the world economic system has been very high since World War II because of its success in achieving favorable trade relations with other countries; this prestige has, in turn, given the dominant political party prestige and legitimacy (at least until the recent economic downturn in Japan). Similarly, leaders enjoyed considerable prestige within Japan during the early phases of World War II when they had military success, but this prestige began to decline with each successive loss in the Pacific and in Asia.

Another force entering this basic relationship between prestige in the world system and the legitimacy of its ruling elites is the level of inequality. Weber recognized that when high levels of inequality exist and when memberships in classes, parties, and status groups are highly correlated (that is, members in upper classes are also members of ruling parties and high prestige status groups, and vice versa), the potential for the emergence of charismatic leadership increases. Thus, a state that engages in diplomatic or military adventurism in relation with other states is more vulnerable if political power has been used in ways to increase the level of inequality within the society. Success in external relations will stave off the emergence of a charismatic leader, but if the state should lose prestige in external relations, then the conflict potential inhering in the stratification system increases dramatically, especially if charismatic leaders can emerge to take advantage of the state's loss of prestige.[56]

Thus, the dynamics of domination are very much tied to geopolitics, and in these, Weber can be considered an early world systems theorist. He saw clearly the connection between legitimation of a stratification order and the state, on one side, and the geopolitical position of the society in relation to its neighbors, on the other side.

WEBER ON CAPITALISM
AND RATIONALIZATION

As we have emphasized, Weber was very much concerned with the process of rationalization. Why had rational-legal domination spread? Part of the answer can be found in Weber's famous analysis of religion, where he argues that a change in religious beliefs (toward Protestantism) was the critical force in tipping western European societies towards capitalism. We will examine this famous and controversial thesis in the next section, but before exploring this thesis, we address what Weber saw as a fundamental relationship between the use of money in free markets and the rationalization of orders in the political and economic arenas.[57] Weber recognized that market forces were an important precondition for the emergence of capitalism; moreover, once in place, they dramatically accelerated the process of rationalization.

Weber argued that when money is introduced into exchanges, it becomes possible to engage in more precise and efficient calculations of value. That is, the worth of a good or commodity is more readily ascertained with money as a common measure of value. The use of money to mediate transactions and social relationships had slowly expanded in human history, primarily as the result of (1) the expansion of markets where money would greatly facilitate transactions, and (2) the growth of the state where liquid revenues could be taxed to expand state power. Money is a generalized medium that can be used to purchase any good or commodity, and so it greatly accelerated the ease of market transactions over barter (where one commodity is exchanged for another). As a result, it gave the state a means to purchase labor and other resources necessary to wield power. Moreover, Weber argued that rational-legal bureaucracy, whether the state's or that of economic actors, would not be possible without free labor willing to sell its services in a labor market for money; thus rationalization depended heavily on the emergence of free markets using money.

Money also facilitates the extension of credit because a debt can be expressed with one measure—the value of money—and the interest rate can similarly be calculated. With credit, economic activity can expand, as can the activities of the state that, like any other actor, can enter credit markets. The use of credit further extends the calculability of utilities, and in so doing, money and credit rationalize market transactions. As this transition occurs, Weber believed, tradition, patronage, and other ways of regulating markets would decline. In their place come rational calculations of price, payments, debts, and interest. As these other ways of organizing economic activity and exchange are pushed aside, productive units become more rational and begin to calculate their costs and profits against the yardstick of money, credit, interest-notes, and market forces. Similarly, the relationship between labor and its employers shifts to one based on rational calculations rather than on patrimony or some other nonrational mechanism for organizing the work force.

Once this level of rationalization has occurred, Weber felt, it will feed on itself, constantly expanding the rationality of the state. As the state becomes dependent on market-driven productivity to finance its operations, it will introduce legal rules, and enforce these rules, to ensure that contracts and agreements are honored. Thus, the rationality of the market becomes rational-legal domination by the state as ever-more rational actors use the law rather than tradition, affect, religion, and other non-rational bases of regulation to organize their affairs. Indeed, one of the most important bodies of law is the tax code, which specifies how, and how much, revenue the state can take from other actors to finance its operations. As the state depends more on this monetary source of revenue, coupled with its access to credit markets, it enacts more laws and expands its administrative structure to ensure a constant flow of revenue. As a result, the bureaucratized state becomes more instrumentally rational as it seeks ways to secure money (e.g., taxes, credit) to expand its operations and, hence, domination.

Similarly, once economic actors and markets are funded by money and credit, these allow markets and production to expand, which further extends the use of money and credit. When the state begins to support markets through law, the use of money and credit in free markets can increase the scale of production and market distribution even more. Once a certain level of rationalization exists in the economy, organizations regulated by law and driven by instrumental rationality (for profits) come to dominate.

Thus, Weber saw a dynamism in capitalism, much as Marx had, but with a very different prediction: the disenchantment of the social world by rational-legal domination in the economic and political arenas. Rather than sowing the seeds of its own destruction, as Marx had argued, Weber felt that rationalization had planted the seeds for its own perpetuation. The capacity to oppose this monolith would be lessened, and ever-more aspects of social life could be calculable, rational, efficient, and dull. Weber thus came to a very different conclusion than Marx about the liberating potential of markets and capitalism. Humans would, in Weber's words, be locked in "the iron cage" of bureaucracy and rational-legal authority.

Weber also disagreed that this rational-legal juggernaut had been inevitable. He did not see history as marching inexorably anywhere, and certainly not to a liberating utopia. Rather, history was often random, moving in response to chance confluence of events, and this had been the case for the rise of capitalism and the spread of rational-legal authority. Why had history made this turn toward increasing rationalization? For Weber, the answer resided in the emergence of Protestantism.

Weber had long been interested in religion, for its own sake, and particularly for an understanding of how it worked to legitimate orders and systems of domination, but he had also been interested in religion for another reason: to understand the rise of capitalism. Perhaps his most famous work, certainly outside of sociology, is the *Protestant Ethnic and the Spirit of Capitalism*. In this controversial work, Weber argued that the key chance event turning the historical tide toward rational-legal forms of domination had been the rise of

Protestantism in the West. This controversial work was only part of a more comprehensive study of religion, but even here, Weber was interested in explaining how religion in other, equally developed societies of the East had inhibited the rise of capitalism. Thus, one of the most important reasons that Weber studied religion was to discover the critical causal event that had unleashed capitalism and rationalization.

WEBER'S STUDY OF RELIGION

The Protestant Ethic and the Spirit of Capitalism is Weber's most famous and prob-ably most important study.[58] Published in two parts, in 1904 and 1905, it was one of his first works to be translated into English, and even more significantly, it was the first application of his mature methodological orientation. As a result, the *Protestant Ethic* is neither a historical analysis nor a politically com-mitted interpretation of history. Rather, it is part of an exercise in historical hypothesis testing in which Weber constructed a logical experiment using ideal types as conceptual tools.

In retrospect, we can see that Weber had three interrelated purposes in writing the *Protestant Ethic*. First, he wanted to refute those forms of Marxist analysis prevalent at the turn of the century. Second, he wanted to explain why the culture of capitalism had emerged in the West and its significance for the development of modern rationalized capitalist economies. Thus, his work is not only an explanation of how the modern world, dominated by instrumen-tal and value rationality, came into being but also a demonstration that an objective sociology can deal with historical topics. Third, he also wanted to demonstrate that cultural values circumscribe social action, primarily by directing people's interests in certain directions.

Although the *Protestant Ethic* is the most well-known of Weber's studies in the sociology of religion, it is nonetheless only a small portion of a much larger intellectual enterprise that he pursued intermittently for about fifteen years. In this grandiose "logical experiment," Weber tried not only to account for the confluence of events that were associated with the rise of modern capitalism in the West but also to explain why capitalism was not likely to have developed in any other section of the world—that is, "why did not the scientific, the artistic, the political, and the economic development [of China, India, and other areas] enter upon that path of rationalization which is peculiar to the occident?"[59] Thus, the *Protestant Ethic* is the first portion of a two-stage quasi-experiment. The second element in Weber's experiment is contained in a series of book-length studies on *The Religion of China* (1913), *The Religion of India* (1916–1917), and *Ancient Judaism* (1917).[60]

Before reviewing either the *Protestant Ethic* or Weber's other work on reli-gion, we should make explicit the implicit research design of these works. In so doing, we will describe another sense in which Weber constructed "quasi-experimental designs" for understanding the causes of historical events. Weber's basic question is: Why did modern capitalism initially occur in the West and not in other parts of the world? To isolate the cause of this monumental historical change, Weber constructed the "quasi-experimental design" diagramed in Figure 11.5.

In Figure 11.5, steps 1–5 approximate the stages of a laboratory situation as it must be adapted to historical analysis, which is why we label it a logical experiment. The West represents the experimental group in that something stimulated capitalism, whereas China and India represent the control groups because their economic systems did not change, even though they were as advanced as the West in technologies and other social forms. The stimulus that caused capitalism in the West was the religious beliefs associated with Protestantism (see step 3 of Figure 11.5), and Weber wrote *The Protestant Ethic and the Spirit of Capitalism* for this reason.

The Protestant Ethic and the Spirit of Capitalism

Weber opened the *Protestant Ethic* with what was a commonplace observation at the end of the nineteenth century: Occupational statistics in those nations of mixed religious composition seemed to show that those in higher socioeconomic positions were overwhelmingly Protestant. This relationship appeared especially true, Weber wrote, "wherever capitalism . . . has had a free hand."[61] Many observers in economics, literature, and history had commented on this phenomenon before Weber, and he cited a number of them.[62] Hence, in the *Protestant Ethic* Weber was not trying to prove that a relationship between Protestantism and economic success in capitalist societies existed because he took its existence as given. In his words, "it is not new that the existence of this relationship is maintained. . . . Our task here is to explain the relation."[63]

To show that Protestantism was related to the origin of the spirit of capitalism in the West, Weber began with a sketch of what he meant by the latter term. Like many of his key concepts, the "spirit of capitalism" is a historical ideal type in that it conceptually accentuates certain aspects of the real world as a tool for understanding historical processes.[64] Although he did not state what he meant very clearly, an omission that helped contribute to the tremendous controversy over the *Protestant Ethic*'s thesis, his concept of the spirit of capitalism appears to have the following components:[65]

First, work is valued as an end in itself. Weber was fascinated by the fact that a person's "duty in a calling [or occupation] is what is most characteristic of the social ethic of capitalistic culture, and is in a sense the fundamental basis of it."

Second, trade and profit are taken not only as evidence of occupational success but also as indicators of personal virtue. In Weber's words, "the earning of money within the modern economic order is, so long as it is done legally, [seen as] the result and the expression of virtue and proficiency in a calling."

Third, a methodically organized life governed by reason is valued not only as a means to a long-term goal, economic success, but also as an inherently proper and even righteous state of being.

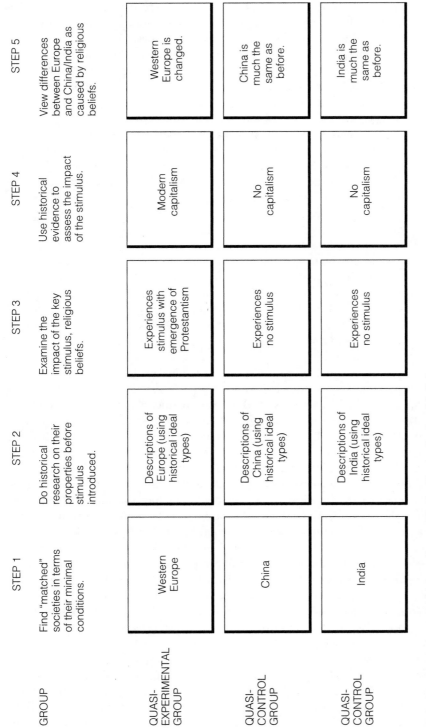

FIGURE 11.5 Weber's Quasi-Experimental Design in the Study of Religion

GROUP	STEP 1 Find "matched" societies in terms of their minimal conditions.	STEP 2 Do historical research on their properties before stimulus introduced.	STEP 3 Examine the impact of the key stimulus, religious beliefs.	STEP 4 Use historical evidence to assess the impact of the stimulus.	STEP 5 View differences between Europe and China/India as caused by religious beliefs.
QUASI-EXPERIMENTAL GROUP	Western Europe	Descriptions of Europe (using historical ideal types)	Experiences stimulus with emergence of Protestantism	Modern capitalism	Western Europe is changed.
QUASI-CONTROL GROUP	China	Descriptions of China (using historical ideal types)	Experiences no stimulus	No capitalism	China is much the same as before.
QUASI-CONTROL GROUP	India	Descriptions of India (using historical ideal types)	Experiences no stimulus	No capitalism	India is much the same as before.

Fourth, embodied in the righteous pursuit of economic success is a belief that immediate happiness and pleasure should be forgone in favor of future satisfaction. As Weber noted, "the *summum bonum* of this ethic, the earning of more and more money, combined with the strict avoidance of all spontaneous enjoyment of life, is above all completely devoid of an eudaemonistic, not to say hedonistic, admixture." In sum, then, these values—the goodness of work, success as personal rectitude, the use of reason to guide one's life, and delayed gratification—reflect some of the most important cultural values in the West because they constitute perceptions of appropriate behavior that are shared by all.

Weber emphasized, however, that the widespread application of such values to everyday life was historically unique and of relatively recent origin. He believed that the greatest barrier to the rise of the spirit of capitalism in the West was the inertial force of traditional values. To varying degrees, European societies before the seventeenth century were dominated by "traditional modes of action." For example, religion rather than science was used as the primary means of verifying knowledge. Bureaucracies composed of technically trained experts were unknown. Patterns of commerce and most other forms of daily life were dominated by status rather than class considerations; that is, people's economic actions were dictated by their membership in religious groups rather than by market factors. Finally, legal adjudication did not involve the equal application of law to all individuals. In short, the choice between instrumentally rational action and value-rational action did not exist. Rather, custom dictated behavior.

In such a context, of course, some individuals tried to make money. As Weber remarks, "capitalism existed in China, India, Babylon, in the classic world, and in the Middle Ages."[66] But it was traditional capitalism in which the ideal was simply to acquire enough money without spending too much time doing it so one could live as one was "accustomed to live." Weber described the traditional enterprises of the nobility as "adventurer capitalism," referring to investment in long-distance trade (for example, in spices or silk) from which one might obtain a windfall profit sufficient to last a lifetime or the purchase of government offices (for example, tax collector) from which one skimmed off a portion of the revenue. "Capitalistic acquisition as an adventure has been at home in all types of economic society which have known trade with the use of money."[67] Such activities, however, were not the center of these people's lives. The existence of individual adventurer capitalists, moreover, differs from a rationalized capitalist economy, based on the mass production of consumer goods, in which the entire population is oriented to making money.

In his *General Economic History,* written some years later, Weber identified the major structural changes that he believed, taken together, had caused the development of rationalized capitalist economies in Western Europe rather than elsewhere: (1) the process of industrialization through which muscle power was supplanted by new forms of energy, (2) the rise of a free labor force whose members had to work for wages or starve, (3) the increasing use of systematic accounting methods, (4) the rise of a free market unencumbered by religious restrictions, (5) the gradual imposition and legitimation of a system of calculable law, (6) the increasing commercialization of economic life through the use of stock certificates and other paper instruments, and (7) the rise of the spirit of capitalism. Although all these historical developments were to varying degrees unique to the West, in the *General Economic History* Weber still regarded the last factor as the most decisive. Thus, in attempting to understand the origin of the economic differences between Protestants and Catholics, he was not trying to deny the fundamental significance of these structural changes but to show the significance of the culture of capitalism and its unintended relationship to Protestant ethical teachings.

Weber believed that the cultural values of traditional society were destroyed by Puritanism and the other Protestant sects, although this was not the intent of those who adopted the new faiths and could not have been predicted in advance. In the *Protestant Ethic* Weber focused mainly on Calvinism, with much shorter discussions of Pietism, Methodism, and Baptism appended to the main analysis. His strategy was to describe Calvinist doctrines by quoting extensively from the writings of its various theologians, then to impute the psychological consequences those doctrines had on people who organized their lives in these terms, and finally to show how they resulted in specific (and historically new) secular values and ways of living. In Weber's words, he was interested in ascertaining "those psychological sanctions which, originating in religious belief and the practice of religion, gave a direction to practical conduct and held the individual to it."[68] In this way, he believed, he could give a powerful example of the manner in which cultural phenomena influence social action and, at the same time, rebut the vulgar Marxists who thought economic factors were the sole causal agents in historical change.

Based on an analysis of Calvinist writings, such as the *Westminster Confession of 1647,* from which he quoted extensively, Weber interpreted Calvinist doctrine as having four consequences for those who accepted its tenets:

First, because the Calvinist doctrine of Predestination led people to believe that God, for incomprehensible reasons, had divided the human population into two groups, the saved and the damned. A key problem for all individuals was to determine the group to which they belonged. Second, because people could not know with certainty whether they were saved, they inevitably felt a great inner loneliness and isolation. Third, although a change in one's relative state of grace was seen as impossible, people inevitably began to look for signs that they were among the elect. In general, Calvinists believed that two clues could be used as evidence: (1) faith, for all had an absolute duty to consider themselves chosen and to combat all doubts as temptations of the devil, and (2) intense worldly activity, for in this way the self-confidence necessary to alleviate religious doubts could be generated. Fourth, all believers were expected to lead methodical and ascetic lives unencumbered by irrational emotions, superstitions, or desires of the flesh. As Weber put it, the good Calvinist was expected to "methodically supervise his own state of grace in his own conduct, and thus to penetrate it with asceticism," with the result that each person engaged in "a rational planning of the whole of one's life in accordance with God's will."[69] The significance of this last doctrine is that in

Calvinist communities worldly asceticism was not restricted to monks and other "religious virtuosi" (to use Weber's phrase) but required of all as they conducted their everyday lives in their mundane occupations, or callings.

To show the relationship between the worldly asceticism fostered by the Protestant sects and the rise of the spirit of capitalism, Weber chose to focus on the Puritan ministers' guidelines for everyday behavior, as contained in their pastoral writings. The clergy's teachings, which were set forth in such books as Richard Baxter's *Christian Directory,* tended to reflect the major pastoral problems they encountered. As such, their writings provide an idealized version of everyday life in the Puritan communities. Although it must be recognized that patterns of social action do not always conform to cultural ideals, such values do provide a general direction for people's behaviors. Most people try, even if imperfectly, to adhere to those standards of appropriate behavior dominant in their communities, and the Puritans were no exceptions. Further, the use of this sort of data suggests the broad way in which Weber interpreted the idea of "explanatory understanding" (*verstehen*): Because people's own explanations of their actions often involve contradictory motives that are difficult to reconcile, he was perfectly willing to use an indirect means of ascertaining the subjective meaning of social action among the Puritans.

Based on Weber's analysis, it appears that the Puritan communities were dominated by three interrelated dictums, which, although a direct outgrowth of Puritan theology, eventuated over the long run in a secular culture of capitalism.

The first of these pronouncements is that God demands rational labor in a calling. As Weber noted, Puritan pastoral literature is characterized "by the continually repeated, often almost passionate preaching of hard, continuous bodily or mental labour."[70] From this point of view, there can be no relaxation, no relief from toil, for labor is an exercise in ascetic virtue, and rational, methodical behavior in a calling is taken as a sign of grace. Hence, from the Puritan's standpoint, "waste of time is . . . the first and in principle the deadliest of sins," because "every hour lost is lost to labour for the glory of God." As Weber observed, this dictum not only provides an ethical justification for the modern division of labor (in which occupational tasks are divided up efficiently) but also reserves its highest accolades for those sober, middle-class individuals who best exemplify the methodical nature of worldly asceticism. As an aside, it should be noted that members of this stratum became the primary carriers of Puritan religious beliefs precisely because they garnered immense economic, social, and political power as a result. In general, Weber emphasized that those who were able to define and sanctify standards of appropriate behavior also benefit materially.[71]

The second directive states that the enjoyment of those aspects of social life that do not have clear religious value is forbidden. Thus, from the point of view of the Puritan ministers, secular literature, the theater, and nearly all other forms of leisure-time activity were at best irrelevant and at worst superstitious. As a result, they tried to inculcate in their parishioners an extraordinarily serious approach to life, for people should direct their attention toward the practical problems dominating everyday life and subject them to rational solutions.

The third guideline specifies that people have a duty to use their possessions for socially beneficial purposes that redound to the glory of God. Thus, the pursuit of wealth for its own sake was regarded as sinful, for it could lead to enjoyment, idleness, and "temptations of the flesh." From this point of view, those who acquire wealth through God's grace and hard work are mere trustees who have an obligation to use it responsibly.

Even allowing for the usual amount of human imperfection, Weber argued, the accumulation of capital and the rise of the modern bourgeoisie were the inevitable results of whole communities sharing the values dictating hard work, limited enjoyment and consumption, and the practical use of money. Hence, modern capitalism emerged. Weber's causal argument is diagramed in Figure 11.6. Over time, of course, Puritan ideals gave way under the secularizing influence of wealth because people began to enjoy their material possessions. Thus, although the religious roots of the spirit of capitalism died out, Puritanism bequeathed to modern people "an amazingly good, we may even say a pharisaically good, conscience in the acquisition of money."[72]

The predominance of such a value in Western societies is historically unique. It led Weber to some rather pessimistic observations in the concluding paragraphs of the *Protestant Ethic*. He believed that the rise of modern capitalism reflected the process of rationalization we referred to earlier. People are now taught to lead methodical lives, using reason buttressed by knowledge to achieve their goals. As he put it, "the idea of duty in one's calling prowls about in our lives like the ghost of dead religious beliefs," with the result that whereas "the Puritan wanted to work in a calling; we are forced to do so." In effect, the culture of capitalism, combined with capitalist social and economic institutions, places people in an "iron cage" from which there appears to be no escape and for which there is no longer a religious justification. This recognition leads to Weber's last, sad lament: "specialists without spirit, sensualists without heart; this nullity imagines that it has attained a level of civilization never before achieved."[73]

This conclusion remains controversial. In considering it, think again about Weber's procedure in conducting his "logical experiment": He began with a research finding (the relationship between Protestantism and business success in Germany at the turn of the century) and attempted to account for it. This strategy reverses that normally taught in courses on quantitative methodology, in which observers (presumably) form hypotheses and then seek data to explain them. In contrast, Weber's strategy is to offer an explanatory hypothesis in which the variables are logically related to one another and the available evidence supports the interpretation. The task is to phrase the explanation and describe the data as clearly as possible and then allow what Seymour Martin Lipset subsequently called the "method of dialogue" to lead to the growth of knowledge.[74] This procedure is tricky, especially with controversial topics. What happens is that observers think about issues, publish their data and interpretations, debate ensues, and—over the long-run—greater understanding results.[75] Such a process implies that sociological explanations are always probabilistic, both substantively and statistically.[76] This means that absolute certainty is impossible, in sociology as in other arenas of life.

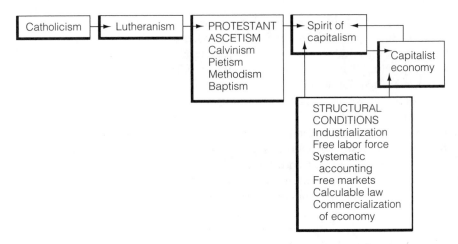

FIGURE 11.6 Weber's Casual Argument for the Emergence of Capitalism

Weber's Comparative Studies
of Religion and Capitalism

During the years following publication of the *Protestant Ethic,* Weber continued his analysis of the relationship between religious belief and social structure to show why it was not very likely that capitalism as an economic system could have emerged anywhere else in the world. Weber's most important works in this regard are *The Religion of China* and *The Religion of India.* They represent the extension of the "logical experiment" begun some years earlier.

Both of these extended essays are similar in format in that Weber began by assessing those characteristics of Chinese and Indian social structure that either inhibited or, under the right circumstances, could have contributed to the development of capitalism in that part of the world (see Figure 11.6 for Weber's implicit experimental design). For purposes of illustration, the examples used here come from *The Religion of China.* Thus, in China during the period when capitalism arose in the West, a number of structural factors existed that could have led to a similar development in the Orient. First, there was a great deal of internal commerce and trade with other nations. Second, because of the establishment and maintenance (for more than 1,200 years) of nationwide competitive examinations, there was an unusual degree of equality of opportunity in the process of status attainment. Third, the society was generally stable and peaceful, although Weber was clearly too accepting of the myth of the "unchanging China." Fourth, China had many large urban centers, and geographical mobility was a relatively common occurrence. Fifth, there were relatively few formal restrictions on economic activity. Finally, a number of technological developments in China were more advanced than those in Europe at the same time (the use of gunpowder, knowledge of astronomy, book printing, and so forth). As Weber noted, all these structural factors could have aided in the development of a Chinese version of modern capitalism.

He emphasized, however, that Chinese social structure also displayed a number of characteristics that had clearly inhibited the widespread development of any form of capitalism in that part of the world. First, although the Chinese possessed an abundance of precious metals, especially silver, an adequate monetary system had never developed. Second, because of the early unification and centralization of the Chinese empire, cities never became autonomous political units. As a result, the development of local capitalistic enterprises was inhibited. Third, Chinese society was characterized by the use of "substantive ethical law" rather than calculable legal procedure. As a result, legal judgments were made considering the particular characteristics of the participants and sacred tradition rather than by the equal imposition of common standards. Finally, the Chinese bureaucracy comprised classically learned persons rather than technically trained experts. Thus, the examinations regulating status attainment "tested whether the candidate's mind was thoroughly steeped in literature and whether or not he possessed the ways of thought suitable to a cultural man."[77] Hence, the idea of the trained expert was foreign to the Chinese experience.

In sum, according to Weber, all these characteristics of Chinese social structure inhibited the development of an oriental form of modern capitalism. Nonetheless, the examples noted do suggest that such a development remained possible. Yet Weber argued that the rise of capitalism as an economic system was quite unlikely in either China or India, for he found no evidence of patterns of religious beliefs that could be compatible with the spirit of capitalism. Moreover, he believed that without the transformative power of religion, the rise of new cultural values was not likely. In China before this century, the religion of the dominant classes, the bureaucrats, was Confucianism. Weber characterized Confucianism by noting that it had no concept of sin but only of faults resulting from deficient education. Further, Confucianism had no metaphysic, and thus no concern with the origin of the world or with the possibility of an afterlife and hence no tension between sacred and secular law. According to Weber, Confucianism was a rational religion concerned with events in this world but with a peculiarly individualistic emphasis. Good Confucians were less interested in the state of society than with their own propriety, as indicated by their development as educated persons and by their pious relations with others (especially their parents). In Weber's words, the educated Chinese person "controls all his activities, physical gestures, and movements as well, with politeness and with grace in accordance with the status mores and the commands of 'propriety.'"[78] Rather than seeking salvation in the next world, the Confucian accepted this world as given and merely desired to behave prudently.

On this basis, Weber asserted that Confucianism was not very likely to result in an Asian form of the culture of capitalism. He made a similar argument about Hinduism four years later in *The Religion of India*. Thus, by means of these comparative studies, Weber tried to show logically not only why Protestantism was associated with the rise of the culture of capitalism in the West but also why other religions could not have stimulated similar developments in other parts of the world.[79]

In sum, then, beginning from what he thought was a relatively simple empirical fact—income differences between Protestants and Catholics in nineteenth-century Europe—Weber asserted that even after taking into account the unique occurrence of a number of important structural phenomena (free labor, rational law, political independence of cities, early industrialization, and so on), the most important factor accounting for the rise of capitalism as an economic system was the spread of new cultural values, which he called the spirit of capitalism. These values originated as the inadvertent consequence of Puritan religious beliefs, which over the long run produced economic behavior (among both believers and nonbelievers) that was uniquely compatible with other historical developments occurring during that period. Over time, of course, the religious origin of modern values disappeared from view, leaving a secular legacy of beliefs in the innate goodness of work, acquisitiveness, the methodical organization of one's life, and delayed gratification. Weber believed that without the influence of Puritan religious beliefs, however, modern society would be fundamentally different than it is today. In his words, "it was the power of religious influence, not alone, but more than anything else, which created the differences of which we are conscious today."[80]

Weber's Outline of the Social System

In all his work, Weber employed, at least implicitly, a vision of society as a social system that consists of three analytically separable dimensions: (1) culture, (2) social structure (patterns of social action), and (3) psychological orientations (see Figure 11.7). Cultural values and beliefs, patterned ways of acting in the world, and the psychological states are all reciprocally related.[81] *The Protestant Ethic and the Spirit of Capitalism* provides one illustration of this model. The components of the spirit of capitalism include an emphasis on work as an end in itself, an emphasis on the legitimacy of acquisition and profit, an emphasis on living a methodical lifestyle to obtain economic success, and a belief that immediate happiness and pleasure should be forgone in expectation of greater satisfaction in the future. Stripped of their religious connection, all these phenomena have become fundamental Western cultural values; that is, they embody shared beliefs about right and wrong, appropriate and inappropriate behavior. The *Protestant Ethic* is, therefore, a book about culture and seeks the origin of the values dominating modern life. Weber's answer is that these values are the secular result of certain peculiar religious movements that began in the sixteenth century.

But the book is about more than cultural phenomena. Indeed, Weber was no more an idealist than Marx was. Because the Puritans' religious beliefs plainly did not become dominant cultural values by themselves, Weber had to consider the extent to which both social structure and people's psychological orientations were reciprocally related to the cultural values he was analyzing. Thus, he argued that the religious beliefs characteristic of the new faiths fundamentally influenced patterns of social action among people—not only among the Puritans but also among those who encountered the underlying

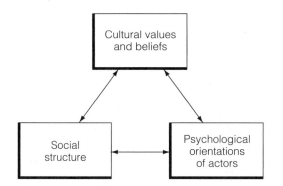

FIGURE 11.7 Weber's Model of the Social System

beliefs of these faiths. Given the rise of industrialization along with several other structural changes, none of which Weber stated explicitly in the *Protestant Ethic* but of which he was clearly aware, Puritan values spread in part because people found them to be congenial to their own secular ambitions as they developed during a time of great change. Hence, as people adopted new beliefs and values, they altered their daily lives; conversely, as individuals began living in new ways (often because they were forced to), they changed their fundamental beliefs and values. Thus, historically new patterns of social action reinforced the new values that had arisen.

Weber, however, concentrated on the Puritans themselves, describing the psychological consequences that their beliefs must have had for their daily lives. Because of their uncertainty and isolation, people looked for signs of their salvation and found them in the ability to work hard and maintain their faith. Hence, the Puritans' psychological needs led them to historically new and unique patterns of social action characterized not only by hard work (medieval peasants certainly worked as hard) but also by a methodical pursuit of worldly goods. The result was the secularization of Puritan religious values and their transformation into what he called the spirit of capitalism. Hence, it is possible to extrapolate from Weber's analysis a set of generic factors—culture, social structure, and psychological orientations—that are taken today as the fundamental features of social organization.

Weber's implicit model of the most general components of social organization has proved to be of tremendous significance in the development of sociology. For example, virtually all introductory sociology textbooks (which we will use here as a rough indicator of the state of the field) now contain a series of chapters, usually located at the beginning, titled something like "Culture," "Social Structure" or "Society," and "Personality and Socialization." The reason for this practice is that these topics provide an essential conceptual orientation to the discipline of sociology. Weber's model, then, serves as a heuristic device rather than as a dynamic analysis of the process of interaction. It can, however, lead to such an analysis—which is all Weber intended. Thus, the model is an essential first step in constructing a set of concepts that would be useful in describing and, hence, understanding modern societies.

CRITICAL CONCLUSIONS

Max Weber might be the most important of the early classical figures in sociology. We have not been able to capture the breadth and depth of his analyses that are, all at once, historical, sociological, methodological, and theoretical. Rather, we have concentrated on the key works that have generated the most attention in contemporary sociology. These works can be seen in the context of Weber's interests in the process of rationalization—that is, why and how rational-legal domination came about. Weber's most famous ideal types revolve around this project, and his most important substantive works on stratification, bureaucracy, law, religion, and change are best interpreted as parts of his analysis of the rationalization of the West and beyond.

Despite Weber's eminence, he can be criticized on many fronts. One is methodology. Weber did not believe that timeless, universal laws could ever be developed because so much that had occurred in history was the result of chance events. Even without the ability to conduct inquiry in the same way as in the natural sciences, however, Weber still wanted to be scientific and objective. Moreover, he sought to do more than write historical descriptions; he also wanted to provide a methodology for more abstract and analytical statements. The result was a strange compromise: the study of historical causes through the vehicle of abstract ideal types. For each cause and effect, Weber constructs an ideal type of its essential characteristics, and some of these are among sociology's more enduring descriptions of basic social forms. In this way, Weber could be more abstract and analytical than historians, but he would not posit laws of human organization. Too much of history, he believed, was the result of random or chance confluence of events. Thus, we are given many rather ponderous descriptions of basic relationships among phenomena portrayed as ideal types, but in following Weber's methodology, we are kept from asking the most interesting question of any science: Are some of these relationships so generic and basic that they might constitute sociological laws? For some, an answer to this question is not essential because they do not believe that the laws of the social universe can ever be uncovered—for many of the same reasons that Weber argued. Sociologists continue to read Weber, however, and they do so for more than the historical descriptions. They sense that Weber was positing some basic laws of the social universe, even though it is often difficult to dig them out of his rather thick descriptions of ideal types.

Methodology aside, Weber's substantive works range widely. We have tried to give them more coherence than they actually reveal by emphasizing the theme of rationalization. Much like Herbert Spencer's work, which also is highly descriptive, Weber's presentation of details is often so convoluted that the main line of his argument becomes difficult to follow. We suspect that Weber came to the theme of rationalization rather late in his work, and he either tried to push and shove earlier works into this theme or, in many cases, did not bother. Although the result was fascinating, though dense, essays on many diverse topics, these do not hang together as did those of Herbert Spencer, Karl Marx, and Émile Durkheim. As a consequence, it is hard to extract a general theory from Weber; rather, what emerges are rich descriptions, ideal types of empirical cases, and complex causal statements on many topics without a general model to guide us. Thus, we will simply have to live with the scattered character of Weberian sociology, taking from it what we find useful.

In the next chapter, we try to generate some general models and laws from Weber's analysis. To do so requires a great deal of inference because Weber never intended to present a general theory. As will become evident, however, it is possible to pull from Weber's writing some very interesting complex causal models and theoretical propositions. We can only criticize Weber for his reluctance to do so on his own, but because he did not believe in a law-driven natural science of society, we cannot be too critical.

NOTES

1. Max Weber, *Economy and Society,* trans. and ed. Guenther Roth and Claus Wittich (New York: Bedminster, 1968).

2. Wolfgang Mommsen, *The Age of Bureaucracy: Perspectives on the Political Sociology of Max Weber* (New York: Harper Torchbooks, 1974), p. 2.

3. Max Weber, "'Objectivity' in Social Science and Social Policy," in *The Methodology of the Social Sciences,* trans. Edward A. Shils and Henry A. Finch (New York: Free Press, 1949), p. 51.

4. Weber, "'Objectivity,'" p. 143.

5. In general, sociologists have followed Weber's lead in making the distinction between "what is" and "ought to be." Nonetheless, some critics of positivism argue that Weber's methodological goal of objective knowledge is not attainable, a position with which we disagree. For a discussion of these issues as they relate to Weber's work, see Alan Scott, "Value Freedom and Intellectual Autonomy," *History of the Human Sciences* 8 (1995), pp. 69–88.

6. Weber used the concept of rationalization in a number of different ways and, hence, subsequent scholars differ

in how to interpret it. See, for example, Randall Collins, *Max Weber: A Skeleton Key* (Beverly Hills, CA: Sage, 1986); Stephen Kahlberg, "Max Weber's Types of Rationality: Cornerstones for the Analysis of Rationalization Processes in History," *American Journal of Sociology* 85 (1980), pp. 1145–1179; John Patrick Diggins, *Max Weber: Politics and the Spirit of Tragedy* (New York: Basic Books, 1996).

7. Weber, "'Objectivity,'" p. 53.

8. Weber, *Economy and Society,* p. 4.

9. Weber, "'Objectivity,'" p. 80.

10. Mommsen, *The Age of Bureaucracy,* p. 5.

11. Weber, *Economy and Society,* p. 4.

12. See Theodore Abel, "The Operation Called Verstehen," *American Journal of Sociology* 54 (1948), pp. 211–218; Peter A. Munch, "Empirical Science and Max Weber's *Verstehen Sociologie,*" *American Sociological Review* 22 (1957), pp. 26–32; Murray L. Wax, "On Misunderstanding Verstehen: A Reply to Abel," *Sociology and Social Research* 51 (1967), pp. 322–333; and Theodore Abel, "A Reply to Professor Wax," *Sociology and Social*

Research 51 (1967), pp. 334–336. A good recent summary is in John Patrick Diggins, *Max Weber,* pp. 114–122.

13. See Thomas Burger, *Max Weber's Theory of Concept Formation: History, Laws, and Ideal Types* (Durham, NC: Duke University Press, 1976), pp. 130–134.

14. Weber, "'Objectivity,'" p. 93 (emphasis in original).

15. Weber, *Economy and Society,* p. 6.

16. As an aside, it should be recognized that *Economy and Society* was left in a highly disorganized state at Weber's death in 1920, and what he intended to do with the fragments that were eventually placed together under that title is not altogether clear. Part 1 is actually the last section he wrote, apparently between 1918 and 1920, whereas part 2 appears to have been written several years earlier, between 1910 and 1914. Titled simply "Conceptual Exposition," Part 1 is essentially an unfinished catalogue of the meaning Weber attached to each of his key concepts. As such, it is quite different in style and tone from the earlier, longer, and more historically oriented Part 2. There is some indication that Weber intended to rewrite the earlier material in terms of the system of concepts he had recently developed.

17. Weber, *Economy and Society,* pp. 24–26. Again, scholars have interpreted the types of social action in various ways. Compare, for example, Collins, *Max Weber,* pp. 42–43, and Raymond Aron, *Main Currents in Sociological Thought, II* (Garden City, NY: Doubleday, 1970), pp. 220–221.

18. Weber, *Economy and Society,* p. 26.

19. Weber's approval of and participation in research projects designed to obtain quantitative data show clearly that he was amenable to the use of such information as one means of "explanatory understanding." See Paul F. Lazarsfeld and Anthony R. Oberschall, "Max Weber and Empirical Social Research," *American*

Sociological Review 30 (April 1965), pp. 185–199.

20. Weber, *Economy and Society,* pp. 22–26.

21. Ibid, pp. 26–28, 40–43.

22. Ibid, 31–39.

23. Considerable controversy exists over the proper translation of *herrschaft.*

24. Weber, *Economy and Society,* p. 946.

25. Ibid., pp. 956–958.

26. Ibid., p. 241.

27. See Reinhard Bendix, "Charismatic Leadership," in *Scholarship and Partnership: Essays on Max Weber,* Reinhard Bendix & Guenther Roth (Berkeley: University of California Press, 1971), pp. 170–187, and Edward A. Shils, "Charisma, Order, and Status," *American Sociological Review* 30 (1965), pp. 199–213.

28. Weber, *Economy and Society,* p. 1148.

29. Ibid., p. 226.

30. It is important to be realistic about this issue. Patterns of social action do not become customary and are not maintained without conflict. In the case of the British monarchy, for example, many struggles over the right of succession occurred, such as the War of the Roses.

31. Weber, *Economy and Society,* pp. 237–241.

32. Quoted in Reinhard Bendix, *Max Weber: An Intellectual Portrait* (London: Heinemann, 1960), pp. 418–419.

33. Weber, *Economy and Society,* pp. 217–220. Our interpretation of the impact of bureaucracy is more benign than is Weber's. He saw bureaucracies as encasing people in an "iron cage" of reason and thereby stifling freedom; as such, bureaucracies are the archetype of the process of rationalization. This is why Diggins subtitles his book on Weber *Politics and the Spirit of Tragedy.*

34. Weber, *Economy and Society,* p. 975.

35. Ibid., p. 951.

36. See Leonard Beeghley, "Social Structure and Voting in the United States: A Historical and International

Analysis," *Perspectives on Social Problems* 3 (1992), pp. 265–287.

37. Weber, *Economy and Society,* pp. 302, 927.

38. Ibid., p. 928.

39. Ibid., p. 927.

40. Ibid., pp. 303, 928.

41. Ibid., p. 937.

42. Ibid., pp. 303, 927.

43. Ibid., p. 304.

44. Ibid., pp. 305, 930–931.

45. Ibid., pp. 305–306, 932.

46. Ibid., p. 932.

47. Ibid., p. 306.

48. Reinhard Bendix, "Inequality and Social Structure: A Comparison of Marx and Weber," *American Sociological Review* 39 (April 1974), pp. 149–161.

49. Weber, *Economy and Society,* pp. 306, 935.

50. Ibid., pp. 342–343.

51. Ibid., p. 933.

52. Karl Marx and Friedrich Engels, *The Birth of the Communist Manifesto* (New York: International, 1975), p. 90.

53. Karl Marx, *The Eighteenth Brumaire of Louis Bonaparte* (New York: International, 1963), and "Critique of the Gotha Program," in *Selected Works,* vol. 3, ed. Karl Marx and Friedrich Engels (New York: International, 1969), pp. 9–30.

54. See Leonard Beeghley, *The Structure of Stratification,* 2nd edition (New York: Allyn & Bacon, 1996).

55. Weber, *Economy and Society,* pp. 901–1372; see in particular, pp. 901–920.

56. Theda Skocpol has pursued this Weberian argument in her *States and Social Revolutions* (New York: Cambridge University Press, 1979).

57. Weber, *Economy and Society,* pp. 63–212.

58. Max Weber, *The Protestant Ethic and the Spirit of Capitalism,* trans. Talcott Parsons (New York: Scribner's, 1958).

59. Max Weber, "Author's Introduction," in *Protestant Ethic,* p. 25. It is impor-tant to recognize that Weber wrote this introduction in 1920 for the German edition of his *Collected Essays in the Sociology of Religion.* Thus, it is an overall view of Weber's work in the sociology of religion rather than an introduction to the *Protestant Ethic.* Scribner's 1976 edition of the book does not make this clear.

60. Max Weber, *The Religion of China,* trans. Hans Gerth (New York: Free Press, 1951), *The Religion of India,* trans. Hans Gerth and Don Martindale (New York: Free Press, 1958), and *Ancient Judaism,* trans. Hans Gerth and Don Martindale (New York: Free Press, 1952). In addition, Part 2 of *Economy and Society* contains a book-length study, "Religious Groups (The Sociology of Religion)," which is also available in paperback under the title *The Sociology of Religion,* trans. Ephraim Fishoff (Boston: Beacon, 1963).

61. Weber, *Protestant Ethic,* p. 25.

62. Ibid., pp. 43–45, 191 (note 23). See also Reinhard Bendix, "*The Protestant Ethic*—Revisited," in *Scholarship and Partisanship: Essays on Max Weber,* ed. Reinhard Bendix and Guenther Roth (Berkeley: University of California Press, 1971), pp. 299–310.

63. Ibid., p. 191.

64. Weber hints at his ideal type strategy but does not bother to explain it in the initial paragraphs of Chapter 2 of the *Protestant Ethic,* p. 47. He refers to the need to develop a "historical indi-vidual"—that is, "a complex of ele-ments associated in historical reality which we unite into a conceptual whole from the standpoint of their cultural significance." This phrasing reveals the influence of Heinrich Rickert, as discussed in Chapter 10 of this book.

65. Weber, *Protestant Ethic,* pp. 53–54.

66. Ibid., p. 52.

67. Ibid., p. 57–58. See also Max Weber, "Anticritical Last Word on the Spirit of Capitalism," trans. Wallace A. Davis, *American Journal of Sociology* 83 (1978), p. 1127.

68. Ibid., p. 97.

69. Ibid., p. 153. As a Protestant, Weber may have been biased and overemphasized the distinction between Protestantism and Catholicism.

70. Ibid., p. 158.

71. Weber, *Economy and Society,* pp. 439–517.

72. Weber, *Protestant Ethic,* p. 176. Although Weber's point was that religious values influenced behavior, he emphasized that values and lifestyles are interrelated. Religiosity permeated every aspect of daily life. See Robert Wuthnow, *Communities of Discourse: Ideology and Social Structure in the Reformation, the Enlightenment, and European Socialism* (Cambridge, MA: Harvard University Press, 1989).

73. Ibid., pp. 180–183.

74. Seymour Martin Lipset, "History and Sociology: Some Methodological Considerations," in *Sociology and History: Methods,* eds. Seymour Martin Lipset and Richard Hofstadter (New York: Basic Books, 1968), pp. 20–58.

75. Weber participated in this process; see his "Anticritical Last Word," as well as the many explanatory footnotes in the *Protestant Ethic.* Most of these footnotes were added around 1920. The debate over Weber's "logical experiment" has continued as observers struggle to understand the origins of modernity. See Robert L. Green (ed.), *Protestantism and Capitalism: The Weber Thesis and Its Critics* (Lexington, MA: D.C. Heath, 1959); S. N.

Eisenstadt (ed.), *The Protestant Ethic and Modernization* (New York: Basic Books, 1968); Gordon Marshall, *In Search of the Spirit of Capitalism: An Essay on Max Weber's Protestant Ethic Thesis* (New York: Columbia University Press, 1982); Hartmut Lehmann and Guenther Roth (eds.), *Weber's Protestant Ethic: Origins, Evidence, Context* (New York: Cambridge University Press, 1993); Stephen Innes, *Creating the Commonwealth: The Economic Culture of Puritan New England* (New York: Norton, 1995).

76. Stanley Lieberson, "Einstein, Renoir, and Greeley: Some Thoughts about Evidence in Sociology," *American Sociological Review* 57 (1992), 1–15.

77. Weber, *Religion of China,* p. 156.

78. Ibid., p. 156.

79. Subsequent events, however, have shown that Confucianism is congenial to capitalism. See Peter M. Berger, *The Capitalist Revolution* (New York: Basic Books, 1986).

80. Weber, *Protestant Ethic,* p. 89.

81. The basis for our interpretation is Talcott Parsons, "Introduction," pp. 3–86 in Max Weber, *The Theory of Social and Economic Organization* (Glencoe, IL: Free Press, 1947). Some scholars agree with Parsons' interpretation; see Mommsen, *The Age of Bureaucracy,* pp. 1–21. Others disagree with it; see Reinhard Bendix, *Max Weber: An Intellectual Portrait* (Garden City, NY: Doubleday, 1962).

12

⁕

Max Weber's
Theoretical Legacy

The first part of Weber's *Economy and Society* begins with an analysis of action, social relationships, and legitimated orders. Each of these phenomena is defined as an ideal type, revolving around the relative amounts of rational, traditional, and affectual orientations of actors. Weber gives the impression that he is building a theory from the micro, moving through more meso structures like legitimated orders, and then on to macro-level societal processes like rationalization. These opening pages are misleading, however, because Weber does not develop a theory in this systematic way. Rather, most of his theorizing is at the meso and macro levels, emphasizing such phenomena as religious and legal systems, patterns of domination, organizational structures like bureaucracies and cities, and stratification systems revolving around classes, status groups, and parties. Weber continually refers to orientations of individuals but he does not *systematically* analyze the relationships among culture, social structures, and individual actions. Probably the most systematic analysis among these various levels of reality can be found in *The Protestant Ethic and the Spirit of Capitalism,* but for most of Weberian sociology, we get only glimpses of a multidimensional approach. How, then, do we get a handle on Weber's sociology when our goal is to convert it into abstract models and general principles?

The answer to this question is that we must examine meso-level structures to get a picture of Weber's more macro views of society as a whole, but this strategy leaves the micro-level of analysis of action and social relationships outside of the models and principles. There is no easy way to include the micro portions of Weber's sociology because a systematic analysis of the connection between micro- and meso-levels of analysis is not undertaken, despite appearances in the early pages of *Economy and Society*. In general, Weber simply stated that the structure of a given meso structure was sustained by a particular type of action, but assertions are not analyses. Thus, in our models of Weberian sociology, we will not explicitly incorporate the analysis of action, except to note where Weber sees certain cultural and structural forces as constraining the type of action orientations among actors, and vice versa.

WEBER'S CAUSAL MODELS
OF SOCIAL ORGANIZATION

Weber's Model on Rationalization

As we have emphasized, Weber organized his work around the theme of rationalization. In Figure 12.1, we have abandoned much of Weber's vocabulary, which is too mired in ideal type distinctions; instead, we offer our views on Weber's general image of the cultural and social forces causing rationalization. As these forces operate, Weber argued, actors are increasingly oriented to instrumental rationality, whereas legitimated orders are structured by rational-legal rules. Thus, for each force listed in the model, there are organizational orders emerging that move away from tradition, affect, and even value rationality toward instrumental rationality; because these forces begin to play off each other and accelerate the dynamics involved, the "iron cage" of rational-legal bureaucracies in politics and the economy pushes society toward a rational-legal form of domination.

In the model in Figure 12.1, we begin at the left with Weber's emphasis that certain material conditions are necessary for capitalism—the ultimate form of rational-legal authority—to emerge. One is inanimate sources of power, or some minimal level of industrial production; another is free markets for exchanging goods and services, including labor; and perhaps the most important is a system of ascetic cultural orientations, as personified by the "Protestant Ethic." Added to these conditions is money, which makes utilities calculable and credit possible. With credit and money, the volume and velocity of market exchanges increases; and as more and more goods and services are bought and sold, the level of profit and wealth increases. The reverse causal loops are particularly important in Weber's argument because once ascetic values, markets, money, credit, and wealth are transformed, these very processes feed back on each other. As the scale of markets increases, so does production; as the use of money and credit expand, the volume and velocity of transactions increases, with the latter feeding back to increase the scale of markets. Once profits and wealth are created, these are reinvested in production, especially when ascetic cultural orientations dominate and encourage people to save. Both ascetic cultural orientations and market transactions operate to shift cultural values to

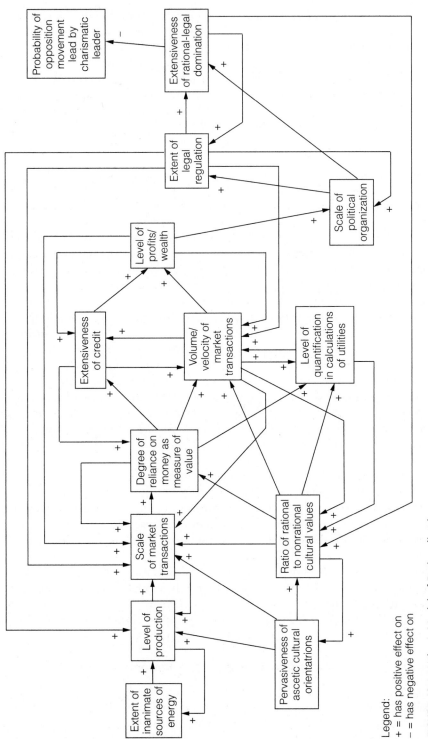

Legend:
+ = has positive effect on
– = has negative effect on

FIGURE 12.1 Weber's Model of Rationalization

emphasize instrumental rationality over value rationality, tradition, and affect, although this process takes time (as is illustrated by Weber in his view that the "spirit of capitalism" emerges out of religious-inspired "Protestant Ethic"). The important point is that all of the arrows in the model—direct, indirect, and reverse—are positive causal chains (except for the one heading to opposition movement), emphasizing that once a certain threshold of capitalism is initiated, economic growth and rationality continually accelerate and decrease the likelihood of revolt.

We have not yet mentioned the political and legal processes involved, but they are critical to rational-legal domination and, hence, the rationalization of society as a whole. Increased wealth allows for the expansion of the state, especially when wealth comes in the form of money that can be used to buy labor (administrators, police, military) and to finance large-scale projects from war to domestic infrastructures (and, of course, elite privilege). Once the larger state becomes dependent on the flow of money, it begins to enact laws that allow it to regulate production and market activities (as indicated by the reverse causal arrows flowing out of legal regulation). Government increasingly has an interest in the productivity of the economy, so it begins to shift the basis of regulation from tradition to law; as this occurs, the centers of power in a society begin to dominate through rational-legal means. Indeed, economic, political, and legal orders become heavily bureaucratized, and as this bureaucratic domination spreads, it reinforces cultural values emphasizing instrumental rationality. In turn, these values encourage expansion of bureaucratization of production and market orders, as well as the state. In the end, these positively reinforcing cycles cause the spread of rational-legal domination; and in this form of domination, opposition movements are less likely to occur because the traditional forms of domination in which such movements had typically arisen have now been pushed aside by rational-legal orders. Indeed, the only negatively signed arrow in the model is from rational-legal domination to opposition movements lead by a charismatic leader, a clear indicator of Weber's views about the iron cage imposed by instrumental rationality.

The model in Figure 12.1 is rather complex, and it would be more so if we inserted a number of more micro- and meso-level factors, including: (1) the action orientations among people (from traditional, affectual, and value-rational to instrumentally rational), (2) the shift in social relations from a communal to associative basis, (3) the movement in organizational forms from patrimonial to bureaucratic bases, and (4) the alteration of stratification systems from traditional status groups, parties, and classes to those increasingly shaped by market forces and the power of actors in the state. We can perhaps recapture some of this detail by examining key elements suggested by the model in more detail.

Weber's Model of Culture and Rationalization

The model in Figure 12.2 simply amplifies elements in Figure 12.1. Here, our goal is to emphasize the relationship among ascetic cultural values emphasizing worldly activity and other key processes of rationalization. The first is the arrow from a chance event, as Weber thought the Protestant Reformation to

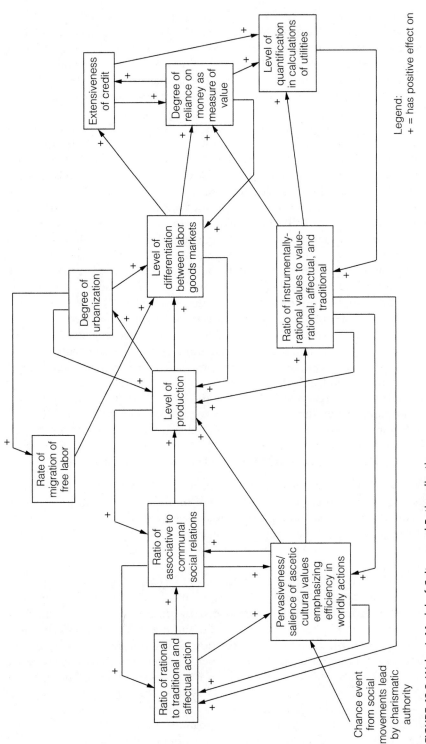

Legend:
+ = has positive effect on

FIGURE 12.2 Weber's Model of Culture and Rationalization

be. Sometimes, by chance, charismatic leaders articulate a set of cultural values that, potentially, can initiate rationalization. As Weber would have emphasized, however, other material conditions need to be present, and these are outlined in the model. Ascetic values will direct action and social relationships in a rational direction, but it is also possible that rational activity at the micro level can, over time, generate a sufficient number of associative relations so that new, more ascetic value orientations gradually emerge in a society. These might or might not have a religious basis, but over time, they can evolve into purely secular values emphasizing instrumental rationality. Once this causal sequence is initiated, ascetic and rational values become the "orientations" that motivate actors to behave rationally to form rational associative relations and to create legitimated orders based on law.

In the model, however, we stress that for this set of causal chains to operate, other conditions (from Weber's model on the rise of capitalism) must be in place. Production must reach a certain threshold, with at least the beginnings of industrialism. Urban areas must grow to concentrate productive capital and labor. A distinct labor market must exist and be differentiated from other markets. The use of money and credit will need to increase the extent to which people quantify the calculation of utilities, particularly through new kinds of accounting and bookkeeping systems. As the direct, indirect, and reverse causal arrows all point, these forces will increase the ratio of instrumentally rational values to those based on affect, value rationality, and tradition. As this process occurs, a secular worldly asceticism begins to provide the orientations that guide the motivations of individuals in all legitimated orders. As a result, people engage in associative social relations in organizational and stratification orders that push production, urbanization, market differentiation, use of money and credit, and quantification further along the road to rational-legal domination.

Weber's Model of Markets, Money, Power, and Law

In Figure 12.3, we have delineated in more detail Weber's underlying argument about the power of markets and the use of money to transform not only the economy but political and legal systems as well. For rational-legal domination to be complete, both the centers of power and production must be organized in bureaucratic organizational forms within institutional orders legitimated by law. Although cultural values are critical in this process, so are markets, money, and credit. Relations among this latter set of forces are arrayed in Figure 12.3.

When money is increasingly used in exchange relations, it has transforming effects, many of which are mediated by the extension of credit. With money, it becomes possible to offer credit, and the two together force a general increase in the quantitative calculation of utilities. Monies lent, monies paid back, and interest accrued must all be calculated when credit is extended; more generally, money orients people to the quantitative nature of exchange relations. All of these forces together will increase the velocity and volume of market activities. As such activities accelerate, they feed back and increase the pervasiveness of the very forces that expanded market activity in the first place—that is, use of money, credit, and quantitative calculations.

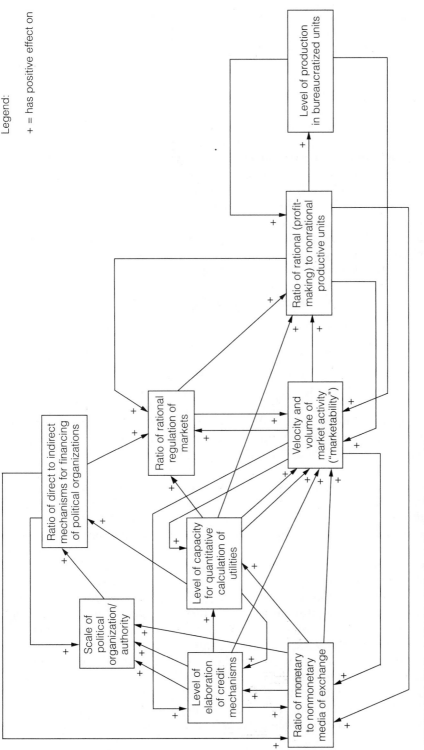

FIGURE 12.3 Weber's Model of Markets, Money, Power, and Law

As markets grow, the ratio of rational, profit-oriented to nonrational forms of organizing production increases. For, once markets are up and running, using money, credit, and new ways of accounting, production becomes increasingly oriented to profits; and as profits become the motive for action, these new orientations of actors feed back and accelerate the processes that shifted the ratio of profit-making to non-profit-making organizations.

The seat of power is never a passive bystander to these processes. With money and credit, government can tax and borrow, and in this way, the scale of political organization and authority can expand. Government can hire more people and do more things. No longer must political actors rely on traditional ways of financing their activities, such as patronage in exchange for soldiers or shares of crops; government can now use more direct mechanisms for securing the necessary resources once money is widely used and credit is readily available. This concern about direct financing of its operations leads the center of power to begin the rational regulation of productive units and markets with laws. Once laws are used to regulate an ever-growing bureaucratic state and bureaucratized economic units, however, rational-legal domination accelerates.

Weber's Model on Stratification and Conflict

Although Weber saw the spread of rational-legal authority as decreasing the potential for conflict, he nonetheless posited a model of conflict that, in general terms, parallels Karl Marx's analysis. Unlike Marx, of course, Weber did not see inequality as inevitably leading to conflict; rather, chance has always played a part in determining if charismatic leaders can emerge to mobilize sentiments against elites. When we recast Weber's model in more abstract terms, we can see a more general theory on the conditions generating conflict between superordinates and subordinates in a system of inequality. Figure 12.4 summarizes the basic model in Weber's analysis.

For Weber, conflict emerges when the legitimacy of existing patterns of domination is questioned by subordinates. Thus, Weber saw the critical question as what conditions will increase the likelihood that this questioning of legitimacy will occur. One condition is the correlation among memberships in class, status groups, and parties. The more one's class position predicts status group membership, and vice versa, and the more one's status group membership and class position predict access to political power, and vice versa, the higher is the correlation among the three dimensions of inequality. The existence of this correlation increases the likelihood that charismatic leaders will emerge to question this situation. Another condition is the discontinuity among classes, status groups, and parties. To some extent, a high correlation will produce big gaps between classes, status groups, and access to power, but this force also operates on its own. If there are large differences between (a) the resources of those in high and low social classes, (b) the prestige and honor of those in high and low status groups, and (c) the respective power of those in different parties, then this kind of discontinuity increases the probability that charismatic leaders will emerge to challenge such high levels of inequality. A

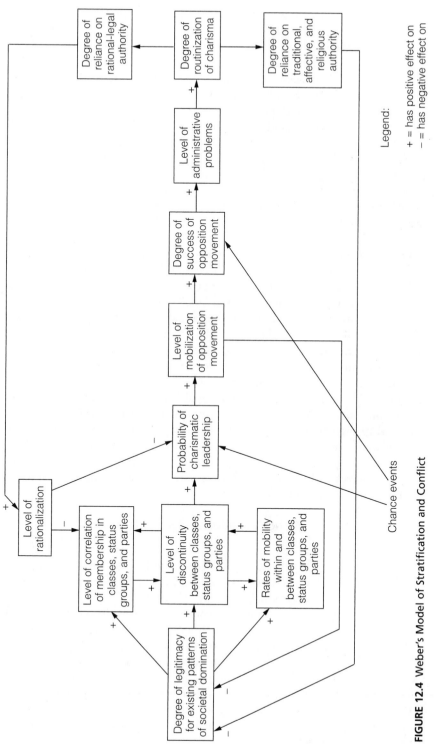

FIGURE 12.4 Weber's Model of Stratification and Conflict

Legend:

+ = has positive effect on
− = has negative effect on

final condition, again often produced by discontinuity, revolves around rates of mobility across different classes, status groups, or parties. If little mobility is possible and individuals are stuck in low positions along all three dimensions of stratification, then people will find charismatic leaders appealing.

When we look at Weber's model, it comes close to Karl Marx's notion of "polarization" of society into two big classes, but Marx simply assumed that as class inequality increases, discontinuity and low mobility rates would follow. As a result, class conflict would be inevitable. In contrast, Weber emphasized that chance and context are important in determining if charismatic leaders will emerge and lead an opposition movement. Moreover, also in contrast to Marx, Weber saw high degrees of rationalization as lowering the probability of charismatic authority by (a) decreasing the correlation among class, status, and party; and (b) decreasing traditional and affectual orientations of actors. If, however, charismatic authority does emerge and is used to mobilize an opposition movement, the outcome of this movement is, once again, subject to chance and contextual events. If the movement is successful, then problems revolving around the "routinization of charisma" emerge. As routinization occurs, the basis for creating a new system of domination is important in determining if this new system creates the conditions for future opposition movements. If routinization moves in the direction of rational-legal domination, then rationalization will decrease the rigidity of the stratification system and lower the potential for the reemergence of opposition movements led by charismatic leaders. In contrast, when routinization involves creating another traditional, religious, or affectual basis of legitimation, stratification is more likely to become rigid once again. This rigidity will increase the probability of a new opposition movement to the stratification system. Again, Weber believed that it is hard to predict which way routinization will go because chance confluence of events are always in play. Similarly, whether charismatic leaders can emerge and be successful is also subject to chance, although the probability of success decreases with rational-legal legitimation and increases with traditional, religious, and affectual legitimation.

Weber's Model of Geopolitics

Weber recognized that the internal stratification system and its potential for generating conflict are very much related to the level of prestige enjoyed by a society vis-à-vis its neighbors. In Figure 12.5, we outline the causal model implied by Weber's analysis of geopolitics. Prestige in the external system is connected to the relations among (a) centers of power, (b) key actors in the economy, and (c) levels of inequality. When economic actors are dependent on the state, they push the state to use coercion in external relations with other societies when these economic actors' interests are tied to the local world system; conversely, when dependence on the state is low, economic actors seek to have the state engage in more co-optive strategies, such as encouraging dependence of other states on the goods provided by the productive sector or creating trade agreements that can facilitate the market relations with other societies.

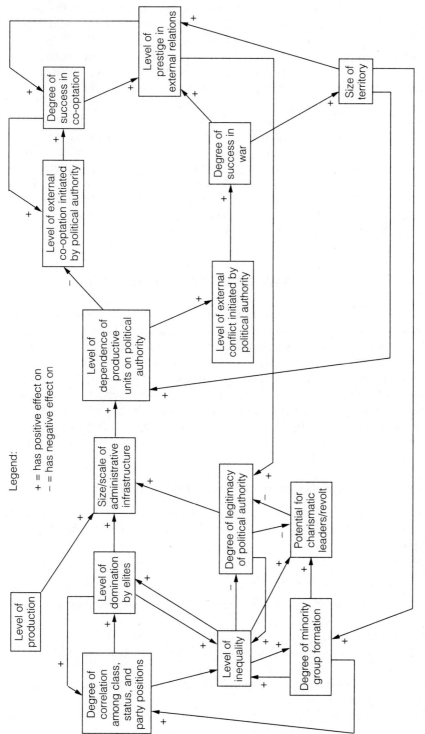

FIGURE 12.5 Weber's Model of Geopolitics

When successful in external relations, the level of prestige of the society and its government increases, and as the long reverse causal arrow from prestige in external relations to degree of legitimacy of political authority suggests, the state will enjoy internal prestige and legitimacy, even under conditions of high inequality. Thus, the effect of inequalities on lowering legitimacy is, to a degree, mitigated by success in the external world system of societies. Although Weber says relatively little about war, he does imply that when a society engages in external conflict and is successful, then its territorial borders are extended, increasing minority groups and creating new classes, status groups, and parties. As these emerge, inequalities may increase as new classes, status groups, and parties are pushed to the bottom of their respective hierarchies, thus increasing the likelihood of an opposition movement. Should a society lose a war, however, the resulting drop in prestige will erode legitimacy independently of inequalities; if high inequality and loss of war are combined, then the potential for opposition movements is dramatically increased.

WEBER'S THEORETICAL PRINCIPLES

We immediately run into problems in trying to convert Weber's sociology into propositions because of two factors: (1) the portrayal of phenomena in nominal categories as ideal types rather than as variables,[1] and (2) the emphasis on chance events and historical context. Weber did not see phenomena as varying by degree but, rather, by type, and when this is the case, it is difficult to construct statements of the form: degree of y varies with levels of x. Similarly, it is hard to make determinative statements when the social universe is seen as very much influenced by chance confluences of events. Nonetheless, it is worthwhile to see what we can do to convert Weber's sociology into a series of abstract theoretical principles.

Principles on the Process of Rationalization

The central theme in Weber's sociology is rationalization, especially its origins and its effects on patterns of domination in society. Thus, we can begin to articulate Weber's theory of society with some elementary principles on this master process:

1. The degree of rational-legal domination in a society increases as
   a. Free markets spread to distribute goods, services, and labor.
   b. Money and credit become more widely available and reorient actors to calculations of utilities.
   c. Ascetic cultural orientations become more prevalent and guide actions of individuals and organizational systems.
   d. The state has interest in taxing and using money and credit to finance its operations increases.
   e. The state shifts the basis of its legitimation toward law.

Under these conditions, rationality will increasingly pervade action, social relationships and legitimated orders, and organizational systems in the economy and state will increasingly become bureaucratic. Weber emphasized the importance of cultural orientations in this process of rationalization, so we might offer a supplemental proposition that clarifies the elements of 1c:

2. The pervasiveness of ascetic cultural values emphasizing efficiency and rationality in worldly actions increases as
    a. The level of production increases, with production growing as
        1. the use of inanimate sources of energy (industrialization) spreads
        2. the size and prevalence of urban centers increase
        3. the movement of free labor into urban areas accelerates
        4. the prevalence of free markets using money and credit increases
    b. The volume and velocity of market exchanges increases, with the volume and velocity of markets increasing as
        1. the level of production increases
        2. the availability of money and credit increases
        3. the availability of money and markets increase quantitative calculation of utilities
    c. The ratio of associative to communal social relations in the society increases

As is evident, these ascetic cultural values are very much tied to market forces in a society, as is the overall process of rationalization. Thus, we should offer a supplemental principle to 1a and 1b:

3. The velocity and volume of market activity increases with the use of money and credit in exchange relations, with this use growing as
    a. The prevalence of ascetic cultural values to guide actions of individuals and organizational systems increases.
    b. The number of profit-oriented organizational systems increases.
    c. The state's interests in taxing and using money for its ends increases.
    d. The state's use of law to regulate exchange relations and to legitimate its actions increases.

The forces enumerated in these propositions are all interconnected. Weber thought the confluence of forces set off rationalization. Certain conditions had to prevail to start the process, but once it got going, the conditions would feed off one another, increasing the weights for each and pushing action, social relations, organizational systems, and stratification toward rational-legal forms.

Principles on De-Legitimation and Conflict

Weber was always concerned with the way patterns of domination in society are legitimated. Because domination generates a stratification system composed of classes, status groups, and parties, this system of inequality is critical in sustaining legitimation. If those high along these three hierarchies of inequality are seen to have the right to be there, then not only is the stratification system legitimated but so is the overall system of domination. Inequality will generate the potential for opposition movements under conditions that increase the likelihood that charismatic leaders will emerge to articulate the grievances of subordinates and to lead them in opposition. These conditions are summarized in principle 4:

4. The more subordinates in a system of inequality withdraw legitimacy from superordinates, the higher the rate of conflict is with superordinates, with the withdrawal of legitimacy increasing as

   a. The correlation among memberships in class, status, and political hierarchies increases.
   b. The discontinuity or degree of inequality along all hierarchies increases.
   c. The chances for mobility of subordinates to higher positions in any or all hierarchies declines.
   d. The opportunities for charismatic leaders who can articulate the grievances of subordinates and lead them in opposition to the current system of domination increase.

If charismatic leaders emerge and are successful in orchestrating an opposition movement, then Weber emphasized that they would face the problem of how to routinize this charisma as the goal shifts from revolution to administration. This problem will become even more acute when a charismatic leader dies without a system for choosing a successor in place, but the problem always exists when opposition movements must transform themselves into systems of administration.

5. The more success charismatic leaders have in de-legitimating a system of domination and in overthrowing elites along each hierarchy of inequality, the more problematic is the establishment of a new system of domination through rules and organizational systems.

6. The stability of this new system in the aftermath of a charismatic revolt will increase to the extent that it is based on rational-legal domination, whereas the stability of the new system will decrease if it is based on traditional domination. The more a rational-legal system of authority is put into place, the more stable is this system and the less the conditions listed in proposition 4 prevail.

De-Legitimation and Geopolitics

Systems of domination are very much connected to geopolitics. A society's prestige in the world system will determine, to some extent, the prestige and legitimacy of political leaders. Thus, the actions of these leaders vis-à-vis other societies in their region and beyond will be critical in determining whether or not they can sustain their power, and especially so when the conditions of inequality portrayed in proposition 4 prevail. We can translate Weber's argument into several interesting propositions:

7. The more those with power can sustain a sense of prestige and success in relations with external societies, the greater will be the capacity of leaders to be viewed as legitimate and the less the conditions listed in proposition 4 prevail.

   a. The more productive sectors of a society depend on political authority for their viability, the more they encourage political authority to engage in military expansion to augment their interests; and the more successful these engagements, the more prestige and legitimacy are given to political authority.

   b. The more productive sectors do not depend on the state for their viability, the more they encourage political authority to rely on co-optation rather than on military engagement; the more successful these efforts at cooptation, the more prestige and legitimacy are given to political authority.

8. The less successful is political authority in external relations, the greater is its loss of prestige vis-à-vis other societies and the greater is its loss of prestige legitimacy within a society.

9. The greater are the loss of prestige in external relations and of legitimacy in internal relations by centers of political authority, the more likely is an opposition movement, especially under the conditions of high inequality summarized in proposition 4.

In sum, then, we can convert many of Weber's critical arguments into theoretical principles, but these do not capture either the breadth or depth of Weberian sociology. If Weber is to endure as a theorist, however, this kind of exercise is important because with each passing decade, the historical context of his sociology becomes ever-more remote. Weber was an historical sociologist as much as a theorist, and although his descriptions of historical contexts may prove interesting to historians, his legacy is best kept alive by extracting the more general models and principles to be found in his work. If we refuse to perform exercises like the one in this chapter, Weber's sociology will become less relevant in the twenty-first century. The power of Weber's ideas resides, we believe, in the more general and abstract arguments that he made. In this chapter, we have tried to emphasize what we see as the most enduring of these arguments.

NOTE

1. The biggest problem of converting Weberian sociology into theory is Weber's tendency to posit variations in phenomena as types rather than as variables. Thus, rather than degrees or levels of phenomena, we have discrete categories, and these kinds of statements are not easily converted into variables in a causal model or set of propositions. For example, how does

one convert Weber's concept of *action* into a variable state that is high or low on some scale? Rather, the fourfold typology of action—value-rational, instrumental-rational, affectual, and traditional—is phrased as four discrete categories, which requires us to have four variables: degree of value-rational, degree of instrumental-rational, degree of affectual, and degree of traditional orientations. Then, if we want to typify an action as a single variable, we must construct a complicated ratio expressing the relative amounts of these four orientations to each other. For example, we might want to say that the ratio of instrumentation action to the other three types of action is high, which is easy enough to say, but how would we express the ratios among the other types of action in this statement (that is, the ratios among value-

rational, affectual, and traditional action)? What is true of action is the case with so much of Weber; he thought in terms of analytical schemes, to use the terminology of Chapter 3, rather than of variables. This tendency is understandable given that Weber was not trying to make causal statements that are not clear because he has one state, described as a typology, causing another state to come about, also presented as a typology. These problems in Weber's methodology make it impossible to construct an overall model of his argument; such a model would become so complex because of the need to convert each element of an ideal type into a variable and, then, to express each of these variable elements as a set of ratios to each other to get a single causal variable from a typology.

13

✳

The Origin and Context of Georg Simmel's Thought

BIOGRAPHICAL INFLUENCES ON SIMMEL'S THOUGHT

Simmel's Marginality

Georg Simmel was born in 1858 at the very center of Berlin.[1] His father, a successful Jewish businessman, had converted to Christianity, yet Simmel's Jewish background haunted him throughout his career. Simmel's father died when Simmel was young, and a friend of the family was appointed his guardian. Simmel appears to have had an emotionally distant relationship with his mother, so it is reasonable to conclude that he never had any strong ties to his family.

Lewis Coser has stressed that Simmel's work was greatly influenced by this marginality not only to his family but also to the academic establishment in Germany.[2] His marginality, as we will see in the next chapter, helps account for the brilliance of Simmel's analysis of the individual in differentiated and cross-cutting social relationships, for there can be little doubt that he was on the edge of various intellectual worlds. It is difficult to know how embittered he was by this borderline existence. Indeed, in contrast with others of this period who tended to see modern and differentiated social structures as harmful to the individual, he stressed the liberating effects of marginality. Simmel believed that the differentiation of structure, the elaboration of markets, and the detached involvement of people in rational bureaucratic structures gave them options, choices, and opportunities not available in traditional societies. Whether he was justifying his own position in this argument can never be known, but one point is clear: He did not conceptualize modern society in the severe pathological terms of Karl Marx (who stressed oppression and alienation), of Émile Durkheim (who worried over anomie and egoism), or of Max Weber (who warned that the process of rationalization trapped individuals in the "iron cage" of bureaucracy). In any case, there is a tragic quality to Simmel's career, despite its many points of success and accomplishment. Let us now turn to this dual character of his career.

Simmel's Intellectual Career

After graduating from the gymnasium, Simmel studied philosophy at the University of Berlin, and he received his doctorate in 1881. Simmel's thesis was on Immanuel Kant,[3] who as we will see shortly, exerted enormous influence on Simmel's approach to sociological analysis. At the time of Simmel's graduation, Germany in general and Berlin in particular were undergoing a remarkable transition. The nation as a whole was industrializing in record time, and within his adult lifetime, Germany surpassed both England and France in productivity, lagging behind only the United States. This rapid industrialization revealed a critical disjuncture: Though the bourgeoisie brought about rapid economic growth, it failed to advance its claims to power, which remained in the hands of the old feudal elite.[4] This tension between the old and the new produced the disastrous policies that led to World War I and its aftermath, which ultimately created the conditions for the rise of Hitler and Nazism.

Within this broader and foreboding national context, however, intellectual life flourished, especially in Berlin, which in five decades grew from a city of five hundred thousand to four million on the eve of World War I. Even in this prosperous intellectual milieu, however, there was a duality between freedom and authoritarian constraint. In the lively counterculture and intellectual life around the university and in the city itself, there was a vibrant mix of activity. Within the university system, which in the earlier decades of the nineteenth century had served as the model for the research-oriented American universities, there was a conservative and at times authoritarian undercurrent. The university system was prosperous and, as a result, caught in the dilemma between encouraging academic freedom and open expression, on one side, and maintaining its comfortable position in a social and political climate charged with the tension between the capitalist bourgeoisie and the semifeudal political system, on the other side.

Many scholars did great work within this system—Max Weber being the best example in sociology. Others, especially Jews, were denied complete access to the academic system or, as was the case with many, were pushed to the provincial universities outside the main urban centers. Not just Jews but also others who revealed more radical political sympathies that might disrupt the status quo suffered this fate.[5]

Simmel was caught in this conflicting current, and early in his career, he decided to stay in Berlin and hope for the best. The result was that he became a *Privatdozent,* or unpaid lecturer, who lived off student fees. The more typical pattern of this time was for academics to move from one university to another, slowly working their way into major university positions.

Perhaps Simmel saw that, as a Jew, he had the cards stacked against him, but the results of his decision in 1885 to assume a marginal position and remain in Berlin were profound. He became a popular lecturer who attracted a large lay and academic following. As a lecturer, he offered a broad range of courses—from sociology and social psychology to logic, philosophy, and ethics. But this popular success appears to have antagonized the academic establishment. His popularity made many jealous, and his breadth and brilliance threatened narrow specialists. The result was that, in addition to the undercurrent of anti-Semitism, a sense of threat, jealousy, and envy as well as an intolerance of cross-disciplinary scholarship worked against Simmel. Moreover, his style affronted the academic establishment in many specific ways. For example, he never documented his works in detail, with footnotes and scholarly quotations; he jumped from topic to topic, never pursuing a subject in great depth (except perhaps his brilliant work on *The Philosophy of Money*). He also wrote essays for popular magazines and newspapers, a tactic that always antagonizes, even now, traditional academics. Indeed, he appears to have gone out of his way to annoy the academic establishment.

In addition to his popularity among the broader intellectual and artistic community, however, Simmel did enjoy much academic success. With Weber and Ferdinand Tönnies, he was a cofounder of the German Society for Sociology, and his works were widely read, cited, and respected by the first generation of sociologists. Despite his marginal academic status, he did not suffer financially because his guardian left him a considerable fortune. After his marriage in 1890, he lived a comfortable upper-middle-class existence. Given his fame as a lecturer and his active association with artists, critics, commentators, journalists, and writers, Simmel enjoyed a stimulating and full life.

There were still the stigma and frustration of being a well-known scholar and intellectual figure without a real academic position. For fifteen years, Simmel remained a private lecturer, despite the efforts of friends such as Weber to secure a full-time position for him. In 1901, he was given the status of honorary adjunct professor at the University of Berlin, which confirmed his position as an outsider who was neither paid nor entitled to take part in the administrative affairs of the university. Such an appointment was, in reality, an insult for Simmel, who was now a scholar of world fame, having written six books and dozens of articles that had been translated into English, French, Italian, Polish, and Russian.

When Simmel finally received a regular academic appointment in 1914, it was at a provincial university in Strasbourg on the border between France and Germany. Moreover, he was now fifty-six years old, well over a decade past the normative time for promotion to full professor. To top off the frustrations in his career, he arrived at Strasbourg just at the outbreak of World War I. Border university life was suspended during the war, thereby denying him the opportunity to lecture. His last efforts to secure a chair at a major university failed in 1915 at the University of Heidelberg. Shortly before the end of the war in 1918, he died of cancer.

This futility and marginality in the face of world fame must surely have contributed to Simmel's style of scholarship: He maintained a foot in both philosophy and sociology while sustaining a commitment to both formal analytical analysis and social commentary on events and typical questions.[6] The result is a lack of in-depth scholarship on sociological topics; instead, he analyzed specific sociological topics with flashes of insight and sophistication, only to move on to yet another, often disconnected topic. His most in-depth works, particularly *The Philosophy of Money,* are so heavily imbued with philosophical commentary that the sociological theory in them can easily go unnoticed.

Despite this topical character to Simmel's work, however, his sociology has two important themes. First, he was concerned, as were all social theorists of this early period in sociology, with the process of differentiation and its effects on the individual. Second, a methodological unity in his work seeks to extract the underlying essence and form of the particular empirical topics; although he often shifted the substantive content of his analysis, he always sought to discover the underlying structure of social interaction and organization that linked diverse substantive areas. These themes emerged from personal and intellectual contact with a number of thinkers—particularly Weber, Herbert Spencer, Kant, and Marx.

INTELLECTUAL INFLUENCES ON SIMMEL'S THOUGHT

A Note on Simmel and Weber

Simmel's sociological writings span the same three decades, 1890 to 1920, as those of his German colleague and friend, Weber. Despite working in similar social and cultural environments, their respective orientations to sociology were rather dissimilar, partly because they were trained differently and, thus, responded to somewhat different influences.

Weber was trained in the law and in economics, with the result that he was forced to react to the *Methodenstreit* (the German methodological controversy, as discussed in Chapter 10). Weber used the works of Wilhelm Dilthey and Heinrich Rickert to fashion a unique sociology that eschewed the development of abstract laws because he believed that such theoretical statements could not uncover the significance of those historical phenomena in which he was most interested. In contrast, Simmel was a philosopher as well as a sociologist; his published works include books and articles on such diverse figures as the philosophers Arthur Schopenhauer, Friedrich Wilhelm Nietzsche, and Kant; the writer Johann Goethe; and the painter Rembrandt. In addition,

Simmel considered morals, ethics, aesthetics, and many other topics from a philosophical vantage point. As might be expected, his brand of sociology is quite different from Weber's. After an early dalliance with Spencer and some elements of Social Darwinism, Simmel's mature works reflect his adaptation of some of Kant's philosophical doctrines to the study of human society. Essentially, Simmel contended that the social processes in which human beings engage constitute organized and stable structures, that these patterns of social organization affect action in systematic ways, and that their consequences can be both observed and predicted independently of the specific objectives of the actors involved. Thus, unlike Weber, he held that sociology should focus on the development of timelessly valid laws of social organization, a point of view that followed from his Kantian orientation.

Simmel and Weber were alike in at least one respect, however, for both felt compelled to react to the ideological and theoretical challenge posed by Marxism as it existed at the turn of the century. Although this interest is more central to Weber's sociology than to Simmel's, the latter's analysis is a sociologically sophisticated rejection of Marx and Marxism.

Herbert Spencer, Social Darwinism, and Simmel's Thought

Like most of the other classical social theorists, Simmel wanted to understand the nature of modern industrial societies. The title of his first sociological treatise, *Social Differentiation,* published in 1890, suggests immediately the fundamental change that he saw: Modern societies are much more differentiated (or complex) than those of the past.[7] Like Spencer (see Chapter 5), Simmel saw this change in evolutionary terms and as an indication of human progress. Indeed, he labeled it an "upward development." Further, just as Spencer often used organismic analogies to illustrate his argument, so did Simmel. For example, in *Social Differentiation,* he argued that just as a more complex organism could save energy in relation to the environment and use that energy to perform more difficult and complex tasks, so could a more highly differentiated society.[8] Like Spencer, Simmel never confused biological analogies with social facts, partly because he also illustrated his arguments with many other kinds of analogies and examples. Although some of Simmel's later works do not reveal much systematic concern with either the problem of evolution or the historical transition to industrialization, they display a lasting interest in understanding the structure of modern, differentiated societies and in showing how people's participation in complex social systems affects their behavior.

The evolutionary discussion in *Social Differentiation* also shows a less attractive side to the young Simmel because he embraced some of the more questionable aspects of Social Darwinism that were current during the late nineteenth century. For example, he insisted on "the hereditary character of the criminal inclination" and even protested against "the preservation of the weak, who will transmit their inferiority to future generations."[9] However, his mature writings betray no trace of such views. In fact, although he generally did not comment on political events in a partisan manner, his analyses of the poor, of women, and of working people all suggest a sympathetic understanding of their plight.[10] In all these cases, his discussion reflects a more general attempt at focusing on the consequences of social differentiation in modern societies. His emphasis on demonstrating the manner in which social structures influence interaction among human beings independently of their specific purposes is one result of his neo-Kantian orientation.

Immanuel Kant and Simmel's Thought

Kant did not finish his most significant work until the age of fifty-one. But that book, *The Critique of Pure Reason,* published in 1781, stimulated a revolution in philosophy.[11] Among other results, it led eventually to G. W. F. Hegel's denial that material phenomena were real (see Chapter 7). However, Hegel is hardly read today, except by Marxists, whereas Kant's influence has endured. In this section, we will begin by sketching Kant's basic ideas and then briefly discuss how Simmel adapted them for use in sociology.

Kant's Basic Ideas In a short discourse on the nature of philosophy, Simmel remarked that individual philosophers often raised what appeared to be a general problem but stated the issue so that its solution conformed to their preconceptions.[12] He undoubtedly had Kant in mind when stating this little aphorism, for the *Critique of Pure Reason* is an investigation of the potential for "pure reason" that exists apart from the mundane and disorganized sense impressions that human beings experience. Kant's definition of *pure reason* is crucial. By this term, he meant that cognitive capacities and forms of thought existed independently of, or before, sense experience, implying that pure reason is thought that is inherent to the structure of the mind. After defining the issue in this way, Kant concluded that all human conceptions of the external world were products of the activity of the mind, which shaped the unformed and chaotic succession of sense impressions into a conceptual unity that could be understood as scientific laws.[13]

Whereas previous philosophers, such as John Locke and David Hume, had assumed that material phenomena were inherently organized and that human beings' sensations of objects and events in the world merely reflected that organization, Kant held that neither supposition was true. Rather, he asserted that people's sensations were intrinsically chaotic, reflecting nothing more than the endless succession of sights, smells, sounds, odors, and other stimuli that, in and of themselves, were disorganized and meaningless. He contended that the mind transformed this chaos of sense impressions into meaningful perceptions through a process he called the *transcendental aesthetic.* Simmel summarized this part of Kant's philosophy by observing that human sensations were "given forms and connections which are not inherent in them but which are imposed on them by the knowing mind as such."[14] This transformation of disorganized sensations into organized perceptions is accomplished by two fundamental "categories," or forms—space and time—that are inherent parts of the mind. In Kant's view, space and time are not things perceived but modes of perceiving that exist prior to, or independently of, our knowledge of the world; they are elements of pure reason. Only by using these categories are human beings able to transform the chaotic sensations that they receive from the external world into systematic perceptions.

Yet merely being able to perceive objects and events is not enough, for people's perceptions are not spontaneously organized either. Rather, like sensations, perceptions are experienced as confused sequences of observations. Thus, Kant emphasized that pure reason aimed at the establishment of higher forms of knowledge: general truths that are independent of experience. These are the laws of science, truths that are abstract and absolute. Hence, in a process Kant called *transcendental logic,* the mind transforms perceptual knowledge into conceptual knowledge—for example, the transformation of the observation of a falling apple into the law of gravity or (on a different level) the transformation of observations of action during conflict into laws stating the consequences of social conflict for human behavior. Kant held that the mind used a set of preconceived "categories," or forms, by which to arrange perceptions. For example, the ideas of cause, unity, reciprocal relations, necessity, and contingency are modes of conceptualizing empirical processes that are inherent to the mind; like space and time, they are elements of pure reason. In Simmel's words, it is only through the activity of the mind that human perceptions "become what we call nature: a meaningful, intelligible coherence in which the diversity of things appears as a principled unity, knitted together by laws."[15] In this regard, Kant also insisted that the manner in which observations were conceptualized always depended on the purposes of the mind. For example, consider a system of thought, such as Charles Darwin's theory of evolution or Marx's theory of revolution. Kant said that these means of conceptualizing empirical data (perceptions) revealed the purposeful activity of the mind, for in neither case were the objects or events in the world prearranged in the manner conceptualized by the theory. Thus, over the long run, scientific knowledge is one result of the existence of pure reason as an intrinsic characteristic of human beings.

In this context, it is important to remember that, unlike Hegel, Kant never denied the existence of the material world; he merely averred that human knowledge of external phenomena occurred through the forms imposed on it by the active mind. Put differently, the empirical world is an orderly place because the categories of thought organize our sensations, organize our perceptions, and organize our conceptions to produce systematic scientific knowledge. Nonetheless, although Kant contended that such knowledge was absolute, he also indicated that it was limited to the field of actual experience. Therefore, he believed that it was impossible to know what objects and events were "ultimately like" apart from the receptivity of human senses. One of the most important implications of this point of view is that attempts at discovering the nature of ultimate reality, either through religion or science, are impossible. In Kant's phrase, "understanding can never go beyond the limits of sensibility."

Simmel's Adaptation of Kant's Ideas The link between Kant and Simmel is most clearly explained in the latter's essay "How Is Society Possible?" which was originally inserted into the first chapter as an excursus in *Sociology: Studies in the Forms of Sociation*.[16] According to Simmel, the basic question in Kant's philosophy is: How is nature possible? That is, how is human knowledge of nature (the external world) possible? As we have said, Kant answered this query by positing the existence of certain a priori categories that observers use to shape the chaotic sensations they receive into conceptual knowledge. This point of view means that when the elements of nature are conceptualized in some manner, their unity depends entirely on the observer.

In contrast, the unity of society is both experienced by the participants and observed by sociologists. In Simmel's words, "the unity of society needs no observer. It is directly realized by its own elements [human beings] because these elements are themselves conscious and synthesizing units."[17] Thus, as people conduct their daily lives, they are absorbed in innumerable specific relationships with one another—economic, political, social, and familial, for example—and these connections give people an amorphous sense of their unity, a feeling that they are part of an ongoing and stable social structure. Weber believed that because people experience and attribute meaning to the social structures in which they participate, neither the methods nor the goals of the natural sciences were appropriate for the social sciences. So he formulated a version of sociology oriented toward the scientific understanding of historical processes.

Simmel, however, took a different view, one that has had lasting consequences for the emergence of sociological theory. Based on his study of Kant, Simmel argued that social structures (which he called *forms of interaction*) systematically influenced people's behavior before and independently of their specific purposes. He used this argument to show that theoretical principles of social action could be adduced, despite the complications inherent because human beings' own experiences are the objects of study. As he put it, the entire contents of *Sociology: Studies in the Forms of Sociation* constitute an inquiry "into the processes—those which, ultimately, reside in individuals—that condition the existence of individuals in society."[18] More generally, as we will show in the next chapter, Simmel's sociology is oriented toward identifying those basic social forms—conflict, group affiliation, exchange, size, inequality, and space—that influence social action regardless of the intentions of the participants. Because these forms constitute the structure (or, as Kant would say, the "categories") within which people seek to realize their goals, knowledge of the manner in which they affect social behavior can lead to theory. In this sense, then, Simmelian sociology is thoroughly Kantian in orientation. Perhaps this philosophical underpinning allowed Simmel to see some of the major pitfalls in Marx's work.

Karl Marx and Simmel's Thought

Some of Simmel's sociology constitutes a critique of Marx's major work, *Capital,* as well as a rejection of his basic revolutionary goal: the establishment of a cooperative society where people would be free to develop their human potential.[19] *Capital* is an attempt at demonstrating that the value of commodities (including human beings) results from the labor power necessary to produce them; Marx called this the labor theory of value. In *The Philosophy of Money* (1900), Simmel rejected the labor theory of value by arguing more generally that people in all societies placed value on items based on their relative desirability and scarcity.[20] By following Kant rather than Hegel, Simmel believed he could better account for the value that individuals attribute to commodities in different societies (capitalist as well as socialist) by showing how cultural and structural phenomena systematically influence what is both scarce and desired. In this way, then, he undercut the theoretical basis for Marx's analysis. Remember, however, that Marx had a rather different definition of science and theory than did Simmel—or any other classical social theorist.

Simmel also rejected Marx's argument by focusing on the importance of money as a medium of exchange. In *Capital,* Marx tried to show that one of the necessary consequences of capitalism was people's alienation from one another and from the commodities that they produced. In this regard, he emphasized that actors had no control over those activities that distinguish them, as human beings, from other animals. Simmel approached the problem of alienation by simply recognizing that in any highly differentiated society, some people are inevitably going to be alienated. Indeed, he saw the decline in personal and emotional contact among humans as more fundamental than the lack of control by people of their own activities. Apparently, he believed that in most societies, most individuals lacked control over their daily lives. Having recognized the inevitability of alienation, however, Simmel went on to oppose Marx by arguing that the dominance of money as a medium of exchange in modern social systems reduced alienation. Although we have identified some of the positive consequences of money here, see the next chapter for a much fuller analysis of *The Philosophy of Money.* In contrast with Marx, Simmel argued that the widespread use of money allowed exchanges between people who were spatially separated from one another, thereby creating multiple social ties and lowering the level of alienation. In addition, he suggested that the generalized acceptance of money in exchanges increased social solidarity because it signified a relatively high degree of trust in the stability and future of the society. Finally, he concluded that the dominance of money allowed individuals to pursue a wider diversity of activities than was possible in barter or mixed economies and hence gave them vastly increased options for self-expression. The result of this last factor, of course, is that people have greater control over their daily lives in money economies. As an aside, Simmel's explanation of the consequences of money in modern societies is a good example of how social processes "condition the existence of individuals in society" independently of their specific purposes. In this case, the specific ways in which money is used are less important, sociologically, than are its effects on the general nature and form of human relationships in systems where money is the modal medium of social exchange.

Finally, Simmel also discussed Marx's formulation of the problem of alienation in his essay on the functions of social conflict. In this context, Simmel reasoned that individuals probably had the best chance of developing their full human capacities in a competitive rather than a cooperative society:

> Once the narrow and naive solidarity of primitive social conditions yielded to decentralization (which was bound to have been the immediate result of the quantitative enlargement of the group), man's effort toward man, his adaptation to the other, seems possible only at the price of competition, that is, of the simultaneous fight *against* a fellow-man *for* a third one—*against* whom, for that matter, he may well compete in some other relationship *for* the former. Given the breadth and individualization of society, many kinds of interest, which eventually hold the group together throughout its members, seem to come alive and stay alive only when the urgency and requirements of the competitive struggle force them upon the individual.[21]

Simmel did not deny, of course, that competition could have "poisonous, divisive, destructive effects"; rather, he simply noted that these liabilities must be evaluated in light of the positive consequences of competition. Like money, Simmel contended, competition gives people more freedom to satisfy their needs, and in this sense, citizens of a society that is competitively organized are probably less alienated. In addition, as indicated in the quotation, he believed that competition was a form of conflict that promoted social solidarity in differentiated social systems because people established ties with one another that involved a relatively constant "concentration on the will and feeling of fellow-men"—an argument that also implies a lessening of alienation. Finally, Simmel indicated that competition was an important means of creating values in society, a process that occurs as human beings produce objective values (commodities, for example) for purposes of exchange; in this way they attain satisfaction of their own subjective needs and desires. This argument not only suggests that competitive societies display less alienation than do noncompetitive ones, it also implies that Marx's revolutionary goal—a cooperative society—is impractical in modern, industrialized, highly differentiated social systems. On this basis, Simmel rejected socialist and communist experiments inspired by Marxist thought. He believed that they were attempts at institutionalizing, indeed enforcing, cooperative relationships among people to prevent the waste of energy and inequalities that inevitably occur in a competitive environment. However, he insisted that even though such results seemed positive, they could only be "brought about through a central [political] directive which from the start organizes all [people] for their mutual interpenetration and supplementation."[22] He implied that an end to alienation would not and could not occur in such an authoritarian social context. Thus, although his works do not represent a long-term debate with Marx, as was perhaps true for Weber, Simmel addressed the same issues as Marx and other early theorists—issues such as the properties of social differentiation, inequality, power, conflict, cooperation, and the procedures for understanding the nature of the social world.

THE ENIGMATIC SIMMEL

In many ways, Simmel's work remains an enigma in modern sociology. Bits and pieces have exerted enormous influence on modern sociological theory, yet it is difficult to view him as inspiring a "school" of thought, as had Marx, Weber, Durkheim, and Mead. Perhaps Simmel's outsider role in the German academic establishment kept him from developing cohorts of students who could carry on his "formal sociology." The result is that his sociology is not, even now when the early masters are the subject of much commentary, fully appreciated for its breadth and brilliance. We try to correct this situation in the next chapter by reviewing his work.

NOTES

1. This sketch of Georg Simmel's biography draws from Lewis A. Coser's *Masters of Sociological Thought* (New York: Harcourt Brace Jovanovich, 1971), pp. 177–217. See also Lewis A. Coser, ed., *Georg Simmel* (Englewood Cliffs, NJ: Prentice-Hall, 1965); Kurt H. Wolff, "Introduction," in *The Sociology of Georg Simmel* (New York: Free Press, 1950); Kurt H. Wolff, ed., *Georg Simmel, 1858–1918* (Columbus: Ohio State University Press, 1959); and Nicholas Spykman, *The Social Theory of Georg Simmel* (Chicago: University of Chicago Press, 1925).

2. Coser, *Masters of Sociological Thought*.

3. The title of his dissertation was "The Nature of Matter according to Kant's Physical Monadology," and in this early confrontation with Kant's ideas can be seen the seeds of Simmel's "formal sociology."

4. Ralf Dahrendorf, "The New Germanys—Restoration, Revolution, and Reconstruction," *Encounter* 23 (April 1964), p. 50.

5. Wilhelm Dilthey, Heinrich Rickert, and Wilhelm Windelband were among those who blocked Simmel's advancement. Despite the differences in their approach to sociology, Weber strongly supported Simmel. See John Patrick Diggins, *Max Weber: Politics and the Spirit of Tragedy* (New York: Basic Books, 1995), p. 138.

6. We have not mentioned Simmel's sudden burst of patriotism during the war because it is so embarrassing: Gone is the cool and analytical Simmel, and in his place is the passionate patriot. As Coser emphasizes, the latter part of Simmel's career was marked by a romantic emotionalism some-what similar to that of Auguste Comte, who, near the end, suffered much the same fate as Simmel (see Chapter 2 of this book).

7. Georg Simmel, *Über soziale Differenzierung* (Leipzig: Duncker and Humbolt, 1890). Although most of this book remains untranslated, two chapters, "Differentiation and the Principle of Saving Energy" and "The Intersection of Social Spheres," do appear in *Georg Simmel: Sociologist and*

European, trans. Peter Laurence (New York: Barnes & Noble, 1976).

8. Simmel, "Differentiation and the Principle of Saving Energy."

9. Paul Honigsheim, "The Time and Thought of the Young Simmel," in *Essays on Sociology, Philosophy and Aesthetics by Georg Simmel et al.,* ed. Kurt Wolff (New York: Harper & Row, 1965), p. 170.

10. See George Simmel, "The Poor," in *Georg Simmel on Individuality and Social Forms* (Chicago: University of Chicago Press, 1971), pp. 150–178. On working people and women, see Georg Simmel, *Conflict and the Web of Group Affiliations* (New York: Free Press, 1955); also Lewis A. Coser, "Georg Simmel's Neglected Contributions to the Sociology of Women," *Signs* 2 (Summer 1977), pp. 869–876.

11. Immanuel Kant, *The Critique of Pure Reason* (New York: Macmillan, 1929). For a more complete analysis, see T. E. Wilkerson, *Kant's Critique of Pure Reason: A Commentary for Students* (New York: Oxford University Press, 1976).

12. Georg Simmel, "The Nature of Philosophy," in Wolff (ed.), *Essays on Sociology,* pp. 282–310.

13. Ibid.

14. Ibid., p. 291.

15. Ibid., p. 291.

16. Georg Simmel, "How Is Society Possible?" in Wolff (ed.), *Essays on Sociology,* pp. 337–356. We have taken the liberty of translating the German title into English.

17. Ibid., p. 338.

18. Ibid., p. 340.

19. Karl Marx, *Capital* (New York: International, 1967).

20. Georg Simmel, *The Philosophy of Money* (Boston: Routledge & Kegan Paul, 1978).

21. Simmel, *Conflict,* p. 6.

22. Ibid., pp. 72–73.

14

✳

The Sociology
of Georg Simmel

Until late in his career, Georg Simmel was never able to hold a regular
academic position. As a Jew, he was subject to discrimination, and,
despite efforts by Max Weber, most of his career was spent as a private
scholar, lecturing to lay audiences for a fee. This marginal position between the
lay and academic intellectual worlds may have prevented him from developing
a coherent theoretical system. Instead, what emerges are flashes of insight into
the basic dynamics of a wide variety of phenomena. Moreover, Simmel had a
tendency to deal with the same topics repeatedly, each time revising and
updating his thinking. If much of his work appears to be a series of lectures,
that is just what it often was: Lectures that prod and stimulate, often without
detailed and scholarly annotation. Yet, with each passing decade since the
1950s, the importance of Simmel's vision for sociology has been increasingly
recognized. For despite the somewhat disjointed character of his life's work, his
methodology for developing sociological theory and the substance of his
insights into the form of modern society present a reasonably coherent program of sociological analysis.

SIMMEL'S METHODOLOGICAL APPROACH
TO THE STUDY OF SOCIETY

In an essay titled "The Problem of Sociology," Simmel concluded as early as 1894 that an exploration of the basic and generic forms of interaction offered the only viable subject for the nascent discipline of sociology.[1] In Chapter 1 of *Sociology: Studies in the Forms of Sociation,* written in 1908, he reformulated and reaffirmed his thoughts on this issue.[2] In 1918, he revised his thinking again in one of his last works, *Fundamental Problems of Sociology.*[3] In what follows, we rely most heavily on this final brief sketch because it represents his most mature statement.

Simmel began the *Fundamental Problems of Sociology* by lamenting that "the first difficulty which arises if one wants to make a tenable statement about the science of sociology is that its claim to be a science is not undisputed." In Germany after the turn of the century, many scholars still denied that sociology constituted a legitimate science; and to retain their power within the university system, these critics wanted to stop sociology from becoming an academic field. Partly for these reasons, it was proposed that sociology should be merely a label to refer to all the social sciences dealing with specific content areas—such as economics, political science, and linguistics. This tactic was a ruse, of course, for Simmel (and many others) recognized that the existing disciplines had already divided up the study of human life and that nothing would be "gained by throwing their sum total into a pot and sticking a new label on it: 'sociology.'"[4] To combat this strategy and to justify sociology as an academic field of study, Simmel argued that it was necessary for the new discipline to develop a unique and "unambiguous content, dominated by one, methodologically certain, problem idea."[5] His discussion is organized around three questions: What is society? How should sociology study society? What are the problem areas of sociology?

What Is Society?

Simmel's answer to the first question is very simple: "Society" exists when "interaction among human beings" occurs with enough frequency and intensity so that people mutually affect one another and organize themselves into groups or other social units. Thus, he used the term *society* rather loosely to refer to any pattern of social organization in which he was interested. As he put it, society refers to relatively "permanent interactions only. More specifically, the interactions we have in mind when we talk about 'society' are crystallized as definable, consistent structures such as the state and the family, the guild and the church, social classes and organizations based on common interests."[6]

The significance of defining society in this way lies in the recognition that patterns of social organization are constructed from basic processes of interaction. Hence, interaction, per se, becomes a significant area of study. Sociology, in his words, is founded on "the recognition that man in his whole nature and in all his manifestations is determined by the circumstances of living in interaction with other men."[7] Thus, as an academic discipline, "sociology asks what happens to men and by what rules do they behave, not insofar as they unfold their understandable individual existences in their totalities, but insofar as they form groups and are determined by their group existence because of interaction."[8] With this statement, Simmel gave sociology a unique and unambiguous subject matter: the basic forms of social interaction.

How Should Sociology Study Society?

Simmel's answer to the second question is again very simple: Sociologists should begin their study of society by distinguishing between *form* and *content*. Subsequent scholars have often misunderstood his use of these particular terms, mainly because their Kantian origin has been ignored.[9] What must be remembered to understand these terms is that Simmel's writings are pervaded by analogies, with the distinction between form and content being drawn from an analogy to geometry. Geometry investigates the spatial forms of material objects; although these spatial forms might have material contents of various sorts, the process of abstraction in geometry involves ignoring their specific contents in favor of an emphasis on the common features, or forms, of the objects under examination. Simmel simply applied this geometric distinction between form and content to the study of society to suggest how sociology could investigate social processes independently of their content. The distinction between the forms and contents of interaction offers the only "possibility for a special science of society" because it is a means of focusing on the basic processes by which people establish social relations and social structures, while ignoring for analytical purposes the contents (goals and purposes) of social relations.

Thus, forms of interaction refer to the modes "of interaction among individuals through which, or in the shape of which, that content attains social reality."[10] Simmel argued that attention to social forms led sociology to goals that were fundamentally different from those of the other social scientific disciplines, especially in the Germany of his time. For example, sociology tries to discover the laws influencing small-group interaction rather than describing particular families or marriages; it attempts to uncover the principles of formal and impersonal interaction rather than examining specific bureaucratic organizations; it seeks to understand the nature and consequences of class struggle rather than portraying a particular strike or some specific conflict. By focusing on the basic properties of interaction, per se, Simmel believed that sociology could discover the underlying processes of social reality.[11] Although social structures might reveal diverse contents, they can have similar forms:

> Social groups, which are the most diverse imaginable in purpose and general significance, may nevertheless show identical forms of behavior toward one another on the part of individual members. We find superiority and subordination, competition, division of labor, formation of parties, representation, inner solidarity coupled with exclusiveness toward the outside, and innumerable similar features in the state, in a religious community, in a band of conspirators, in an economic association, in an

art school, in the family. However diverse the interests are that give rise to these sociations, the *forms* in which the interests are realized may yet be identical.[12]

On this basis, then, Simmel believed that it was possible to develop "timelessly valid laws" about social interaction. For example, the process of competition or other forms of conflict can be examined in many different social contexts at different times: within and among political parties, within and among different religious groups, within and among businesses, among artists, and even among family members. The result can be some theoretical insight into how the process of competition (as a form of conflict) affects the participants apart from their specific purposes or goals. Thus, even though the terminology has changed over the years, Simmel's distinction between form and content constitutes one of his most important contributions to the emergence of sociological theory. However, the next task he faced was identifying the most basic forms of interaction; in his words, sociology must delineate its specific problem areas. Sadly, his inability to complete this task represents the most significant flaw in his methodological work.

What Are the Problem Areas of Sociology?

Unlike his responses to the other two questions, Simmel's answer to the third query has not proven to be of enduring significance for the development of sociological theory. In his initial attempts at conceptualizing the basic social forms with which sociology ought to be concerned, he referred to "a difficulty in methodology." For the present, he felt, the sociological viewpoint can be conveyed only by means of examples because only later will it be possible "to grasp it by methods that are fully conceptualized and are sure guides to research."[13]

Both the title and the organization of Simmel's *Fundamental Problems of Sociology* (1918) suggest that the major impetus for writing this last little book was his recognition that the "difficulty in methodology" remained unresolved. Unfortunately, this final effort at developing systematic procedures for identifying the generic properties of the social world studied by sociology was not very successful either. In this book, Simmel identified three areas that he said constituted the fundamental problems of sociology. First is the sociological study of historical life and development, which he called *general sociology*. Second is the sociological study of the forms of interaction independent of history, which he called *pure,* or *formal, sociology*. Third is the sociological study of the epistemological and metaphysical aspects of society, which he called *philosophical sociology*. In *Fundamental Problems,* which has only four chapters, Simmel devoted a separate chapter to each of these problem areas.

General Sociology Simmel began by noting that "general sociology" was concerned with the study "of the whole of historical life insofar as it is formed societally"—that is, through interaction. The process of historical development can be interpreted in a number of ways, however, and Simmel believed that it was necessary to distinguish the sociological from the nonsociological approach. For example, he indicated that Émile Durkheim saw historical development "as a process proceeding from organic commonness to mechanical simultaneousness," whereas Auguste Comte saw it as occurring through three distinct stages: theological, metaphysical, and positive.[14] Although both claims are reasonable, Simmel remarked, neither constitutes a justification for the existence of sociology. Rather, the historical development of those observable social structures studied by the existing disciplines (politics, economics, religion, law, language, and others) must be subjected to a sociological analysis by distinguishing between social forms and social contents. For example, when the history of religious communities and labor unions is studied, it is possible to show that the members of both are characterized by patterns of self-sacrifice and devotion to ideals. These similarities can, in principle, be summarized by abstract laws.

What Simmel was apparently arguing, although this is not entirely clear, is that studies of the contents of interaction can yield valid theoretical insights only when attention is paid to the more generic properties of the social structures in which people participate. However, his chapter on general sociology, which deals with the problem of the development of individuality in society, proceeds in ways that are, at best, confusing.[15] Thus, the overall result is that readers are left wondering just what the subject of general sociology is and how it relates to the other problem areas.

Pure, or Formal, Sociology   For Simmel, "pure, or formal, sociology" consists of investigating "the societal forms themselves." Thus, when "society is conceived as interaction among individuals, the description of this interaction is the task of the science of society in its strictest and most essential sense."[16] Simmel's problem was thus to isolate and identify fundamental forms of interaction. In his earlier work, he attempted to do this by focusing on a number of less observable but highly significant social forms, which can be divided (roughly) into two general categories, although he did not use these labels: (1) generic social processes, such as differentiation, conflict, and exchange, and (2) structured role relationships, such as the role of the stranger in society. Nearly all of his substantive work consists of studies of these less observable social forms. For example, a partial listing of the table of contents of *Sociology: Studies in the Forms of Sociation* reveals that the following topics are considered:[17]

1. The quantitative determinateness of the group
2. Superordination and subordination
3. Conflict
4. The secret and the secret society
5. Note on adornment
6. The intersection of social circles (the web of group affiliations)
7. The poor
8. The self-preservation of the group

9. Note on faithfulness and gratitude
10. Note on the stranger
11. The enlargement of the group and the development of the individual
12. Note on nobility

Yet Simmel's description of pure, or formal, sociology suffers from a fundamental defect: It does not remedy the "methodological difficulty" referred to.[18] In *Fundamental Problems,* Simmel failed to develop a precise method for either identifying the most basic forms of interaction or analyzing their systematic variation.

Philosophical Sociology Simmel's "philosophical sociology" is an attempt to recognize the importance of philosophical issues in the development of sociology as an academic discipline. As he put it, the modern scientific attitude toward the nature of empirical facts suggests a "complex of questions concerning the fact 'society.'" These questions are philosophical, and they center on epistemology and metaphysics. The epistemological problem has to do with one of the main cognitive presuppositions underlying sociological research: Is society the purpose of human existence, or is it merely a means for individual ends?[19] Simmel's explanatory chapter on philosophical sociology deals with this question by studying the relationship between the individual and society in the eighteenth and nineteenth centuries.[20] As with the other chapters in *Fundamental Problems,* however, this material is so confusing that it is of little use. Apparently, Simmel wanted to argue that questions about the purpose of society or the reasons for individual existence could not be answered in scientific terms, but even this reasonable conclusion is uncertain.[21] Ultimately, then, his vision of philosophical sociology has simply been ignored, mainly because his analysis is both superficial and unclear.

In the end, Simmel had to confess that he had failed to lay a complete methodological foundation for the new discipline. This failure stems from his uncertainty about his ability to isolate truly basic or generic structures and processes. Thus, both *Sociology* and *Fundamental Problems* contain disclaimers suggesting that his analysis of specific topics—such as the significance of group affiliations, the functions of social conflict, and the process of social exchange—can only demonstrate the potential value of an analysis of social forms.[22] We now examine Simmel's three most important studies in formal, or pure, sociology.

THE WEB OF GROUP AFFILIATIONS

"The Web of Group Affiliations" is a sociological analysis of how patterns of group participation are altered with social differentiation and the consequences of such alterations for people's everyday behavior. Simmel first dealt with this topic in his *Social Differentiation* (1890).[23] However, this early version is not very useful, and the text explicated here is taken from *Sociology: Studies in the Forms of Sociation*. Like all the classical sociologists, Simmel saw a general historical tendency toward increasing social differentiation in modern industrialized societies. Rather than tracing this development either chronologically or by increased functional specialization, he focused on the nature and significance of group memberships. In this way, he was able to identify a unique social form.

The Web of Group Affiliations as a Social Form

Social forms refer to the modes of interaction through which people attain their purposes, or goals. In "The Web of Group Affiliations" Simmel was interested in the extent to which changes in the network of social structures making up society affect people. Indeed, he believed that the number of groups a person belongs to and the basis on which they are formed influence interaction apart from the interests that the groups are intended to satisfy.[24]

One of the most important variables influencing the number of groups to which people belong, as well as the basis of their attachment to groups, is the degree of *social differentiation,* or the number of different activities or structures organizing these activities. For example, in a hunting and gathering society, almost all tasks are done in and by the family (gathering and producing food, educating children, worshiping gods, making law, and the like). Thus, people have only a few roles in an undifferentiated society. This simplicity means that everyone they know is just like them because the accident of birth means that people interact with one another and establish ties within a homogeneous group. By contrast, in an industrial society, many important tasks are divided up. This increase in complexity, which sociologists call differentiation, affects interaction. People still produce goods, worship, educate, and adjudicate, but they do so differently. This disparity occurs because people increasingly choose which groups to belong to based on "similarity of talents, inclinations, activities," and other factors over which they have some control.[25] Thus, people play a far greater number and variety of roles and, in so doing, often interact with others different from themselves. The remainder of "The Web of Group Affiliations" explores the structural changes that result. In this way, Simmel demonstrated how a sociological analysis can reveal what happens to people "insofar as they form groups and are determined by their group existence because of interaction." He also demonstrated some implications of modernity.

Structural Changes
Accompanying Social Differentiation

Simmel observed that the process of social differentiation produced two fundamental changes in patterns of interaction. First, the principle underlying group formation changed, in his words, from *organic* to *rational* criteria. As Simmel uses it, the term "organic" is a biological metaphor suggesting that a family or village is like a living organism in which the parts are inherently connected.[26] Thus, when groups have an "organic" basis, people belong to them based on birth—into a family, a religion, village—and they are so strongly identified with the group to which they belong that they are not seen as individuals in their own right. In Shakespeare's play, *Romeo and Juliet,* for example, Romeo did not have an identity apart from his family and village; they constituted who he was. This is why his banishment was so devastating. In contrast, the term "rational" suggests the use of reason and logic. Thus, as Simmel uses the term, when groups have a "rational" basis, people belong by choice. For example, Simmel noted that English trade unions had originally "tended toward local exclusiveness" and had been closed to workers who came from other cities or regions,[27] but over time workers ended their dependence on local relationships, choosing to build and join national unions to pursue their interests.[28]

Second, social differentiation also leads to an increase in the number of groups that people can join. When groups have an "organic" basis, people can only belong to a few primary groups (that is, small, intimate, face-to-face groups): their family, their village, and not much more. In contrast, when groups have a "rational" basis, people can join a greater number and variety of them, based on skill, mutual interests, money, and other types of commonality. Simmel observed a trend in modern societies for people to join many groups and for such affiliations to be based on conscious reflection. This tendency applies even to intimate relationships, such as marriages. Many of these groups, however, are larger and more formal and are called *secondary groups.* Thus, people can also belong to occupational groups of various sorts, purely social groups, and a virtually unlimited number of special-interest groups. Further, individuals might also identify themselves as members of a social class and a military reserve unit. Finally, they might see themselves as citizens of cities, states, regions, and nations. Not surprisingly, Simmel concluded,

> This is a great variety of groups. Some of these groups are integrated. Others are, however, so arranged that one group appears as the original focus of an individual's affiliation, from which he then turns toward affiliation with other, quite different groups on the basis of his special qualities, which distinguish him from other members of his primary group.[29]

Put differently, group affiliations in differentiated societies are characterized by a superstructure of secondary groups that develops beyond primary group membership. From Simmel's point of view, the most important sociological characteristics of these groups—both primary and secondary—are that individuals choose to affiliate, everyone belongs to different groups, and people are often treated as individuals with unique experiences. These attributes mean that in many important respects, every person differs from every other.

The Consequences of Differentiation

The implications of this change are profound. Simmel suggests, for example, that when groups are formed by choice and people belong to a large number of them, the possibility of role conflict arises because membership in groups places competing demands on people. "As the individual leaves his established position within one primary group, he comes to stand at a point at which many groups 'intersect.'" As a result, "external and internal conflicts arise through the multiplicity of group affiliations, which threaten the individual with psychological tensions or even a schizophrenic break."[30] Thus, it is now common for people to have multiple obligations: as spouse, parent, employee, club member, student, and so forth. Sometimes these duties lead to hard choices; this happens, for example, when obligations to one's employer compete with obligations to one's family. Usually, Simmel says, people try to balance their competing responsibilities by keeping them spatially and temporally separate. Nonetheless, the impact of conflicting expectations can lead to psychological stress and, hence, influence behavior.

One example of the impact of role conflict and the resulting stress occurs in gender relations—although Simmel did not say much about this topic. For example, employed wives often have two roles that seem incompatible: keeping house and earning a living. One way married women resolve this conflict is by lowering their aspirations, working part-time, and other schemes designed to make their lives a little easier. Another way is to get a divorce. After all, Simmel's analysis implies that a marriage is simply one group among many in modern societies; it remains special to be sure, and intimate, but one that people choose to form and sometimes choose to leave—as they do other groups.[31]

The problem of role conflict in gender relations seems to be a negative consequence of modernity, a perception that leads some observers to wishful thinking. If, it is said, we could return to a time when men's and women's roles were assigned at birth and expectations were clear, we would not have so many problems today. There might be fewer divorces, for example. This kind of argument suggests that modern societies are demoralized in two senses: We lack moral guidance and we lack morale.[32] Regardless of whether you agree with such judgments, note that our understanding of the source of role conflict in today's world stems from a simple theoretical proposition that specifies the impact of differentiation: increased choice in both group affiliation and the number of groups to which people belong. Moreover, the connection between differentiation and the divorce rate is interesting and not obvious—until one thinks through the implications of Simmel's ideas.

Simmel's analysis leads to other insights about modern societies, insights that reveal some of the positive consequences of modernity. For example, if it is accurate, people now play many different roles in contexts that are spatially and temporally separated: spouse, parent, son or daughter, athlete, employee, and political activist, among other roles. This list, which could be extended, gives each of them a distinct identity in relationship to other people. These others also have a unique set of characteristics (roles) that make them distinct. Thus, Simmel's theory implies that the changes produced by social differentiation lead to greater individuality (what he called a "core of inner unity") that makes each person discrete. In Simmel's words, "the objective structure of society provides a framework within which an individual's non-interchangeable and singular characteristics may develop and find expression."[33] Such a result is impossible when everyone resembles everyone else. Ironically, then, modernity not only produces role conflict and psychological stress but also creates the conditions under which individuality emerges.

Such individuality emerges precisely because people in modern societies can, indeed must, make choices. Moreover, they must adjust their behavior to different people in different situations—an insight that carries many implications. For example, as people choose and become aware of their uniqueness they enjoy greater personal freedom. As Simmel put it, although "the narrowly circumscribed and strict custom of earlier conditions was one in which the social group as a whole . . . regulated the conduct of the individual in the most varied ways," such regulation is not possible in differentiated societies because people belong to so many different groups.[34] It is not accidental, from this point of view, that the ideology of personal freedom as an inalienable right of every adult arose during the last two centuries. Its structural basis, Simmel said, lies in social differentiation.

These insights lead to others. For example, when people play many roles and face conflicting expectations, they develop the capacity for empathy—the ability to identify with and understand another's situation or motives.[35] This capacity can sometimes reduce the level of conflict between people. Thus, the increasing complexity of modern societies provides a structural basis for an important personality characteristic. Note two implications of this argument: First, role conflict now appears to be a positive feature of modern societies. Second, the distribution of psychological characteristics in a population (people's sense of individuality and empathy, for example) does not happen by chance; they reflect the social structure. In addition, although role conflict burdens individuals, it also forces them to make choices and thereby encourages creativity.[36] After all, in a complex society, roles cannot be taken for granted, they must be negotiated; so people have to consider both their own and others' situations and be creative. Moreover, the aptitude for thinking imaginatively and originally extends to all arenas of life as people confront problems. Thus, "the development of mental ability takes place together with the grasp of the complexity of social roles" in modern societies.[37] The logic of this analysis suggests that modernity results from and, at the same time, produces a spiral effect such that as societies became more differentiated, more people became creative, and as more people became creative, societies became more complex. (See Figure 14.1.)

CONFLICT

Although his initial sketch of "The Sociology of Conflict" appeared in 1903, the basis for our commentary is a much-revised version that was included as a chapter in *Sociology: Studies in the Forms of Sociation*.[38] Simmel began the latter essay by remarking that although the social "significance of conflict has in principle never been disputed," it has most commonly been seen as a purely

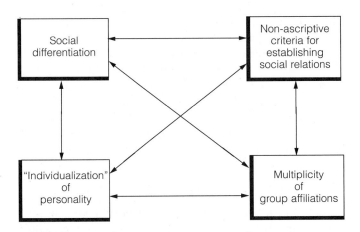

FIGURE 14.1 Model 1: Simmel's Image of Group Affiliations

destructive factor in people's relationships, one that should be prevented from occurring if possible. He believed that this orientation stemmed from an emphasis on exploring the contents of interaction; people observe the destructive consequences of conflict on other individuals (both physically and psychologically) and assume that it must have a similar effect on collectivities. In Simmel's view, this emphasis is shortsighted because it fails to recognize that conflict often serves as a means of maintaining or increasing integration within groups. In his words, "it is a way of achieving some kind of unity." For example, people's ability to express their hostilities toward one another can give them a sense of control over their destinies and thereby increase social solidarity within a group.

Conflict as a Social Form

Human beings, Simmel observed, have an "*a priori* fighting instinct"; that is, they have an easily aroused sense of hostility toward others. Although this fighting instinct is probably the ultimate cause of social conflict, he said, humans are distinguished from other species because, in general, conflicts are means to goals rather than merely instinctual reactions to external stimuli. This fundamental principle in Simmel's discussion means that conflict is a vehicle by which individuals achieve their purposes in innumerable social contexts, such as marriage, work, play, politics, and religion. As such, conflict reveals certain common properties in all contexts, and hence it can be viewed as a basic social form.

Moreover, conflict is nearly always combined with cooperation: People agree on norms that regulate when, where, and how to fight with one another, and this is true in marriage, business, games, war, and theological disputes. As Simmel wrote, "there probably exists no social unit in which convergent and divergent currents among its members are not inseparably interwoven. An absolutely centripetal and harmonious group . . . not only is empirically unreal, it could show no real life process."[39] The importance of this fusion of conflict and cooperation can be seen most clearly in those instances where a cooperative element appears to be lacking, for example, interaction between muggers and their victims or when conflict is engendered exclusively by the lust to fight. Simmel believed that these examples were clearly limiting cases, however, for if "there is any consideration, any limit to violence, there already exists a socializing factor, even though only as the qualification of violence."[40] This is why he emphasized that social conflict was usually a means to a goal; its "superior purpose" implies that people can change or modify their tactics depending on the situation.

In his essay on conflict, Simmel sketched some of the alternative forms of conflict, the way in which they are combined with regulatory norms, and the significance that this form of interaction has for the groups to which people belong. He first examined how conflict within groups affects the reciprocal relations of the parties involved, then he turned to the consequences that conflict with an outgroup has for social relations within a group. The following sections deal with each of these topics.

Conflict within Groups

Simmel's investigation of the sociological significance of conflict within groups revolves around three forms: (1) conflicts in which the opposing parties possess common personal qualities, (2) conflicts in which the opposing parties perceive each other as a threat to the existence of the group, and (3) conflicts in which the opposing parties recognize and accept each other as legitimate opponents.

Conflict Among Those with Common Personal Qualities[41] Simmel noted that "people who have many common features often do one another worse or 'wronger' wrong than complete strangers do," mainly because they have so few differences that even the slightest conflict is magnified in its significance. As examples, he referred to conflict in "intimate relations," such as marriages, and to the relationship between renegades and their former colleagues. In both cases, the solidarity of the group is based on the parties' possessing many common (or complementary) characteristics. As a result, people are involved with one another as whole persons, and even small antagonisms between them can be highly inflammatory, regardless of the content of the disagreements. Thus, when conflict does occur, the resulting battle is sometimes so intense that previous areas of agreement are forgotten. Most of the time, Simmel observed, participants develop implicit or explicit norms that keep conflicts within manageable bounds. When emotions run high or when group members see the conflict as transcending their individual interests, however, the fight can become violent. At that point, he suggested, the very existence of those who differ might be taken as a threat to the group.

Conflict as a Threat to the Group[42] Conflict sometimes occurs among opponents who have common membership in a group. Simmel argued that this type of conflict should be treated as a distinct form because when a group is divided into conflicting elements, the antagonistic parties "hate each other not only on the concrete ground which produced the conflict but also on the sociological ground of hatred for the enemy of the group itself." Such antagonism is especially intense and can easily become violent, Simmel argued, because each party identifies itself as representing the group and sees the other as a mortal enemy of the collective.

Conflicts Among Recognized and Accepted Opponents Simmel distinguished two forms of conflict among parties who recognize and accept each other as opponents. When conflict is "direct," the opposing parties act squarely against eachother to obtain their goals.[43] When conflict is "indirect," the opponents interact only with a third party to obtain their goals. Simmel referred to this latter form of conflict as *competition*.[44] Yet both forms share certain distinguishing characteristics that differentiate them from the forms of conflict noted earlier: Opponents are seen to have a right to strive for the same goal, conflict is pursued mercilessly yet nonviolently, personal antagonisms and feelings of hostility are often excluded from the conflict, and the opponents either develop agreements among themselves or accept the imposition of overriding norms that regulate the conflict.

The purest examples of direct conflict are antagonistic games and conflicts over causes. In the playing of games, "one *unites* [precisely] in order to fight, and one fights under the mutually recognized control of norms and rules."[45] Similarly, in the case of conflicts over causes, such as legal battles, the opponents' essential unity is again the underlying basis for interaction because to fight in court, the opponents must always follow agreed-upon normative procedures. Thus, even as parties confront each other, they affirm their agreement on larger principles. The analysis of direct conflict within groups was, however, of less interest to Simmel, with the result that he did not devote much space to it. Rather, he emphasized the sociological importance of competition because this form of fighting most clearly illustrates how conflict can have positive social consequences. By proceeding indirectly, competition functions as a vital source of social solidarity within a group.

Although recognizing the destructive and even shameful aspects of competition to which Marx and other observers had pointed, Simmel argued that even after all its negative aspects were taken into account, competition had positive consequences for the group because it forced people to establish ties with one another, thereby increasing social solidarity. Because competition between parties proceeds by the opponents trying to win over a third party, each of them is implicated in a web of affiliations that connects them with one another.[46]

With some exceptions, Simmel noted, the process of competition is restricted because unregulated conflict can too easily become violent and lead to the destruction of the group itself.[47] Hence, all collectivities that allow competition usually regulate it in some fashion, either through interindividual restrictions, in which regulatory norms are simply agreed on by the participants, or through super-individual restrictions, in which laws and other normative principles are imposed on the competitors.[48] Indeed, the existence of competition often stimulates normative regulation, thereby providing a basis of social integration.

Finally, Simmel recognized instances in which groups or societies try to eliminate competition in the name of a higher principle. For instance, in socialist or communist societies, competition is suspended in favor of an emphasis on organizing individual efforts in such a way as to (1) eliminate the wasted energy that accompanies conflict and (2) provide for the common good. Nonetheless, Simmel appears to have regarded a competitive environment as more useful than a noncompetitive one in modern, highly differentiated societies, not only in economic terms but also in most other arenas of social life. He believed that such an environment provided an outlet for people's "fighting instincts" that redounded to the common good and provided a stimulus for regulatory agreements that also contributed to the common good.

Conflict Between Groups

In the final section of his essay, Simmel examined the consequences that conflict between groups has "for the inner structure of each party itself."[49] Put differently, he was concerned with understanding the effect that conflict has on social relationships within each respective party to the conflict. To make his point, Simmel identified the following consequences of conflict between groups: (1) it increases the degree of centralization of authority within each group; (2) it increases the degree of social solidarity within each group and, at the same time, decreases the level of tolerance for deviance and dissent; and (3) it increases likelihood of coalitions among groups having similar opponents.

Conflict and Centralization[50] Just as fighters must psychologically "pull themselves together," Simmel observed, so must a group when it is engaged in conflict with another group. There is a "need for centralization, for the tight pulling together of all elements, which alone guarantees their use, without loss of energy and time, for whatever the requirements of the moment may be." This necessity is greatest during war, which "needs a centralistic intensification of the group form." In addition, Simmel noted, the development and maintenance of a centralized group is often "guaranteed best by despotism," and he argued that a centralized and despotic regime was more likely to wage war precisely because people's accumulated energies (or "hostile impulses") needed some means of expression. Finally, Simmel remarked that centralized groups generally preferred to engage in conflict with groups that were also centralized. For despite the conflict-producing consequences of fighting a tightly organized opponent, conflict with such an opponent can be more easily resolved, not only because the boundaries separating each side are clearly demarcated but also because each party "can supply a representative with whom one can negotiate with full certainty." For example, in conflicts between workers and employers or between nations, Simmel argued, it is often "better" if each side is organized so that conflict resolution can proceed in a systematic manner.

Conflict, Solidarity, and Intolerance[51] Simmel argued that conflict often increased social solidarity within each of the opposing groups. As he phrased it, a "tightening of the relations among [the party's] members and the intensification of its unity, in consciousness and in action, occur." This is especially true, he asserted, during wars or other violent conflicts. Moreover, increasing intolerance also accompanies rising solidarity, for whereas antagonistic members can often coexist during peacetime without harm to the group, this luxury is not possible during war. As a result, "groups in any sort of war situation are not tolerant" of deviance and dissent because they often see themselves as fighting for the existence of the group itself and demand total loyalty from members. Thus, in general, conflict between groups means that members must develop solidarity with one another, and those who cannot are often either expelled or punished. As a result of their intolerance toward deviance and dissent, Simmel remarked, groups in conflict often become smaller, as those who would compromise are silenced or cast out. This tendency can make an ongoing conflict more difficult to resolve, because "groups, and especially minorities, which live in conflict and persecution, often reject approaches or tolerance from the other side." The acceptance of such overtures would mean that "the closed nature of their opposition without which they cannot fight on would be blurred." Finally, Simmel suggested that the internal solidarity of many groups depends on their continued conflict with other parties and that their complete victory over an opponent could result in a lessening of internal social solidarity.

Conflict, Coalitions, and Group Formation[52] Under certain conditions, Simmel wrote, conflict between groups can lead to the formation of coalitions and ultimately to new solidarities among groups where none had existed before. In his words, "each element in a plurality may have its own opponent, but because this opponent is the same for all elements, they all unite—and in this case, they may, prior to that, not have had anything to do with each other." Sometimes such combinations are only for a single purpose, and the allies' solidarity declines immediately at the conclusion of the conflict. However, Simmel argued, when coalitions are engaged in wars or other violent conflicts and when their members become highly interdependent over a long period, more cohesive social relations are likely to ensue. This phenomenon is even more pronounced when a coalition is subjected to an ongoing or relatively permanent threat. As Simmel wrote, "the synthetic strength of a common opposition may be determined, not [only] by the number of shared points of interest, but [also] by the duration and intensity of the unification. In this case, it is especially favorable to the unification if instead of an actual fight with an enemy, there is a permanent *threat* by him."

Like so much of Simmel's work, the essay on conflict does not embody a unified conceptual perspective; rather, we get a series of provocative insights. In addition, as always with discursive writings, problems arise in presenting his insights. Nonetheless, we can extrapolate a model of the process of differentiation in Figure 14.2. As societies differentiate (become more complex), the number of organized units and their potential for conflict increases. Increased numbers of units, per se, create pressures for regulation of social relations by such mechanisms as centralization of power, laws, courts, mediating agencies, and coalitions among varying social units. Conflict escalates these pressures while unifying or consolidating social units structurally (centralization of authority, normative clarity, increased sanctioning) and ideologically (increased salience of beliefs and values). If conflicts are sufficiently frequent, low in intensity, and regulated, they release tensions, thereby encouraging further differentiation and elaboration of regulative structures. Such structures also encourage further differentiation by providing the capacity to coordinate increased numbers of units, manage tensions among them, and reduce their respective sense of threat when in potential conflict.

THE PHILOSOPHY OF MONEY

Simmel's *The Philosophy of Money* is a study of the social consequences of exchange relationships among human beings, with special emphasis on those forms of exchange in which money is used as an abstract measure of value. Like all his other work, *The Philosophy of Money* is an attempt at exposing how the forms of interaction affect the basic nature of social relations independently of their specific content. Although Simmel first considered this issue as early as 1889 in an untranslated article titled "The Psychology of Money," the final formulation of his ideas did not appear until the second edition of *The Philosophy of Money* was published in 1907.[53] Unlike the works reviewed earlier, *The Philosophy of Money* is both a sociological and philosophical treatise,[54] forcing us to extract the more sociological ideas from a philosophical text.

Exchange as a Social Form

The Philosophy of Money represents Simmel's effort to isolate another basic social form. Not all interaction is exchange, but exchange is a universal form of interaction.[55] In analyzing social exchange, Simmel concentrated on "economic exchange" in general and on money exchanges in particular. Although not all economic exchanges involve the use of money, money has historically come into increasing use as a medium of exchange. This historical trend, Simmel emphasized, reflects the process of social differentiation. But it does much more: Money is also a major cause and force behind this process. Thus, the sociological portions of *The Philosophy of Money* are devoted to analyzing the transforming effects on social life of the ever-increasing use of money in social relations.

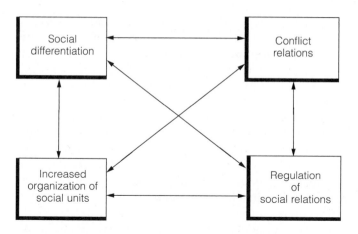

FIGURE 14.2 Model 3: Simmel's Image of Social Conflict

In analyzing differentiation from an exchange perspective, Simmel developed a number of philosophical assumptions and linked these to a sociological analysis of the modern world. Much like his friend and intellectual defender, Max Weber, Simmel was interested in understanding not just the forms of modern life but also their historical origins.[56] But unlike Weber, Simmel did not engage in detailed historical analyses, nor was he interested in constructing elaborate taxonomies. Rather, his works always sought to link certain philosophical views about humans and the social universe to understanding the properties of a particular social form. Thus, before explicating Simmel's specific analysis of money and exchange, it is necessary to place his analysis in philosophical context.

Simmel's Assumptions about Human Nature

In *The Philosophy of Money,* Simmel presented a vision of human nature that is implicit but less visible in his sociological works. He began by asserting that people are teleological beings; that is, they act on the environment in the pursuit of anticipated goals. In the essay on conflict, Simmel emphasized that this characteristic made human conflict different from that occurring among other animals. In *The Philosophy of Money* Simmel took the position that although people's goals would vary in accordance with their biological impulses and social needs, all action reflected human's ability to manipulate the environment in an attempt to realize goals. In so doing, individuals use a variety of "tools," but not just in the obvious material sense. Rather, people use more subtle, symbolic tools, such as language and money, to achieve their goals. In general, Simmel argued that the more tools people possessed, the greater would be their capacity to manipulate the environment, and hence, the more they could causally influence the flow of events. Moreover, the use of tools allows many events to be connected in chains that can form more extended social relations, as when money is used to buy a good. Money, for example, pays the salary of the seller, becomes profit for the manufacturer, and is transformed into wages for the worker, and so on, in a chain of social relations. Thus, Simmel thought all action reveals the properties presented in Figure 14.3. (As an interesting aside, compare Simmel's model with George Herbert Mead's analysis of the phases of "the act," reviewed in Chapter 23.)

Money, Simmel asserted, is the ultimate social tool because it is generalized; that is, people can use it in many ways to manipulate the environment to obtain their goals. This means that money can potentially connect many events and persons who would not otherwise be related. In an indirect way, then, the use of money allows a vast increase in the number of groups to which individuals can belong; thus, it is a prime force behind social differentiation.

A related assumption is that humans have the capacity to divide their world into an internal, subjective state and an external, objective state. This division occurs only when impulses are not immediately satisfied—that is, when the environment presents barriers and obstacles. When such barriers exist, humans separate their subjective experiences from the objects of the environment that are the source of need or impulse satisfaction. As Simmel emphasized,

> We desire objects only if they are not immediately given to us for our use and enjoyment; that is, to the extent that they resist our desire. The content of our desire becomes an object as soon as it is opposed to us, not only in the sense of being impervious to us, but also in terms of its distance as something not enjoyed.[57]

Value inheres in this subject-object division. In contrast with Marx, Simmel stressed that the value of an object existed not in the "labor power" required to produce it but in the extent to which it was both desired and unattainable; that is, value resides in the process of seeking objects that are scarce and distant. Value is thus tied to humans' basic capacity to distinguish a subjective from an objective world and in the relative difficulty in securing objects. Patterns of social organization, Simmel emphasized, perform much of this subject-object separation: They present barriers and obstacles, they create demands for some objects, and they determine how objects will circulate. The economic production of goods and their sale in a market is only a special case of the more general process of subject-object division among humans. Long before money, markets, and productive corporations existed, humans desired objects that were not easily obtainable. Thus, whether in the economic marketplace or the more general arena of life, value is a positive function of the extent to which an object of desire is difficult to obtain.[58]

Money, as Simmel showed, greatly increases the creation and acceleration of value because it provides a common yardstick for a quick calculation of values (that is, "how much" this or that commodity or service is "worth"). Moreover, as a "tool," money greatly facilitates the acquisition of objects; as money circulates and is used at each juncture to calculate values, all objects in the environment become assessed by their monetary value. Unlike Marx, Simmel did not see this as a perverse process but as a natural reflection of

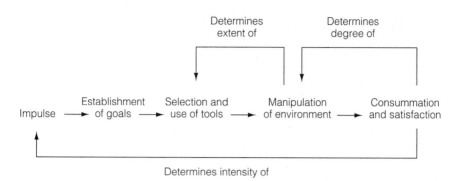

FIGURE 14.3 Simmel's Model of the Dynamics of Human Action

humans' innate capacity and need to create values for the objects of their environment.

Another assumption about human nature is to be found in Simmel's discussion of "world view."[59] People naturally seek stability and order in their world, he argued. They seek to know the place of objects and of their relationship to these objects. For example, Simmel observed, humans develop totems and religious rituals to regularize their relations to the supernatural; similarly, the development of money as a standardized measurement of value is but another manifestation of this tendency for humans to seek order and stability in their view of the world. By developing money, they can readily compare objects by their respective value and can therefore develop a "sense of order" about their environment.

In sum, then, Simmel believed that the development of money is an expression and extension of basic human nature. Money is a kind of tool in teleological acts, it is a way to express the value inherent in humans' capacity for subject-object division, and it is a means for attaining stability and order in people's worldview. All these innate tendencies are the driving force behind much human action, and this is why exchange is such a basic form of social interaction. For exchange is nothing more than the sacrificing of one object of value for the attainment of another. Money greatly facilitates this process because it provides a common reference point for calculating the values of objects that are exchanged.

Money in Social Exchange

For Simmel, social exchange involves the following elements:

1. The desire for a valued object that one does not have

2. The possession of the valued object by an identifiable other

3. The offer of an object of value to secure from another the desired object

4. The acceptance of this offer by the possessor of the valued object[60]

Contained in this portrayal of social exchange are several additional points that Simmel emphasized. First, value is idiosyncratic and is ultimately tied to an individual's impulses and needs. Of course, what is defined as valuable is typically circumscribed by cultural and social patterns, but how valuable an object is will be a positive function of both the intensity of a person's needs and the scarcity of the object. Second, much exchange involves efforts to manipulate situations so that the intensity of needs for an object is concealed and the availability of an object is made to seem less than it actually is. Inherent in exchange, therefore, is a basic tension that can often erupt into other social forms, such as conflict. Third, to possess an object is to lessen its value and to increase the value of objects that one does not possess. Fourth, exchanges will occur only if both parties perceive that the object given is less valuable than the one received.[61] Fifth, collective units as well as individuals participate in exchange relations and, hence, are subject to the four processes listed. Sixth, the more liquid the resources of an actor are in an exchange—that is, the more resources that can be used in many types of exchanges—the greater that actor's options and power will be. If an actor is not bound to exchange with any other and can readily withdraw resources and exchange them with another, that actor has considerable power to manipulate any exchange.

Economic exchange involving money is only a special case of this more general social form. But it is a very special case. When money becomes the predominant means for establishing value in social relationships, the properties and dynamics of social relations are transformed. This process of displacing other criteria of value, such as logic, ethics, and aesthetics, with a monetary criterion is precisely the long-term historical trend in societies. This trend is, as we mentioned earlier, both a cause and an effect of money as the medium of exchange. Money emerged to facilitate exchanges and to realize even more completely humans' basic needs. Once established, however, the use of money has the power to transform the structure of social relations in society. In seeking to understand how money has this power to alter social relations, Simmel's *The Philosophy of Money* becomes distinctly sociological.

Money and Its Consequences for Social Relations

In much of Simmel's work, there is an implicit functionalism. He often asked: What are the consequences of a social form for the larger social whole, or what functions does it perform? This functionalism is most evident in Simmel's analysis of conflict, but it is also found in his analysis of money. He asked two related questions in tracing the consequences of money for social patterns: (1) What are the consequences of money for the structure of society as a whole? (2) What are the consequences of money for individuals?

In answering these two questions, Simmel added to his lifelong preoccupation with several issues. We mention these to place his specific analysis of the consequences of money for society and the individual into context. One prominent theme in all Simmel's work is the dialectic between individual attachments to, and freedom from, groups. On the one hand, he praised social relations that allow individuals freedom to choose their options, but on the other hand, he was somewhat dismayed at the alienation of individuals from the collective fiber of society (although not to the extent of other theorists during his time). This theme is tied to another prominent concern in his work: the growing rationalization of society, or as he phrased the matter, the "objectification" of social life. As social relations lose their traditional and religious content, they become mediated by impersonal standards—law, intellect, logic, and money. The application of these standards increases individual freedom and social justice, but it also makes life less emotional and involving. It reduces relations to rational calculations, devoid of the emotional bonds that come with attachments to religious symbols and long-standing traditions. Simmel's analysis of the "functions" of money for individuals and the social whole must be viewed in the context of these two themes.

Money and the Social Whole Much like Weber, but in a less systematic way, Simmel was concerned with the historical trend toward rationalization, or objectification, of social relations. In general, humans tend to symbolize their relations, both with one another and with the natural environment. In the past, this was done with religious totems and then with laws. More recently, Simmel believed, people expressed their relationships with physical entities and with one another in monetary terms, with the result that they have lost intimate and direct contact with others as well as with the objects in their environment. Thus, money represents the ultimate objective symbolization of social relations—unlike material entities, money has no intrinsic value. Money merely represents values, and it is used to express the value of one object in relation to another. Although initial forms of money, such as coins of valuable metals and stones that could be converted into jewelry, possessed intrinsic value, the evolutionary trend is toward the use of paper money and credit, which merely express values in exchanges. As paper money and credit dominate, social relations in society are profoundly altered, in at least the following ways:

1. The use of money enables actors to make quick calculations of respective values. People do not have to bargain and haggle over the standards to be used in establishing the respective values of objects—whether commodities or labor. As a result, the "velocity" of exchange dramatically increases. People move through social relations more quickly.[62]

2. Because money increases the rate of social interaction and exchange, it also increases value. Simmel felt that people did not engage in exchange unless they perceived that they would get more than they gave up. Hence, the greater the rate of exchange is, the greater people's accumulation of value will be—that is, the more they will perceive that their needs and desires can be realized.[63]

3. The use of money as a liquid and nonspecific resource allows for much greater continuity in social relations. It prevents gaps from developing in social relations, as is often the case when people have only hard goods, such as food products or jewelry, to exchange in social relations. Money

gives people options to exchange almost anything because respective values can be readily calculated. As a result, there is greater continuity in social relations because all individuals can potentially engage in exchanges.[64]

4. In a related vein, money also allows the creation of multiple social ties. With money, people join groups other than those established at birth and thereby interact with many more others than is possible with a more restrictive medium of exchange.[65]

5. Money also allows exchanges among human beings located at great distances. As long as interaction involves exchange of concrete objects, there are limits to how distant people can be from one another and how many actors can participate in a sequence of exchanges. With money, these limitations are removed. Nations can engage in exchanges; individuals who never see one another—such as a factory worker and consumers of goods produced in the factory—can be indirectly connected in an exchange sequence (because some of the payment for a good or commodity will ultimately be translated into wages for the worker). Thus, money greatly extends the scope of social organization; money allows organization beyond face-to-face contact or beyond the simple barter of goods. With money, more and more people can become connected through direct and indirect linkages.[66]

6. Money also promotes social solidarity, in the sense that it represents a "trust"; that is, if people take money for goods or services, they believe that it can be used at a future date to buy other goods or services. This implicit trust in the capacity of money to meet future needs reinforces people's faith in and commitment to society.[67]

7. In a related argument, money increases the power of central authority, for the use of money requires that there be social stability and that a central authority guarantees the worth of money.[68] As exchange relations rely on government to maintain the stability of money, government acquires power. Moreover, money makes it much easier for a central government to tax people.[69] As long as only property could be taxed, there were limitations on the effectiveness of taxation by a remote central government because knowledge of property held would be incomplete and because property, such as land, is not easily converted into values that can be used to increase the power of central government. (How can, for example, property effectively buy labor services in the army of the administrative staff of a government?) As a liquid resource, however, tax money can be used to buy those services and goods necessary for effective central authority.

8. The creation of a tax on money also promotes a new basis of social solidarity. Because all social strata and other collectivities are subject to a monetary taxation system, they have at least one common goal: control and regulation of taxes imposed by the central government. This

commonality laces diverse groupings together because of their common interest in the taxing powers of government.

9. The use of money often extends into virtually all spheres of interaction. As an efficient means for comparing values, money replaces other, less efficient ways to calculate value. As money begins to penetrate all social relations, resistance to its influence in areas of personal value increases. Efforts to maintain the "personal element" in transactions increase, and norms about when it is inappropriate to use money become established. For example, traditions of paying a bride price vanish, using money to buy influence is considered much more offensive than personal persuasion, paying a price as punishment for certain crimes decreases, and so on.[70]

10. While these efforts are made to create spheres where the use of money declines, there is a general "quantification" and "objectification" of social relations.[71] Interactions become quantified as their value is expressed as money. As a result, moral constraints on what is possible decrease because anything is possible if one just has the money. Money releases people from the constraints of tradition and moral authority; money creates a system in which it is difficult to restrain individual aspirations and desires. Deviance and "pathology" are, therefore, more likely in systems where money becomes the prevalent medium of interaction.[72]

Money and the Individual For Simmel, the extensive use of money in social interaction has several consequences for individuals. Most of these reflect the inherent tension between individual freedom from constraint, on the one hand, and alienation and detachment from social groups, on the other hand. Money gives people new choices and options, but it also depersonalizes their social milieu. Simmel isolated the following consequences of money for individuals:

1. As a "tool," money is nonspecific and thus gives people an opportunity to pursue many diverse activities. Unlike less liquid forms of expressing value, money does not determine how it can be used. Hence, individuals in a society that uses money as its principal medium of exchange enjoy considerably more freedom of choice than is possible in a society that does not use money.[73]

2. In a similar vein, money gives people many options for self-expression. To the degree that individuals seek to express themselves through the objects of their possession, money allows unlimited means for self-expression. As a result, the use of money for self-expression leads to, and indeed encourages, diversity in a population that is no longer constrained in the pursuit of its needs (except, of course, by the amount of money its members have).[74]

3. Yet, money also creates a distance between one's sense of self and the objects of self-expression. With money, objects are easily acquired and discarded, and hence long-term attachments to objects do not develop.[75]

4. Money allows a person to enter many different types of social relations. One can, for example, buy such relationships by paying membership dues in organizations or by spending money on various activities that ensure contacts with particular types of people. Hence, money encourages a multiplicity of social relations and group memberships. At the same time, however, money discourages intimate attachments. Money increases the multiplicity of individuals' involvements, but it atomizes and compartmentalizes their activities and often keeps them from emotional involvement in each of their segregated activities. This trend is, Simmel felt, best personified by the division of labor that is made possible by money wages but that also compartmentalizes individuals, often alienating them from others and their work.[76]

5. Money also makes it less often necessary to know people personally because their money "speaks" for them. In systems without money, social relations are mediated by intimate knowledge of others, and adjustments among people are made through the particular characteristics of each individual. As money begins to mediate interaction, the need to know another personally is correspondingly reduced.

Thus, in Simmel's analysis of consequences, money is a mixed blessing for both the individual and society. Money allows greater freedom and provides new and multiple ways for connecting individuals. Money also isolates, atomizes, and even alienates individuals from the persons and objects in their social milieu. As a result, money alters the nature of social relations among individuals in society, and therefore an analysis of its consequences is decidedly a sociological topic.

Embedded in this descriptive analysis of the consequences of money is a more general model of exchange, differentiation, and individualization of the person. Some of the key elements of this model are delineated in Figure 15.3 (see p. 298). Social differentiation increases the volume, rate, velocity, and potential scope of social ties among individuals and groups because there are more different kinds of units and hence more opportunities for multiple and varied social contacts. Increases in the number of social relations create pressures for the use of objective or rational symbolic media, such as money, to facilitate exchange transactions; reciprocally, the use of money allows an ever-increasing volume of social ties because money makes it easy to determine the value of each actor's resources and to conduct social transactions. Increases in social exchanges mediated by money feed back on differentiation, encouraging further differentiation, which in turn increases the volume, rate, velocity, and scope of social ties mediated by money. Such processes cause ever more individualization of people—that is, increased involvement of only small parts of one's personality in groups, increased group affiliations, and greater potential alienation from society. Yet, these trends toward individualization are important contributors to the increased volume and rates of interaction, as well as the escalated use of money, on which social differentiation depends.

CRITICAL CONCLUSIONS

In evaluating Simmel's work as a whole, his major theoretical contribution to sociology resides in his concern with the basic forms of interaction. By looking beyond differences in the "contents" of diverse social relations and by attempting to uncover their more generic forms, he was able to show that seemingly diverse situations reveal basic similarities. He implicitly argued that such similarities could be expressed as abstract models or laws, although we can criticize Simmel for not explicitly stating these laws.

Thus, although Simmel did not employ the vocabulary of abstract theory, his many essays on different topics reveal a commitment to formulating abstract statements about basic forms of human relationships. This orientation, however, is not always clear because Simmel tended to argue by example. His works tend to focus on a wide variety of empirical topics, and even when he explored a particular type of social relation, such as conflict and exchange, the discussion proceeds with many illustrations. He would, for instance, talk about conflicts among individuals and wars among nation-states in virtually the same passage. Such tendencies give his work an inductive and descriptive flair, but a more careful reading indicates that he clearly held a deductive view of theory in sociology.[77] For example, if conflict between such diverse entities as two individuals and two nations reveals certain common forms, diverse empirical situations can be understood by the same abstract law, or principle.

Simmel's work is often difficult to read and understand because he jumps from topic to topic, from the micro to the macro, and from the historical past to contemporary situations in his time. If we keep in mind, however, that this seeming lack of focus represents an effort to use abstract models and principles to explain many diverse empirical cases, much of the confusion surrounding his work recedes. His goal is similar to that of all theorists: to explain many empirical events with a few highly abstract models and principles.

In the next chapter, we make more explicit these theoretical principles, but before moving on, we must note what he failed to do. He promised a pure sociology, but what emerges are only snippets of insight. We are not given a coherent theme guiding his work, although one can be extracted (as will be shown in the next chapter). Rather, Simmel jumps from topic to topic, never really stating what the generic properties of the social universe are. Without such statements on the fundamental properties of pure sociology, it is hard to know what forces in the social universe are to be the subject matter of sociological theory. Of course, Simmel's need to lecture to lay audiences contributed to this great failing, but still, in several decades of sociological inquiry he never produced a formal sociology that could guide subsequent generations of theorists. He had time, but he did not deliver.

Many might criticize Simmel for his implicit functionalism. He tended to ask: What are the consequences of a phenomenon—for example, differentiation, conflict, exchange, money—for the social whole? Such questions are functional because they analyze social processes in terms of their outcomes.

Simmel did not fall into the functionalist trap of seeing outcomes as the causes of these very outcomes, but his work does tend to emphasize the positive outcomes. True, he recognized the atomizing and alienating effects of differentiation of structure and objectification (rationalization) of social patterns, but in general, he tends to see conflict, money, and differentiation in terms of their positive outcomes. In some ways, this orientation is refreshing because most German theorists tended to see modernity as evil and as doing harmful things to people. In contrast, Simmel argued that the great events that were making society more complex, impersonal, and objectified could free individuals from constraints and give them options not available in simpler societies. Moreover, he saw the potential for low-intensity, frequent, and regulated conflicts in differentiated societies as potentially increasing their integration. In a sense, then, Simmel's work stands as a corrective to the rather dreary prognosis of Max Weber about rationalization or to the polemical views of Karl Marx on the evils of capitalism.

Simmel has enjoyed a great re-birth in late twentieth century sociology, spilling over into the twenty-first century, because he recognized historical trends that have been picked up by scholars within contemporary "postmodern" theory. Simmel saw that differentiation and the spread of exchanges using money created a new kind of person, one with potentially as many identities as affiliations in diverse groups. This theme has been used to condemn late capitalist society as destroying a unified self; Simmel recognized this potential, but unlike postmodernists, he saw the liberating effects of being able to fashion one's own group affiliations and, hence, one's identity. Simmel more than any other theorist of the classical founders saw the transforming effects of money and markets on society. For postmodernists, everything is "commodified"—people, self, group culture, sacred symbols, affiliations—and they see this power of money and markets to create a world of unstable groups structures whose culture is marketed and bought by people seeking to purchase an identity. Simmel saw this potential, but again, he came down on the more positive side, emphasizing that people are freed of the oppressive constraints associated with traditional, communal societies. Thus, because Simmel addressed the issues of interest to postmodernism, he has moved from a more minor place in sociology's pantheon to a plane just below that of Marx, Weber, and Durkheim.

NOTES

1. The translation appeared the following year. See Georg Simmel, "The Problem of Sociology," *Annals of the American Academy of Political and Social Science* 6 (1895), pp. 412–423.

2. Georg Simmel, "The Problem of Sociology," in *Essays on Sociology, Philosophy and Aesthetics by Georg*

Simmel et al., ed. and trans. Kurt Wolff (New York: Harper & Row, 1959), pp. 310–336. This is chapter 1 of Simmel's *Sociology: Studies in the Forms of Sociation* (1908). This book, which is Simmel's major work, has not been translated into English, but portions appear in various edited collections. See note 17.

3. Simmel, *Fundamental Problems of Sociology,* appears as part 1 of *The Sociology of Georg Simmel,* trans. Kurt Wolff (New York: Free Press, 1950), pp. 3–86.

4. Ibid., p. 4.

5. This remark is from the preface to *Sociology: Studies in the Forms of Sociation;* it is quoted in Kurt Wolff's introduction to *The Sociology of Georg Simmel,* p. xxvi. For another review and analysis of Simmel's methodology, see Donald N. Levine, "Simmel and Parsons Reconsidered," *American Journal of Sociology* 96 (1991), pp. 1097–1116, "Simmel as a Resource for Sociological Metatheory," *Sociological Theory* 7 (1989), pp. 161–174, and "Sociology's Quest for the Classics: The Case of Simmel," in *The Future of Sociological Classics,* ed. Buford Rhea (London: Allen & Unwin, 1981), pp. 60–80.

6. Simmel, *Fundamental Problems,* p. 9.

7. Ibid., p. 12.

8. Ibid., p. 11.

9. The most well-known criticism of Simmel's presumably excessive "formalism" are by Theodore Abel, *Systematic Sociology in Germany* (New York: Octagon, 1965), and Pitirim Sorokin, *Contemporary Sociological Theories* (New York: Harper & Row, 1928). The best defenses of Simmel against this spurious charge are those by F. H. Tenbruck, "Formal Sociology," in *Essays in Sociology,* pp. 61–69, and Donald N. Levine, "Simmel and Parsons Reconsidered," as well as his *Simmel and Parsons: Two Approaches to the Study of Society* (New York: Arno, 1980).

10. Simmel, "Problem of Sociology," p. 315.

11. Simmel, *Fundamental Problems,* p. 18.

12. Ibid., p. 22 (emphasis in original). See Levine, "Simmel and Parsons Reconsidered," for elaboration on this point of emphasis in Simmel's work.

13. Simmel, "Problem of Sociology," pp. 323–324.

14. Simmel, *Fundamental Problems,* pp. 19–20. In general, Simmel does not cite his sources. On these two pages, however, his references are relatively clear, even though neither Durkheim nor Comte is mentioned by name.

15. Ibid., pp. 26–39.

16. Ibid., p. 22.

17. Items 1, 2, 4, 5, and 9 are available in *The Sociology of Georg Simmel.* Items 3 and 6 are in *Conflict and the Web of Group Affiliations,* trans. Reinhard Bendix (New York: Free Press, 1955). Item 10 is in *Essays on Sociology.* Item 8 is in the *American Journal of Sociology* 3 (March 1900), pp. 577–603. Items 7, 11, and 12 are in *Georg Simmel on Individuality and Social Forms,* trans. Donald Levine (Chicago: University of Chicago Press, 1971). The remaining chapters, about one-fourth of the book, are still untranslated. They deal with such topics as social psychology, hereditary office holding, the spatial organization of society, and the relationship between psychological and sociological phenomena.

18. Simmel, *Fundamental Problems,* pp. 40–57. See Levine, "Simmel and Parsons Reconsidered," p. 1107, and "Sociology's Quest for the Classics," pp. 69–71, for an effort to extract the basic forms from Simmel's work.

19. This same issue was dealt with ten years earlier in Georg Simmel, "Note on the Problem: How Is Society Possible?" in *Essays on Sociology,* pp. 337–356.

20. Simmel, *Fundamental Problems,* pp. 58–86.

21. Ibid., p. 25.

22. Ibid., p. 18.

23. See Georg Simmel, "The Intersection of Social Spheres," in *Georg Simmel: Sociologist and European,* trans. Peter Lawrence (New York: Barnes & Noble, 1976), pp. 95–110.

24. The remainder of this section draws on Leonard Beeghley, "Demystifying Theory: How the Theories of Georg

Simmel (and Others) Help Us to Make Sense of Modern Life," Chapter 34 in Jon Gubbay, Chris Middleton, and Chet Ballard (eds.), *The Blackwell Companion to Sociology* (New York: Blackwell, 1997).

25. Simmel, "Web of Group Affiliations," p. 127 in *Conflict and the Web of Group Affiliations* (New York: Free Press, 1955).

26. The classical theorists were often groping for how to communicate. Partly for this reason, they often used the same or similar concepts with quite different meanings. Thus, Simmel's notion of "organic" is the opposite of Durkheim's (considered in Chapter 17).

27. Simmel, "Web of Group Affiliations," p. 129.

28. Ibid., p. 137.

29. Ibid., p. 137.

30. Ibid., p. 141.

31. See Leonard Beeghley, *What Does Your Wife Do? Gender and the Transformation of Family Life* (Boulder, CO: Westview, 1996).

32. See Gertrude Himmelfarb, *The Demoralization of Society: From Victorian Virtues to Modern Values* (New York: Knopf, 1995).

33. Simmel, "Web of Group Affiliations," p. 150; see also pp. 139, 149, 151.

34. Ibid., p. 165.

35. Robert K. Merton, *Social Theory and Social Structure* (New York: Free Press, 1968), p. 436.

36. Rose Laub Coser, *In Defense of Modernity* (Stanford, CA: Stanford University Press, 1991).

37. Ibid., p. 7.

38. Georg Simmel, "The Sociology of Conflict," *American Journal of Sociology* 9 (1903–1904), pp. 490–525, 672–689, 798–811.

39. Simmel, "Conflict," in *Conflict and the Web of Group Affiliations,* p. 15. See Levine, "Sociology's Quest for the Classics," p. 68, for an effort to extract

the formal properties of conflict in Simmel's essay.

40. Simmel, "Conflict," p. 26

41. Ibid., pp. 43–48.

42. Ibid., pp. 48–50.

43. Ibid., pp. 34–43.

44. Ibid., pp. 57–86.

45. Ibid., p. 35 (emphasis in original).

46. Ibid., p. 62.

47. Ibid., pp. 68–70. Simmel recognized that within families and to some extent within religious groups, the interests of the group often dictate that members refrain from competing with one another.

48. Ibid., p. 76.

49. Ibid., p. 87.

50. Ibid., pp. 88–91.

51. Ibid., pp. 17–19, 91–98.

52. Ibid., pp. 98–107.

53. Georg Simmel, "Psychologie des Geldes," *Jahrbücher für Gesetzgebung, Verwaltung und Volkswirtschaft* 23 (1889), pp. 1251–1264. *The Philosophy of Money*, 2nd ed., trans. Tom Bottomore and David Frisby (Boston: Routledge, 1990).

54. It is often forgotten that Simmel was a philosopher as well as a sociologist. As noted in Chapter 10, he wrote books and articles on the works of Kant, Goethe, Schopenhauer, and Nietzsche and considered more general philosophical issues and problems as well.

55. Simmel, *Philosophy of Money*, p. 82.

56. As noted in Chapter 13, Simmel was excluded from senior academic positions for much of his career, and his work was often attacked. Weber was one of his most consistent defenders and apparently helped him maintain at least a marginal intellectual standing in Germany. But Weber revealed some ambivalence toward Simmel; see Weber's "George Simmel as a Sociologist," with an introduction by Donald N. Levine, *Social Research* 39 (1972), pp. 154–165.

57. Simmel, *Philosophy of Money*, p. 66.

58. Ibid., pp. 80–98.

59. Ibid., pp. 102–110.

60. Ibid., pp. 85–88.

61. Surprisingly, Simmel did not explore in any detail the consequences of unbalanced exchanges, in which people are forced to give up a more valuable object for a less valuable one. He simply assumed that at the time of exchange, one party felt that an increase in value had occurred. Retrospectively, a redefinition might occur, but the exchange will not occur if at the moment people do not perceive that they are receiving more value than they are giving up.

62. Simmel, *Philosophy of Money*, pp. 143, 488–512.

63. Ibid., p. 292.

64. Ibid., p. 124.

54. Ibid., p. 307.

66. Ibid., pp. 180–186.

67. Ibid., pp. 177–178.

68. Ibid., pp. 171–184.

69. Ibid., p. 317.

70. Ibid., pp. 369–387.

71. Ibid., p. 393.

72. Ibid., p. 404. Many analysts of Simmel emphasize these pathologies, especially when factoring in his analysis in other works, such as those translated by Peter Etzkorn in *The Conflict in Modern Culture and Other Essays* (New York: Teacher's College Press, 1980) and such famous essays as "The Metropolis and Mental Life." See, for example, David Frisby, *Sociological Impressionism: A Reassessment of Simmel's Social Theory* (London: Heinemann, 1981), and *Georg Simmel* (Chichester, England: Ellis Horwood, 1984). For a balanced assessment that corresponds to the one offered here, see Donald R. Levine, "Simmel as Educator: On Individuality and Modern Culture," *Theory, Culture and Society* 8 (1991), pp. 99–117.

73. Simmel, *Philosophy of Money*, p. 307.

74. Ibid., pp. 326–327.

75. Ibid., p. 297.

76. Ibid., p. 454.

77. That is, explanation occurs by deduction to empirical cases from abstract laws, which are universal and context-free.

15

✳

Simmel's Theoretical Legacy

I f there is a central theme in Simmel's work, it is the effects of the process of differentiation on forms of social relations, and vice versa. Like his fellow countryman, Max Weber, Simmel saw the growing reliance on rationality and calculation over ascription and tradition in establishing social relations in differentiated societies. In particular, Simmel saw this shift in the basis of social relations as related to the use of money in market exchanges. With differentiation, money, and markets, the number and diversity of ties increases, and this shift in patterns of group affiliations increases the level of individuality in society. People can now tailor their relations to their unique needs and tastes, and they can carry varying identities in different groups. This individualization gives individuals more freedom and choice, but there is a cost to pay in the increased potential for marginality, atomization, and alienation.

Like Karl Marx, Simmel examined conflict in complex differentiated societies. Unlike Marx, Simmel saw conflict can potentially be an integrative force, especially if it is frequent and of low intensity. Indeed, in complex societies where diverse groups reveal different interests, conflict is inevitable, but if conflict can release tensions and be regulated by government and law, then it can help sustain society rather than rip it apart.

These are the general themes that run through Simmel's main core of work. There are many essays in Simmel's sociology, but those focusing on differentiation and the changing basis of integration in society for both individuals and groups mark Simmel's enduring theoretical legacy. Let us see if we can present these themes in a general model.

SIMMEL'S CAUSAL MODELS
OF SOCIAL ORGANIZATION

Simmel's Model of Social Organization

In Figure 15.1, we present the key elements in Simmel's model of social organization. For Simmel, the major transformation of modern society was the increasing level of differentiation among individuals, their roles, and social relations as well as the groupings to which they belong. Thus, the model begins at the left with the level of social differentiation, and like Simmel's argument, it moves to the right enumerating the outcomes of differentiation. As Simmel would have emphasized, these outcomes feed back as reverse causal effects (moving from right to left) influencing the values for the forces that cause them. In Simmel's approach, all of the direct, indirect, and reverse causal effects are positive, signaling his recognition that differentiation had unleashed forces that dramatically changed the form of social relations in society.

Increased differentiation has several consequences, the most important of which is to rationalize social relations and give individuals the ability to choose affiliations. This shift away from ascription as a basis for establishing social relations has the effect of increasing the multiplicity of social ties among individuals as well as the diversity of secondary groups to which individuals can belong. The introduction of money accelerates this shift in the basis for ties because it gives individuals a generalized marker of value to choose and buy affiliations. More generally, money alters people's worldviews and mindsets toward the calculation of utilities to be gained in social relations. The diversity of secondary groups, the multiplicity of ties, and the use of money as a marker of value all directly increase individualization of people. People can have different identities in various groups, and they have the ability to realize their idiosyncratic needs. The lower portions of the model denote some of the dynamics that accelerate individualization. When money circulates widely, diversity of demand increases because money can be used to buy whatever a person chooses; moreover, once individuals can express their tastes with money, this change encourages the further use of money and the rational calculation of utilities, both of which increase the level of differentiation. With diverse demand, markets can grow and differentiate, and as markets increase the volume and velocity of exchanges, reliance on money as the medium of exchange will become even greater. These self-reinforcing cycles give individuals further options to express their own needs, and as these processes feed back and increase the level of differentiation, rational bases for establishing social ties and group memberships allow individuals to express distinctive but also multiple identities.

Differentiation of secondary groups also increases rates of conflict. Differentiated groups have different interests, and these interests can lead to competition, one form of conflict. Conflict will increase the degree of organization within units (e.g., authority systems, boundaries, normative clarity, intolerance for deviance), and it often leads to the formation of coalitions among groups in larger networks and confederations. Most important, conflict

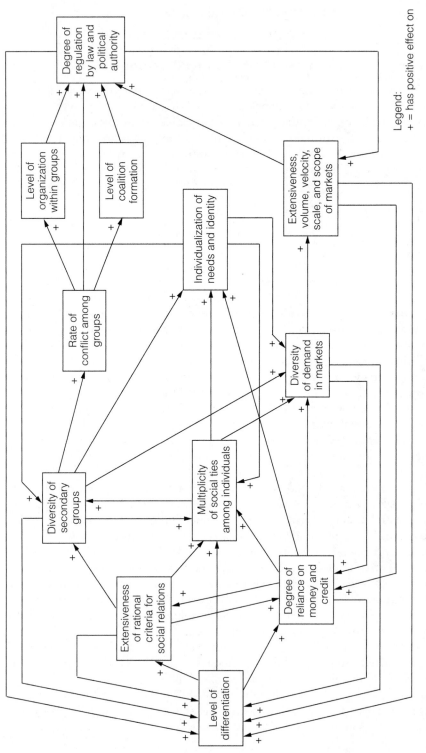

FIGURE 15.1 Simmel's Model of Social Organization

will increase the extent to which law and government begin to regulate social relations, and, if frequent and low-intensity conflicts can be consistently regulated, further differentiation is possible (as the long reverse causal chain across the top of the model underscores).

Political-legal regulation is also influenced by markets. Simmel argued that dynamic markets using money require political authority and laws to regulate exchanges; as markets become more extensive, they draw political authority into the economy. Reciprocally, with some degree of regulation, markets can expand further and set into motion the reverse causal paths back to differentiation. Thus, both conflict and markets will expand the use of power and law to regulate a society, and when this regulation can mitigate conflicts and facilitate exchanges, it creates the conditions for a new level of differentiation (and the outcomes of this differentiation that are enumerated in the model).

Thus, it is possible to pull a number of the prominent themes in Simmel's work into a general model. Perhaps more than any theorist of his time, Simmel recognized the power of reverse causal effects—that is, the outcomes of differentiation will accelerate each other as well as the general process of differentiation. The model in Figure 15.1 is, however, very general, and it only emphasizes the positive effects among forces. Simmel understood that there are pathologies inhering in these forces, and we can see these better by breaking particular paths in Figure 15.1 into separate models.

Simmel's Model of Differentiation and Group Affiliations

Simmel argues that social differentiation increases the proliferation of non-ascriptive, or "rational," criteria for group membership. Such non-ascriptive, rational criteria allow individuals greater freedom and choice in establishing social relations. This increased choice stimulates the proliferation of secondary groups, and conversely, the existence of such groups encourages the use of rational or choice-based criteria. For Simmel, then, social differentiation dramatically changes the nature of the individual's relationship to groups, and in turn, these changes in social relations alter the nature of individuals. The basic model employed by Simmel is outlined in Figure 15.2.

As secondary groups proliferate and non-ascriptive criteria are used in establishing ties, individuals can have multiple group affiliations. Although primary group affiliations still exist, an increasing proportion of social relations occur within secondary groups created for particular goals and ends (what Simmel called "contents"). Members of these groups make explicit calculations about their needs and interests in joining. As the number of secondary groups increases and as the proportion of attachments to them increases, *individualization* of personality also becomes more frequent. By individualization, Simmel meant that people can have (1) multiple identities tied to their unique web of groups and (2) choice in their affiliations tailored to their particular needs. Without differentiation, this kind of individualization is not possible; rather, people remain embedded in a smaller number of more primary groups on the basis of ascription rather than choice.

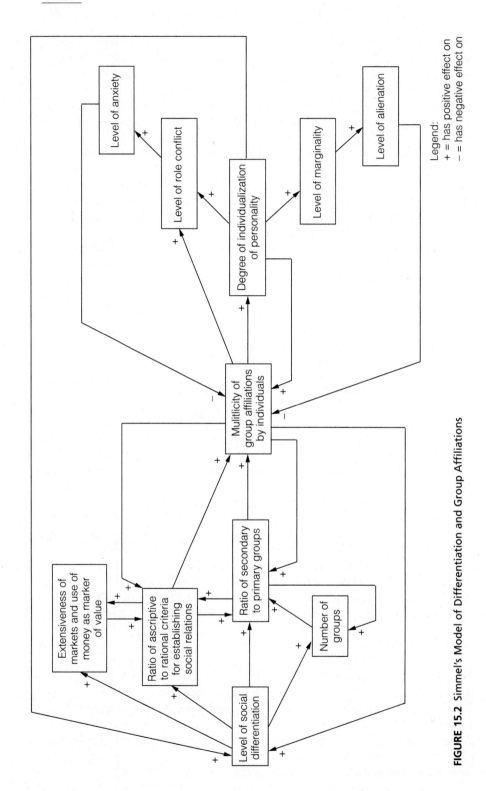

FIGURE 15.2 Simmel's Model of Differentiation and Group Affiliations

However, there are potential problems for individuals that reverberate back and, potentially, influence the multiplicity of ties and social differentiation. One problem is role conflict stemming from too many obligations attached to too many roles in too many groups. Under these conditions, performing a role or maintaining a social relation makes it difficult to do the same in other roles and relations, thereby generating stress and anxiety in the individual. Under these conditions, individuals will, if they can, cut back on affiliations (as is underscored by the negatively signed arrow from anxiety (generated by role conflict) to multiplicity of group affiliations). Another problem is marginality whereby the individual does not feel highly integrated into any one group. Rather, people stand between groups, being a part of many groups but also a stranger to each. This can lead individuals to cut back on secondary group memberships and seek more primary group affiliations.

Simmel argues that as non-ascriptive criteria are used in establishing social relations, *objectification* of the world ensues. This idea parallels Weber's notion of *rationalization,* but it focuses on the criteria that people use to affiliate with groups rather than on the structure of the groups themselves. Much as Weber did, Simmel saw this objectification as inevitable, once initiated. Simmel's assessment of rationalization is just the opposite of Weber's, however. Rationalization does not lead to an "iron cage"; on the contrary, there is now freedom of choice not possible in more traditional patterns of group affiliation, and this freedom can enhance the quality of people's lives.

Simmel's Model of Money, Markets, and Differentiation

Simmel's *The Philosophy of Money* contains a model of market dynamics and how these increase the level of differentiation in society. His model is very similar to the one developed by Weber, except Simmel sees markets and the rationality that they encourage as more liberating than constraining. In Figure 15.3, we outline this model (its similarity to Weber's can be seen by comparing it to Figure 12.3 on page 241). The model begins with increasing penetration of money into social exchanges, although the long reverse causal arrow from differentiation on the far right emphasizes that differentiation is one of the causes behind the use of money in exchanges. With money, it becomes possible to calculate values rapidly. In turn, this ability to make rapid calculations increases the velocity of social exchanges. An increase in the velocity of exchanges has a positive effect on social differentiation because it becomes possible to specialize and still secure resources from other specialists by virtue of high-volume, high-velocity markets. Once markets exist using money, these processes all feed back and accelerate the use of money and the velocity of exchanges, as is indicated by the positive signs on the arrows marking these reverse causal effects.

The row of causal effects running across the bottom of the figure recapitulates in the context of market dynamics Simmel's analysis of affiliations presented in the previous section. Money gives individuals freedom to choose, but it lowers the level of intimacy of relations. At the same time, money increases the potential for

298

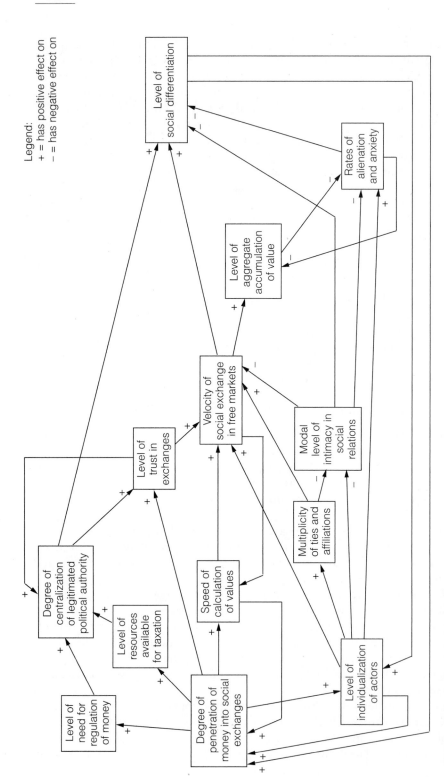

Legend:
+ = has positive effect on
− = has negative effect on

FIGURE 15.3 Simmel's Model of Markets, Money and Differentiation

anxiety and alienation to the extent that it encourages a high ratio of secondary to primary group affiliations; and anxiety and alienation stemming from multiple group affiliations can work against further differentiation.

A countervailing force to these problems is the effect of markets in increasing aggregate sense of value experienced by individuals. As Simmel noted, when people use money in markets, including markets for group memberships, they will do so because they perceive the bargain to be worthwhile. They engage in exchange because they think that they will receive increased value; and as many individuals do so, the aggregate sense of value in a society rises. When people sense that they are accumulating value (or rewards and utilities), this accumulated value can mitigate against anxiety or alienation (as is indicated by the negatively signed arrow from value to these potential problems). Thus, Simmel was criticizing both Karl Marx and Max Weber in their respective views that money, rational calculation, and markets are dehumanizing; rather, Simmel argued these forces allow individuals to enhance their sense of value or well being.

Across the top of Figure 15.3 is Simmel's theory of how money influences the growth of the state. The use of money requires regulation, if only to coin the money and to sustain its value in markets. Moreover, money also represents a form of liquid revenue to the state, and as a result, the government soon learns how to tax the flows of money in order to expand government. But Simmel recognized something more: The value of money becomes a critical force in legitimating government. If the state can sustain the stability of currency (that is, prevent inflation), then the value of money is preserved, and people will have positive sentiments toward political authority.

Moreover, money increases the level of trust in a society, particularly when the value of money remains stable. This enhanced sense of trust further legitimatizes the political system and the operation of markets, as is denoted by the positive arrows leaving level of trust in exchanges. This trust is generated because money has no intrinsic value by itself, especially paper money. As a result, when money is accepted in exchange, there is an implied trust: The money has value in future purchases by those who take it. The more money is used and trusted, the greater will be the aggregate sense of trust in the society. This trust encourages individuals to use their money in markets, and because the state is involved in maintaining the value of money, this trust is reflected back on government in the form of enhanced legitimacy. Thus, for Simmel, the basis of trust shifts from ascription, tradition, and emotional attachments to a more diffuse sense that the means of exchange—money—carries and sustains value. Should money lose its value because of an increase in the rate of inflation, the process works in the opposite way: people withdraw legitimacy from government; they do not trust others using money; markets become stagnant or even revert back to older forms of exchange such as barter (where goods are exchanged without the use of money); and, as this occurs, the level of differentiation in a society is arrested, or even reversed. Indeed, people may begin to use ascriptive criteria in forming fewer affiliations with more primary groups to recreate a sense of trust.

Simmel's Model of Social Differentiation, Conflict, and Societal Integration

Unlike his fellow Germans, Karl Marx and Max Weber, Simmel tended to focus on the positive outcomes, or functions, of conflict for societal integration. Integration can occur at two levels, within the parties to a conflict and across the broader society. Figure 15.4 outlines the model developed by Simmel on these two aspects of conflict and integration, although the forces denoted in the model give more attention to societal integration. Any conflict will increase the level of organization among the units engaged in conflict, but Simmel was most interested in frequent conflicts of low intensity because these kinds of conflicts offered the most potential for societal integration. Such conflicts are likely, Simmel argued, in differentiated systems, where they would not polarize all members into a few warring camps. Differentiated systems generate multiplicity of ties in secondary groups, and these groups are less likely to arouse the same emotions as more primary groups engaged in conflict. Moreover, given the many ties involved among a large number of groups, conflicts would be more frequent, but this frequency would mitigate against their intensity, as is emphasized by the negatively signed causal arrow connecting frequency to intensity of conflict.

Frequent conflicts of lower intensity would allow for their regulation by political authority and law. Indeed, conflict would be a principle force behind the expansion of government, and once legitimated political authority is in a position to manage conflicts, it can work to reduce their intensity. Thus, as the negatively signed causal arrows from frequency of conflict and extensiveness of political regulation to level of accumulated hostility emphasize, frequent conflicts of low intensity regulated by political authority and law release tensions and, thereby, decrease the level of accumulated hostility among units.

Coalitions among groups are likely under all conditions of conflict, and such coalitions will increase the organization of social units and create larger confederations of groups. But, the relationship between coalitions and societal integration is more complex: A certain amount of confederation promotes integration among units, but when a society becomes confederated into just a few flocks standing in conflict, societal integration decreases. Thus, the relationship between coalitions and integration depends on the number of coalitions; the more coalitions embracing a diversity of units, the more these confederations will work for integration, whereas the fewer are the coalitions and the more they embrace all units, the more potential for highly disruptive conflict that can destroy societal integration. This is why we have phrased the variable as *number of coalitions,* since this gives us a rough indicator of whether a society is composed of just a few confederated camps or many cross-cutting points of conflict that do not have the power to polarize a society.

In sum, then, these four models capture Simmel's general theory. We have pieced together somewhat disparate elements of Simmel's sociology, but this exercise demonstrates that Simmel did have an image of the key dynamics operating in differentiating social systems. Although he never presented a unified

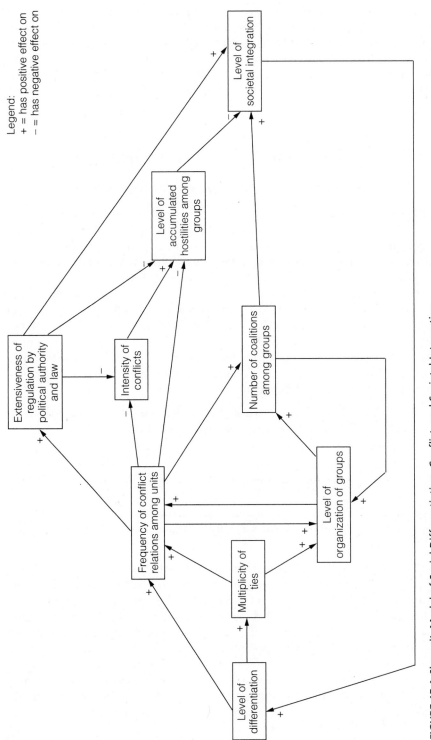

FIGURE 15.4 Simmel's Model of Social Differentiation, Conflict, and Societal Integration

theory, the models reveal a certain coherence and consistency in the variables presented. Differentiation is the master force, both as an outcome and cause of other forces. Within differentiated systems, a variety of forces are unleashed—money, markets, expansion of political authority and law, conflicts, coalitions, individualization, alienation and anxiety, and trust—and each of these forces reverberates off the others, while increasing or decreasing the potential for further differentiation.

SIMMEL'S THEORETICAL PRINCIPLES

Simmel's Analysis of the Process of Differentiation

Although many would not see his work in quite this way, we believe that Simmel was a theorist of differentiation. Like so many in his time, he was fascinated by the dramatic changes that were occurring with "modernity," and as emphasized earlier, he was concerned with the consequences of differentiation for forms of social relations. Moreover, he believed that as the forms of social relations changed with differentiation, these new forms would encourage further differentiation, at least to the point where people felt too marginal and alienated or too conflicted and anxious. Simmel did not examine the causes of differentiation in detail because he tended to see the outcomes of differentiation as accelerating differentiation. Still, we should try to express Simmel's argument on the causes of differentiation before moving onto the consequences or outcomes of differentiation:

1. The level of differentiation will increase when non-ascriptive media of social exchange, like money, are available, with the availability of non-ascriptive criteria increasing as
   a. Free markets using money as their medium of exchange proliferate.
   b. Secondary groups using non-ascriptive criteria for membership proliferate.
   c. Individualization of preferences increases.
   d. Centers of power increasingly rely on money to sustain their operations.

For Simmel, then, differentiation is a result of a breakdown of ascriptive criteria—such as tradition, community, family, and religion—as a basis for group membership and for determining individuals' preferences. When individuals are no longer controlled by ascription and, instead, can use non-ascriptive criteria like money or educational credentials for joining groups and for expressing preferences, the potential for greater diversity in ties and affiliations exists. This potential can be realized, however, only when markets operate as the mechanism for distribution goods, services, and even memberships. Markets rely on money, and together, money and markets allow people to calculate values (their preferences) and to make decisions based on these calculations. Yet, markets alone cannot drive differentiation; secondary groups, as both a consequence and cause of differentiation, must also become widespread. If individuals have few options beyond primary groups where membership is often ascribed at birth, their affiliations will remain in these primary groups, thus shutting down demand for memberships in diverse secondary groups (and hence, for differentiation of groups). Not only must markets and secondary groups prevail, but also some degree of individualization of people's tastes and preferences must have already occurred to make them interested in using markets and money to express these preferences and in making calculations of their interests in joining new kinds of groups. Finally, centers of power must have an interest in using liquid resources like money for their ends, and they must come to have an interest in maintaining the stability or purchasing power of money to sustain legitimacy. Under these conditions, differentiation will increase, but the converse is also true; as differentiation increases, it will push for reliance on non-ascriptive media of exchange, for the expansion of markets, for the multiplication of secondary groups, for individualizing tastes and preference expressed as demands in markets, for government regulation of relations among diverse groups, and for the maintenance of the value of money.

Differentiation depends upon the use of generalized media of exchange, most particularly money; and for Simmel, the most interesting aspect of this relationship is how social relations are changed as money becomes the dominant medium of exchange:

2. The greater is the use of money in exchanging relations, the greater will be

    a. The volume of exchange relations.
    b. The rate of social exchange.
    c. The scope of social exchange.
    d. The accumulation of value in social exchange.
    e. The accumulation of trust based on the value of money.
    f. The calculation of utilities.
    g. The multiplicity of social ties and exchanges.
    h. The differentiation of power to regulate exchanges and the value of money.
    i. The options for individuals.
    j. The individualization of people.

These propositions read much like a "laundry list," but it is how Simmel argued. Simmel wanted to emphasize that the form or nature of social relations changes with differentiation, and he set out to "list the ways" this process occurred. He saw most of these outcomes as positive, but he also recognized the potential for pathological situations:

3. The greater the use of money in exchange relations, the greater is the multiplicity of ties and affiliations in secondary groups, and hence, the greater is the potential for

    a. Anxiety-generating role conflicts arising from too many diverse ties and group affiliations.

    b. Alienation, stemming from weak attachments to groups and a sense of standing between rather than within groups.

Yet, unlike Weber and Marx, Simmel did not seem to see such outcomes as a chronic state of affairs. Role conflict and marginality are indeed potentials, and individuals frequently experience these states, but Simmel did not see them as overwhelming the mass of people in society or as posing threats to the stability and integration of society. Rather, he generally saw the outcomes of differentiation and use of money as positive because they gave individuals more freedom, increased options, and enhanced sense of value. We can summarize Simmel's argument as follows:

4. The greater is the level of differentiation and the greater is the reliance on non-ascriptive media of exchange in a society, the more choice individuals have in their group affiliations and the greater will be the number of affiliations in diverse groups; the more choice individuals have in establishing their multiple group affiliations, the more they experience

    a. Enhanced value or utility because their affiliations better match their preferences.
    b. Confirmation of identities from each group with which they chose to affiliate.

Simmel on Conflict

Differentiated social systems will reveal more secondary groups, thus increasing not only the diversity of groups but also the potential for conflicts of interest among various groups. Even in less differentiated societies, conflict is an ever-present force in social relations; indeed, conflict will tend to be most violent in these less differentiated systems. Simmel's essay on conflict is much like all of his sociology in that it emphasizes the consequences of conflict rather than its causes. We can divide his theoretical principles accordingly: (1) those on the conditions affecting the violence of conflict, (2) those on the consequences of conflict for the overall social system within which the conflict occurs, and (3) those on the consequences of conflict for the units or parties involved.

Simmel's principle on the violence of conflict is incomplete but it does contain some important insights:

5. The level of violence in conflict increases as

    a. Parties to the conflict see it as transcending their individual interests and, hence, as a moral cause.
    b. Parties to the conflict have a high degree of emotional involvement in the conflict, with such involvement increasing as
       1. parties to the conflict perceive the conflict to be a moral cause.
       2. parties to the conflict each reveal high solidarity.
       3. parties to the conflict once had harmonious relations, and each perceives the breaking of this relation as a moral violation of trust and obligation.

6. The level of violence in conflict will decrease to the degree that parties to the conflict see it as a means to well-defined ends or goals.

What is perhaps the most interesting aspect to these propositions on the violence of conflict is the corrective they pose to Karl Marx's analysis. Simmel is saying that when conflicts are instrumental—that is, they are a means to explicit goals—they are less likely to be violent. His reasoning is that when clear goals exist, leaders are likely to negotiate and compromise rather than risk the high costs of violent conflict. Marx thought that the more clear-cut the goals of conflict are, the more violent it will be, or in his terms, when a class was transformed from a class "of itself" to one "for itself," a revolution is more likely. Marx would argue that the revolution is a moral cause by workers with high solidarity who perceive that their superordinates have violated their obligations, thus leading to his prediction of a violent overthrow of the bourgeoisie. But Simmel counters that as the subordinates get increasingly organized and have a clearer understanding of their purposes and goals, the conflict will be seen in more instrumental than emotional terms, thus leading to less violence. For Simmel, then, modern society was not polarizing into warring camps, but just the opposite was occurring: there was increasing differentiation of many diverse secondary groups, each with more clearly defined purposes and, hence, interests. Under these conditions, conflict would be more instrumental and, thereby, less violent. The less violent and the more frequent conflicts are, the more they have positive outcomes for the larger social whole in which they occur:

7. The less violent and more frequent the conflict among social units in a differentiated social system, the more likely is the conflict to

    a. Allow units to release hostilities before they accumulate to extremely high levels.
    b. Encourage the creation of norms to regulate the conflict.
    c. Encourage the development of authority and judiciary systems to regulate the conflict.

Similarly, when analyzing the consequences of conflict for the respective parties involved, Simmel stressed its integrative impact, especially when the conflict is violent. Thus, for Simmel, although violent conflict can have disintegrative consequences for the social whole, it can also promote integration within the participants, as is evident in the following propositions.

8. The more violent are intergroup hostilities and the more frequent are the conflicts among groups, the less likely are groups' boundaries to disappear.

9. The more violent is the conflict, the more likely is centralization of power within the conflict groups.

10. The more violent is the conflict, the greater will be the internal solidarity of conflict groups, especially as

    a. The size of the conflict groups decreases.
    b. The conflict group represents a minority position.
    c. The conflict group is engaged in self-defense.

11. The more violent and prolonged is the conflict between groups, the more likely is the formation of coalitions among previously unrelated groups in a system.

12. The more prolonged is the threat of violent conflict between groups, the more enduring are the coalitions of each of the conflict parties.

We have made Simmel's analysis of society more coherent than it really is. His work is a series of essays; in his books, he never pulled the contents of the essays together into a coherent theory. Simmel argued strongly for a "formal sociology" that would examine basic forms of social relations and extract the laws explaining the operation of these forms. As we have seen, he offered some interesting principles, but he did not develop a general theory of differentiation, group affiliations, or conflict. We get glimpses of insight, but not a general theory.

16

✳

The Origin and Context of Émile Durkheim's Thought

BIOGRAPHICAL INFLUENCES ON DURKHEIM'S THOUGHT[1]

Émile Durkheim was born in Épinal, France, in 1858. Because his Jewish family was deeply religious, the young Durkheim studied Hebrew, the Old Testament, and the Talmud, apparently intending to follow his father's example and become a rabbi. He began to move away from religion in his early teens, however, and he eventually abandoned personal religious involvement and proclaimed himself an agnostic. As will become evident in the next chapter, however, he never lost interest in religion as a topic of intellectual inquiry; perhaps equally important, his passion for creating a new "civil morality" in France was fueled by the high sense of morality instilled by his family and early religious training.

Durkheim was an excellent student, and in 1879, he was admitted to the École Normale Supérieure, the traditional training ground for the intellectual elite of France in the nineteenth century. In the new environment he became indifferent, however, apparently finding the literary, esthetic, and rhetorical thrust of the instruction unappealing. Instead, he preferred the disciplined logic of philosophical arguments and the hard facts and findings of the sciences. Several teachers at the École did influence Durkheim, however. The great French historian Fustel de Coulanges provided him with a firm appreciation for careful assessment of historical causes, and the philosopher Émile

Boutroux instilled in him an understanding of how reality consists of discontinuous levels that reveal emergent properties that distinguish them from one another. As we will see, these concerns for historical cause and emergent realities became central to Durkheim's sociology, and he acknowledged his debt by later dedicating his two doctoral theses to these teachers at the École.

Between 1882 and 1887, Durkheim taught in various schools around Paris, except for a year in Germany, where he studied German academic life and wrote a series of reports on German sociology and philosophy. These reports gave him some visibility in academic circles and promoted contacts with important officials in the educational establishment in France. More significantly, in the 1880s, his sociological orientation took on a more coherent form. This orientation represented a mixture of moral commitment to creating an integrated and cohesive society, on the one hand, and the application of rigorous analysis of social processes, on the other. Much like Auguste Comte before him, Durkheim believed that the observations of facts and the development of theories to explain these facts would lead to a body of knowledge that could be used to create a "better society." To achieve this goal Durkheim had to recreate sociology in an era when Comte's ideas were not highly regarded and the traditional academic structure was hostile to any "science of society." Thus, it is to Durkheim's credit that he could use his powers of persuasion as well as his personal contacts to secure a position at the University of Bordeaux in the Department of Philosophy, where he was allowed to teach a social science course, heretofore an unacceptable subject in French universities.

During this period at Bordeaux, Durkheim wrote the three works that made him famous and placed him in a position to change the structure and content of the French educational system from primary schools to the universities themselves. Indeed, at no other time in the history of sociology has a sociologist exerted this degree of influence in a society.

At Bordeaux, Durkheim wrote *The Division of Labor in Society, The Rules of the Sociological Method,* and *Suicide.*[2] These three books established the power of sociological analysis, generating enormous controversy and begrudging respect for Durkheim as a scholar. Perhaps more significant for his ultimate influence on French intellectual thinking was his creation of *L'Année Sociologique* in 1898. This journal soon became the centerpiece of an intellectual movement revolving around his approach to sociology. Each annual issue contained contributions by Durkheim and a diverse group of young and creative scholars who, though from varying disciplines, were committed to defending his basic position.

In 1902, Durkheim's stature allowed him to move to the Sorbonne in Paris, and in 1906, he became a professor of science and education. In 1913, by a special ministerial decree, the name of his chair was changed to Science of Education *and Sociology*. As Lewis Coser notes of this event,[3] "after more than three-quarters of a century, Comte's brainchild had finally gained entry at the University of Paris." In Paris, Durkheim continued to edit *L'Année* and inspire a new generation of gifted scholars. Moreover, he was able to help reform the French educational system: At the time he rose to prominence, the government had embarked on a difficult process of secularizing the schools and creating a state system of public education that rivaled and then surpassed the Catholic school system, which until the early decades of this century dominated the education of children. Through his contacts in high-level government positions, Durkheim created a new kind of curriculum in the public schools and a revolutionary program of teacher education. Emphasis was on secular topics, with the schools serving as a functional sub-stitute for the church. In essence, the school was to teach reverence for "soci-ety," and the teacher was to be the "priest" who guided this worship of civil society. Durkheim's desire was to create a "civil morality" under which stu-dents became committed to the institutions of society and, at the same time, developed the secular skills and knowledge to analyze and change society for the better. Although he was very cautious in dictating the precise nature of the school curriculum and in proposing the desirable direction of society, no social scientist has ever exerted more influence on the general profile of such a major institutional structure in a society.

This concern with a civil morality was, of course, inherited from Comte, who in turn had merely carried forth the banner of earlier French philoso-phers—Charles Montesquieu and Jean Jacques Rousseau being the most prominent. For Durkheim, the central question of all sociological analysis is this: What forces integrate society, especially as it undergoes rapid change and differentiation? Durkheim believed that integration will always involve a "morality" or set of values, beliefs, and norms that guide the cognitive orien-tations and behaviors of individuals. Durkheim approached the analysis of moral integration in many different ways, but the need for a common moral-ity permeates his work—from his first great work, *The Division of Labor in Society,* to his last major book, *The Elementary Forms of Religious Life.*[4] Thus, the young boy who was to have been a rabbi developed into the secular academic who was to preach for societal integration.

World War I disrupted the "Année School," as it had come to be known. Indeed, it killed off many of its most promising members, including Durkheim's son, André, who would no doubt have had a distinguished career as a sociological linguist. Durkheim never recovered emotionally from this blow because he had hoped his son would carry on his work in the social sci-ences. In 1917, two years after his son's death, Durkheim died at the age of fifty-nine. Emotionally drained and physically declining, he simply did not care to live any longer.

Durkheim thus died at the height of his intellectual and political promi-nence. As we will see in the next two chapters, he left an intellectual legacy that is as influential on sociological theorizing today as it was at the turn of the century. Durkheim's work represents the culmination of the French intellec-tual tradition that began with the Enlightenment (see Chapter 1). At the same time, his sociology is a response to both the perceived strengths and weaknesses of German and English sociology. The result is an approach that is true to its French pedigree—especially the works of Charles Montesquieu, Jean Jacques Rousseau, Auguste Comte, and Alexis de Tocqueville. Yet, the pedigree is con-ditioned by Durkheim's reaction against Herbert Spencer and, to a lesser extent, Karl Marx. Let us now turn to this list of influential thinkers and observe how they influenced Durkheim's thinking. In this way, we can place into broader intellectual and historical context the works to be examined in the next chapter.

MONTESQUIEU AND DURKHEIM

As mentioned in our analysis of Auguste Comte's work, Montesquieu marked the beginning of a French intellectual line that came to a climax with Durkheim. To appreciate many of Durkheim's concepts and points of emphasis and his methodological approach, we must return to Montesquieu, one of the giant intellects of the eighteenth century.

Montesquieu as the First Social Scientist

Montesquieu introduced an entirely new approach to the study of society. If we look at any number of scholars whose thought was prominent in his time, we can immediately observe dramatic differences between their approach to the study of society and his. Many scholars of the eighteenth century were philosophers who were primarily concerned with the question: What is the ultimate origin of society? Their answer to this question was more philosophical than sociological and tended to be given in two parts. First, humans once existed in a "natural state" before the first society was created. Theory about society thus began with speculations about "the state of nature"—whether this state be warlike (Thomas Hobbes), peaceful (John Locke), or idyllic (Rousseau). Second, in this state of nature humans formed a "social contract" and thereby created "society." People agreed to subordinate themselves to government, law, values, beliefs, and contracts.

In contrast with these philosophical doctrines, Montesquieu emphasized that humans have never existed without society. In his view, humans are the product of society, and thus speculation about their primordial state does not represent an analysis of the facts of human life. He was an empiricist, concerned with actual data rather than speculation about the essence of humans and the ultimate origins of their society. In many ways, he was attracted to the procedures employed by Newton in physics: Observe the facts of the universe, and from these make statements about their basic properties and the lawlike relations. Although it was left to Comte in the following century to trumpet the new science of "social physics," Montesquieu was the first to see that a science of society, molded after the physical sciences, was possible.

Durkheim saw Montesquieu as positing that society was a "thing" or "fact" in the same sense that physical matter constitutes a thing or fact. Montesquieu was the first to recognize, Durkheim believed, that "morals, manners, customs" and the "spirit of a nation" are subject to scientific investigation. From this initial insight, it is a short step to recognizing, as Comte did, that a discipline called sociology can study society. Durkheim gave explicit credit to Montesquieu for recognizing that a "discipline may be called a science only if it has a definite field to explore. Science is concerned with things, realities. Before social science could begin to exist, it had to be assigned a subject matter."[5]

Montesquieu never completely carried through on his view that society could be studied in the same manner as phenomena in the other sciences, but his classic book, *The Spirit of Laws,* represents one of the first sociological works with a distinctly scientific tone. Although he had become initially famous for other works, *The Spirit* had the most direct influence on Durkheim.[6] Indeed, Durkheim's Latin doctoral thesis was on *The Spirit of Laws* and was published a year before his famous French thesis, *The Division of Labor in Society.*[7] From *The Spirit,* Durkheim took both methodological and substantive ideas, as is emphasized in the following review of Montesquieu's work and its influence on Durkheim.

Montesquieu's View of "Laws"

The opening lines of *The Spirit of Laws* read,

> Laws, in their most general signification, are the necessary relations aris-
> ing from the nature of things. In this sense all beings have their laws: the
> Deity His laws, the material world its laws, the intelligences superior to
> man their laws, the beasts their laws, man his laws.[8]

There is an ambiguity in this passage that is never clarified, for Montesquieu used the term *law* in two distinct senses: (1) law as a command-ment, or rule, created by humans to regulate their conduct and (2) law as a sci-entific statement of the relations among properties of the universe in its physical, biological, and social manifestations. The first is a substantive con-ception of law—that of the jurist and political scientist. The second is a con-ception of scientific laws that explains the regularities among properties of the natural world. Durkheim incorporated this distinction implicitly in his own work. On the one hand, his first great work, on *The Division of Labor in Society,*[9] is about law, for he used variations in laws and the penalties for their violation as concrete indicators of integration in the broader society. On the other hand, *The Division of Labor* involves a search for the scientific laws that explain the nature of social integration in human societies.

Montesquieu also revealed an implicit "hierarchy of laws," an idea that may have suggested to Comte the hierarchy of the sciences (see Chapter 3). For Montesquieu, "lower-order" phenomena, such as physical matter, cannot devi-ate from the scientific laws that govern their operation, but higher-order beings with intelligence can violate and transgress laws, giving the scientific laws of society a probabilistic rather than absolute character. Durkheim, who was to view statistical rates as social facts in many of his works, also adopted this idea of probabilistic relations among social phenomena.

Montesquieu's Typology of Governments

To search for the scientific laws of the social world, Montesquieu argued, classification and typology are necessary. The enormous diversity of social patterns can easily obscure the common properties of phenomena unless the underlying type is exposed. The first thirteen books of *The Spirit* are thus devoted to Montesquieu's famous typology of governmental forms: (1) republic, (2) monarchy, and (3) despotism. Both methodological and substantive facets of Montesquieu's typology influenced Durkheim. On the methodological side, Durkheim saw as significant the way in which Montesquieu went about constructing his typology. Durkheim stressed the more strictly methodological technique of using "number, arrangement, and cohesion of their component parts" for classifying social structures. In many ways, he was making assumptions about Montesquieu's views on this matter because Montesquieu was never very explicit. Yet, Durkheim's lifelong advocacy of typology and his use of the number, arrangement, and cohesion of parts as the basis for constructing typologies were apparently inspired by Montesquieu's classification of governmental forms.

Another methodological technique that Durkheim appears to have borrowed from Montesquieu is the notion that laws enacted by governments will reflect not only the "nature" (structural form) of government but also its "principle" (underlying values and beliefs). Moreover, laws will reflect the other institutions that the nature and principle of government influence. Law is thus a good indicator of the culture and structure of society. This premise became the central methodological tenet of Durkheim's first major work on the division of labor.

On the substantive side, Montesquieu's view of government as composed of two inseparable elements, nature and principle, probably influenced Durkheim more than he acknowledged. For Montesquieu, each government's nature, or structure, is a reflection of both who holds power and how power is exercised. Each government also reveals a principle, leading to the classification of governments by structural units and cultural beliefs. For a republic, the underlying principle is "virtue," in which people have respect for law and for the welfare of the group; for a monarchy, the guiding principle is respect for rank, authority, and hierarchy, and for despotism, the principle is fear. The specifics of Montesquieu's political sociology are less important than the general insight they illustrate: Social structures are held together by a corresponding system of values and beliefs that individuals have internalized. Moreover, as Montesquieu emphasized, when a government's nature and principle are not in harmony—or, more generally, when social structures and cultural beliefs are in contradiction—social change is inevitable. Durkheim wrestled with these theoretical issues for his entire intellectual career. Yet, one finds scarce notice in his thesis of Montesquieu's profound insight into this aspect of social reality.

Another substantive issue, for which Montesquieu is most famous, is the "balance of powers" thesis. The basic argument is that a separation, or division, of powers among elements of government is essential to a stable government.

Power must be its own corrective, for only counterpower can limit the abuse of power. Thus, Montesquieu saw the two branches of the legislature (one for the nobility, the other for commoners) as they interact with each other and with the monarch as providing checks and balances on each other.[10] The judiciary, the third element of government, was not considered by Montesquieu to be an independent source of power, as it became in the American governmental system. Several points in this analysis no doubt influenced Durkheim. First, Montesquieu's distrust of mass democracy, in which the general population directly influences political decisions, was to be retained in Durkheim's analysis of industrial societies. In Durkheim's eye, representation is always to be mediated to avoid instability in political decisions. Second, Durkheim shared Montesquieu's distrust of a single center of power. Montesquieu feared despotism, whereas Durkheim distrusted the monolithic and bureaucratized state, but both recognized that to avoid the danger of highly centralized power, counterpower must be created.

The Causes and Functions of Governments

In addition to the notion of social types and scientific laws, the most conspicuous portions of Durkheim's Latin doctoral thesis are those on the "causes" and "functions" of government, a distinction that became central to his sociology.

Montesquieu's *The Spirit* is often a confused work, and commentators have frequently misunderstood his analysis of causes. After the typology of governments in the first thirteen books of *The Spirit,* Montesquieu suddenly launched into a causal analysis that could appear to undermine his emphasis on the importance of "the principle" in shaping the "nature," or structure, of government. For suddenly, in books 14 through 25, a variety of physical and moral causes of governmental forms is enumerated: Climates, soil fertility, manners, morals, commerce, money, population size, and religion are introduced one after another as causes of governmental and social forms.

The confusion often registered in this discussion of causes can be mitigated by recognizing Montesquieu's underlying assumptions. First, these causes do not directly affect governmental forms. Rather, each affects people's behavior, temperament, and disposition in ways that create a "general spirit of the nation," an idea that was not far from Durkheim's conceptualization of the "collective conscience" and "collective representations" or Comte's similar notions. Thus, "physical causes," such as climate, soil, and population size, as well as "moral causes," such as commerce, morals, manners, and customs, all constrain how people act, behave, and think. From the collective life constrained and shaped by these causes comes the "spirit of a nation," which is a set of implicit ideas that bind people to one another and give them a sense of their common purpose.[11]

Once it is recognized that these causes do not operate directly on the structure of government, a second point of clarification is possible. According to Montesquieu, the underlying principle of government is linked to the general spirit that emerges from the actions and thoughts of people as they are constrained by the list of causes. Montesquieu was ambiguous on this issue, but this interpretation is the most consistent with how Durkheim probably viewed his argument. Curiously, despite the similarity of their views on the importance of collective ideas or "spirits" for social relations, Durkheim did not give Montesquieu much credit for this aspect of his sociology.

Durkheim did give Montesquieu explicit credit, however, for recognizing that a society must assume a "definite form" because of its "particular situation" and that this form stems from "efficient causes." In particular, Durkheim indicated that Montesquieu's view of ecological and population variables stimulated his concern with "material density" and how it influences "moral density." Durkheim recognized that to view social structures and ideas as the result of identifiable causes marked a dramatic breakthrough in social thought, especially because many social thinkers of the time were often locked into discussions of human nature and the "origins" of the first social contract.

Durkheim was, however, highly critical of Montesquieu's causal analysis in one respect: He saw Montesquieu as arguing for "final causes." That is, the ends served by a structure such as law cause it to emerge and persist. As Durkheim stressed, "Anyone who limits his inquiry to the final cause of social phenomena loses sight of their origins and is untrue to science. This is what would happen to sociology if we followed Montesquieu's method."[12] Durkheim recognized that Montesquieu had been one of the first scholars to argue for what is now termed *cultural relativism*. Social structures must be assessed not in relation to some absolute, ethnocentric, or moralistic standard but in their own terms and in view of the particular context in which they are found. For example, Montesquieu could view slavery not so much as a moral evil but as a viable institution in certain types of societies in particular historical periods. Implicit in this kind of argument is the notion of "function": A structure must be assessed by its functions for the social whole; if a social pattern, even one like slavery, promotes the persistence and integration of a society, it cannot be deduced to be an evil or good pattern. Durkheim felt that Montesquieu too easily saw the consequence of structures—that is, integration—as their cause, with the result that Montesquieu's functional and causal analyses frequently became confused. Yet Montesquieu might have suggested to Durkheim a critical distinction between causal and functional analysis.

In sum, Durkheim gave Montesquieu credit for many insights that became a part of his sociology. The social world can be studied as a "thing"; it is best to develop typologies; it is necessary to examine the number, arrangement, and relations among parts in developing these typologies; it is important to view law as an indicator of broader social and cultural forces; and it is wise to employ both causal and functional analyses.

Despite Durkheim's praise of Montesquieu, however, it is interesting to note what he did not acknowledge in Montesquieu's work: the view that laws, like those of physics, can be formulated for the social realm (in the thesis, Durkheim gave Comte credit for this insight);[13] the recognition that social morphology and cultural symbols are interconnected; the position that causes of morphological structures are mediated through, and mitigated by, cultural ideas; the notion that causes, and the laws that express relations among events, are probabilistic in nature; and the view that power in social relations must be checked by counterpower. These ideas became an integral part of Durkheim's thinking, and thus much of Montesquieu's theoretical legacy lived in Durkheimian sociology.[14]

ROUSSEAU AND DURKHEIM

Writing his major works in the decade following the 1748 publication of Montesquieu's *The Spirit of Laws,* Jean Jacques Rousseau produced a philosophical doctrine that contains none of Montesquieu's sense for social science but much of his sense for the nature of social order.[15] Although not greatly admired in his time, Rousseau's ideas were, by the beginning of the nineteenth century, viewed in a highly favorable light. Indeed, in retrospect, Rousseau was considered the leading figure of the Enlightenment, surpassing Hobbes, Locke, Voltaire, and certainly Montesquieu. It is not surprising, therefore, that Durkheim read with interest Rousseau's philosophical doctrine and extracted many ideas.

Rousseau's Doctrine

Rousseau's doctrine was a unique combination of Christian notions of the Fall that came with Original Sin and Voltaire's belief in the progress of humans.[16] Rousseau first postulated a presocietal "state of nature" in which individuals were dependent on nature and had only simple physical needs, for "man's"[17] desires "do not go beyond his physical needs; in all the universe the only desirable things he knows are food, a female, and rest." In the "state of nature," humans had little contact with or dependence on one another, and they had only crude "sensations" that reflected their direct experiences with the physical environment.

The great "fall" came from this natural state. The discovery of agriculture, the development of metallurgy, and other events created a new and distinct entity: society. People formed social relations; they discovered private property; they appropriated property; they competed; those with property exploited others; they began to feel emotions of jealousy and envy; they began to fight and make war; and in other ways they created the modern world. Rousseau felt that this world not only deviated from humans' natural state but also made their return to this state impossible.

As an emergent reality that destroys the natural state, modern society poses a series of problems that makes life agonizing misery. In particular, humans feel no limit to their desires and passions; self-interest dominates; and one human exploits another. For Rousseau, society is corrupt and evil, destroying not only the natural controls on passions and self-interest but also the liberty from exploitation by one's fellows that typified the natural state.

Rousseau's solution to this evil was as original as it was naive, and yet it exerted considerable influence on Durkheim. Rousseau's solution was to eliminate self-interest and inequality by creating a situation in which human beings would have the same relation to society as they once had to nature. That is, people should be free from one another and yet equally subject to society. In Rousseau's view, only the political state could ensure individual freedom and liberty, and only when individuals totally subordinated their interest to what he termed the *general will* could inequality, exploitation, and self-interest be eliminated. If all individuals must subjugate themselves equally to the general will and the state, they are equal. If the state can ensure individual freedom and maintain equal dependence of individuals on the general will, then the basic elements of nature are re-created: freedom, liberty, and equal dependence on an external force (society instead of nature).

What is the general will? And how is it to be created? Rousseau was never terribly clear about just what constituted the general will, but it appears to have referred to an emergent set of values and beliefs embodying "individual wills." The general will can be created and maintained, Rousseau asserted, only by several means: (1) the elimination of other-world religions, such as Christianity, and their replacement by a "civil religion" with the general will as the supreme being; (2) the elimination of family socialization and its replacement by common socialization of all the young into the general will (presumably through schools); and (3) the creation of a powerful state that embodies the general will and the corresponding elimination of groups, organizations, and other "minor associations" that deflect the power of general will and generate pockets of self-interest and potential dissensus among people.[18]

Specific Influences on Durkheim

Society as an Emergent Reality In his courses, Durkheim gave Rousseau credit for the insight that society constituted a moral reality, *sui generis,* that could be distinguished from individual morality.[19] Although Montesquieu had reached a similar recognition, Rousseau phrased the matter in a way that Durkheim emulated on frequent occasions. For Rousseau as well as for Durkheim, society is "a moral entity having specific qualities [separate] from those of the individual beings who compose it, somewhat as chemical compounds have properties that they owe to none of their elements."[20]

Thus, Durkheim took from Rousseau the view of society as an emergent and moral entity, much like emergent physical phenomena. Like Rousseau, Durkheim abhorred a society in which competition and exchange dominated a common morality. Indeed, for Durkheim, society was not possible without a moral component guiding exchanges among individuals.

Social Pathology Durkheim viewed Rousseau's discussion of the natural state as a "methodological device" that could be used to highlight the pathologies of contemporary society and to provide guidelines for the remaking of society. Although many others in the eighteenth and nineteenth centuries had also emphasized the ills of the social world, Durkheim appeared to be drawn to three central conditions emphasized by Rousseau. Durkheim termed these (1) *egoism,* (2) *anomie,* and (3) *the forced division of labor,* but his debt to Rousseau is clear. For Durkheim, egoism is a situation in which self-interest and self-concern take precedence over commitment to the larger collectivity. Anomie is a state of deregulation in which the collective no longer controls people's desires and passions. The forced division of labor is a condition in which one class can use its privilege to exploit another and to force people into certain roles. Indeed, the inheritance of privilege and the use of privilege by one class to exploit another was repugnant to Durkheim. Like Rousseau, he felt that inequalities should be based on "natural" differences that spring from "a difference of age, health, physical strength, and mental and spiritual qualities."[21]

Thus, Durkheim was highly sympathetic to Rousseau's conception of what ailed society: People force others to do their bidding, they are deregulated, and they are unattached to a larger purpose. Hence, the social order should be structured in ways that mitigate these pathologies.

The Problem of Order Durkheim accepted the dilemma of modern society as Rousseau saw it: How is it possible to maintain individual freedom and liberty, without also releasing people's desires and encouraging rampant self-interest, while creating a strong and cohesive social order that does not aggravate inequality and oppression?

For Rousseau, this question could be answered with a strong political state that ensured individual freedom and a general will that emulated nature. Like Rousseau, Durkheim believed that the state was the only force that would guarantee individual freedom and liberty, but he altered Rousseau's notion of society as the equivalent of the physical environment in the state of nature. For Durkheim, society and the constraints it imposes must be viewed as natural, with the result that people must be taught to accept the constraints and barriers of society in the same way that they accept the limitations of their biological makeup and the physical environment. Only in this way can both egoism and anomie be held in check. Durkheim believed that constraint by the moral force of society is in the natural order of things.

Like Rousseau, Durkheim argued for a view of society as "sacred" and for the transfer of the same sentiments toward civil and secular society that people had traditionally maintained toward the gods (which, Durkheim later emphasized, are only symbolizations of society).[22] Moreover, like Rousseau, Durkheim stressed the need for a moral education outside of the family in which children could be taught in schools to understand and accept the importance of commitment to the morality of the collective. Such a commitment could be achieved, he argued, through a unified "collective conscience" or set of "collective representations" that could regulate people's desires and passions. This view represented a reworking of Rousseau's view of absolute commitment to the general will. In sharp contrast with Rousseau, however, Durkheim came to believe that only through attachment to cohesive subgroups, or what Rousseau had called "minor associations," could egoism be mitigated. Such groups, Durkheim felt, can attach individuals to the remote collective conscience and give them an immediate community of others. Moreover, like Montesquieu, he distrusted an all-powerful state, and hence he came to view these subgroups as a political counterbalance to the powers of the state.

Thus, Durkheim borrowed many ideas from Rousseau. Some of his central concepts about social pathologies—anomie, egoism, and the forced division of labor—owed much to Rousseau's work. Durkheim's vision of society as integrated by a strong state and by a set of common values and beliefs reflected Rousseau's vision of how to eliminate these pathologies. Rousseau also inspired Durkheim's desire to use schools to provide moral education for the young and to rekindle the spirit of commitment to secular society that people once had toward the sacred.

Yet, Durkheim could never accept Rousseau's trust of the state. Durkheim believed that the state's power must be checked and balanced. People must be free to associate and to join groupings that encourage diversity based on common experiences and that create centers of counterpower to mitigate the state's power. Durkheim thus internalized Rousseau's vision of an integrated society in which individual freedom and liberty prevail. Durkheim accepted the challenge of proposing ways to achieve that society, but he could never abide by Rousseau's vision of an all-powerful state and its oppressive general will.[23]

Rousseau's impact on Durkheim was, no doubt, profound, but the extremes of Rousseau's philosophy are mitigated in Durkheim's work. Montesquieu's emphasis on the balancing of power with counterpower, and his emphasis on empirical facts rather than on moral precepts, represented one tempering influence on Durkheim. Still another moderating influence came from Comte, whose work consolidated many intellectual trends into a clear program for a science that could be used to create the "good society."

COMTE AND DURKHEIM

It is difficult to know how much of the French intellectual tradition of the eighteenth century came to Durkheim through Comte, because Durkheim did not always acknowledge his debt to the titular founder of sociology. This difficulty is compounded because, like Durkheim's work, Comte's intellectual scheme represents a synthesis of ideas from Montesquieu, Rousseau, Saint-Simon,[24] and others in the French lineage. Many of the specific features of Durkheimian sociology owed much to Comte's grand vision for the science of society. We reviewed Comte's thought in Chapter 2, and so we will focus only on the specific aspects of his intellectual scheme that appear to have exerted the most influence on Durkheim.

The Science of Positivism

Comte must have reinforced for Durkheim Montesquieu's insistence that "facts" and "data," rather than philosophical speculation, should guide the science of society. Borrowing Comte's vision of a science of "social facts," Durkheim agreed with Comte's view that the laws of human organization could be discovered. These laws, as Montesquieu stressed, will not be as "rigid" or "deterministic" as in sciences lower in the hierarchy, but they will be the equivalent of those laws in physics, chemistry, and biology in that they will allow for the understanding of phenomena. Thus, Comte cemented in Durkheim's mind the dictum that the search for sociological laws must be guided by empirical facts, and conversely, the gathering of facts must be directed by theoretical principles.

The Methodological Tenets of Positivism

Collecting facts requires a methodology, and Comte was the first to make explicit the variety of methods that could guide the new science of society. As he indicated, four procedures are acceptable: (1) "observation" of the social world by the use of human senses (best done, he emphasized, when guided by theory); (2) "experimentation," especially as allowed by social pathologies; (3) "historical" observation, in which regular patterns of change in the nature of society—especially in the nature of its ideas—can be seen; and (4) "comparison," in which human and animal societies, coexisting human societies, and different elements of the same society are compared in order to isolate the effects of specific variables. Durkheim employed all of these methods in his sociology, and hence, we can conclude that Comte's methodological approach influenced Durkheim's methodology.

Another methodological aspect of Comte's thought revolves around the organic analogy. Comte, as we saw in Chapter 3, often compared society to a biological organism, with the result that a part, such as the family or the state, could be understood by what it did for or contributed to the "body social." Montesquieu had made a similar point, although Comte first drew the clear analogy between the social and biological organisms. The functional method developed by Durkheim thus owed much to Comte's biological analogy. Indeed, as Durkheim emphasized, complete understanding of social facts is not possible without assessing their functions for maintaining the integration of the social whole.[25]

Much less prominent in Comte's scheme than in Montesquieu's was the emphasis on typology; yet Comte recognized that the construction of somewhat "idealized" types of social phenomena could help in sociological analysis. Although many intermediate cases will not conform to these extreme types, their deviations from the types allow their comparison against a common yardstick—that is, the idealized type.[26] In his early work, Durkheim developed typologies of societies; thus, we can assume that Montesquieu's emphasis on types, as reinforced by Comte's emphasis on the use of types as an analytical device for comparison, must have shaped Durkheim's approach. Throughout his career, Durkheim insisted that classification of phenomena by their "morphology," or structure, must precede either causal or functional analysis.

Social Statics and Dynamics

Durkheim was also influenced by the substance of Comte's scheme. Comte divided sociology into "statics" and "dynamics," a distinction that Durkheim implicitly maintained. Moreover, Durkheim adopted the specific concepts Comte used to understand statics and dynamics.

With regard to social statics, Durkheim shared Comte's concern with social solidarity and with the impact of the division of labor on this solidarity. In particular, Durkheim asked the same question as Comte: how can *consensus universalis,* or what Durkheim termed the *collective conscience,* be a basis for social integration given the growing specialization of functions in society? How can consensus about ideas, beliefs, and values be maintained at the same time that people are differentiated and pulled apart by their occupational specialization? *The Division of Labor in Society,* Durkheim's first major work, addressed these questions, and though Rousseau in the eighteenth century and a host of others in the nineteenth century had also tried to address these same issues, Durkheim's approach owed more to Comte than to any other thinker.[27]

With respect to social dynamics, Comte held an evolutionary vision of human progress. Societies, especially their ideas, are moving from theological through metaphysical to positivistic modes of thought. Durkheim adopted this specific view of the evolution of ideas late in his career in his work on religion. More fundamentally, however, he retained the evolutionary approach to studying social change held by Comte and a host of other thinkers. Durkheim saw societies as moving from simple to complex patterns of social structure and, correspondingly, from religious to secular systems of ideas. Almost everything he examined was couched in these evolutionary terms, which, to a very great extent, he adopted from Comte.

Science and Social Progress

Like his teacher and collaborator, Saint-Simon, Comte saw the development of sociology as a means to creating a better society. Although Durkheim's sense of pathology in the modern world probably owes more to Rousseau than to either Saint-Simon or Comte, Durkheim accepted the hope of the two latter thinkers for a society based on the application of sociological laws. Durkheim was much less extreme than either Saint-Simon or Comte, who tended to make a religion of science and to advocate unattainable utopias, but Durkheim retained Comte's view that a science of society could be used to facilitate social progress. Indeed, Durkheim never abandoned his dream that applying the laws of sociology could create a just and integrated social order.

In sum, then, Durkheim's debt to Rousseau was mitigated by his exposure to Comte. His view of science as reliant on data and as generating laws of human organization came as much from Comte as from any other thinker, as did his adoption of explicit methodological techniques. His substantive view of society similarly reflected Comte's emphasis: a concern for social integration of differentiated units and for determining how ideas (values, beliefs, and norms) are involved in such integration. Comte's insistence that science be used to promote the betterment of the human condition translated Rousseau's passionate and moralistic assessment of social ills into a more rational concern with constructing an integrated society employing sociological principles.

TOCQUEVILLE AND DURKHEIM

In 1835, Alexis de Tocqueville, a young member of an elite family, published the first two volumes of a book based on his observations of American society. *Democracy in America* was an almost immediate success, propelling Tocqueville into a lifelong position of intellectual and political prominence in France. The third and fourth volumes of *Democracy in America* appeared in 1840,[28] and after a short political career culminating in his abbreviated appointment as foreign minister for France, he retired to write what he saw as his major work, *The Old Regime and the French Revolution,* the first part of which appeared in 1857.[29] His death in 1859 cut short the completion of *The Old Regime,* but the completed volumes of *Democracy in America* and the first part of *The Old Regime* established him as the leading political thinker in France, one who carried the tradition of Montesquieu into the nineteenth century and one whom Durkheim read carefully.

Tocqueville probably never read Comte, but his effort to emulate Montesquieu's method of analysis must have had considerable influence on Durkheim. Indeed, from Tocqueville's analysis of democracy in America, Durkheim got many of the ideas that mitigated the extremes of Rousseau's political solutions to social pathologies.

Tocqueville's *Democracy in America*

Tocqueville saw the long-term trend toward democracy as the key to understanding the modern world. In *Democracy in America,* the young Tocqueville attempted to discover why individual freedom and liberty were being preserved in the United States and, implicitly, why French efforts toward democracy had experienced trouble (a theme more explicitly developed in *The Old Regime and the French Revolution*). In this effort, he isolated two trends that typified democracies:

1. *The trend toward a leveling of social status.* Although they preserve economic and political ranks, democratic societies bestow equal social status on their members—creating, in Tocqueville's eye, an increasingly homogeneous mass.

2. *The trend toward centralization of power.* Democratic governments tend to create large and centralized administrative bureaucracies and to concentrate power increasingly in the hands of legislative bodies.

Tocqueville saw a number of potential dangers in these two trends. First, as differences among people are leveled, the only avenue for social recognition becomes ceaseless material acquisition motivated by blind ambition. Tocqueville felt that traditional status and honor distinctions had kept ambition and status striving in check. As the old hereditary basis for bestowing honor is destroyed, the individual is released and freed from the constraints imposed by traditional social patterns. This point—the releasing of individuals from social control—is reminiscent of Rousseau's analysis and, no doubt, stimulated Durkheim's conceptualization of egoism and anomie.

Second, the centralization of administrative power can become so great that it results in despotism, which then undermines individual freedom and liberty. Moreover, centralized governments tend to rely on external war and to suppress internal dissent in an effort to promote further consolidation of power.

Third, the centralization of decision making in the legislative branch can make government too responsive to the immediate, short-lived, and unreasoned sentiments of the social mass. Under these conditions, government becomes unstable as it is pulled one way, and then another, by public sentiment.

Montesquieu's influence is evident in these concerns. Unlike Montesquieu, however, Tocqueville saw another side of democracy, a side in which individual liberty and freedom could be preserved even with the centralization of power, and where people's ambitions and atomization could be held in check even as the old system of honor and prestige receded. The democratic pattern in America, Tocqueville believed, provided an illustration of conditions that could promote this other side of democracy.

It is not surprising that Tocqueville, as a student of Montesquieu, made references in his analysis of American democracy to historical causes, placing emphasis on geography, unique historical circumstances, the system of laws (in particular the Constitution), and, most important, the "customs, manners, and beliefs" of the American people. From these causes, Tocqueville described several conditions that mitigated the concentration of political power and the overatomization of individuals:

1. The system of checks and balances in government, with power in the federal government divided into three branches

2. The federalist system, in which state and local governments, with their own divisions of power, check each other's power as well as that of the federal government

3. A free and independent press

4. A strong commitment of the people to use and rely on local institutions

5. The freedom to form and use political and civil associations to achieve individual and collective goals

6. A powerful system of values and beliefs stressing individual freedom

The power and subtlety of Tocqueville's description of America cannot be captured with a short list like this. Yet, this list probably best communicates what Durkheim pulled out of Tocqueville's work. Rousseau and Tocqueville had both highlighted the ills of the modern world—unregulated passions and rampant self-interest—but Rousseau's solution to these problems was too extreme for the liberal Durkheim, whereas Tocqueville's analysis of America provided a view of a modern and differentiated social structure in which freedom and individualism could be maintained without severe pathologies and without recourse to a dictatorial state.

Specific Influences on Durkheim

Durkheim viewed modern social structure as integrated when (1) differentiated functions were well coordinated, (2) individuals were attached to collective organizations, (3) individual freedom was preserved by a central state, (4) this central state set broad collective goals and personified common values, and (5) the state's broad powers were checked and balanced by countersources of power.

It is not hard to find these themes in Tocqueville's work. In particular, Durkheim found the idea of "civil and political" associations appealing. These associations can provide people with a basis for attachment and identification, and they can serve as a mechanism for mediating between their members and the state. Durkheim termed these associations *occupational,* or *corporate* groups, and he took much from Tocqueville's analysis of voluntary civil and political associations. Moreover, in adopting Montesquieu's emphasis on customs, manners, and beliefs that promote strong commitments to freedom and liberty, without also promoting atomization, he recognized in Tocqueville's work the importance of general values and beliefs (Rousseau's general will and Comte's *consensus universalis* and general spirit) for promoting integration among the diversified groupings of modern societies. For even if these values and beliefs emphasize individual freedom and liberty, they can be used to unite people by stressing a collective respect for the rights of the individual.

Thus, Tocqueville gave to Durkheim a sense for some of the general conditions that could mitigate the pathologies of modern societies. These conditions became a part of Durkheim's practical program as well as his more strictly theoretical analysis.

SPENCER AND DURKHEIM

Montesquieu, Rousseau, Comte, and Tocqueville represent the French intellectual heritage from which Durkheim took many of his more important concepts.[30] His criticism of these thinkers is not severe, and we can sense that he never reacted against their thought. He took what was useful and ignored obvious weaknesses. Such is not the case with Spencer, for throughout his career, Durkheim singled out Spencer for very special criticism.

Durkheim and Spencerian Utilitarianism

Durkheim reacted vehemently against any view of social order that ignored the importance of collective values and beliefs. Utilitarian doctrines stress the importance of competition and exchange in creating a social order held together by contracts among actors pursuing their own self-interests. Durkheim did not ignore the importance of competition, exchange, and contract, but he saw blind self-interest as a social pathology. A society could not be held together by self-interest and legal contracts alone; there must also be a "moral" component or an underlying system of collective values and beliefs guiding people's interactions in the pursuit of "collective" goals or interests.

Durkheim was thus highly critical of Spencer, who as we saw in Chapters 4 and 5, coined such phrases as "survival of the fittest" and emphasized that modern society was laced together by contracts negotiated from the competition and exchange of self-interested actors. In fact, Durkheim's works are so filled with references to the inadequacies of Spencerian sociology that we might view Durkheimian sociology as a lifelong overreaction to the imputed ills in Spencerian sociology.[31]

Durkheim and Spencerian Organicism

Spencer wore two intellectual hats: (1) the moralist, who was a staunch individualist and utilitarian, and (2) the scientist, who sought to develop laws of both organic and super-organic forms. Moreover, in attempting to realize the latter, Spencer took Comte's organic analogy and converted it into an explicit functionalism: System parts function to meet needs of the "body social." Durkheim clearly drew considerable inspiration from this mode of analysis because one of his major methodological tenets is to stress the importance of assessing the functions of social phenomena. We might even speculate that had Spencer not formulated functionalism, it is unlikely that Durkheim would have adopted this mode of sociological analysis.

Durkheim and Spencerian Evolutionism

Spencer also had an evolutionary view of societies as moving from a simple to a treble-compound state. Although perhaps deficient in some respects, Spencer's description was far more attuned than was Comte's to the structural and cultural aspects of social evolution. Comte's evolutionism had been vague, with references to the movement of systems of thought and the view of the social organism as embracing all of humanity. In contrast, Spencer's analysis was far more sociological and emphasized explicit variables that could distinguish types of societies from one another and that could provide a view of the dimensions along which evolutionary change could be described. Thus, it is difficult to imagine that Durkheim was unimpressed with Spencer's analysis of the broad contours of social evolution.[32] Indeed, Durkheim's first major work was to explore social evolution from simple to complex societies, a task that had initially occupied Spencer in Volume 1 of his *Principles of Sociology.*

MARX AND DURKHEIM

Durkheim analyzed socialist and communist doctrines in his courses, especially in a course on the history of social thought. He was often critical, as can be seen in posthumously published essays taken from his lectures.[33] Some evidence indicates that Durkheim wanted to devote a full course to Marx's thought, but apparently he never found the time. Durkheim was thus aware of Marx but was generally dismayed by socialism's "working-class bias" and by the emphasis on revolution and conflict. He felt that the problems of alienation, exploitation, and class antagonism were relevant to all sectors of society and that revolution caused more pathology than it resolved. Yet in his first work he discussed the forced division of labor, the value theory of labor, and the problems of exploitation—points highly reminiscent of Marx's conceptualization.

On balance, however, Marx's influence was negative. Durkheim reacted against Marx's insistence that integration in capitalist societies could not be achieved because of its "internal contradictions." What for Marx were the "normal" conflict-generating forms of capitalism were for Durkheim "abnormal forms," which could be eliminated without internal revolution. Indeed French social thought in the aftermath of the French Revolution and the lesser revolution of 1848 was decidedly conservative and did not consider revolutionary conflict as a productive and constructive way to bring about desired change. Thus, although Marx's influence on Durkheim is evident, it is not profound. Unlike Weber, for whom the "ghost of Marx" was ever present, Durkheim considered Marx's thought, reacted against Marx's ideas in his first works, and eventually rejected and ignored Marx in later works.

ANTICIPATING DURKHEIMIAN SOCIOLOGY

A scholar's ideas are the product of multiple influences, some obvious and others more subtle. We have mentioned some biographical influences on Durkheim's thought, but our emphasis has been on those scholars from whom he took concepts and methods. Our view is that simply looking at the key elements of Durkheim's thought and then examining the major figures of his intellectual milieu make the sources of his basic concepts and concerns rather clear.

The influence of various scholars on Durkheim's sociology is evident at different points in his career, which will become clear in the next chapter. By way of anticipating this discussion, we close this chapter with a brief listing of the elements of his sociology. All these elements derive from the scholars discussed in this chapter, but his unique biography led him to combine them in ingenious ways, creating a distinctive sociological perspective. His sociology can be seen as (1) a series of methodological tenets, (2) a theory-building strategy, (3) a set of substantive topics, and (4) a host of practical concerns. Each is briefly summarized in an effort to anticipate the detailed analysis of the next chapter.

Methodological Tenets

From Montesquieu and Comte, Durkheim came to view a science of society as possible only if social and moral phenomena were considered as distinct realities. Moreover, a science of the social world must be like that of the physical and biological worlds; it must be based on data, or facts. Montesquieu initially emphasized this point, but Comte articulated the methods to be employed by the science of society. Historical, comparative, experimental, and observational techniques must all be used to discover the social facts that can build a theory of society.

Theoretical Strategy

Durkheim took from Montesquieu and Comte the vision that sociological laws could be discovered. In particular, causal analysis came to be an integral part of Durkheim's approach. Like Montesquieu, Durkheim believed that theory should seek the general causes of phenomena, for only in this way can the abstract laws of social organization be uncovered. Yet without a corresponding, but nonetheless separate, analysis of the functions served by social phenomena, these laws will remain hidden, a point implicit in Montesquieu's and Comte's work that became explicit in Spencer's sociology. Thus, for Durkheim, the laws of sociology will come from the causal and functional analysis of social facts.

Substantive Interests

Durkheim believed the basic theoretical question in sociology was what forces hold society together. At a substantive level, this question involved the examination of (1) social structures, (2) symbolic components, such as values, beliefs, and norms, and (3) the complex relations between 1 and 2. From Montesquieu, Tocqueville, and Spencer, Durkheim acquired a sense for social structure. From Montesquieu's spirit of a nation, Rousseau's general will, and Comte's *consensus universalis,* Durkheim came to understand the significance of cultural symbols for integrating social structures. The specific topics of most concern to Durkheim—religion, education, government, the division of labor, intermediate groups, and collective representations—come from all the scholars discussed in this chapter and from specific intellectual and academic concerns of his time. The emphasis on symbolic and structural integration connected his examination of specific topics, which will become increasingly clear in the next chapter.

Practical Concerns

Like Rousseau and Comte, Durkheim wanted to create a well-integrated society. Such a goal could be achieved only by recognizing the pathologies of the social order, which were first articulated with a moral passion in Rousseau's work and then reinforced in Tocqueville's more dispassionate analysis of American democracy. As Durkheim came to view the matter, the solution to these pathologies involved the creation of a system of constraining ideas

(Comte, Rousseau, and Montesquieu), integration in intermediate subgroups (Tocqueville), coordination of differentiated functions through exchange and contract (Comte and Spencer), and the creation of a central state that provided overall coordination while maintaining individual freedom (Rousseau, Tocqueville, and Comte).

In sum, then, these methodological, theoretical, substantive, and practical concerns mark the critical elements of Durkheim's sociology. We now explore the specific works that can yield further insight into his thought.

NOTES

1. In this section, we have drawn heavily from Lewis A. Coser's *Masters of Sociological Thought* (New York: Harcourt Brace Jovanovich, 1977); Steven Lukes's *Émile Durkheim: His Life and Work* (London: Allen Lane, 1973); and Robert Alun Jones's *Émile Durkheim* (Newbury Park, CA: Sage, 1986).

2. Émile Durkheim, *The Division of Labor in Society* (New York: Free Press, 1947; originally published in 1893), *The Rules of the Sociological Method* (New York: Free Press, 1938; originally published in 1895), and *Suicide* (New York: Free Press, 1951; originally published in 1897).

3. Coser, *Masters of Sociological Thought,* p. 147.

4. Émile Durkheim, *The Elementary Forms of Religious Life* (New York: Free Press, 1947; originally published in 1912).

5. Émile Durkheim, *Montesquieu and Rousseau* (Ann Arbor: University of Michigan Press, 1960), p. 3.

6. Montesquieu's major works include *The Persian Letters* (New York: Meridian, 1901; originally published in 1721), and *Considerations on the Grandeur and Decadence of the Romans* (New York: Free Press, 1965; originally published in 1734). In many ways, these two early books represented a data source for the more systematic analysis in *The Spirit of Laws,* 2 vols. (London: Colonial, 1900; originally published in 1748).

7. In academic circles of Durkheim's time, two doctoral dissertations were required, one in French and another in Latin. The Latin thesis, on Montesquieu, was published in 1892, and the French thesis, on the division of labor, was published in 1893.

8. Montesquieu, *Spirit of Laws,* p. 1.

9. Durkheim, *Division of Labor.*

10. If one computes these balances, they consistently work out in favor of the nobility and against the common people. The nobility and commoners unite to check the monarch, and the nobility and monarch check the commoners. But the monarch cannot, as an elevated figure, unite with the commoners. Montesquieu's aristocratic bias is clearly evident.

11. This argument is derived from Louis Althusser, *Politics and History* (Paris: Universities of Paris, 1959).

12. Durkheim, *Montesquieu and Rousseau,* p. 44.

13. As Durkheim noted, "No further progress could be made until it was recognized that the laws of societies are no different from those governing the rest of nature and that the method by which they are discovered is identical with that of the other sciences. This was Auguste Comte's contribution (*Montesquieu and Rousseau,* pp. 63–64).

14. For further commentary on Montesquieu's work, see Althusser, *Politics and History,* pp. 13–108; Raymond Aron, *Main Currents in Sociological Thought,* vol. 1 (Garden

City, NY: Doubleday, 1968), pp. 13–72; W. Stark, *Montesquieu: Pioneer of the Sociology of Knowledge* (Toronto: University of Toronto Press, 1961); and Thomas L. Pangle, *Montesquieu's Philosophy of Liberalism: A Commentary on "The Spirit of Laws"* (Chicago: University of Chicago Press, 1973).

15. Jean-Jacques Rousseau, *The Social Contract and Discourses,* trans. G. D. H. Cole (New York: Dutton, 1950). This book is a compilation of Rousseau's various *Discourses* and *The Social Contract*—his most important works—which were written separately between 1750 and 1762. Durkheim analyzed *The Social Contract,* and it appears, along with his Latin thesis on Montesquieu, in Durkheim, *Montesquieu and Rousseau,* pp. 65–138.

16. J. H. Broome, *Rousseau: A Study of His Thought* (New York: Barnes & Noble, 1963), p. 14.

17. Rousseau's phraseology uses the term "the natural state of man," which is retained here.

18. For more detailed analyses of Rousseau's doctrines, see Broome, *Rousseau;* Ernst Cassirer, *The Question of Jean-Jacques Rousseau* (Bloomington: Indiana University Press, 1963); Ronald Grimsley, *The Philosophy of Rousseau* (New York: Oxford University Press, 1973); John Charuet, *The Social Problem in the Philosophy of Rousseau* (Cambridge University Press, 1974); and David Cameron, *The Social Thought of Rousseau and Burke: A Comparative Study* (Toronto: University of Toronto Press, 1973).

19. Durkheim's essay on Rousseau in *Montesquieu and Rousseau* was drafted from a course he taught at Bordeaux. It was published posthumously in 1918.

20. Durkheim, *Montesquieu and Rousseau,* p. 82. Durkheim took this quote from Rousseau.

21. Ibid., p. 86.

22. Many of the specifics of Durkheim's ideas about religion as the symboliza-tion of society were borrowed from Roberton Smith. See Lukes, *Émile Durkheim,* p. 450. For a further documentation of the influences on Durkheim's sociology of religion, see Robert Alun Jones and Mariah Evans, "The Critical Moment in Durkheim's Sociology of Religion," paper read at the meeting of the American Sociological Association, September 1978.

23. For a discussion of Durkheim's differences with Rousseau, see Lukes, *Émile Durkheim,* chap. 14.

24. Many have noted how much Comte took from his teacher, Saint-Simon. We can argue, however, that Saint-Simon's more scientific concerns reached Durkheim via Comte's reinterpretation, although Durkheim rejected Saint-Simon's utopian socialism.

25. For a more complete analysis of Comte's organicism and its impact on Durkheim's functionalism, see Jonathan H. Turner and Alexandra Maryanski, *Functionalism* (Menlo Park, CA: Benjamin/Cummings, 1979).

26. This approach obviously anticipated by a half-century Max Weber's ideal type method.

27. Durkheim, *Division of Labor.*

28. Alexis de Tocqueville, *Democracy in America* (New York: Knopf, 1945; originally published in 1835 and 1840).

29. Alexis de Tocqueville, *The Old Regime and the French Revolution* (Garden City, NY: Doubleday, 1955; originally published in 1857).

30. This is not to deny the influence of specific teachers and less well-known scholars. But we think the degree to which Durkheim took from the giants of French thought has been underemphasized in commentaries. There is too much similarity in the combined legacy of Montesquieu, Rousseau, Comte, and Tocqueville, on the one hand, and Durkheim's thought, on the other, for the impact of these prominent social thinkers to be ignored.

31. For a more detailed analysis of Durkheim and Spencer, see Turner and Maryanski, *Functionalism,* Chapter 1. Indeed, if Durkheim's and Spencer's actual theories of social differentiation are compared, side by side, they are virtually identical. We can see this by examining the closing portions of Chapters 6 and 18 of this book. For further analysis and commentary, see Jonathan H. Turner, *Herbert Spencer: Toward a Renewed Appreciation* (Newbury Park, CA: Sage, 1985); "Durkheim's and Spencer's Principles of Social Organization: A Theoretical Note," *Sociological Perspectives* 27 (January 1984), pp. 21–32, and "The Forgotten Giant: Herbert Spencer's Theoretical Models and Principles," *Revue Européene des Sciences Sociales* 29 (no. 59, 1981), pp. 79–98.

32. Some commentaries, surprisingly those by British scholars, have tended to underemphasize Spencer's impact on Durkheim. Although all commentaries note the positive reaction of Durkheim to the German organicist Albert Schäffle, they fail to note that Schäffle was simply adopting Spencer's ideas. We suspect that Durkheim knew he was restating Spencer's ideas.

33. See, for example, Émile Durkheim, *Socialism and Saint-Simon* (Yellow Springs, OH: Antioch, 1959; originally published in 1928).

17

✳

The Sociology
of Émile Durkheim

Six decades after Auguste Comte proposed a field of inquiry called sociology, Émile Durkheim pulled together the long French lineage of social
thought into a coherent theoretical approach. Throughout Durkheim's
illustrious career, his theoretical work revolved around one fundamental question: What is the basis for integration and solidarity in human societies?[1] At
first, he examined this question from a macro perspective, looking at society
as a whole. Later, he shifted his attention to the micro bases of solidarity, examining ritual and interaction of people in face-to-face contact. In all his works,
he not only brought past theorizing together into a coherent scheme but also
stimulated a number of twentieth-century intellectual movements that persist
to this day.[2]

THE DIVISION OF LABOR IN SOCIETY

Durkheim's first major work was the published version of his French doctoral
thesis, *The Division of Labor in Society*.[3] The original subtitle of this thesis was
A Study of the Organization of Advanced Societies.[4] On the surface, the book is
about the causes, characteristics, and functions of the division of labor in modern societies, but, as we will explore, the book presents a more general theory
of social organization, one that can still inform sociological theorists.[5] In this
great work, Durkheim stressed a number of issues that will guide our review:

(1) social solidarity,[6] (2) the collective conscience, (3) social morphology,[7] (4) mechanical and organic solidarity, (5) social change, (6) social functions, and (7) social pathology.

Social Solidarity

The Division of Labor is about the shifting basis of social solidarity as societies evolve from an undifferentiated and simple profile[8] to a complex and differentiated one.[9] Today this topic would be termed *social integration* because the concern is with how units of a social system are coordinated. Durkheim posited that the question of social solidarity, or integration, turns on several related issues: (1) How are individuals made to feel part of a larger social collective? (2) How are their desires and wants constrained in ways that allow them to participate in the collective? (3) How are the activities of individuals and other social units coordinated and adjusted to one another? These questions, we should emphasize, not only dominated *The Division of Labor* but also guided all of Durkheim's subsequent substantive works.

These questions take us into the basic problem of how patterns of social organization are created, maintained, and changed. It is little wonder, therefore, that Durkheim's analysis of social solidarity contains a more general theory of social organization; we should explore those concepts that he developed to explain social organization in general. One of the most important of these concepts is the *collective conscience*.

The Collective Conscience

Throughout his career, Durkheim was vitally concerned with "morality," or *moral facts*. Although he was often somewhat vague on what constituted a moral fact, we can interpret the concept of morality to embody what sociologists now call culture. That is, Durkheim was concerned with the systems of symbols—particularly the norms, values, and beliefs—that humans create and use to organize their activities.

Durkheim had to assert the legitimacy of the scientific study of moral phenomena because other academic disciplines, such as law, ethics, religion, philosophy, and psychology, all laid claim to symbols as their subject matter. Thus, he insisted, "Moral facts are phenomena like others; they consist of rules of action recognizable by certain distinctive characteristics. It must, then, be possible to observe them, describe them, classify them, and look for laws explaining them."[10]

We should emphasize, however, that Durkheim in his early work often used the concept of *moral facts* to denote structural patterns (groups, organizations, and so on) as well as systems of symbols (values, beliefs, laws, norms). In *The Division of Labor*, however, we can find clear indications that he wanted to separate analytically the purely structural from the symbolic aspects of social reality. This isolation of cultural or symbolic phenomena can best be seen in his formulation of another, somewhat ambiguous, concept that suffers in translation: the *collective conscience*. He later dropped extensive use of this term in favor of *collective representations,* which, unfortunately, adds little clarification. But we can begin to understand his meaning with the formal definition provided in *The Division of Labor:* "The totality of beliefs and sentiments common to average citizens of the same society forms a determinate system which has its own life; one may call it the *collective* or *common conscience.*"[11] He went on to indicate that although the terms *collective* and *common* were "not without ambiguity," they suggested that societies reveal a reality independent of "the particular conditions in which individuals are placed." Moreover, people are born into the collective conscience or the culture of a society, and this culture regulates their perceptions and behaviors. What Durkheim was denoting with the concept of collective conscience, then, is that aspects of culture—systems of values, beliefs, and norms—constrain the thoughts and actions of individuals.

In the course of his analysis of the collective conscience, Durkheim conceptualized its varying states as having four variables: (1) volume, (2) intensity, (3) determinateness, and (4) religious versus secular content.[12] *Volume* denotes the degree to which the values, beliefs, and rules of the collective conscience are shared by the members of a society; *intensity* indicates the extent to which the collective conscience has power to guide a person's thoughts and actions; *determinateness* denotes the degree of clarity in the components of the collective conscience; and *content* pertains to the ratio of religious to purely secular symbolism in the collective conscience.

Social Morphology

Durkheim saw social structure (or as he termed it, *morphology*) as involving an assessment of the "nature," "number," "arrangement," and "interrelations" among parts, whether these parts were individuals or corporate units, such as groups and organizations. Their nature is usually assessed by such variables as size and functions (economic, political, familial, and so on). Arrangement concerns the distribution of parts in relationship to one another; interrelations deal with the modes of communication, movement, and mutual obligations among parts.

Although Durkheim's entire intellectual career involved an effort to demonstrate the impact of social structures on the collective conscience as well as on individual cognitions and behaviors, he never made explicit use of these variables—that is, nature, number, arrangement, and interrelations—for analyzing social structures. In his more methodological statements, he argued for the appropriateness of viewing social morphology by nature, size, number, arrangement, and interrelations of specific parts. Yet his actual analysis of social structures in his major substantive works left these more formal properties of structure implicit.[13]

Mechanical and Organic Solidarity

With these views on the collective conscience and structural morphology, Durkheim developed a typology of societies based on their modes of integration or solidarity. One type is termed *mechanical,* and the other, *organic.*[14] As we will show later, each of these types rests on different principles of social integration, involving different morphologies, different systems of symbols, and different relations between social and symbolic structures. Durkheim's distinction between mechanical and organic is both a descriptive typology of traditional and modern societies and a theoretical statement about the changing forms of social integration that emerge with increasing differentiation of social structure.

At a descriptive level, mechanical solidarity is based on a strong collective conscience regulating the thought and actions of individuals located within structural units that are alike. Of the four variables by which Durkheim conceptualized the collective conscience, the cultural system is high in volume, intensity, determinateness, and religious content. Legal codes, which in his view are the best empirical indicator of solidarity, are repressive, and sanctions are punitive. The reason for such repressiveness is that deviation from the dictates of the collective conscience is viewed as a crime against all members of the society and the gods. The morphology, or structure, of mechanical societies reveals independent kinship units that organize relatively small numbers of people who share strong commitments to their particular collective conscience. The interrelations among kin units are minimal, with each unit being like the others and autonomously meeting the needs of its members. Not surprisingly, then, individual freedom, choice, and autonomy are low in mechanical societies. People are dominated by the collective conscience, and their actions are constrained by its dictates and by the constraints of cohesive kin units.

In contrast, organically structured societies are typified by large populations, distributed in specialized roles in many diverse structural units. Organic societies reveal high degrees of interdependence among individuals and corporate units, with exchange, legal contracts, and norms regulating these interrelations. The collective conscience becomes "enfeebled" and "more abstract," providing highly general and secular value premises for the exchanges, contracts, and norms regulating the interdependencies among specialized social units. This alteration is reflected in legal codes that become less punitive and more "restitutive," specifying nonpunitive ways to redress violations of normative arrangements and to reintegrate violators back into the network of interdependencies that typify organic societies. In such societies, individual freedom is great, and the secular and highly abstract collective conscience becomes dominated by values stressing respect for the personal dignity of the individual.

This descriptive contrast between mechanical and organic societies is summarized in Table 17.1.[15] At the more theoretical level, Durkheim's distinction between mechanical and organic solidarity posits a fundamental relationship in the social world among "structural differentiation," "value–generalization," and "normative specification." Let us explore this relationship in more detail. As societies differentiate structurally, values become more abstract.[16] The collective conscience changes its nature as societies become more voluminous. Because these societies are spread over a vaster surface, the common conscience or culture rises above local diversities and consequently becomes more abstract. Only by becoming general can culture be common to distinctive environments.[17]

Table 17.1 Descriptive Summary of Mechanical and Organic Societies

Morphological (structural) Features	Mechanical Solidarity	Organic Solidarity
1. Size	Small	Large
2. Number of parts	Few	Many
3. Nature of parts	Kinship based	Diverse: Dominated by economic and governmental content
4. Arrangement	Independent, autonomous	Interrelated, mutually interdependent
5. Nature of interrelations	Bound to common conscience and punitive law	Bound together by exchange, contract, norms, and restitutive law
Collective Conscience (culture)	**Mechanical Solidarity**	**Organic Solidarity**
1. Volume	High	Low
2. Intensity	High	Low
3. Determinateness	High	Low
4. Content	Religious, stressing commitment and conformity to dictates of sacred powers	Secular, emphasizing individuality

As basic values lose their capacity to regulate the specific actions of large numbers of differentiated units, normative regulations arise to compensate for the inability of general values to specify what people should do and how individuals as well as corporate units should interact:

> If society no longer imposes upon everybody certain uniform practices, it takes greater care to define and regulate the special relations between different social functions and this activity is not smaller because it is different.[18]
>
> It is certain that organized societies are not possible without a developed system of rules which predetermine the functions of each organ. In so far as labor is divided, there arises a multitude of occupational moralities and laws.[19]

Thus, in his seemingly static comparison of mechanical and organic societies, Durkheim was actually proposing lawlike relationships among structural and symbolic elements of social systems.

Social Change

Durkheim's view of social change revolves around an analysis of the causes and consequences of increases in the division of labor:

The division of labor varies in direct ratio with the volume and density of societies, and, if it progresses in a continuous manner in the course of social development, it is because societies become regularly denser and generally more voluminous.[20]

Some translation of terms is necessary if this "proposition," as Durkheim called it, is to be understood. Volume refers to population size and concentration; density pertains to the increased interaction arising from escalated volume. Thus, the division of labor arises from increases in the concentration of populations whose members increasingly come into contact. Durkheim also termed the increased rates of interaction among those thrust into contact dynamic and moral density. He then analyzed those factors that increase the material density of a population. Ecological boundaries (rivers, mountains, and so on), migration, urbanization, and population growth all directly increase volume and thus indirectly increase the likelihood of dynamic density (increased contact and interaction). Technological innovations, such as new modes of communication and transportation, directly increase rates of contact and interaction among individuals. But all these direct and indirect influences are merely lists of empirical conditions influencing the primary explanatory variable, dynamic, or moral density.

How, then, does dynamic density cause the division of labor? Dynamic density increases competition among individuals who, if they are to survive the "struggle," must assume specialized roles and then establish exchange relations with each other. The division of labor is thus the mechanism by which competition is mitigated:

> Thus, Darwin says that in a small area, opened to immigration, and where, consequently, the conflict of individuals must be acute, there is always to be seen a very great diversity in the species inhabiting it. . . . Men submit to the same law. In the same city, different occupations can co-exist without being obliged mutually to destroy one another, for they pursue different objects.[21]

Figure 17.1 delineates these causal connections. To recapitulate, Durkheim saw migration, population growth, and ecological concentration as causing increased *material density,* which in turn caused increased *moral* or *dynamic density*—that is, escalated social contact and interaction. Such interaction could be further heightened by varied means of communication and transportation, as is illustrated in the model in Figure 17.1. The increased rates of interaction characteristic of a larger population within a confined ecological space cause increased competition, or "struggle," among individuals. Such competition allows those who have the most resources and talents to maintain their present positions and assume high-rank positions, whereas the "less fit" seek alternative specialties to mitigate the competition. From this competition and differentiation comes the division of labor, which, when "normal," results in organic solidarity.

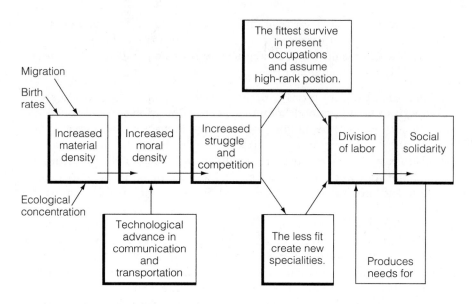

FIGURE 17.1 Durkheim's Casual Model of the Division of Labor

The major problem with the model is the implicit argument by "final causes": The function of the division of labor is to promote social solidarity. Durkheim implied that the need for social solidarity caused the struggle to be resolved by the division of labor; yet, he never specified how the needs met by the division of labor (that is, social solidarity) caused it to emerge. Without clear specification of this causal connection, the model becomes an illegitimate teleology in which the end state (social solidarity) causes the very thing (the division of labor) that brings about this end state. As is denoted by the feedback arrow moving from right to left in the model, Durkheim implied that the model was also circular, or tautological, in that cause and effect become difficult to separate: The division of labor causes social solidarity, and the need for social solidarity causes the division of labor. In such an argument, just what causes what becomes unclear.

The model in Figure 17.1 thus contains some suggestive ideas, particularly the notions that material density causes moral density, that moral density causes competition, that competition causes differentiation, and that differentiation causes new mechanisms of integration. On the other hand, without specifying the conditions under which these causal connections are generated, the model is vague. These causal statements still have great power, and in our view, they do constitute one of the lawlike properties of the social universe. The problems emerged as Durkheim sought to perform functional analysis. Let us explore, then, why he embraced functionalism.

Social Functions

As we observed in the discussion of Spencer's sociology in Chapter 5, he clearly formulated the notions of structure and function, with functions assessed by the needs of the social organism being met by a structure. Durkheim appears to have borrowed these ideas and, indeed, opened *The Division of Labor* with an assessment of its functions.[22] The function of the division of labor is to promote social solidarity, or societal integration. Such functional analysis, Durkheim argued, must be kept separate from causal analysis. It will be recalled from the previous chapter that he was highly critical of Montesquieu's analysis of "final causes," or functions. Thus, in his first major work, Durkheim attempted to keep his functional and causal analyses separated. As the model in Figure 17.1 illustrates, however, he was less than successful in maintaining such separation.

We might now ask a more fundamental question: Why did Durkheim engage in functional analysis at all? That is, why did he feel it necessary to determine the consequences for society of the division of labor? In his Latin doctoral thesis, he had been highly critical of Montesquieu for not separating his scientific observations from his personal feelings about what society should be. However, he was more sympathetic to Comte, who explicitly argued that the science of society could be used to create a better society. It was thus in the French tradition—Montesquieu, Rousseau, Saint-Simon, and Comte—to seek a better society. Beginning with Comte and culminating with Durkheim, then, scholars felt that an understanding of the laws of society could be used to create a new and better society.

Durkheim wanted to keep the science of sociology separated from such disciplines as moral philosophy and ethics, however. He had utter contempt for what he saw as the idle speculation and moral imperialism of these and related disciplines. Rather, he viewed sociology as the applied side of biology; that is, sociologists could be the "physicians of society."

The concept of function was central to this vision, especially in conjunction with the related notions of societal types. By assessing what a structure "does for" a society of a particular type or at a specific stage of evolution, one is in a position to determine what is "normal" and "abnormal" for that society—a point that Comte had first made in his advocacy of the experimental method as it could be used in sociology. The concept of function allowed Durkheim to judge whether a structure, such as the division of labor, was functioning normally for a particular type of society. Hence, to the degree that the division of labor fails to promote societal integration or social solidarity in a society, he viewed that society as in a "pathological" state and in need of alterations to restore "normality" to the "body social." These considerations led him to analyze "abnormal forms" of the division of labor at the close of this first major sociological work because the abnormality of structures can be determined only in reference to their "normal functions."

Pathology and Abnormal Forms

Durkheim opened his discussion of abnormal forms with the following statement:

> Up to now, we have studied the division of labor only as a normal phenomenon, but, like all social facts, and, more generally, all biological facts, it presents pathological forms which must be analyzed. Though normally the division of labor produces social solidarity, it sometimes happens that it has different, and even contrary results.[23]

Durkheim isolated three abnormal forms: (1) the anomic division of labor, (2) the forced division of labor, and (3) the inadequately coordinated division of labor. In discussing these abnormal forms, he drew considerable inspiration from his French predecessors, particularly Jean-Jacques Rousseau and Alexis de Tocqueville, while carrying on a silent dialogue with Karl Marx and other socialists. Thus, his analysis of abnormal forms represents his effort to address issues that had been discussed and contested for several previous generations of intellectuals. Indeed, individuals' isolation, their detachment from society, their sense of alienation, their exploitation by the powerful, and related issues had been hotly debated in both intellectual and lay circles. Yet, although Durkheim's selection of topics is not unique, his conclusions and their theoretical implications are highly original.

The Anomic Division of Labor The concept of anomie was not well developed in *The Division of Labor*. Only later, in the 1897 work *Suicide,* did this concept become theoretically significant. Durkheim's discussion in *The Division of Labor* is explicitly directed at Comte, who had noted the essence of the basic dilemma confronting organic social systems. As Comte stated,

> From the moral point of view, while each individual is thus made closely dependent on the mass, he is naturally drawn away from it by the nature of his special activity, constantly reminding him of his private interests, which he only very dimly perceives to be related to the public.[24]

For Durkheim, this dilemma was expressed as maintaining individuals' commitment to a common set of values and beliefs, while allowing them to pursue their specialized interests. At this stage in his thinking, anomie represented insufficient normative regulation of individuals' activities, with the result that individuals do not feel attached to the collectivity.

Anomie is inevitable, Durkheim believed, when the transformation of society from a mechanical to an organic basis of social solidarity is rapid and causes the "generalization," or "enfeeblement," of values. With generalization, individuals' attachment to, and regulation by, values are lessened. The results of this anomic situation are diverse. One result is that individuals feel alienated because their only attachment is to the monotony and crushing schedule dictated by the machines of the industrial age. Another is the escalated frustrations and the sense of deprivation, manifested by increased incidence of revolt, that come in a state of underregulation.

Unlike Marx, however, Durkheim did not consider these consequences inevitable. He rejected the notion that there were inherent contradictions in capitalism, for "if, in certain cases, organic solidarity is not all it should be . . . [it is] because all the conditions for the existence of organic solidarity have not been realized."[25] Nor would he accept Comte's or Rousseau's solution to anomie: the establishment of a strong and somewhat dictatorial central organ, the state.

Yet in the first edition of *The Division of Labor*, Durkheim's own solution is vague; the solution to anomie involves reintegration of individuals into the collective life by virtue of their interdependence with other specialists and the common goals that all members of a society ultimately pursue.[26] In many ways this argument substitutes for Adam Smith's invisible hand the "invisible power of the collective" without specifying how this integration into the collective is to occur.

Durkheim recognized the inadequacy of this solution to the problem of anomie. Moreover, his more detailed analysis of anomie in *Suicide* (1897) must have further underscored the limitations of his analysis in *The Division of Labor*. Thus, the second edition of *The Division of Labor*, published in 1902, contained a long preface that sought to specify the mechanism by which anomie was to be curbed. This mechanism is the "occupational," or "corporate," group.[27]

Durkheim recognized that industrialism, urbanization, occupational specialization, and the growth of the bureaucratized state all lessened the functions of family, religion, region, and neighborhood as mechanisms promoting the integration of individuals into the societal collectivity. With the generalization and enfeeblement of the collective conscience, coupled with the potential isolation of individuals in an occupational specialty, Durkheim saw that new structures would have to evolve to avoid anomie. These structures promote social solidarity in several ways: (1) they organize occupational specialties into a collective; (2) they bridge the widening gap between the remote state and the specific needs and desires of the individual; and (3) they provide a functional alternative to the old loyalties generated by religion, regionalism, and kinship. These new intermediate structures are not only occupational but also political and moral groupings that lace together specialized occupations, counterbalance the power of the state, and provide specific interpretations for the more abstract values and beliefs of the collective conscience.

Durkheim had taken the idea of "occupational groups" from Tocqueville's analysis of intermediate organizations in America (see Chapter 16). He extended the concept considerably, however, and in so doing, he posited a conception of how a society should be economically, politically, and morally organized.[28] Economically, occupational groups would bring together related occupational specialties into an organization that could set working hours and wage levels and that could bargain with management of corporations and government.

Politically, the occupational group would become a kind of political party whose representatives would participate in government. Like most French scholars in the postrevolutionary era, Durkheim distrusted mass democracy, feeling that short-term individual passions and moods could render the state helpless in setting and reaching long-range goals. He also distrusted an all-powerful and bureaucratized state on the ground that its remote structure was too insensitive and cumbersome to deal with the specific needs and problems of diverse individuals. Moreover, Durkheim saw that unchecked state power inevitably led to abuses, an emphasis that comes close to Montesquieu's idea of a balance of powers in government. Thus, the power of the state must be checked by intermediate groups, which channel public sentiment to the state and administer the policies of the state for a particular constituency.[29]

Morally, occupational groups are to provide many of the recreational, educational, and social functions formerly performed by family, neighborhood, and church. By bringing together people who are likely to have common experiences because they belong to related occupations, occupational groups can provide a place where people feel integrated into the society and where the psychological tensions and monotony of their specialized jobs can be mitigated. Moreover, these groups can make the generalized values and beliefs of the entire society relevant to the life experiences of each individual. Through the vehicle of occupational groups, then, an entire society of specialists can be reattached to the collective conscience.

Inequality and the Forced Division of Labor Borrowing heavily from Rousseau, Claude-Henri de Saint-Simon, and Comte, but reacting to Marx,[30] Durkheim saw inequalities based on ascription and inheritance of privilege as "abnormal." He advocated an inheritance tax that would eliminate the passing of wealth across generations, and indeed, he felt that in the normal course of things, this change would come about. Unlike Marx, however, Durkheim had no distaste for the accumulation of capital and privilege, as long as it was earned and not inherited.

What Durkheim desired was for the division of labor and inequalities in privilege to correspond to differences in people's ability. For him, it was abnormal in organic societies for wealth to be inherited and for this inherited privilege to be used by one class to oppress and exploit another. Such a situation represents a "forced division of labor," and in the context of analyzing this abnormality Durkheim examined explicitly Marxian ideas: (1) the labor theory of value and exploitation and (2) the domination of one class by another. Let us examine each briefly:

1. Durkheim felt that the price one pays for a good or service should be proportional to the "useful labor which it contains."[31] To the degree that this is not so, he argued, an abnormal condition prevails. What is necessary, and inevitable in the long run, is for buyers and sellers to be "placed in conditions externally equal"[32] in which the price charged for a good or service corresponds to the "socially useful labor" in it and where no

seller or buyer enjoys an advantage or monopoly that would allow prices to exceed socially useful labor.

2. Durkheim recognized that as long as there is inherited privilege, especially wealth, one class can exploit and dominate another. He felt that the elimination of inheritance was inevitable, because people could no longer be duped by a strong collective conscience into accepting privilege and exploitation (a position that parallels Marx's notion of *false consciousness*). For as religious and family bonds decrease in salience and as individuals are liberated from mechanical solidarity, people can free themselves from the beliefs that have often been used to legitimate exploitation.

Durkheim was certainly naive in his assumption that these aspects of the forced division of labor would, like Marx's state, "wither away." What he saw as normal was a situation that sounds reminiscent of Adam Smith's utilitarian vision.[33]

Lack of Coordination Durkheim termed the lack of coordination *another abnormal form* and did not devote much space to its analysis.[34] At times, he noted, specialization of tasks is not accompanied by sufficient coordination, creating a situation where energy is wasted and individuals feel poorly integrated into the collective flow of life. In his view, specialization must be "continuous," with functions highly coordinated and individuals laced together through their mutual interdependence. Such a state, he argued, will be achieved as the natural and normal processes creating organic solidarity become dominant in modern society.

On this note, *The Division of Labor* ends. Durkheim's next major work, published two years after *The Division of Labor,* sought to make more explicit assumptions and methodological guidelines that were implicit in *The Division of Labor. The Rules of the Sociological Method* (1895), as we will show, represents a methodological interlude in his efforts to understand how and why patterns of social organization are created, maintained, and changed.

THE RULES
OF THE SOCIOLOGICAL METHOD

The Rules of the Sociological Method is both a philosophical treatise and a set of guidelines for conducting sociological inquiry.[35] Durkheim appears to have written the book for at least three reasons.[36] First, he sought intellectual justification for his approach to studying the social world, especially as evidenced in *The Division of Labor.* Second, he wanted to persuade a hostile academic community of the legitimacy of sociology as a distinctive science. Third, because he wanted to found a school of scholars, he needed a manifesto to attract and guide potential converts to the science of sociology. The chapter titles of *The Rules* best communicate Durkheim's intent: (1) "What Is a Social

Fact?" (2) "Rules for the Observation of Social Facts," (3) "Rules for Distinguishing Between the Normal and the Pathological," (4) "Rules for the Classification of Social Types," (5) "Rules for the Explanation of Social Facts," and (6) "Rules Relative to Establishing Sociological Proofs." We will examine each of these.

What Is a Social Fact?

Durkheim was engaged in a battle to establish the legitimacy of sociology. In *The Division of Labor* he proclaimed "moral facts" to be sociology's subject, but in *The Rules* he changed his terminology to that employed earlier by Comte and argued that "social facts" were the distinctive subject of sociology. For Durkheim, a social fact "consists of ways of acting, thinking, and feeling, external to the individual, and endowed with power of coercion, by which they control him."[37]

In this definition, Durkheim lumped behaviors, thoughts, and emotions together as the subject of sociology. The morphological and symbolic structures in which individuals participate are thus to be the focus of sociology, but social facts are, by virtue of transcending any individual, "external" and "constraining." They are external in two senses:

1. Individuals are born into an established set of structures and an existing system of values, beliefs, and norms. Hence, these structural and symbolic "facts" are initially external to individuals, and as people learn to play roles in social structures, to abide by norms, and to accept basic values, they feel and sense "something" outside of them.

2. Even when humans actively and collaboratively create social structures, values, beliefs, and norms, these social facts become an emergent reality that is external to their creators.

This externality is accompanied by a sense of constraint and coercion. The structures, norms, values, and beliefs of the social world compel certain actions, thoughts, and dispositions. They impose limits, and when deviations occur, sanctions are applied to the deviants. Moreover, social facts are "internalized" in that people want and desire to be a part of social structures and to accept the norms, values, and beliefs of the collective. In the 1895 edition of *The Rules,* this point had been underemphasized, thus in the second edition Durkheim noted,

> Institutions may impose themselves upon us, but we cling to them; they compel us, and we love them.[38]
>
> [Social facts] dominate us and impose beliefs and practices upon us. But they rule us from within, for they are in every case an integral part of ourself.[39]

Durkheim thus asserted that when individuals come into collaboration, a new reality consisting of social and symbolic structures emerges. This emergent reality cannot be reduced to individual psychology, because it is external to, and constraining on, any individual. And yet, like all social facts, it is registered on the individual and often "rules the individual from within." Having established that sociology has a distinct subject matter—social facts—Durkheim devoted the rest of the book to explicating rules for studying and explaining social facts.

Rules for the Observation of Social Facts

Durkheim offered several guidelines for observing social facts: (1) Personal biases and preconceptions must be eliminated. (2) The phenomena under study must be clearly defined. (3) An empirical indicator of the phenomenon under study must be found, as was the case for "law" in *The Division of Labor*. (4) Social facts must be considered "things." Social facts are things in two different, although related, ways. First, when a phenomenon is viewed as a thing, it is possible to assume "a particular mental attitude" toward it. We can search for the properties and characteristics of a thing, and we can draw verifiable conclusions about its nature. Such a position was highly controversial in Durkheim's time because moral phenomena—values, ideas, morals—were not often considered proper topics of scientific inquiry, and when they were, they were seen as a sub-area in the study of individual psychology. Second, Durkheim asserted, phenomena such as morality, values, beliefs, and dogmas constitute a distinctive metaphysical reality, not reducible to individual psychology. Hence, they can be approached with the same scientific methods as any material phenomenon in the universe.[40]

Rules for Distinguishing Between the Normal and the Pathological

Throughout his career, Durkheim never wavered from Comte's position that science is to be used to serve human ends: "Why strive for knowledge of reality if this knowledge cannot serve us in life? To this we can make reply that, by revealing the causes of phenomena, science furnishes the means of producing them."[41] To use scientific knowledge to implement social conditions requires knowledge of what is normal and pathological. Otherwise, one would not know what social facts to create and implement, or one might actually create a pathological condition. To determine normality, the best procedure, Durkheim argued, is to discover what is most frequent and typical of societies of a given type or at a given stage of evolution. That which deviates significantly from this average is pathological.

Such a position allowed Durkheim to make some startling conclusions for his time. In regard to deviance, for example, a particular rate of crime and some other form of deviance could be normal for certain types of societies. Abnormality is present only when rates of deviance exceed what is typical of a certain societal type.

Rules for the Classification of Social Types

Durkheim's evolutionary perspective, coupled with his strategy for diagnosing normality and pathology in social systems, made inevitable a concern with social classification. Although specific systems reveal considerable variability, it is possible to group them into general types on the basis of (1) the "nature" and "number" of their parts and (2) the "mode of combination" of parts.

In this way, Durkheim believed, societies that reveal superficial differences can be seen as belonging to a particular class, or type. Moreover, by ignoring the distracting complexities of a society's "content" and "uniqueness," it is possible to establish the stage of evolutionary development of a society.

Rules for the Explanation of Social Facts

Durkheim emphasized again a point he had made in *The Division of Labor:* "When the explanation of social phenomena is undertaken, we must seek separately the efficient cause which produces it and the function it fulfills."[42] Causal analysis involves searching for antecedent conditions that produce a given effect. Functional analysis is concerned with determining the consequences of a social fact (regardless of its cause) for the social whole or larger context in which it is located. Complete sociological explanation involves both causal and functional explanations, as Durkheim had sought to illustrate in *The Division of Labor* (see Figure 17.1).

Rules for Establishing Sociological Proofs

Durkheim advocated two basic procedures for establishing "sociological proofs"—proofs being documentation that causal and, by implication, functional explanations are correct. One procedure involves comparing two or more societies of a given type (as determined by the rules for classification) to see if one fact, present in one but not the other(s), leads to differences in these otherwise similar societies.

The second procedure is the method of concomitant variation. If two social facts are correlated and one is assumed to cause the other, and if all alternative facts that might also be considered causative cannot eliminate the correlation, it can be asserted that a causal explanation has been "proved." If an established correlation, and presumed causal relation, can be explained away by the operation of another social fact, the established explanation has been disproved and the new social fact can, until similarly disproved, be considered "proved." The essence of Durkheim's method of concomitant variation, then, was similar in intent to modern multivariate analyses: to assert a relation among variables, controlling for the impact of other variables.[43]

The Rules marks a turning point in Durkheim's intellectual career. It was written after his thesis on the division of labor and while he was pondering the question of suicide in his lectures. Yet it was written before his first public course on religion.[44] He had clearly established his guiding theoretical interests: the nature of social organization and its relationship to values, beliefs, and other symbolic systems. He had developed a clear methodology: asking causal and functional questions within a broad comparative, historical, and evolutionary framework. He had begun to win respect in intellectual and academic circles for the fact that social organization represents an emergent reality, *sui generis,* and that it is the proper subject matter for a discipline called sociology.

Durkheim's next work appears to have been an effort to demonstrate the utility of his methodological and ontological advocacy. For he sought to understand sociologically a phenomenon that, at the time, was considered uniquely psychological: suicide. In this work, he attempted to demonstrate the power of sociological investigation for seemingly psychological phenomena, employing social facts as explanatory variables. Far more important than the specifics of suicide, we believe, is his extension of concepts introduced in *The Division of Labor.* In so doing, he presented additional theoretical principles concerning why patterns of social organization were created, maintained, and changed.

SUICIDE

In *Suicide,* Durkheim appears to follow self-consciously the "rules" of his sociological method.[45] He was interested in studying only a social fact, and hence he did not study individual suicides but rather the general pervasiveness of suicide in a population—that is, a society's aggregate tendency toward suicide. In this way, suicide could be considered a social rather than an individual fact, and it could be approached as a "thing." Suicide is clearly defined as "all causes of death resulting directly or indirectly from a positive or negative act of the victim himself which he knows will produce this result."[46] The statistical rate of suicide is then used as the indicator of this social fact.[47] Suicide is classified into four types: egoistic, altruistic, anomic, and fatalistic. The cause of these types is specified by the degree and nature of individual integration into the social collective. A variant of modern correlational techniques is employed to demonstrate, or "prove," that other hypothesized causes of suicide are spurious and that integration into social and symbolic structures is the key explanatory variable.

The statistical manipulations in *Suicide* are important because they represent the first systematic effort to apply correlational and contingency techniques to causal explanation. Our concern, however, is with the theoretical implications of this work, and hence the following summary will focus on theoretical rather than statistical issues.

Types of Suicide

As noted, Durkheim isolated four types of suicide by varying causes. We should emphasize that despite his statistical footwork, isolating types by causes and then explaining these types by the causes used to classify them is a suspicious, if not spurious, way to go about understanding the social world. These flaws aside, Durkheim's analysis clarifies notions of social integration that are somewhat vague in *The Division of Labor.* Basically, Durkheim argued that suicides could be classified by the nature of an individual's integration into the social fabric. There are, in Durkheim's eye, two types of integration:

1. *Attachment* to social groups and their goals. Such attachment involves the maintenance of interpersonal ties and the perception that one is a part of a larger collectivity.

2. *Regulation* by the collective conscience (values, beliefs, and general norms) of social groupings. Such regulation limits individual aspirations and needs, keeping them in check.

In distinguishing these two bases of integration, Durkheim was explicitly recognizing the different "functions" of the structural and cultural elements of the social world. Interpersonal ties that bind individuals to the collective keep them from becoming too "egoistic"—a concept borrowed from Tocqueville and widely discussed in Durkheim's time. Unless individuals can be attached to a larger collective and its goals, they become egoistic, or self-centered, in ways that are highly destructive to their psychological well being. In contrast, the regulation of individuals' aspirations, which are potentially infinite, prevents anomie. Without cultural constraints, individual aspirations, as Rousseau[48] and Tocqueville had emphasized, escalate and create perpetual misery for individuals who pursue goals that constantly recede as they are approached. These two varying bases of individual integration into society, then, form the basis for Durkheim's classification of four types of suicide— egoistic, altruistic, anomic, and fatalistic.

Egoistic Suicide When a person's ties to groups and collectivities are weakened, there is the potential for excessive individualism and, hence, egoistic suicide. Durkheim stated this relation as a clear proposition: "Suicide varies inversely with the degree of integration of social groups of which the individual forms a part."[49] And as a result,

> The more weakened the groups to which he belongs, the less he depends on them, the more he consequently depends only on himself and recognizes no other rules of conduct than what are founded on private interest. If we agree to call this state egoism, in which the individual ego asserts itself to excess in the face of the social ego and at its expense, we may call egoistic the special type of suicide springing from excessive individualism.[50]

Altruistic Suicide If the degree of individual integration into the group is visualized as a variable continuum, ranging from egoism on the one pole to a complete fusion of the individual with the collective at the other pole, the essence of Durkheim's next form of suicide can be captured. Altruistic suicide is the result of individuals being so attached to the group that, for the good of the group, they commit suicide. In such a situation, individuals count for little; the group is paramount, with individuals subordinating their interests to those of the group. Durkheim distinguished three types of altruistic suicide:

1. *Obligatory altruistic suicide,* in which individuals are obliged, under certain circumstances, to commit suicide

2. *Optional altruistic suicide,* in which individuals are not obligated to commit suicide, but it is the custom for them to do so under certain conditions

3. *Acute altruistic suicide,* in which individuals kill themselves "purely for the joy of sacrifice, because, even with no particular reason, renunciation in itself is considered praiseworthy"[51]

In sum, then, egoistic and altruistic suicides result from either over-integration or under-integration into the collective. Altruistic suicide tends to occur in traditional systems—what Durkheim termed *mechanical* in *The Division of Labor*—and egoistic suicide is more frequent in modern, organic systems that reveal high degrees of individual autonomy. At the more abstract level, Durkheim is positing a critical dimension of individual and societal integration: the maintenance of interpersonal bonds within coherent group structures.

Anomic Suicide In *The Division of Labor,* Durkheim's conceptualization of anomie was somewhat vague. In many ways, he incorporated both anomie (deregulation by symbols) and egoism (detachment from structural relations in groups) into the original definition of anomie. In *Suicide,* Durkheim clarified this ambiguity: Anomic suicide came to be viewed narrowly as the result of deregulation of individuals' desires and passions. Although both egoistic and anomic suicide "spring from society's insufficient presence in individuals,"[52] the nature of the disjuncture or deficiency between the individual and society differs.[53]

Fatalistic Suicide Durkheim discussed fatalistic suicide in a short footnote. Just as altruism is the polar opposite of egoism, so fatalism is the opposite of anomie. Fatalistic suicide is the result of "excessive regulation, that of persons with futures pitilessly blocked and passions violently choked by oppressive discipline."[54] Thus, when individuals are overregulated by norms, beliefs, and values in their social relations, and when they have no individual freedom, discretion, or autonomy in their social relations, they are potential victims of fatalistic suicide.

Suicide and Social Integration

These four types of suicide reveal a great deal about Durkheim's conception of humans and the social order. The study of suicide allows us a glimpse at how he conceived of human nature. Reading between the lines in *Suicide,* the following features are posited:

1. Humans can potentially reveal unlimited desires and passions, which must be regulated and held in check.

2. Total regulation of passions and desires creates a situation where life loses all meaning.

3. Humans need interpersonal attachments and a sense that these attachments connect them to collective purposes.

4. Excessive attachment can undermine personal autonomy to the point where life loses meaning for the individual.[55]

These implicit notions of human nature, it should be emphasized, involve a vision of the "normal" way in which individuals are integrated into the structural and cultural structures of society. Indeed, Durkheim was unable even to address the question of human nature without also talking about the social order. Durkheim believed that the social order is maintained only to the degree that individuals are attached to, and regulated by, patterns of collective organization. This belief led him later in his career to explore in more detail an essentially social-psychological question: In what ways do individuals become attached to society and become willing to be regulated by its symbolic elements?

Suicide and Deviance

Durkheim made an effort to see if other forms of deviance, such as homicide and crime, were related to suicide rates, but the details of his correlations are not as important as the implications of his analysis for a general theory of deviance. As he had in *The Division of Labor,* he recognized that a society of a certain type would reveal a "typical," or "average," level of deviance, whether of suicide or some other form. However, when rates of suicide, or deviance in general, exceed certain average levels for a societal type, a "pathological" condition might exist.

Durkheim's great contribution is his recognition that deviance is caused by the same forces that maintain conformity in social systems. Moreover, he specified the two key variables in understanding both conformity and deviance: (1) the degree of group attachment and (2) the degree of value and normative regulation. Thus, excessive or insufficient attachment and regulation will cause varying forms of deviance in a social system. Moreover, the more a system reveals moderate degrees of regulation and attachment, the less likely are pathological rates of deviance and the greater is the social integration of individuals into the system.

Thus, Durkheim's analysis in *Suicide* is much more than a statistical analysis of a narrowly defined topic. It is also a venture into understanding how social organization is possible. This becomes particularly evident near the end of the book, where Durkheim proposes his solution to the high rates of suicide and other forms of deviance that typify modern or "organically" structured societies.

Suicide and the Social Organization
of Organic Societies[56]

At the end of *Suicide,* Durkheim abandoned his cross-sectional statistical analysis and returned to the evolutionary perspective contained in *The Division of Labor.* During social change, as societies move from one basis of social solidarity, deregulation (anomie) and detachment (egoism) of the individual from society can occur, especially if this transition is rapid. Deregulation and detachment create not only high rates of deviance but also problems in maintaining the social order. If these problems are to be avoided and if social "normality" is to be restored, new structures that provide attachment and regulation of individuals to society must be created.

In a series of enlightening pages, Durkheim analyzed the inability of traditional social structures to provide this new basis of social integration. The family is an insufficiently encompassing social structure, religious structures are similarly too limited in their scope and too oriented to the sacred, and government is too bureaucratized and hence remote from the individual. For Durkheim, the implications are that modern social structures require intermediate groups to replace the declining influence of family and religion, to mediate between the individual and state, and to check the growing power of the state. He saw the occupational group as the only potential structural unit that could regulate and attach individuals to society.

Thus, in *Suicide,* the ideas that were later placed in the 1902 preface to the second edition of *The Division of Labor* found their first forceful expression. The analysis in suicide allowed Durkheim to explore further the concept of social integration, and for this reason *Suicide* represents both an application of the method advocated in *The Rules* and a clarification of substantive ideas contained in *The Division of Labor.* It also represents an effort to incorporate social psychology[57] into structural sociology.

THE ELEMENTARY FORMS
OF THE RELIGIOUS LIFE

Although Durkheim turned to the study of religion in his last major work, it had been an important interest for a long time. Indeed, his family background ensured that religion would be a central concern, and from 1895 on, he had taught courses on religion.[58] Regardless of any personal reasons for his interest, we suspect that Durkheim pursued the study of religion through most of his career because it allowed him to gain insight into the basic theoretical problem that guided all of his work: the nature of symbols and their reciprocal effects with social organization. In *The Division of Labor,* he had argued that in mechanical societies the collective conscience or culture is predominately religious in content and that it functions to integrate the individual into the collective. He had recognized that in organic systems the collective conscience becomes "enfeebled" and that religion as a pervasive influence recedes. The potential pathologies that can occur with the transition from mechanical to organic solidarity—particularly anomie—became increasingly evident to Durkheim. Indeed, the naive optimism that these pathologies would "spontaneously" wither away became increasingly untenable, and as is evident in *Suicide,* he began to ponder how to create a social system in which individuals are both regulated by a general set of values and attached to concrete groups. As he came to view the matter, these concerns revolve around the more general problem of "morality."

Durkheim never wrote what was to be the culmination of his life's work: a book on morality. In many ways, however, his study of religion represents the beginning of his formal work on morality. Although he had lectured on morality in his courses on education[59] and had written several articles on morality,[60] he saw in religion a chance to study how interaction among individuals leads to the creation of symbolic systems that (1) lace together individual actions into collective units, (2) regulate and control individual desires, and (3) attach individuals to both the cultural (symbolic) and structural (morphological) facets of the social world. Given the rise of anomie and egoism in modern societies, he thought that an understanding of religious morality in primitive social systems would help explain how such morality could be created in modern, differentiated systems. Thus, we could retitle *The Elementary Forms of the Religious Life* "the fundamental forms of moral integration" and be close to his purpose in examining religion in aboriginal societies, particularly the Arunta aborigines of Australia.[61]

In the course of writing what was his longest work, however, Durkheim introduced many other intellectual issues that occupied his attention over the years. Thus, *Elementary Forms* is more than a study of social integration; it is also an excursion into human evolution, the sociology of knowledge, functional and causal analyses, the origin and basis of thought and mental categories, the process of internalization of beliefs and values, and many other issues. Between the long descriptive passages on life among the Australian tribes, a myriad of ideas burst forth and give evidence of the wide-ranging concerns of Durkheim's intellect.

Elementary Forms is thus a long, complex, and—compared with earlier works—less coherently organized book. This requires that our analysis be divided into a number of separate topics. After a brief overview of the argument in *Elementary Forms,* we will examine in more detail some of its implications.

An Overview of Durkheim's Argument

By studying the elementary forms of religion among the most primitive[62] peoples, it should be possible, Durkheim felt, to understand the essence of religious phenomena without the distracting complexities and sociocultural overlays of modern social systems.[63] As dictated in *The Rules,* a clear definition of the phenomenon under study was first necessary. Thus, Durkheim defined religion as "a unified system of beliefs and practices relative to sacred things, that is to say, things set apart and forbidden—beliefs and practices which unite into one single moral community called a Church, all those who adhere to them."[64]

Durkheim believed that religiosity had emerged among humans when they occasionally assembled in a larger mass. From the mutual stimulation and "effervescence" that comes from animated interaction, people came to perceive a force, or "mana," that seemed superior to them. The mutual stimulation of primitive peoples thus made them "feel" an "external" and "constraining" force above and beyond them.[65] This force seemed to be imbued with special significance and with a sense that it was not part of this world. It was, then, the first notion of a "sacred" realm distinct from the routine or "secular" world of daily activities. The distinction between sacred and secular was thus one of the first sets of mental categories possessed by humans in their evolutionary development.

As humans came to form more permanent groupings, or clans, the force that emerged from their interaction needed to be more concretely represented.[66] Such representation came with "totems," which are animals and plants that symbolize the force of mana. In this way, the sacred forces could be given concrete representation, and groups of people organized into "cults" could develop "ritual" activities directed toward the totem and indirectly toward the sacred force that they collectively sensed.

Thus, the basic elements of religion are (1) the emergence of beliefs in the sacred, (2) the organization of people into cults, and (3) the enactment of rituals, or rites, toward totems that represent the forces of the sacred realm. What the aboriginals did not recognize, Durkheim argued, is that in worshipping totems, they were worshipping society. Totemic cults are nothing but the material symbolization of a force created by their interaction and collective organization into clans.

As people first became organized into clans and associated totemic cults, and as they perceived a sacred realm that influenced events in the secular world, their first categories of thought were also formed. Notions of causality, Durkheim argued, could emerge only after people perceived that sacred forces determined events in the secular world. Notions of time and space could exist only after the organization of clans and their totemic cults. According to Durkheim, the basic categories of human thought—cause, time, space, and so on—emerged after people developed religion. Thus, in an ultimate sense, science and all forms of thought have emerged from religion, an argument, we might note, reminiscent of Comte's law of the three stages. Before religion, humans experienced only physical sensations[67] from their physical environment, but with religion, their mental life became structured by categories. In Durkheim's view, mental categories are the cornerstone of all thought, including scientific thought and reasoning. In looking back on *Elementary Forms* a year after its publication, Durkheim was still moved to conclude,

> The most essential notions of the human mind, notions of time, of space, of genus and species, of force and causality, of personality, those in a word, which the philosophers have labeled categories and which dominate the whole logical thought, have been elaborated in the very womb of religion. It is from religion that science has taken them.[68]

For Durkheim, the cause of religion is the interaction among people created by their organization into the simplest form of society, the clan. The functions of religion are (1) to regulate human needs and actions through beliefs about the sacred and (2) to attach people, through ritual activities (rites) in cults, to the collective. Because they are internalized, religious beliefs generate needs for people to belong to cults and participate in rituals. As people participate in rituals, they reaffirm these internalized beliefs and, hence, reinforce their regulation by, and attachment to, the dictates of the clan. Moreover, the molding of such basic mental categories as cause, time, and space by religious beliefs and cults functions to give people a common view of the world, thus facilitating their interaction and organization.

This argument is represented in Figure 17.2, which delineates Durkheim's model on the origins of functions of religion. Regarding origins, he had an image of aboriginal peoples periodically migrating and concentrating themselves in temporary gatherings. Once they have gathered, increased interaction escalates collective emotions, which produce a sense that there is something external and constraining to each individual. This sense of constraint is given more articulate expression as a sacred force, or mana. This causal sequence occurs, Durkheim maintained, each time aboriginals gather in their periodic festivals. Once they form more permanent groupings, called clans, the force of mana is given more concrete expression as a sacred totem. The creation of beliefs about, and rituals toward, the totem function to promote clan solidarity.

This model is substantively inaccurate, as are all of Durkheim's intellectual expeditions into the origins of society. For example, the clan was not the first kinship structure, and many hunter-gatherers like the aborigines do not worship totems. These errors can be attributed to Durkheim's reliance on Australian aborigine kinship and religious organization, which in many ways deviate from modal patterns among hunting-and-gathering peoples. Aside from these factual errors, the same problems evident in the model of the division of labor resurface. First, the conditions under which any causal connection holds true are not specified. Second, the functions of religious totems (for social solidarity) are also what appear to promote their very creation. In addition, a psychological need—the "primitive need" to make concrete and symbolize "mana"—is invoked to explain why totems emerge.

Because of these problems, it must be concluded that the model does not present any useful information in its causal format. But as a statement of relationships among rates of interaction, structural arrangements, emotional arousal, and symbolic representation, Durkheim's statements are suggestive and emphasize: (1) highly concentrated interactions increase collective sentiments, which mobilize actors' actions, (2) small social structures tend to develop symbols to represent their collective sentiments, and (3) these structures evidence high rates of ritual activity to reinforce their members' commitment.

Some Further Implications of *Elementary Forms*

Practical Concerns Durkheim's analysis of pathologies in *Suicide,* along with other essays, forced the recognition that a more active program for avoiding egoism and anomie might be necessary to create a normal "organic" society. Religion, he thought, offers a key to understanding how this can be done. Early in his career, however, he rejected the idea that religion could ever again assume major integrative functions. The modern world is too secular and

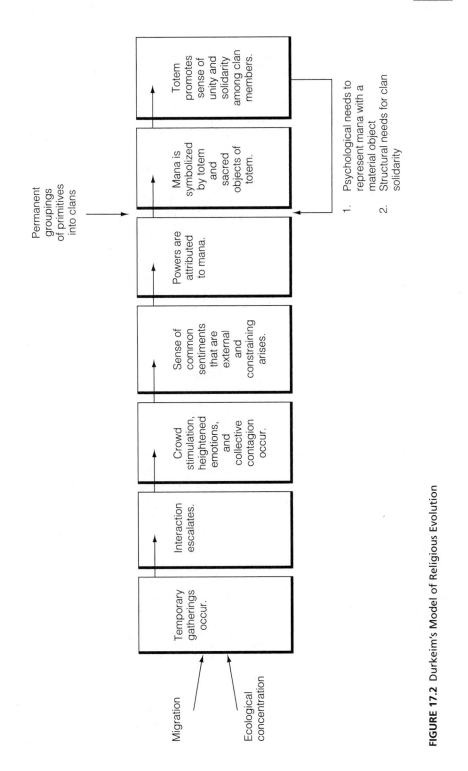

FIGURE 17.2 Durkeim's Model of Religious Evolution

individualistic for the subordination of individuals to gods. He also rejected, to a much lesser degree, Saint-Simon's and Comte's desire to create a secular religion of humanity based on science and reason. Although Durkheim saw a need to maintain the functions and basic elements of religion, he had difficulty accepting Comte's ideal of positivism, which, as Robert Nisbet noted, was "Catholicism minus Christianity." For Comte, the Grand Being was society, the church was the hierarchy of the sciences, and the rites were the sacred canons of the positive method.[69] Durkheim also rejected Max Weber's pessimistic view of a secular, rational world filled with disenchantment and lacking in commitments to a higher purpose.

The "solution" implied in *Elementary Forms* and advocated elsewhere in various essays is for the re-creation in secular form of the basic elements of religion: feelings of sacredness, beliefs and values about the sacred, common rituals directed toward the sacred, and cult structures in which these rituals and beliefs are reaffirmed. Because society is the source and object of religious activity anyway, the goal must be to make explicit this need to "worship" society. Occupational groups and the state would become the church and cults; nationalistic beliefs would become quasi-sacred and would provide underlying symbols; and activities in occupational groups, when seen as furthering the collective goals of the nation, would assume the functions of religious ritual in (1) mobilizing of individual commitment, (2) reaffirming beliefs and values, and (3) integrating individuals into the collective.

Theoretical Concerns Contained in these practical concerns are several important theoretical issues. First, integration of social structures presupposes a system of values and beliefs that reflects and symbolizes the structure of the collective. Second, these values and beliefs require rituals directed at reaffirming them as well as those social structures they represent or symbolize. Third, large collectivities, such as a nation, require subgroups in which values and beliefs can be affirmed by ritual activities among a more immediate community of individuals. Fourth, to the degree that values and beliefs do not correspond to actual structural arrangements and to the extent that substructures for the performance of actions that reaffirm these values and beliefs are not present, a societal social system will experience integrative problems.

We can see, then, that Durkheim's practical concerns follow from certain theoretical principles he had tentatively put forth in *The Division of Labor.* The study of religion seemingly provided him with a new source of data to affirm the utility of his first insights into the social order. There are, however, some noticeable shifts in emphasis, the most important of which is the recognition that the "conscience collective" cannot be totally "enfeebled"; it must be general but also strong and relevant to the specific organizations of a social system. Despite these refinements, however, *Elementary Forms* affirms the conclusion contained in the preface to the second edition of *The Division of Labor.*

The most interesting aspect of the analysis is perhaps the social psychological emphasis of *Elementary Forms.* Although Durkheim, in courses and essays, had begun to feel comfortable with inquiry into the social-psychological dynamics of social and symbolic structures, these concerns are brought together in his last major book.

Social-Psychological Concerns *Elementary Forms* contains the explicit recognition that morality—that is, values, beliefs, and norms—can integrate the social order only if morality becomes part of an individual's psychological structure. Statements in *Elementary Forms* mitigate the rather hard line taken in the first edition of *The Rules,* where social facts are seen as external and constraining things. With the second edition of *The Rules,* Durkheim felt more secure in verbalizing the obvious internalization of values, beliefs, and other symbolic components of society into the human psyche. In *Elementary Forms,* he revealed even fewer reservations:

> For the collective force is not entirely outside of us; it does not act upon us wholly from without; but rather, since society cannot exist except in and through individual consciousnesses, this force must also penetrate us and organize itself within us, it thus becomes an integral part of our being.[70]

Durkheim hastened to add in a footnote, however, that although society was an "integral part of our being," it could not ever be seen as reducible to individuals.

Another social-psychological concern in *Elementary Forms* is the issue of human thought processes. For Durkheim, thought occurs in categories that structure experience for individuals:

> At the roots of all our judgments there are a certain number of essential ideas which dominate all our intellectual life; they are what philosophers since Aristotle have called the categories of the understanding: ideas of space, class, number, cause, substance, personality, etc. They correspond to the most universal properties of things. They are like the solid frame which encloses all thought.[71]

Durkheim sought in *Elementary Forms* to reject the philosophical positions of David Hume and Immanuel Kant. Hume, the staunch empiricist, argued that thought was simply the transfer of experiences to the mind and that categories of thought were merely the codification of repetitive experiences. In contrast with Hume, Kant argued that categories and mind were inseparable; the essence of mind is categorization. Categories are innate and not structured from experience. Durkheim rejected both of these positions; in their place he wanted to insert the notion that categories of thought—indeed, all thinking and reflective mental activity—were imposed on individuals by the structure and morality of society. Indeed this imposition of society becomes a critical condition not just for the creation of mind and thought but also for the preservation of society.[72] Thus, Durkheim believed that the basic categories of thought, such as cause, time, and space, were social products, in that the structure of society determines them in the same way that values and beliefs also structure human "will," or motivations. For example, the idea of a sacred force, or mana, beyond individuals became, in the course of human evolution, related to ideas of causality as rituals and beliefs came to concern the effects of sacred acts in the secular world. Similarly, the idea of time emerged among humans as they developed calendrical rituals and related them to solar and lunar rhythms. The conception of space was shaped by the structure of villages, so that if the aboriginal village is organized in a circle, the world will be seen as circular and concentric in nature. These provocative insights were at times taken to excessive extremes in other essays, especially in one written with his nephew and student, Marcel Mauss, on "primitive classification."[73] Here mental categories are seen to be exact representations of social structural divisions and arrangements. Moreover, Durkheim and Mauss appear to have selectively reported data from aboriginal societies to support their excessive claims.[74]

Durkheim is often viewed as the "father of structuralism," a school of thought in the twentieth century that embraced social science, linguistics, and literature. In Durkheim's *Elementary Forms* and other works of this period can be found an implicit model that appears to have inspired this structuralist reasoning. Figure 17.3 outlines the contours of this model. In Durkheim's view, the morphology of a society (the number, nature, size, and arrangement of parts—see Table 17.1) determines the structure of the collective conscience or culture (the volume, density, intensity, and content of values, beliefs, and norms). Reciprocally, the collective conscience reinforces social morphology. Both morphology and collective conscience circumscribe each individual's cognitive structure by determining the nature of basic categories of thought with respect to time, space, and causality. In turn, these categories mediate between morphology and the collective conscience, on the one hand, and the nature of rituals that individuals emit in face-to-face interaction, on the other. The reverse causal loops in the model are crucial to Durkheim's argument: The enactment of rituals reinforces not only cognitive categories but also the structure and idea systems of society. In this way the macrostructural features of society—morphology and idea systems—are conceptually tied to the microstructural dimensions of reality—that is, the internal psychological structure of thought and the face-to-face interactions among individuals in concrete settings.

A SCIENCE OF "MORALITY"

As early as *The Division of Labor,* Durkheim defined sociology as the science of "moral facts," and he always wanted to write a book on morality. In light of this unfulfilled goal, perhaps we should close our analysis by extracting from his various works what would have been the core ideas of this uncompleted work.[75]

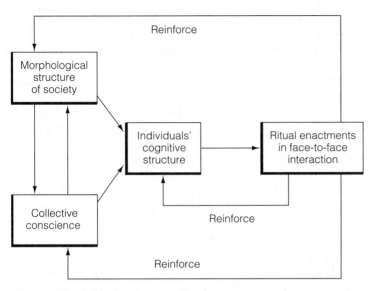

FIGURE 17.3 Durkheim's Structuralism

What Is Morality?

In only two places did Durkheim provide a detailed discussion of morality.[76] For him, morality consists of (1) rules, (2) attachment to groups, and (3) voluntary constraint.

Rules Morality is ultimately a system of rules that guide the actions of people. For rules to be moral, they must reveal two additional elements:

1. *Authority.* Moral rules are invested with authority; that is, people feel that they ought to obey them, and they want to abide by them. Moral rules are a "system of commandments."

2. *Desirability.* Moral rules also specify the "desirable" ends toward which a collectivity of people should direct its energies. They are more than rules of convenience; they carry conceptions of the good and desirable and must therefore be distinguished from strictly utilitarian norms.

Attachment to Groups Moral rules attach people to groups. They are the product of interactions in groups, and as they emerge, they bind people to groups and make individuals feel a part of a network of relations that transcends their individual being.

Durkheim termed these two facets of morality *the spirit of discipline.* Morality provides a spirit of self-control and a commitment to the collective. In terms of the concepts developed in *Suicide,* morality reduces anomie and egoism to the degree that it regulates desires and attaches people to the collective. True morality in a modern society must do something else, how-ever: It must allow people to recognize that the constraints and restraints imposed on them are in the "natural order of things."

Voluntary Constraint Modern morality must allow people to recognize that unlimited desires (anomie) and excessive individualism (egoism) are pathologi-cal states. Durkheim felt that these states violate the nature of human society and can be corrected only by morality. In simple societies, morality seems to oper-ate automatically, but "the more societies become complex, the more difficult [it becomes] for morality to operate as a purely automatic mechanism."[77] Thus, morality must be constantly implemented and altered to changing conditions. Individuals must also come to see that such alteration is necessary and essential because to fail to establish a morality and to allow people to feel free of its power is to invite the agonies of anomie and egoism.

Durkheim then resumed an argument first made by Rousseau: Morality must be seen as a natural constraint in the same way that the physical world constrains individuals' options and actions. So it is with morality; humans can no more rid themselves of its constraint than they can eliminate the physical and biological world on which their lives depend. The only recourse is to use science, just as we use the physical and biological sciences, to understand how morality works.[78]

Thus, Durkheim never abandoned his original notion, first given forceful expression in *The Division of Labor,* that sociology is the science of moral facts. His conception of morality had become considerably more refined, however, in three senses:

1. Morality is a certain type of rule that must be distinguished from both the morphological aspects of society and other, nonmoral, types of nor-mative rules.

2. Morality is therefore a system of rules that reflects certain underlying value premises about the desirable.

3. Morality is not only external and constraining; it is also internal. It calls people to obey from within. For although morality "surpasses us it is within us, since it can only exist by and through us."[79]

By the end of Durkheim's career, the study of morality involved a clear separation between two types of norms and rules: those vested with value premises and those that simply mediate and regularize interactions. Moreover, an understanding of these types of rules could come only by visualizing their relationship to the morphological or structural aspects of society—nature, size, number, and relations of parts—and to the process by which internalization of symbols or culture occurs. Durkheim had thus begun to develop a clear con-ception of the complex relations among normative systems, social structures, and personality processes of individuals.

What would Durkheim have said in his last work—the book on morality—if he had lived to write it? *Moral Education,* when viewed in the context of his other published books, can perhaps provide some hints about the direction of his thought. For *Moral Education* offers a view of how a new secular morality can be instilled.

For a new secular morality to be effective, the source of all morality must be recognized: society. This means that moral rules must be linked to the goals of the broader society, but they must be made specific through the participation of individuals in occupational groups. The commitment to the common morality must be learned in schools, where the teacher operates as the functional equivalent of the priest. The teacher gives young students an understanding of and a reverence for the nature of the society and the need to have a morality that regulates passions and provides attachments to groupings organized to pursue societal goals. Such educational socialization must assure that the common morality is a part of students' motivational needs (their "will," in Durkheim's language), their cognitive orientations ("categories of mind"), and their self-control processes ("self-mastery").

A modern society that cannot meet these general conditions, Durkheim would have argued in this unwritten work, is a society that will be rife with pathologies revolving around (1) the failure to limit individual passions, desires, and aspirations and (2) the failure to attach individuals to groups with higher purposes and common goals.

Durkheim must have felt that the implicit theory of social organization contained in this argument had allowed him to realize Comte's dream of a science that could create "the good society." Although Durkheim was cautious in implementing his proposals, they were often simplistic, if not somewhat reactionary. At the same time, there is the germ of a theory of human organization. This theory, we believe, marks Durkheim's enduring legacy, as we will see in the next chapter.

CRITICAL CONCLUSIONS

Émile Durkheim is, along with Karl Marx and Max Weber, one of the "holy trinity" of sociology's early masters. He enjoys this high place in sociology's pantheon because he addressed issues that have long fascinated sociologists. Although he borrowed a great deal from Herbert Spencer in his early work, he nonetheless presented what is now termed an ecological model, emphasizing population growth, competition for resources, and differentiation—a model that is widely used in sociology today. More fundamentally, he isolated in *The Division of Labor in Society* some of the key mechanisms by which complex systems sustain integration: structural interdependence, abstract and general values and beliefs, more specific beliefs and norms to regulate relations within and between differentiated groups and organizations, networks of subgroups forming larger coalitions and confederations with common interests. Moreover, his analysis of the pathologies, such as anomie, that arise from the failure to achieve integration, are some of sociology's more enduring concepts.

Durkheim's later work, where questions of how individuals become integrated into society became ever-more important, has also had an enormous impact on sociology. The analysis in *Suicide,* emphasizing the individual's integration into social structures and culture, has been widely used in sociological studies of deviance, crime, and other social "pathologies." Perhaps more significant is Durkheim's recognition in his last major work, *The Elementary Forms of the Religious Life,* that rituals directed at symbolic representations of groups constitute the basis of integration at the micro level of social organization.

Finally, in his insistence on analyzing one set of social facts by another, Durkheim made a very strong case for sociology as a distinctive kind of enterprise. He was trying to make a place for sociology in the academic and broader intellectual worlds, and he argued for sociology in a way reminiscent of Comte's advocacy. Perhaps he argued too much, but he did gain an academic beachhead for sociology in France during the last decade of the nineteenth century.

Still, there are problems in Durkheim's approach. Functional reasoning almost always gets a theorist into trouble, and Durkheim is no exception. He often argued that, in seemingly mysterious ways, the need for social integration brought about the cultural and structural arrangements that would meet this need for integration. Arguing for a separate causal analysis (as distinct from a functional analysis) did not obviate this problem of seeing outcomes as the cause of these very outcomes. Such reasoning usually becomes rather circular, explaining very little.

Durkheim was at his best when making causal arguments, but even here, there are problems. First, he never gave Spencer much credit for presenting the key ideas on the causes on the division of labor some twenty years before *The Division of Labor.* Second, he never really indicated the causal sequences by which new forms of integration are to be achieved with differentiation; he simply assumes that these forms will emerge without specifying causality. As with Spencer, we can invoke a selection argument to help the argument; that is, problems of integration generate selection pressures for new types of cultural symbols and social structures, but this too is vague.

More substantively, Durkheim ignored the importance of power and stratification in society. He assumed that the "forced division of labor" would simply go away, and he never addressed adequately the conflict potential in systems of stratification, assuming that this too would go away as new bases of integration were achieved. Like Marx before him, but in the opposite direction, Durkheim's ideological commitments to finding a new basis of integration led him to mistake what he wanted to occur from what would actually transpire in differentiated societies. Complex societies always reveal points of tension and conflict, but through Durkheim's rose-colored glasses, we would hardly know that this was the case.

Another substantive problem is Durkheim's overemphasis on cultural forces to the detriment of recognizing the importance of power and mutual interdependence as mechanisms of integration. Durkheim addressed these topics, to be sure, but one really does not get a sense for how power integrates complex societies or how markets and other mediators of interdependence operate.

Rather, values, beliefs, and norms seem to do most of the integrative work, and although this point of emphasis has added a great deal to sociology's understanding of cultural processes, the more structural dimensions of integration are underemphasized—a rather remarkable conclusion given Durkheim's emphasis on studying "social facts."

Yet, as we will see in the next chapter, some of the most important models and principles in sociology today owe their origins to Durkheim's analysis. For all their problems, then, Durkheim's collective works continue to inspire sociology in the twenty-first century.

NOTES

1. Commentators have disagreed about whether Durkheim's work changed fundamentally from a macro perspective to a micro one, or from structural to social-psychological, between 1893 and 1916. For relevant commentary on this issue and on Durkheim's approach in general see Anthony Giddens, *Capitalism and Modern Theory: An Analysis of the Writings of Marx, Durkheim, and Max Weber* (Cambridge: Cambridge University Press, 1971); Anthony Giddens, ed. and trans., *Émile Durkheim: Selected Writings* (Cambridge: Cambridge University Press, 1972); Talcott Parsons, *The Structure of Social Action* (New York: McGraw-Hill, 1937); Steven Lukes, *Émile Durkheim, His Life and Work: A Historical and Critical Study* (London: Allen Lane, 1973); and Robert Alun Jones, *Émile Durkheim* (Newbury Park, CA: Sage, 1985).

2. For an extensive bibliography of Durkheim's published works, see Lukes, *Émile Durkheim,* pp. 561–590. See also Robert A. Nisbet, *The Sociology of Émile Durkheim* (New York: Oxford University Press, 1974), pp. 30–41, for an annotated bibliography of the most important works forming the core of Durkheim's theoretical system.

3. Émile Durkheim, *The Division of Labor in Society* (New York: Free Press, 1947; originally published in 1893).

4. See Lukes, *Émile Durkheim,* Chapter 7, for a detailed discussion.

5. Our view of Durkheim's *The Division of Labor in Society* underemphasizes the social evolutionism contained in this work because we think that too much concern is placed on the model of social change and not enough is placed on the implicit theory of social organization.

6. Alternatively, social integration.

7. Or the nature of social structure.

8. Durkheim described such simple societies as based on mechanical solidarity. *Mechanical* was a term intended to connote an image of society as a body in which cohesion is achieved by each element revealing a similar cultural and structural form.

9. Such societies were seen as based on organic solidarity. *Organic* was intended to be an analogy to an organism in which the elements are distinctive in form and operate independently, but for the welfare of the more inclusive social organism.

10. Durkheim, *Division of Labor,* p. 32. This idea owes its inspiration to Comte. As Durkheim noted in his Latin thesis on Montesquieu: "No further progress could be made until it was recognized that the laws of societies are no different from those governing the rest of nature. . . . This was Auguste Comte's contribution." Émile Durkheim, *Montesquieu and Rousseau* (Ann Arbor: University of Michigan Press, 1960; originally published in 1892), pp. 63–64.

11. Durkheim, *Division of Labor,* pp. 79–80 (emphasis in original).

12. Ibid., p. 152 for 1, 2, and 3 and throughout book for 4. For interesting secondary discussions, see Lukes, *Émile Durkheim,* and Giddens, *Selected Writings.*

13. The concern for "social morphology" was, no doubt, an adaptation of Comte's idea of social statics, as these were influenced by German organicist Albert Schäffle, with whom Durkheim had been highly impressed. See Lukes, *Émile Durkheim,* pp. 86–95.

14. Such typologizing was typical in the nineteenth century. Spencer distinguished societal types, but more influential was Tönnies's distinction between *Gemeinschaft* and *Gesellschaft.* Durkheim spent a year in Germany as a student in 1885–1886, and although Tönnies's famous work had not yet been published, his typology was well known and influenced Durkheim's conceptualization of mechanical and organic solidarity.

15. This table is similar to one developed by Lukes, *Émile Durkheim,* p. 151.

16. Durkheim, *Division of Labor,* p. 171.

17. Ibid., p. 287.

18. Ibid., p. 205.

19. Ibid., p. 302.

20. Ibid., p. 262.

21. Ibid., pp. 266–267.

22. For a more detailed analysis of Durkheim's debt to Spencer and of his contribution to functionalism, see Jonathan H. Turner and Alexandra Maryanski, *Functionalism* (Menlo Park, CA: Benjamin/Cummings, 1979); and Jonathan H. Turner, "Durkheim's and Spencer's Principles of Social Organization," *Sociological Perspectives* 9 (1984), pp. 283–291.

23. Durkheim, *Division of Labor,* p. 353.

24. Quoted in Lukes, *Émile Durkheim,* p. 141.

25. Durkheim, *Division of Labor,* pp. 364–365.

26. Ibid., pp. 373.

27. Ibid., pp. 1–31.

28. We are supplementing Durkheim's discussion of occupational groups with additional works; see Émile Durkheim, *Professional Ethics and Civil Morals* (Boston: Routledge & Kegan Paul, 1957), and his *Socialism and Saint-Simon* (Yellow Springs, OH: Antioch, 1958).

29. Durkheim, *Division of Labor,* p. 28.

30. Durkheim rarely addressed Marx directly. Though he wanted to devote a special course to Marx's ideas in addition to his course on Saint-Simon and socialism, he never got around to doing so. Much as with Weber, however, one suspects that Durkheim's discussion of "abnormal forms" represented a silent dialogue with Marx.

31. Durkheim, *Division of Labor,* p. 382.

32. Ibid., p. 383.

33. Ibid., p. 377.

34. Ibid., pp. 389–395; see also Charles H. Powers, "Durkheim and Regulatory Authority," *Journal of the History of the Behavioral Sciences* 21 (1985), pp. 26–36.

35. Émile Durkheim, *The Rules of the Sociological Method* (New York: Free Press, 1938; originally published in 1895).

36. Lukes, *Émile Durkheim,* Chapter 10.

37. Durkheim, *The Rules,* p. 3.

38. Ibid., footnote 5, p. 3.

39. Ibid., p. 7.

40. Many commentators, such as Giddens, *Capitalism and Modern Theory,* and Lukes, *Émile Durkheim,* emphasize that Durkheim was not making a metaphysical statement. We think that he was making both a metaphysical and a methodological statement.

41. Durkheim, *The Rules,* p. 48.

42. Ibid., p. 95.

43. Durkheim made other assertions: A social fact can only have one cause, and this cause must be another social fact (rather than an individual or psychological fact).

44. Lukes, *Émile Durkheim,* p. 227.

45. Émile Durkheim, *Suicide: A Study in Sociology* (New York: Free Press, 1951; originally published in 1897).

46. Ibid., p. 44.

47. Ibid., p. 48. It should be emphasized that suicide had been subject to extensive statistical analysis during Durkheim's time, and thus he was able to borrow the data compiled by others.

48. This view of humans, we should note, is very similar to that of Marx.

49. Durkheim, *Suicide,* p. 209.

50. Ibid., p. 209.

51. Ibid., p. 223.

52. Ibid., p. 258.

53. Ibid., p. 258.

54. Ibid., p. 276, in footnote.

55. See also Lukes, *Émile Durkheim,* Chapter 9, for a somewhat different discussion.

56. Durkheim dropped the term *organic societies,* but we have retained it here to emphasize the continuity between *Suicide* and *The Division of Labor.*

57. Durkheim would, of course, not admit to this label.

58. Émile Durkheim, *The Elementary Forms of the Religious Life* (New York: Free Press, 1947; originally published in 1912).

59. The work on "moral education" will be examined later in this chapter in a discussion of Durkheim's more general concern with "morality."

60. See, for example, Émile Durkheim, "The Determination of Moral Facts," in *Sociology and Philosophy,* trans. D. F. Poccock (New York: Free Press, 1974); this article was originally published in 1906.

61. Baldwin Spencer and F. J. Gillian, *The Native Tribes of Central Australia* (New York: Macmillan, 1899), presents the first collection of "accounts" of these aboriginal peoples, which was in itself fascinating to urbane Europeans. Sigmund Freud, in *Totem and Taboo* (New York: Penguin Books, 1938; originally published in 1913), and two anthropologists, Bronislaw Malinowski, in *The Family among the Australian Aborigines* (New York: Schocken, 1963, originally published in 1913), and A. R. Radcliffe-Brown, in "Three Tribes of Western Australia," *Journal of Royal Anthropological Institute of Great Britain and Ireland* 43 (1913), were all preparing works on the aborigines of Australia at the same time that Durkheim was writing *The Elementary Forms of the Religious Life.*

62. Obviously Durkheim was wrong on this account, but this was one of his assumptions.

63. This strategy was the exact opposite of that employed by Max Weber, who examined the most complex systems of religion with his ideal-type methodology.

64. Durkheim, *Elementary Forms,* p. 47. His earlier definition of religious phenomena emphasized the sacred—beliefs and ritual—but did not stress the morphological units of community and church. For example, an early definition read, "Religious phenomena consist of obligatory beliefs united with divine practices which relate to the objects given in the beliefs" (quoted in Lukes, *Émile Durkheim,* p. 241). His exposure to the compilation in Spencer and Gillian, *Native Tribes,* apparently alerted him to these morphological features.

65. Durkheim clearly borrowed the ideas of crowd behavior developed by Gustave LeBon and Gabriel Tarde, even though the latter was his lifelong intellectual enemy.

66. Durkheim, in both *The Division of Labor* and *The Rules,* had stressed that the segmental clan was the most elementary society. He termed the presocietal "mass" from which the clan emerges the horde.

67. As will be recalled, Durkheim took this idea from Rousseau and his description of the "natural state of man."

68. Quoted in Lukes, *Émile Durkheim,* p. 445 (taken from *L'Année Sociologique,* 1913). This line of thought is simply Comte's idea of the movement of thought from the theological through the metaphysical to the positivistic.

69. Nisbet, *Émile Durkheim,* p. 159.

70. Durkheim, *Elementary Forms,* p. 209.

71. Ibid., p. 9.

72. Ibid., pp. 17–18.

73. Émile Durkheim and Marcel Mauss, *Primitive Classification,* trans. Rodney Needham (Chicago: University of Chicago Press, 1963; originally published in 1903).

74. See the introduction to the translation for documentation of this fact.

75. See, in particular, Émile Durkheim, *Moral Education: A Study in the Theory and Application of the Sociology of Education,* trans. E. K. Wilson and H. Schnurer (New York: Free Press, 1961; originally published in 1922). This is a compilation of lectures given in 1902–1903; the course was repeated in 1906–1907.

76. One is Durkheim, *Moral Education;* the other is an article, published in 1906, on "The Determination of Moral Facts." Reprinted in Émile Durkheim, *Sociology and Philosophy* (New York: Free Press, 1974), from papers originally collected and translated in 1924.

77. Durkheim, *Moral Education,* p. 52.

78. Ibid., pp. 119–120.

79. Durkheim, "Determination of Moral Facts," p. 55.

18

✳

Durkheim's
Theoretical Legacy

É mile Durkheim's sociology is, ultimately, a search for the conditions generating social integration or solidarity during the course of societal differentiation. As societies move from simple (what he called "mechanical")
to more complex ("organic") forms, the basis of integration changes.
Durkheim's goal was to discover the causes of differentiation and the mechanisms for reintegrating society around new kinds of cultural and structural
systems.

The problem with Durkheim's analysis is that it mixes causal, functional, and
ideological statements. The causal arguments are relatively easy to isolate, but the
functional statements are problematic, as are all functional arguments. For
Durkheim, there is one functional requisite—the need for integration—and all
social and cultural processes in society should be examined with reference to
how they meet this master need. As we emphasized in the conclusion to the last
chapter, it is difficult to sort out causality in functional arguments because the
need for integration appears to be causing the very cultural and structural forms
meeting this need. Causality seems backwards, but as with Herbert Spencer, we
can obviate this backward causality by positing a "selectionist" argument,
emphasizing that differentiation sets into motion selection pressures for actors to
find new ways of integrating the social system. If such efforts are successful, then
integration is achieved and the society is viable; if they are not, the society dis-
integrates. Unlike Spencer, however, Durkheim never holds out the option of
disintegration. He believed that pathologies like anomie, poor coordination,
forced divisions of labor, and the like are temporary aberrations that will be

eliminated as the new, organic basis of solidarity falls into place. Here, Durkheim turns into a moral preacher who has faith that integration will naturally emerge as differentiation proceeds; what he wanted to transpire is conflated with statements about what would inevitably happen. Thus, Durkheim tended to assume that the new structural and cultural forms emerging with differentiation would automatically reintegrate society, perhaps after a brief period where pathological forms are evident. Even if we assume that Durkheim implicitly made a selectionist argument that pathological conditions alerted actors to problems and set them to find solutions to anomie, forced divisions of labor, and poorly coordinated structures, the causal argument is difficult to disentangle from the moral functionalism that pervades Durkheimian sociology. Yet, despite these problems, we can pull from Durkheim's sociology causal models and abstract principles of social organization in differentiated social systems.

DURKHEIM'S
UNDERLYING CAUSAL MODEL

In Figure 18.1, a complex causal model of Durkheim's analysis of differentiation and integration is delineated. In this model, we summarize the arguments in *The Division of Labor in Society* with elements of those in *Suicide* and *The Elementary Forms of the Religious Life* as these pertain to the bases for integrating complex societies. Moving across the top of the model from left to right is the argument presented in *The Division of Labor*. At the right side of the model are the ideas from *Suicide* and the preface to the second edition of *The Division of Labor*, as these bear on the bases for integrating differentiated social systems. On the bottom and middle portions of the model are the key micro-level ideas from *The Elementary Forms* that are easily plugged into Durkheim's more macro analysis of differentiation.

Ecological concentration is a key force in Durkheim's theory because it sets into motion (1) the competition that leads to the division of labor and (2) the co-presence of individuals that leads to those micro-interpersonal rituals generating the (religious) morality of the group. The power of ecological concentration is highlighted by the positive arrows from this force to competition for resources at the top and to co-presence at the bottom. At the far left are listed the causal relations among those forces—population size, migration and growth as well as communication and transportation technologies—that reduce space between individuals and, thereby increase rates of interaction and competition. Pursuing the argument in *The Division of Labor* first, competition for resources initiates Darwinian selection among individuals and collective actors for resources. Durkheim very explicitly draws an analogy from Darwin's explanation for speciation, seeing ecological concentration of individuals and collective units as putting pressure on resources, forcing actors to compete, with this competition leading to specialization of activities (as opposed to speciation). This competition, Durkheim assumed, need not lead to the actual

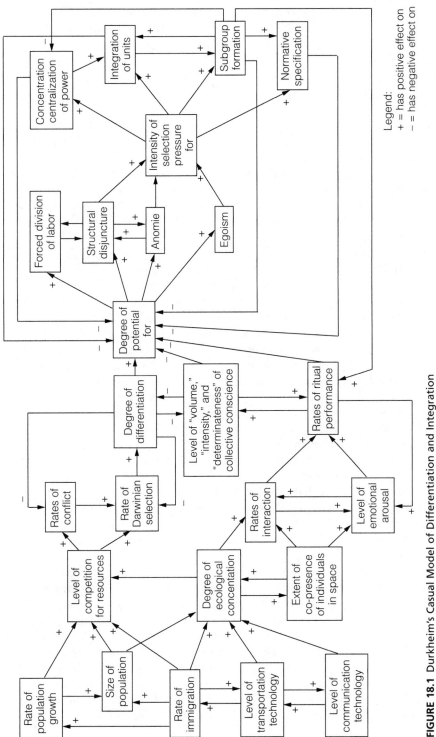

FIGURE 18.1 Durkheim's Casual Model of Differentiation and Integration

Legend:
+ = has positive effect on
– = has negative effect on

"extinction" or "death" of actors; rather, those less successful in one niche would simply move to another resource niche and differentiate themselves from those against whom they could not successfully compete. From such competition, then, came increased specialization and, hence, a more complex division of labor, or as we have phrased the matter in the model, a higher degree of differentiation. In turn, increasing differentiation would reduce conflict and struggle emerging from competition for resources, as is denoted by the negative reverse causal arrow connecting differentiation to conflict/struggle as well as Darwinian selection. The negative arrow from differentiation to the volume, intensity, and determinateness of the collective conscious recapitulates Durkheim's view that the power of the collective conscious to regulate diversely situated actors must decline. That is, differentiated actors having somewhat diverse experiences could no longer share the exact same values and beliefs ("volume") as in mechanical societies; they could no longer be subject to precise regulation by the collective conscious ("intensity"); and they could not experience the collective conscious with the same degree of clarity ("determinateness"). Moreover, the secular to religious "content" of the collective conscious would increase with differentiation.

Differentiation, coupled with more abstract tenets in values and beliefs, increases the potential for the pathologies that emerge before the new basis of organic solidarity is in place. The forced division of labor, where resources are not distributed in line with people's talents, may increase. Coordination of structures through mutual interdependence and through regulation by laws and other norms may prove difficult. A generalized, and in Durkheim's words, "enfeebled" collective conscience may open the door to anomie or the lack of regulation of people's desires and passions and of their relations to each other and the units organizing their activities. Egoism may become rampant as differentiated structures force individuals to stand between structures rather than embed in them. Durkheim often assumes that these pathological conditions will simply go away, but we have converted his argument, much like we did for Spencer, into a selectionist statement. These pathologies will alert people and collective actors to problems, forcing actors to try to find ways to deal with these problems, or face disintegration. Thus, the pathological forms of the division of labor can be conceptualized as the axes along which selection pressures are activated, pushing actors to find solutions. The more prevalent anomie, disjuncture among social units, forced division of labor, and egoism, the more intense are the selection pressures, as is emphasized by the positive signs on the arrows going into selection pressures. The positive signs going out of selection pressures signal Durkheim's faith that solutions to these pathological forms in the division of labor would be found.

The solutions represent Durkheim's views on the bases for integrating complex societies. We can say little about the forced division of labor because Durkheim simply assumed that it would disappear—a naive conviction. Therefore, we cannot draw a chain of causal events leading to the demise of this pathology. For the other pathologies, however, we can construct a causal argument from Durkheim's writings.

Structural disjunctions increase selection pressures for interdependence of units and for normative specification of relations between units so that their relations are better coordinated. Moreover, if units could be collated into larger subgroups or federations, or what Durkheim termed "occupation groups" in the preface to the second edition of *The Division of Labor,* smaller structural units could be better coordinated as their interests converge (as is indicated by the positively signed arrow from subgroup formation to integration of units). Finally, as the positive arrow underscores, the centralization of power also would facilitate the mutual interdependence because these centers provide the mechanism (regulation of money, systems of courts, and lawmaking legislative bodies) for creating and sustaining relations of mutual interdependence, although Durkheim does not offer much detail on these processes. In Durkheim's sociology, power and confederation of units into larger subgroups or "occupational groups" are important integrating forces. Although Durkheim did not develop this idea very far, he also argued that centralization of political authority (in the state) was necessary to enact and enforce many of the norms (as laws) essential to coordinating diverse social units, and hence, critical to avoiding structural disjuncture and poor coordination of diverse units. Durkheim distrusted centralized power, and so, he sought a source of counterpower in occupation groups or, in terms of the model in Figure 18.1, subgroups among members whose interests and experiences converge. With common experiences and interests, members of these subgroups could form political parties that would push for the common interests of broad sectors in the division of labor and pose a basis of counterpower to the centralized state. A negatively signed arrow from subgroup formation to the centralization and concentration of power emphasizes this counterbalancing effect of occupational groups mobilized as political parties.

Anomie is, of course, the most famous pathology in Durkheim's sociology, and he denoted several aspects of "regulation" with this term. One is the lack of regulation in the sense of coordinating activities within and between structural units. This was the meaning to be gleaned from *The Division of Labor.* Another meaning comes from *Suicide:* the lack of regulation of people's desires, needs, wants, and passions to the point of increasing rates of aberrant or deviant behavior beyond what is "normal" for a society. The first meaning of anomie is managed by the same forces that reduce structural disjuncture—power, interdependence, normative specification, and confederation of individuals and groups into larger subgroups. The second meaning emphasizes the effects of developing norms—what is termed "normative specification" in the model—to fill in the gap between the generalized and "enfeebled" collective conscience for the whole society and the desires and passions of individuals. Durkheim felt that such norms would emerge and regulate people's actions and passions so that anomie would decline. This process is more likely to take place when normative specification occurs within group structures, thus providing a kind of group-based "collective conscience" that would translate the highly generalized tenets of the societywide collective conscience into specific instructions and mandates that would provide regulation.

Another potential pathology in highly differentiated societies is "egoism," a force discussed by Durkheim in *Suicide* but one that can also be applied to his analysis of social differentiation. As Durkheim argued, egoism comes when individuals do not feel attached to groups or cultural symbols, and Durkheim's analysis of "occupational groups" in the preface to the second edition of *The Division of Labor* provides his answer to this problem. Individuals and collective actors located at similar places in the division of labor will have common interests and experiences, thereby leading them to develop common norms that attach them to the more general collective conscious and lower their sense of isolation or egoism. The negative arrow from subgroups and normative specification back to the box denoting the degree of potential for egoism underscores Durkheim's view that the expansion of normatively regulated subgroups would reduce rates of egoism and the deviant behaviors associated with the lack of integration of people into groups. Durkheim also felt that egoism can generate self-interest that is at odds with collective interests, and so, he implicitly argued that as normatively regulated subgroups eliminate egoism, they would transform self-interest into group or collective interests more amenable to societal integration and political democracy.

Durkheim's discussion in *The Elementary Forms* can supplement the macro-level analysis thus far. Co-presence of actors increases rates of interaction and emotional arousal leading to the performance of rituals directed at the perceived power or "mana" of the collective conscience or culture. This is indicated by the positively signed arrows in the lower middle of the model in Figure 18.1. As rituals are performed, individuals "worship" the group or subgroups to which they belong, and consequently, the group becomes more salient and powerful in regulating individuals' actions. This relationship of micro-level rituals on subgroup formation is emphasized by the negative effects of rates of ritual performance on the degree of potential for egoism. Moreover, these ritual performances strengthen the collective conscience by generating more consensus ("volume"), regulatory power ("intensity"), and clarity ("determinateness") in cultural values and (as is indicated by the positive arrows from ritual performance to these variable dimensions marking the power of the collective conscious). Thus, Durkheim's "discovery" of the basis of solidarity in elementary religious forms allowed him to find the mechanisms by which normative specificity and subgroup formation would overcome the problems, respectively, of anomie and egoism. Moreover, once subgroups with more specific norms exist, the positive reverse causal arrow from subgroup to ritual performance underscores Durkheim's argument that these rituals would sustain group boundaries and a more powerful set of cultural symbols for regulating activities in these subgroups.

Durkheim underemphasized power and its effects on integrating complex social systems, and he did not address the modes of interdependence among differentiated actors in any detail. The usefulness of Durkheim's analysis, then, resides more in the dynamics occurring to values and beliefs as they generalize and then are made more specific in normatively regulated subgroups. Durkheim did not, therefore, provide a very complete model of integration in complex social systems, especially when his ideas are placed into a robust causal model. Yet, despite these shortcomings, Durkheim did present to sociology some powerful theoretical principles that state a number of fundamental relationships among forces in the social universe.

DURKHEIM'S THEORETICAL PRINCIPLES

Principles of Social Differentiation

For Durkheim, the level of social differentiation among members of a population and among the social units organizing their activities is related to competition for resources. This competition for resources is, in turn, the result of increased "moral density" in which rates of interaction among members of a population are high. This ecological view of the causes of differentiation was borrowed from Charles Darwin's analysis of speciation of life forms, although Durkheim clearly borrowed heavily from Herbert Spencer's analysis of differentiation. The basic relationship among density, rates of interaction, competition, and differentiation can be stated as a single principle:

1. The greater is the level of competition for resources among members of a population, the higher are rates of interaction and levels of differentiation.
2. The level of competition and rates of interaction increase as
   a. The concentration of the population in space increases, with this concentration rising with
      1. the absolute size of the population
      2. the existence of ecological barriers
      3. the existence of political, social, and cultural barriers
   b. The extent to which transportation and communication technologies and infrastructures reduce the social space among individuals

Durkheim saw the concentration of the population as both a physical fact created by ecological barriers (mountains, bodies of water, and the like) and by large numbers of individuals who, inevitably, must be more concentrated than small numbers, although the absolute size of the territory would determine, to some extent, how concentrated a large population would be. Political, cultural, and social barriers also influence concentration of the population. When political organization is strong, geopolitical boundaries are likely to be sustained, and people will generally stay within them. Similarly, when cultural symbols and social structures are powerful and constraining, they will tend to hold members in the space. Density is also, Durkheim recognized, related to the level of development in transportation and communication technologies. Even when people, structural units, and cultural symbols are more dispersed, the ability to move in space and to communicate efficiently will increase rates of interaction and moral density. Indeed, although Durkheim could not have anticipated the world system of the twenty-first century, where time and space are compressed and where cultural symbols circulate in fast-moving communication channels, he understood the basic relationship between transportation/communication technologies and moral density. Thus, any forces that increase moral density will escalate competition for resources, and this competition will increase the level of differentiation as actors seek viable resource niches in which to sustain themselves.

Principles of System Integration

For Durkheim, integration, or as he phrased it, *social solidarity,* can be defined only by references to what he saw as "abnormal." Anomie, egoism, poor coordination, and the forced division of labor all represented to Durkheim instances of mal-integration. Thus, "normal" integration would represent the converse of these conditions, allowing us to formulate the following definition: Integration occurs when individual passions are regulated by cultural symbols, when individuals are attached to the social collective, when actions are regulated and coordinated by norms, and where inequalities correspond to the distribution of talents and, hence, are considered legitimate.

Differentiating systems face a dilemma, first given forceful expression by Adam Smith and, more immediately to Durkheim, by Auguste Comte: The compartmentalization of actors into specialized roles also partitions them from each other, driving them apart and decreasing their common sentiments. In *The Division of Labor,* Durkheim recognized that differentiation is accompanied by the growing abstractness and generalization of cultural symbols, especially at the level of values and beliefs. Or, in terms of the specific variables he used to describe the collective conscience, there is less consensus over ("volume"), less regulatory power of ("intensity"), less clarity of ("determinateness"), and less religious content in values and beliefs. Increasing generality of the collective conscience is essential if actors in specialized and secularized roles hold common values and beliefs. If values and beliefs are too specific, too rigid, too intense, and too sacred, they cannot be relevant to the diversity of actors' secular experiences and orientations in differentiated roles, nor can they allow the flexibility that comes with the division of labor in society. Hence, moral imperatives become more abstract and general. Durkheim thus saw a fundamental relationship between system differentiation and the generalization of cultural values and beliefs, as is summarized here:

3. The more differentiated a social system is, the more generalized are its values, beliefs, and other evaluational symbols.

The term *generalized* in this principle is used to summarize Durkheim's view of the collective conscience becoming less determinate, voluminous, and intense. Such a process of value and belief generalization presents differentiating systems with an integrative problem: Given the generalization of evaluational symbols and high levels of role specialization, what is to provide unity? In answering this question, Durkheim often inserted what he thought should occur. Yet, even with his moralistic bias, he saw certain fundamental integrative tendencies in differentiated systems. These can be summarized by a series of additional principles:

4. The greater is the level of structural differentiation and the more abstract and general are values and beliefs, the more norms emerge to specify evaluational premises and to regulate relations within and between social units.

Durkheim recognized that social integration in organic societies depends on the development among social units of functional interrelations regulated by "contract," and on regularization of behaviors within social units through concrete norms. In reacting to Spencer and the utilitarians, Durkheim emphasized that there is always a "moral component" or "noncontractual" basis of contract and that norms are more than utilitarian and instrumental "conveniences." They are tied, or at least he hoped that they were, to general values and beliefs, giving actors a common set of premises and assumptions. Principle 3 extracts the essence of Durkheim's argument from its moralistic trappings and argues that there is a fundamental relationship in the social world among structural differentiation, value generalization, and normative specification.

5. The greater is the level of structural differentiation and the more abstract and general are values and beliefs, the more subgroups form around similar or related role activities sharing common interests.

Durkheim's discussion of occupational groups was, in many ways, an expression of his hopes and desires. At a more abstract level, however, he appears to have captured the essence of an important structural principle: As roles become differentiated and specialized and as common value premises generalize, clusters and networks of positions—or, in the terms of principle 4, subgroups—form among those engaged in similar activities. Such subgroup formation reinforces the process of normative specification delineated in principle 3.

In Durkheim's later work *The Elementary Forms of the Religious Life,* he emphasized the power of rituals toward symbols of groups as a key integrative mechanism. If we transpose this more micro view onto his analysis of the bases of integration in differentiated systems, then the following principle is suggested:

6. The greater is the level of structural differentiation and the more abstract and general are values and beliefs, the more normative specification and subgroup formation depend on the enactment of rituals to sustain commitments to norms and subgroups.

From Rousseau, Durkheim appears to have absorbed the idea that the "state" could personify the "collective conscience" and coordinate activities in pursuit of collective goals and purposes. If we seek a more abstract principle to translate Durkheim's somewhat moralistic vision, it can be seen that in any system experiencing differentiation and value generalization, there are increased probabilities for not only normative specification (principle 4) and subgroup formation (principle 5), but also for centralization of authority.

7. The greater is the level of structural differentiation and the more abstract and general are values and beliefs, the more centers of power will coordinate diverse activities.

Durkheim also distrusted centralized authority that went unchecked by counter-authority—an idea that he took from Montesquieu and Tocqueville. It is not clear that centralized authority automatically engenders counter-authority, but if we are to be true to Durkheim's argument, this process must be stated as a general principle. Durkheim felt that subgroups in a differentiating system with centralized authority tend to become sources of power that check and balance the power of the central authority. Thus, we can formulate the following principle:

8. The more centralized is authority in a differentiated social system revealing subgroups, the more subgroups become centers of counterauthority mitigating the power of centralized authority.

For Durkheim, as systems differentiate and "naturally" tend to form subgroups around specialized roles and functions as well as common interests (principle 5), these groups will resist the arbitrary use of power by the centralized authority, thereby creating a "balance of powers" in a system.

9. The greater is the level of structural differentiation and the more abstract and general are values and beliefs, the more the distribution of scarce resources will correspond to the unequal distribution of talents.

This is a highly questionable Durkheimian principle. What defines talent? Is it innate intelligence, training, or cunning? Why would those who have resources at one point in time not pass them on to designated others, creating an ascriptive system in which talent and rewards become poorly correlated. Despite these problems with the principle, however, resources in highly differentiated systems have a slight tendency to be distributed in accordance with the ability of actors to contribute to the system. This is not a strong tendency, however, and ascriptive processes weaken it. Thus, principle 10, more than principles 1 through 9, reflects Durkheim's hope for the future rather than a clear assessment of structural tendencies.

10. The greater is the level of structural differentiation and the more abstract and general are values and beliefs, the more sanctions against deviance will be restitutive rather than punitive.

Durkheim incorrectly overestimated the degree of punishment in mechanical (simple) social systems, and he underestimated the punitive sanctions in organic (complex) systems. Yet, he might have been correct in his view that the ratio of restitutive to punitive sanctions increases with differentiation and value generalization. Because differentiated systems have less-immediate value premises to offend and because they require coordination of parts, Durkheim might be correct in his view that there is a fundamental relationship in social systems among differentiation, value generalization, and restitutive sanctions.

Principles 3 through 10 summarize Durkheim's vision of the basis of integration in differentiating systems. These principles were developed by Durkheim for understanding whole societies, but we suspect that they apply to any differentiating social system, whether a group, organization, or community. Although these principles are not without ambiguities, we believe that they can still inform modern sociological theorizing and provide a promising place from which to begin further work.

Principles of Deviance

Durkheim's analysis of suicide is much more than a discussion of suicide rates; it is also an exploration into the structural and cultural sources of deviance in general. The basic issue in Durkheim's discussion of suicide is the question: What is the nature of individual integration into social systems? His answer is that there are two bases of individual integration: (1) regulation of individual desires and passions by values, beliefs, and norms and (2) attachment of individuals to collective units of organization. Regulation and attachment must be balanced so that there is not too much or too little of either. If the balance is disturbed, then suicide-prone individuals are likely to commit suicide, thereby raising the rate of suicide in society.

If we abstract away from suicide, which is only one type of deviance, we have a perspective for visualizing its general characteristics. Because anomie, or deregulation, and egoism, or detachment, are Durkheim's primary concern in *Suicide* and elsewhere in his work, we focus on these two conditions and their relation to deviance in general. Durkheim saw deviance as intimately connected to the principles of differentiation already presented. He viewed increases in the rates of deviance as "normal" in differentiating systems, but under certain conditions, the rates increase to a point where a pathological condition prevails. If we ignore this distinction between normal and abnormal rates (for no objective criterion exists for determining what is normal and abnormal in social systems), and focus on his statements about the conditions increasing rates of deviance in general, then the following two principles are evident in Durkheim's work:

11. The greater is the level of structural differentiation and the more abstract and general are values and beliefs without a corresponding degree of normative specification, the greater is the level of anomie, and hence, the higher are rates of deviance.

12. The greater is the level of structural differentiation and the more abstract and general are values and beliefs without a corresponding increase in subgroup formation, the greater is the level of egoism, and hence, the higher are rates of deviance.

These two principles summarize Durkheim's view that deregulation of individual passions (anomie) and detachment of people from collective goals and purposes of groups (egoism) are related to basic processes of social differentiation. When the inevitable process of value generalization in differentiating systems is not accompanied by the processes denoted in principle 4 (normative specification), in principle 5 (subgroup formation), and in principle 6 (rituals), then anomie and egoism are likely. With either or both of these states, rates of deviance are likely to increase. Thus, Durkheim was arguing that a fundamental relationship exists in social systems among structural differentiation, value–belief generalization, normative specification, and subgroup formation, on the one hand, and rates of deviance, on the other. Such an insight was truly revolutionary for Durkheim's time, and it can still inform contemporary theorizing on deviation in social systems.

Principles of System Mal-Integration

Durkheim's view of social pathology represented a moral view of the world. Pathologies were simply defined as those processes that deviated from Durkheim's conception of "normal" for a given type of society. A scientific theory of the social should try to keep such moralistic evaluations out of the analysis. Yet much of Durkheim's view of the world concerns mal-integration in social systems, and despite the moral connotations of his analysis, he presented a number of principles that might be useful to modern theory.

In many respects, Durkheim's statements on mal-integration are the converse of principles of 3 through 10 on the conditions of integration, or they are extensions of those on deviance. In this view, mal-integration of a social system can occur when (1) anomie (deregulation) and egoism (detachment) are great, (2) coordination of functions is low, and (3) inequalities in the distribution of resources create tensions between those with and without resources.

13. The greater is the level of structural differentiation and the more abstract and general are values and beliefs, and the less is the degree of normative specification, the greater is the level of anomie, and hence, the less is the level of the integration of individuals into society.

14. The greater is the level of structural differentiation and the more abstract and general are values and beliefs, and the less is the formation of subgroups, the greater is the level of egoism, and hence, the less is the level of integration of individuals into society.

15. The greater is the level of structural differentiation and the more abstract and general are values and beliefs, and the less is the degree of the normative specification of relations among social units, the less is the coordination of units, and hence, the less integrated is a society.

16. The greater is the level of structural differentiation and the more abstract and general are values and beliefs, and the less centralized is authority, the less coordinated are social units, and hence, the less integrated is a society.

17. The greater is the level of structural differentiation, the more abstract and general are values and beliefs, and the centralization of authority, and the less effective the countervailing power of subgroups, the greater are the

tensions between those with and those without power, and hence, the less integrated is a society.

18. The greater is the level of structural differentiation and the more abstract and general are values and beliefs, and the less is the correlation between distribution of scarce resources and talents, the greater are tensions between those with and without resources, and hence, the less integrated is a society.

Principles 13 and 14 concern individual integration into the system; principles 15 and 16 deal with coordination among system units, whether individuals or corporate units; and principles 17 and 18 deal with the mal-integrative impact of tensions between those with, and those without, scarce resources, including power. These last two principles are as close to Marx as Durkheim comes, for despite his acceptance of Rousseau's abhorrence of inherited inequalities, Durkheim was always more concerned with the dual issues of individual integration into the social system and coordination among differentiated system parts than he was with social inequality.

Yet, principles 13 through 18 provide interesting theoretical leads about those forces that are involved in either the change or the breakdown of social systems. Of particular importance is the fact that these principles employ the same variables that are contained in those principles explaining social order and integration. Thus, the same variables explain both order, disorder, and change in social systems. Principles 1 through 17 provide clear evidence that Durkheim took the task of sociological theorizing seriously.

Durkheim's sociology is consistently re-read because of the power in several of his theoretical principles. His ideas on population size, ecological concentration, and differentiation are not original to Durkheim; these ideas were developed in much more detail 20 years earlier by Herbert Spencer. The usefulness of Durkheim's theory thus resides in his analysis of the pathologies of highly differentiated structures and the mechanisms by which these pathologies are overcome. Again, the moralistic and functional statements are often difficult to disentangle, but we can at least see the strong points in his argument. First, Durkheim recognized that there is a fundamental relationship between differentiation and generalization of cultural symbols, especially values and beliefs. Second, as this generalization occurs, disintegrative pressures, or what he termed "pathological forms" of the division of labor, increase. These pressures mount along two interrelated fronts: (a) coordinating activities through power, mutual interdependence, norms, and confederations of differentiated units into subgroups; (b) specifying generalized evaluative symbols in the societywide culture with more specific norms attached to subgroupings of individuals with common interests. Third, Durkheim understood the fundamental micro foundations of these more macro processes: rates of interaction and ritual performances directed toward the collective conscience. From these rituals, norms are generated and commitments to these norms and the groups in which they are embedded are sustained.

The principles Durkheim developed to explain these fundamental relationships among micro and macro forces have, over the last 100 years, represented a rich source of inspiration for sociological theory. Durkheim's theoretical legacy resides in these relationships, summarized in the principles enumerated.

19

✳

The Origin and Context of Vilfredo Pareto's Thought

BIOGRAPHICAL INFLUENCES ON PARETO'S THOUGHT

In this chapter we will present a personal profile of Vilfredo Pareto. Biographical influences are especially important in Pareto's case because his substantive concerns and theoretical style so directly reflect his early experiences and training.[1]

Family Background

Pareto's ancestors came from the vicinity of Genoa, Italy. They were prominent merchants and citizens of good standing. Pareto's great-great-great-grandfather was ennobled as a marquis in 1729. This rank, which Pareto inherited, is just below that of a duke.

Americans automatically expect the nobility to favor autocratic government. But Genoa was a small city-state run on quasi-republican principles by a commercial elite, and it was fiercely antagonistic to expansionist monarchies, such as Austria. Ardent republicans, Pareto's grandfather and grand-uncle occupied important government positions during the Napoleonic period.

After Napoleon's defeat, the Republic of Genoa was given as a war prize to the hereditary rulers of Piedmont, and local interests suffered a setback. But republicanism did not wane among members of the Pareto family. When Giuseppe Mazzini mobilized republican resistance against the monarchy in the 1830s, Pareto's future father (Marquis Raffaele Pareto) was forced to flee to

France. Several other members of the family were imprisoned or otherwise punished.

Raffaele Pareto was a well-known hydrological engineer and was able to support himself in France. He eventually married a French Calvinist by the name of Marie Mattenier, and they had two daughters and one son. Vilfredo Frederico Damaso Pareto was born on July 15, 1848, in Paris.

Pareto never really escaped his republican activist roots, for his later writings were always critical of the ruling class.[2] A lifelong opponent of autocratic regimes, he made his major social scientific contributions by identifying the ways in which self-aggrandizing regimes first gain and then lose political power. This latent republicanism greatly influenced the substance of his sociological theories.[3]

Early Life Experiences

Pareto enjoyed a reasonably affluent upper-middle-class French upbringing for the first ten years of his life. By the late 1850s, his father was able to return to Italy, and the family settled in Turin. Vilfredo completed his undergraduate degree in engineering at the Polytechnic Institute of Turin in 1869.

At the institute Pareto became enthralled with the concept of equilibrium, which would mark so much of his later work. His senior thesis examines how expansion and contraction operate as countervailing forces to determine the volume of solid substances.[4]

Many students of sociological theory overlook Pareto's thesis because it deals purely with engineering and has nothing whatsoever to say about society. But this undergraduate thesis may be more important than any other single work if we are to understand his sociological theory, for it lays his scientific epistemology open to view. Many of his writings take on new meaning when one understands how he conceived complex phenomena, how he sought to conduct scientific inquiry, and what he viewed as the role of theory in science.

Years later Pareto applied the same equilibrium strategy to the study of sociology, examining the ways in which countervailing forces of economic expansion and contraction, political centralization and decentralization, and liberalization and conservatism in public sentiment interact to effect changes in the overall character of the society.[5] If Pareto had not had engineering training, or if he had not used an equilibrium model in his senior thesis in 1869, it is unlikely that he would ever have produced the kind of sociological theory for which he eventually became famous. His training as an engineer had more influence than anything else on the *form* of explanation he employed as a sociologist.

Experiences in the Corporate World

After graduation from college, Pareto held several engineering positions. He worked, first in Rome and then in Florence, as a civil engineer with the Italian railway. In 1874, he left the railroad for a management position with Societá Ferriere Italiana, which operated mining and industrial concerns. While he was with this firm, Pareto traveled several times to England and Scotland on business, where he became enchanted with laissez-faire economic doctrine and the apparent success of British government policies promoting free trade.

His preoccupation with European trade issues marked another important development in Pareto's thinking, for he began to focus on the latent effects that government policy can have on volume of trade. This subject was tailor-made for the kind of theorizing he enjoyed: specifying the nature of interdependence among elements in complex systems. At this point, he stopped thinking like a technician and began to think like a social scientist.

Indeed, Pareto might have been the first real administrative scientist. Approaching managerial problems sociologically led him to make a number of discoveries that have had lasting impact on the corporate world. For example, his 80/20 rule suggests that 20 percent of the items in an inventory account for 80 percent of sales volume and that most of the supply problems a corporation faces are associated with the 80 percent of the inventory that is rarely used. People often misplace seldom-used items or fail to restock such items as they run out. Pareto introduced fixed slots for stock items along with other methods of inventory control, marking a significant organizational advance.[6]

Another of our intellectual debts to Pareto stems from his realization that management style is crucial to organizational success. This theme has been prominent in the administrative science literature since Chester Barnard, borrowing from Pareto, demonstrated the pivotal role of management in maintaining a cooperative environment in which people try to work together as part of a team.[7]

Seventy years after his death in 1923, Pareto's contributions to the field of business administration continue to be of some importance. In his early management work, however, we see only a glimmer of his greatness, for his managerial concerns were very narrowly focused and highly concrete. There is no sign here of the ambition that would eventually characterize his overarching sociological theory. Nonetheless, his management experience allowed him to apply systems analysis to a reasonably complex form of social organization: the industrial corporation.

Experiences as a Social and Political Commentator

As a member of the managerial elite, Pareto made influential friends of many people in artistic, commercial, and intellectual circles, and he participated in a number of social clubs and discussion groups in Florence, such as the Adam Smith Society and the Academy of Geography. He taught himself to read Greek and, by studying the classics, began to see patterns of emergence and decline in Western civilization. These studies provided much of the data he used decades later for serious sociological research and theorizing.

Fired by his analysis of political economy, Pareto ran for parliament in 1881. His loss convinced him that voters hear only what they want to hear and ignore obvious truths. This realization had tremendous impact on his later sociological work, convincing him that whatever people might say to the contrary, they are certainly not governed by logic. This failed candidacy left him with the conviction that sociological theory should be predicated on the analysis of human sentiment. Indeed, as we will see in Chapter 20, a theory of sentiment is the cornerstone of his sociology.

Pareto retreated somewhat after the death of his father in 1882. He did not even marry until after the death of his mother in 1889. By the time of this first marriage (to Dina Bakunin, a Russian who ran off with another man around 1901), Pareto was actively engaged in writing political commentary for newspapers and magazines.[8] Most of this work was polemical. He argued in favor of free trade and attacked the government for establishing protective tariffs, granting monopolies, and pursuing other interventionist policies. As a mark of his effectiveness, Pareto was harassed by the police. But he was more than a polemicist. By 1891, he was putting his engineering skills to use by translating the discursive economic theories of his day into mathematical formulas. European economists took immediate note, and he was appointed professor of political economy at the University of Lausanne, filling a vacancy created by the retirement of Léon Walras.

Experiences as an Academic

In the years that followed, Pareto produced a number of important books, including two classics in economics: *Course in Political Economy* (1896–1897) and *Manual of Political Economy* (first edition published in 1906, revised edition in 1909).[9] Pareto's combined contributions were so important and so distinctive that he came to be known as "the father of mathematical economics." Although the specific content of these economic works played little part in his later sociology, this time spent as an economist was nevertheless an essential part of his development as a sociologist. For in his capacity as professor of political economy, he began his rigorous study of society as a holistic system. The economy gave him some tangible reference points for applying equilibrium models to aggregate patterns of change in the society at large. Thus, the stage was set for him to make a lasting contribution to sociology.

Pareto was just breaking into his stride as an academic economist when he inherited a small fortune from an uncle in 1898. Purchasing a quiet country villa in Céligny, a village west of Lausanne on Lake Geneva, he was able to give full concentration to his work. After being abandoned by his wife in 1901, he was joined by Jane Régis (born in 1877), who remained his trusted companion until his death. (Pareto and Régis were married early in 1923, when he was finally able to procure a divorce from his first wife.) Céligny may truly be one of the most beautiful and tranquil spots on the earth. Most of what we remember Pareto for was written here, where he could work in quiet contemplation.

By this time Pareto's stature as an economist was unsurpassed. Yet, he was deeply troubled by the course being charted within that discipline, for he was convinced that economic events could be adequately understood only within the broader sociopolitical context. He was rebuffed for expressing these views, drifted somewhat from the mainstream of opinion among economists, and entered partial retirement. This gave him the time he needed to work on sociological manuscripts, including his one-million-word tome, *Treatise on General Sociology* (1916).[10] The final years of his life were spent in ill health, but he continued writing and produced *The Transformation of Democracy* (1921), which adds some final touches to his overall theory of society.[11] Vilfredo Frederico Damaso Pareto died on August 19, 1923.

Psychological Profile

First among Pareto's qualities was confidence in his own intellectual ability. What kind of person thinks he can unlock the secrets of the social universe? The kind of person whose idea of entertainment is teaching himself to read Greek and whose idea of a ten-year intellectual project is to formalize and mathematize most of the extant literature in the entire field of economics. Pareto was completely confident in his own intellectual ability, so confident that he could be curt and intolerant of people holding opinions differing from his own.

One of Pareto's other prominent qualities was a love of freedom. He was a steadfast advocate of freedom of speech and an unwavering opponent of autocratic government as well as foreign imperialism. Although he was conservative in his own tastes and inclinations, he was also a libertarian who saw nothing wrong with people exercising high levels of personal autonomy. He loathed taking orders from others, particularly people he did not respect.

Pareto was clever and witty. However, he was deeply disillusioned with humanity. One would have to say that he was a cynic who readily found fault with the world, which is one reason he had few admirers and many detractors. He was confident that he had indeed unlocked some of the secrets of the social universe, but he was also certain that his contemporaries would fail to recognize the scope of his contribution.

Pareto had a compelling sense of obligation. He remained true to his principles, even when doing so carried great personal cost. Last but not least, he was an insomniac, which is one reason he was able to accomplish so much. With limitless time and without distractions of the modern era, he stayed up late into the night reading, analyzing, speculating, researching, and writing.

NINETEENTH-CENTURY INTELLECTUAL CURRENTS AND PARETO'S THOUGHT

Pareto was an intellectual maverick and trailblazer. This description is most accurate for the period when he worked on his sociological manuscripts in the relative tranquility and isolation of Céligny. Although he worked alone in his last years, he was very much the product of powerful intellectual currents of the nineteenth century. This must be recognized if his work is to be properly understood.

In seeking to unravel the intellectual milieu in which Pareto's economic, political, and sociological analysis developed, we begin with the recognition that he was a child of the Newtonian revolution and its view of science. At the same time, he was greatly influenced by three dominant schools of thought—utilitarianism, positivism, and historicism—as well as by the Italian intellectual tradition.[12] The influence on Pareto of each intellectual current is briefly examined in the following pages.

Pareto and the Newtonian Revolution

Pareto was formally trained as a mathematician and civil engineer, and as a result, he was greatly influenced by the promise of Newtonian physics: Through observation of the empirical world, the basic and fundamental properties of this world can be isolated and their lawlike relations discovered. Throughout his varied intellectual career, whether as an engineer, social polemicist, academic economist, or sociologist, he never wavered from the position that his "sole interest is the quest for social uniformities, social laws."[13]

From Pareto's view, then, the ultimate goal of all reflection on the social world is the development of universal laws. Thus, metatheoretical speculation, philosophical schemes, and concrete empirical observations are useful only to the extent that they help develop abstract laws of the social universe. Unbridled philosophical speculation, Pareto maintained, can remove discourse from the actual properties of the world, and the mindless accumulation of empirical facts can impede the process of abstraction that is so essential to the development of universal laws. Hence, social theory will emerge when facts and philosophy are harnessed to the goals of all science: the discovery of uniformities and the articulation of laws that make these uniformities understandable.[14]

Positivism and Pareto's Thought

As we observed in Chapter 3 on Auguste Comte, the Newtonian vision fostered the development of positivism in the social sciences. Although Pareto read and admired Comte's work, he rejected many of the points contained in positivist doctrines: that analogies to the biological realm are useful, that social systems reveal stages of progress and evolution, that structures can be analyzed by their functions, and that the laws of sociology can be used to reconstruct society.[15]

Rather, Pareto accepted only aspects of the Newtonian vision as Comte and other positivists had reformulated them. That is, the general principles of the social realm can be discovered through the direct observation of social facts, through experimentation, through comparisons of different types of societies, and through the analysis of historical records. Thus, Pareto absorbed from positivism the view that diverse methods of empirical inquiry can be used to uncover the laws governing the operation of empirical regularities in the social world.

Utilitarianism and Pareto's Thought

As we observed in our discussion of Herbert Spencer, utilitarian thinkers of the last century, inspired by Adam Smith, tended toward an atomistic view of humans and an evolutionary view of society in which order and progress ensue from people's pursuit of individual self-interest. From such pursuits, individuals find their place, or niche, in society, with the character of society being determined by the qualities of its members. From this perspective, a science of society must be based on the study of individuals. As such, rationality and pursuit of happiness are thought by utilitarians to be the major forces motivating human behavior, and like the principle of attraction in astronomy, rationality and the pursuit of happiness are seen to occupy a place of central theoretical importance.

In his early works, Pareto accepted the utilitarian position that unimpaired free market conditions lead to the optimum collective good, but by the middle of his career, he became aware that government intervention, corporate monopolies, and labor unions all violated precepts of classical economics and made policy decisions based on the assumption of free market operations invalid. After spending years trying to justify laissez-faire and free trade policies as the best possible economic system, he eventually realized that those who tried to "discover" what form of society was "best" were simply disguising their own sentiments in the cloak of pseudoscientific investigation.[16]

In reacting to what he perceived to be the failings of utilitarianism, Pareto brought into clearer focus properties essential to understanding patterns of social organization: power, interest, and ideological rationalization. He rejected much of the substance of utilitarian doctrines, but he retained elements of the utilitarian mode of analysis. In particular, notions of cycles, supply and demand, and equilibrium became a prominent part of his sociological system.

Historicism and Pareto's Thought

Historicists of the nineteenth century tended to view a given society as the product of unique events rather than as a manifestation of certain lawlike relations among properties of the social world. Even more analytical historicists, such as Georg Hegel and Karl Marx, tended to confine their notions of "social laws" to specific historical epochs, rejecting the idea of universal laws applicable to all times and places. In discussing this tradition, Pareto saw as unfortunate the unwillingness of historicists to broaden their vision and to adopt the Newtonian premise.[17]

At the same time, however, Pareto's positivism led him to view historical events as a major source of data, especially because he was most interested in the rhythmic and cyclic dynamics of social systems over time. Thus, he remained sympathetic to historical inquiry—indeed, his work is filled with historical illustrations—but he rejected what he saw as the atheoretical bias of most historicists.

A Note on the Italian Tradition and Pareto's Thought

The established schools of thought that influenced Pareto were largely tied to particular countries: positivism to France, utilitarianism to England and Scotland, and historicism to Germany. In contrast with predominant modes of thinking in these nations, social thought in Italy was more eclectic, drawing inspiration from many diverse sources.

Yet Italy did reveal some unique intellectual traditions, most notably, the concern with social power and its use. Although German scholars, such as Max Weber, were also concerned with power, the work of Niccolò Machiavelli set the tone of much Italian scholarship. Indeed, Pareto felt that criticisms of Machiavelli's *The Prince* had been unjust, for Machiavelli had described not so much his personal ideals as a paramount reality of the social world: the use of power to create more power.[18] Another trend of thought in Italian intellectual circles concerned the way values and beliefs were used to control and manipulate populations. For example, Giambattista Vico's work on the importance of cycles in belief systems exerted considerable influence on Pareto.

Thus, to the extent that it was distinctive, the Italian intellectual tradition focused on two related issues: the use of power and the impact of cyclical changes in beliefs on social arrangements. Both issues became prominent in Pareto's sociology.

SPECIFIC INFLUENCES
ON PARETO'S THOUGHT

Working within the broad intellectual traditions of Pareto's time were a number of scholars from whom he borrowed key assumptions and concepts. To fully appreciate the genesis of his social theories, then, we need to review the influences of several immediate intellectual predecessors as well as some of his contemporaries.

Comte and Pareto

Although Pareto rejected Comte's organismic analogy, his concern for normalcy and social planning, and his moralistic pronouncements, Pareto accepted—indeed, he embraced—Comte's concern with "social facts" and his proposed methodology. Moreover, Pareto adopted Comte's definition of social facts as widespread patterns of observable behavior and the patterns of beliefs guiding such behavior.

Adam Smith and Pareto

Although Pareto eventually rejected Smith's advocacy of free and open competition, he did embrace Smith's insights into the laws of supply and demand. Pareto advanced the science of economics significantly in his formulation of structural equations to describe the dynamics of supply, demand, and other economic forces. Moreover, Pareto adopted and altered the equilibrium processes implied in Smith's economic analysis in a way that allowed analysis of sociopolitical phenomena. In particular, Pareto took from Smith's *The Wealth of Nations* the notion that the social world can be viewed as a system of interdependent properties, tending toward equilibrium points but also subject to change with alterations in the value of any one property.[19]

Maffeo Pantaleoni, Léon Walras, and Pareto

Even after publication of *The Wealth of Nations,* economics remained a largely discursive and inexact discipline. The role Maffeo Pantaleoni played in creating a science of economics is seldom recognized, but in *Pure Economics,* Pantaleoni attempted to formalize verbal propositions on the relationships among such major economic concepts as cost, supply, demand, interest, wages, rent, profit, utility, and value.[20] Pareto read Pantaleoni's book in 1891 and was excited by the prospects it offered for the development of scientific economics. Within a short time, he established correspondence with Pantaleoni and directed his own effort away from political commentary to the development of mathematical equations corresponding to Pantaleoni's propositions.

Pantaleoni suggested that Pareto reread the works of Léon Walras, who had articulated a theory of marginal utility and developed a general theory of equilibrium that characterized economic activity as the result of understandable competitive market adjustments and responses to events occurring within a unified socioeconomic system. Walras had sought to develop his equilibrium model into a general framework for the scientific study of economics, complete with equations specifying relationships among aspects of the economic system. When he retired from his professorial chair in political economy at the University of Lausanne, he followed Pantaleoni's recommendation and requested that Pareto replace him. In his new position Pareto spent several years clarifying and formalizing Walras's economic equilibrium theory and, from his efforts, founding mathematical economics.

Equally important, as Pareto worked with the idea of equilibrium, he soon realized the limitations of the approach when it included purely economic variables. Thus, his exposure to economics gave him a profound appreciation for the analytical power of formal equilibrium models, but at the same time, he came to recognize their limitations. This recognition led him to examine more carefully the sociological works of Spencer, the dominant utilitarian social thinker of the nineteenth century.

Spencer and Pareto

Following publication of several of his works in economics, Pareto attracted considerable attention, but he became increasingly disillusioned with two shortcomings of economic analysis: (1) many important factors that are known to vary are assumed constant in economic models, and (2) the analysis of human motivation in economic models tends to be simplistic. Spencer's early treatment of these two subjects encouraged Pareto to broaden his equilibrium theory, changing it fundamentally in the process.[21]

First, Pareto found Spencer's essays on the interdependent nature of social systems appealing, and thus, Pareto concluded that sociological and economic phenomena that were part of the same system must be studied by the same methods of analysis and treated within a common theoretical framework. Indeed, he came to emphasize that studying social and economic phenomena separately merely obscured the most fascinating theoretical questions about the nature of their interdependence. Second, he credited Spencer's early work with the critical insight that much human behavior is nonlogical and therefore not subject to economic models assuming the rationality of behavior.

Marx and Pareto

There was a natural affinity between Marx's and Pareto's analyses of social systems. Both men recognized the importance of economic interests, both saw the connection between economic and political processes, both realized the significance of cultural symbols in legitimating social conditions, and both saw inequality in the distribution of resources as a driving force behind social change. Indeed, Pareto gave much credit to Marx for demonstrating the connections among economic interests, political power, cultural beliefs, and patterns of inequality.

Pareto disagreed, however, with the specifics of Marx's analysis of capitalism. He regarded Marx's belief in the intrinsic value of labor as a vestige of outmoded economic theory. The importance of surplus value in Marx's analysis led Pareto to the conclusion that Marx had built a misguided theoretical edifice on false assumptions. Moreover, Pareto regarded Marx's analysis of the expansion and collapse of capitalism as flawed and felt that Marx maintained that consumption was the driving force behind capitalist expansion (money begets money through the circulation of commodities). In contrast, Pareto argued that high levels of consumption were associated with capital depletion and economic downturn. Even more fundamental was Pareto's criticism of Marx's tendency to infuse his theory with ideology and to make his doctrine a religious faith, a criticism similar to that leveled against Comte's "religion of humanity." Despite these sources of disagreement, Pareto reinforced Marx's critical insight that societies constitute systems and that the key properties of such systems are economic interests, power, inequality, and cultural symbols.[22]

Georges Sorel and Pareto

As a Marxist who became increasingly disillusioned with the Communist party, Georges Sorel provided some of the most penetrating attacks on the self-serving tendencies of elites who sought to use power to create additional power and, hence, to increase their capacity to exploit others. For whatever their ideological position, Sorel argued, the leaders of political parties are driven by the paramount interest to preserve their privileged position. Sorel's analysis supported Pareto's insights on the "circulation of elites" and encouraged Pareto to continue refining the theory for which he is best known.[23] Coupled with his observation that history is the "graveyard of elites" and his reinterpretation of Marx, Pareto recognized that elites came into power, exploited others, created conditions for their own downfall, and were then replaced by others who initiated the cycle again.

RECURRENT THEMES IN PARETO'S WORK

Family background, personal experience, intellectual training, personal disposition, and the intellectual climate of the time all subtly direct a person's work. Keeping these things in mind, we can see why certain themes reappear throughout Pareto's career. These themes are like the woof and warp of a tapestry. They constitute the basic fabric of Pareto's worldview and give shape to his sociological endeavors.

To begin with, Pareto had an elitist orientation. This should come as no surprise when we recall his aristocratic background, academic achievements, and professional mastery. He held expertise in high regard and showed great disdain for ignorance and lack of cultivation, especially in members of the privileged elite, who enjoyed society's advantages.

It might sound paradoxical, but Pareto was an elitist and also an egalitarian. He understood that members of the privileged classes were innately no more capable than were members of the laboring classes and, in fact, were often slovenly. Unfortunately, elites tend to block upward mobility of the most capable and energetic members of subordinate groups while exercising the power at their disposal for personal benefit rather than collective good. Pareto was an early advocate of meritocracy.[24]

Intellectually, Pareto was a positivist. In his estimation, Sir Isaac Newton was the greatest thinker who had ever lived. Newtonian inspiration was visible as early as 1869, when Pareto tried to formalize the laws of expansion and contraction influencing the volume of solids. Pareto's quest to isolate fundamental laws governing the social universe become increasingly apparent after 1900, when he wrote his sociological works. He summed it up nicely: "The principle end of my studies has been to apply to the social sciences, of which economics is only a part, the experimental method which has given such brilliant results in the natural sciences."[25] Unfortunately, Pareto was never able to capture in any clear terms the principles of sociology for which he so strenuously searched, but he did provide all the ingredients (in a somewhat chaotic way) in his *Treatise on General Sociology*. At least one version of his theory has been articulated in a set of succinct interrelated propositions and translated into simultaneous equations as we imagine Pareto would have appreciated.[26]

In his search for sociological laws, Pareto focused on interdependence and mutual determination among the structural components of social systems. Harking back to his engineering training, he came to understand that every event had potential for reinforcing or undermining the status quo. Any occurrence can trigger a chain of events, either returning a system to some semblance of its previous state or propelling the system to an entirely different structural configuration. Thus, Pareto used the concept of equilibrium expressly for studying change. He never implied that the world was static. As a sociologist, he consistently used this analytical strategy, examining the complex interplay of socioeconomic and political forces to understand how and why societies change over long periods.[27]

As a final note, it would be hard to understand Pareto without recognizing him as a disaffected liberal. He came to believe that powerful people manipulated the government to serve their own selfish interests and then used rhetoric to disguise their own greed under the cloak of national interest. What he found most disconcerting is that common people are reluctant to recognize the deceptiveness of their leaders. People hear only what they want to hear, ignoring the truth unless it happens to be in harmony with their short-term interest

NOTES

1. This chapter borrows heavily from Charles Powers, "The Life and Times of Vilfredo Pareto," in Vilfredo Pareto, *The Transformation of Democracy,* ed. Charles Powers and trans. Renata Girola (New Brunswick, NJ: Transaction, 1984), pp. 1–23; and from Charles Powers, *Vilfredo Pareto* (Newbury Park, CA: Sage, 1987). Biographical insights have been collected from a number of sources, including Norberto Bobbio, *On Mosca and Pareto* (Geneva: Librairie Droz, 1972); Placido Bucolo, *The Other Pareto* (New York: St. Martin's Press, 1980); Giovanni Busino, ed., *Correspondence 1890–1923,* 2 vols. (Geneva: Librairie Droz, 1975); S. E. Finer, "Pareto and Pluto-Democracy: The Retreat to Galapagos," *American Political Science Review* 62 (1968), pp. 440–450; Arthur Livingston, "Vilfredo Pareto: A Biographical Portrait," *Saturday Review,* May 25, 1935; Maffeo Pantaleoni, "Vilfredo Pareto," *Economic Journal* 33 (September 1923), pp. 582–590; Joseph Schumpeter, "Vilfredo Pareto, 1848–1923," *Quarterly Journal of Economics* 63 (May 1949), pp. 147–173; and Vincent Tarascio, *Pareto's*

Methodological Approach to Economics (Chapel Hill: University of North Carolina Press, 1966).

2. One does not have to look very hard at Pareto's work to realize that despite his conservatism, he was a champion of the underdog. See, for example, Vilfredo Pareto, "The Parliamentary Regime in Italy," *Political Science Quarterly* (1893), pp. 677–721.

3. Pareto taught his first sociology course—the first in Switzerland—in 1898. His first sociology publication, "An Application of Sociological Theory," deals explicitly with the demise of corrupt government and lays out the sociological agenda Pareto would spend the rest of his life following. See Vilfredo Pareto, *The Rise and Fall of the Elites,* intro. Hans Zetterberg (Totowa, NJ: Bedminster, 1968; originally published in 1901).

4. Vilfredo Pareto, "Principi fondamentali della teoria della elasticité dé corpi solidi e ricerche sulla intergrazione delle equazioni differenziali che ne definiscono l'equilibrio" (1869). Reprinted in Vilfredo Pareto, *Scritti teorici* (Milan: Malfasi, 1952), pp. 593–639.

5. This approach receives its fullest development in Vilfredo Pareto, *Treatise on General Sociology* (first translated into English as *The Mind and Society*), ed. Arthur Livingston and trans. A. Bongiorno and A. Livingston with J. Rogers (New York: Harcourt Brace Jovanovich, 1935; reprinted under the original title by Dover in 1963 and AMS in 1983). Pareto presents a much briefer and somewhat more refined analysis in *Transformation of Democracy.*

6. See, for example, *Encyclopedia of Professional Management* (New York: McGraw-Hill, 1978).

7. Chester Barnard, *The Functions of the Executive* (Cambridge, MA: Harvard University Press, 1938).

8. To get the flavor of Pareto's commentary, see Vilfredo Pareto, *La liberté économique et les événements d'Italie* (New York: Burt Franklin, 1968; originally compiled in 1898 as a collection of previously published newspaper and magazine articles).

9. Vilfredo Pareto, *Cours d'économie politique* (Geneva: Librairie Droz, 1964; originally published in 1896–1897); and Vilfredo Pareto, *Manual of Political Economy*, 2nd ed., ed. Ann Schwier and Alfred Page and trans. Ann Schwier (New York: August M. Kelley, 1971; translated from the revised edition, originally published in 1909).

10. Pareto, *Treatise on General Sociology.*

11. Pareto, *Transformation of Democracy.*

12. A brief but interesting review of the intellectual milieu in which Pareto matured is provided in Chapter 1 of Tarascio, *Pareto's Methodological Approach to Economics.* Comments interspersed throughout Pareto's work indicate the comparative importance of a variety of intellectual influences.

13. Pareto, *Treatise*, p. 86.

14. Pareto, *Manual,* pp. 47–50; *Treatise,* pp. 2, 102, 144.

15. Pareto, *Treatise,* pp. 217, 287–288, 827–828.

16. Pareto, *Manual,* pp. 268–269.

17. Vilfredo Pareto, "Introduction to Marx," in *Marxisme et économie pure* (Geneva: Librairie Droz, 1966; originally published in 1893). Also see *Treatise,* p. 1790.

18. Niccolò Machiavelli, *The Prince* (New York: Heritage, 1954; originally published in 1532).

19. Adam Smith, *An Inquiry into the Nature and Causes of the Wealth of Nations* (New York: Random House, 1937; originally published in 1776–1784).

20. Maffeo Pantaleoni, *Pure Economics* (New York: Macmillan, 1898; originally published in 1889).

21. Pareto was particularly impressed with Herbert Spencer's *The Classification of the Sciences* (New York: Appleton-Century-Crofts, 1864). Pareto was less impressed with Spencer's other works, although he seems to have read them with interest.

22. Pareto, "Introduction to Marx."

23. Many people credit the theory of circulating elites to Gaetano Mosca, who published on the subject before Pareto. However, similarities between their theories seem to have been the result of independent invention. To the extent that others influenced Pareto's theory, the intellectual debt is probably to Sorel. See, for example, *From Georges Sorel: Essays in Socialism and Philosophy,* ed. and intro. John Stanley and trans. John Stanley and Charlotte Stanley (New York: Oxford University Press, 1976).

24. For example, Pareto did not like people referring to him by his hereditary title of marquis. However, he was comfortable when people referred to him by the achieved title of professor.

25. Bernard DeVoto, "Sentiments and the Social Order," *Harper's Monthly Magazine* 167 (October 1933), pp. 569–581.

26. Charles Powers, "Pareto's Theory of Society," *Cahiers Vilfredo Pareto* 19 (1981), pp. 99–119; and Charles

Powers and Robert Hanneman, "Pareto's Theory of Social and Economic Cycles: A Formal Model and Simulation," ed. Randall Collins, *Sociological Theory* 1 (1983), pp. 59–89.

27. Barbara Heyl, "The Harvard 'Pareto Circle,'" *Journal of the History of the Behavioral Sciences* 4 (1968), pp. 316–334; Joseph Lopreato and Sandra Rusher, "Vilfredo Pareto's Influence on U.S. Sociology," *Cahiers Vilfredo Pareto* 69 (1983), pp. 69–122; Andrzej Kojder, "The Mixed Reputation of Vilfredo Pareto as a Classic of Sociology," *Polish Sociological Bulletin* 10 (1993).

20

✳

The Sociology
of Vilfredo Pareto

During the course of his life, Pareto was a practicing engineer who wrote a baccalaureate dissertation on molecular mechanics, a political and social commentator who published nearly two hundred articles, a business manager and consultant who made a number of noteworthy contributions that still inform the conduct of business in the private sector, an academic who made major breakthroughs in economics and political science, and finally, a retired academic who wrote a major treatise on sociology. Thus, in approaching Pareto's work, we are faced with the immediate problem of selecting the most sociologically important pieces. This task of selection is particularly difficult because each varied stage in his career provided him with certain key concepts that became integral to his culminating work in sociology.

Thus, we should initially approach Pareto's sociology by analyzing how various nonsociological works set the conceptual stage for his purely sociological efforts. We will therefore discuss his work as the product of five distinct stages: (1) engineering, (2) management, (3) commentary, (4) academic life, and (5) sociological work.[1]

PARETO AS AN ENGINEER

Pareto's dissertation for the School of Applied Engineering in Turin examines a theoretical topic, "Fundamental Principles of the Theory of Elasticity in Solid Bodies and Research Concerning the Integration of the Differential Equations Defining Their Equilibrium."[2] The details of his analysis are not particularly important, but his approach to scientific inquiry was to be extended to his study of sociological phenomena.

Pareto's thesis presents an integrated set of equations defining elasticity in solids. Because every system is composed of interdependent parts, any event affecting some elements has repercussions for the system as a whole. Pareto tried to identify principles governing the ways in which change reverberates through a system. These "equilibrium dynamics" became his trademark. In his engineering thesis, the volume of a solid is determined by countervailing forces of expansion and contraction. A relatively stable balance, or equilibrium, can be reached when a system is undisturbed by outside shocks. When the external environment changes, however, the precarious balance among countervailing forces is altered, and the system changes.

This analysis has a number of interesting features. First, Pareto made a clear distinction between the dynamic processes internal to a system and shocks from the external triggering internal processes. Second, he made no effort to explain why external shocks occur. They are simply taken as given. The important question is, What happens *inside* a system when it confronts a changing environment? The only way to answer this question is to understand the dynamics of balance among components internal to the system. Third, depending on specific empirical conditions, countervailing forces can either amplify or retard change. Hence, the same theoretical framework can be used to study both stability and change. Fourth, by understanding the nature of interdependence among system elements, it is possible to predict how the system will be transformed when it confronts changes in its environment.

The equilibrium model employed by Pareto is inherently dynamic. It implies neither that the world is static nor that the status quo is good. All "equilibria" undergo change because no system exists in a vacuum. Hence, the internal configuration of system elements is modified whenever some of the system's interdependent components are influenced by external events. Because change is constant, the goal of science should be to reveal the general dynamics of balance among countervailing forces and the way changes in that balance produce internal structural alterations when a system is confronted with changing exigencies. Later stages in Pareto's work can be viewed as a slow process of clarifying concepts and discovering principles that would allow him to apply equilibrium analysis to the study of social systems.

After graduating from college in 1869, Pareto went to work as a civil engineer with the government-owned railroad. In 1874, he entered the private sector and occupied a variety of posts in which he could apply his engineering training and skills. As time went by, he assumed greater managerial responsibilities.

PARETO AS A MANAGER

In 1874, Pareto left government service and assumed a post with Societá Ferriere Italiana. This position involved complex managerial responsibilities, dealing with issues as diverse as movement of freight, shifting foreign exchange rates and customs duties, labor problems, inventory control, maintenance of equipment, and technical problems in the manufacturing process. Involved as he was in industrial management, Pareto saw the broad picture of a corporate system in which events transpiring in one location influenced the overall production process. This marked a significant milestone in his life, for he had the perfect opportunity for examining the inner workings of a complex social organization.[3]

Pareto's first "social scientific" efforts yielded concrete results. For instance, having a fixed slot for items of inventory and supply significantly reduced the number of work stoppages by making it simpler to locate essential parts and easier to recognize when those parts need to be restocked. During the late nineteenth century, developments of this kind constituted major innovations in administrative science, and they contributed significantly to organizational productivity and helped managers limit operating costs.[4]

Perhaps even more important were Pareto's discoveries about the decision-making process. Contrary to common opinion, there is rarely a single optimal solution to any given business problem. Technical constraints limit what is feasible, but one usually finds a range of possible alternatives involving roughly similar cost-benefit ratios for the actor. Thus, it is possible to select options having the least fallout for others, and Pareto-like strategies can be used to make decisions as important as where to locate power plants or how to select among possible responses to air and water pollution.[5] The implication for decision science is that managers have a certain number of options and should consider factors like morale that might not always be tangible but are nonetheless critically important.[6]

PARETO AS A POLITICAL COMMENTATOR

Pareto's life changed completely in 1889. His mother died, he married his first wife, and he gave up regular employment for work as a private consultant. This freed him to spend much of his time writing political commentary for newspapers and magazines. Working at a fever pitch, he published approximately 160 articles during the next few years.[7]

Pareto understood what the methods and goals of science should be, but the social sciences were yet undeveloped. Therefore, the articles he wrote during this period are more like insightful journalism than sophisticated social science. This stage was critical to his development as a sociologist, however, for without a well-developed body of social scientific literature from which to draw, he had to identify for himself the most important features of the social world. The basic ingredients for his later sociological works—economic interests, political power, social mobility, inequality, and sentiment—all began to appear in his journalistic commentary.

Most of Pareto's work at this time was highly polemical. He was an outspoken advocate of free trade and a critic of government intervention in the economy. He argued that forms of government involvement such as protective tariffs, government-granted monopoly rights, and the use of subsidies and loans to protect corporations all promoted the interests of the rich while undermining general prosperity by supporting inefficient enterprise and discouraging modernization. One of his most damning observations was that governments tend to give special assistance to big corporations, which can hardly claim to be foundling industries. He was also a constant opponent of militarization and colonial expansion.[8]

Gradually, Pareto began to isolate the social structural dynamics essential to understanding society. This new focus to his thought would remain incomplete for many years, but by the early 1890s, he had come to recognize the importance of a number of processes:

1. Powerful economic lobbies promote their interests by exerting influence on political elites to intervene on behalf of the rich.

2. Political elites seek to consolidate their positions by transferring wealth from the nonelite classes to the elite classes.

3. Economic and political elites create ideologies to legitimize their activities while attempting to provide the masses with some benefits to maintain their allegiance.

4. At some point, elites lose their vitality and capacity to control nonelites, setting into motion processes that lead to radical change.

Just as Pareto had adopted an approach to scientific inquiry while an engineer and had begun to focus on the interdependent aspects of social systems while engaged in industrial management, the substantive ingredients for a theory of society emerged from his political commentary. Nevertheless, the task of integrating these insights into a sophisticated theory of social systems remained unfinished.

PARETO'S ACADEMIC WORKS

After reading Maffeo Pantaleoni's *Pure Economics,*[9] Pareto became convinced that social science was possible and that economics would be its cutting edge. Drawing on the view of equilibrium developed in this engineering phase, as it became reinforced by Léon Walras's equilibrium theory (see Chapter 19), he set out to formalize economics. As he did so, he apparently had a broader vision that involved applying the same analytical approach to political phenomena.

On Walras's retirement, Pareto replaced him in the professorial chair of political economy at the University of Lausanne. Thus began Pareto's academic phase, during which he wrote two great works in economics: *Course in Political Economy* (1896–1897)[10] and *Manual of Political Economy* (1906–1909).[11] In the decade between publication of these works, he began to extend his analysis to noneconomic phenomena. The result was his *The Rise and Fall of the Elites* (1901),[12] which contains the ideas for which he is perhaps best known, and his *Les systèmes socialistes* (1902–1903).[13] Written between the first and last revisions of his purely economic work, these books sensitized him to the complex interconnections among social, economic, political, and ideological phenomena. By his retirement from academia in 1907, he had become convinced that

> Human society is the subject of many researches. Some of them constitute specialized disciplines: law, political economy, political history, the history of religions, and the like. Others have not yet been distinguished by special names. To the synthesis of them all, which aims at studying society in general, we may give the name of *sociology.*[14]

To appreciate how Pareto came to this conclusion and why he chose to analyze social systems in his own distinctive way, we need to summarize the substance, style, and strategy evident in his academic stage. Hence, we will first analyze the two great economic works and then those dealing with political phenomena; his sociology emerged from these works after his retirement from academia.

Course in Political Economy and *Manual of Political Economy*

Course in Political Economy is, in many ways, a defense of classical economics, and ideological biases are evident. The basic contribution of *Course* is its application of the equilibrium concept to major economic functions—production, capital formation and movement, and economic cycles. In particular, Pareto demonstrated considerable methodological sophistication, employing formal equations as well as occasionally using longitudinal and cross-cultural data.

Besides the formalization of equilibrium analysis as it applies to the economy, *Course* also contains the initial statement of Pareto's "law of income distribution."[15] He argued that the distribution of wealth tends to be relatively stable in any given society and that efforts to alter substantially the distribution of wealth, making it either more equitable or less equitable, stimulates powerful countervailing forces, resulting in a return to the norm for that society.[16] No matter what politicians may find it convenient to say, drastic reductions in inequality are unlikely because those in power usually find ways of protecting their own interests. This law has an important policy implication: If relative shares of the "economic pie" remain essentially constant, the best way to relieve poverty and help the poor is by "baking a bigger pie."

More important than *Course in Political Economy* is *Manual of Political Economy,* in which Pareto rejected his past polemics and sought to sharpen his analytical edge. The result is a classic in economics in which a theory of maximum efficiency is developed, indifference curves are employed, and a refined statement on the equilibrium dynamics of supply and demand is presented. Drawing renewed inspiration from his early engineering works, Pareto reasserted that theory must seek general principles by isolating generic properties of systems from the mass of available empirical data and, then, attempt to specify the conditions under which these principles hold true. The concept of equilibrium is also expanded to admit political, social, and cultural variables.[17]

Pareto's equilibrium model takes on special clarity in *Manual.* He treated economic decisions as choices reflecting a balance between countervailing *tastes* (desires people want to satisfy) and *obstacles* (factors standing in the way of satisfaction of desires). Tastes are not at all logical. Once a given nonlogical taste develops, however, the individual rationally calculates how to maximize the satisfaction of that taste despite known obstacles.

Pareto did not argue, as some have maintained, that all existing structural features are the products of equilibrium. For example, conditions of exchange are often set by government regulation (for example, price controls). Nor did Pareto maintain that the status quo, even if conceived as a product of equilibrium, is necessarily the best of all possible states.

Further refinements in the concept of equilibrium took on importance in *Manual,* as Pareto outlined two kinds of equilibrium movements. *Stable equilibrium* occurs when a change in one component of a system stimulates modifications within the system that tend to minimize or reverse the original change. For example, an increase in consumer demand can lead to price increases that dampen demand. *Unstable equilibrium* occurs when change in one compartment results in modifications that further add to the initial change. For instance, increasing demand can lead to an increase in the number of competing suppliers, changes in economy of scale, or technological innovations, each of which can result in lower prices and stimulate still greater demand. The U.S. hand calculator industry serves as a good illustration; one had to be affluent to purchase a hand calculator before 1965, whereas by 1980 everybody could afford to own one.

It is important to recognize that stable and unstable states are both types of equilibria and are understandable as balances among those social, cultural, and political variables that influence the economy. To mistake Pareto's discussions of stable periods (which he generally refers to simply as *equilibria*) for his entire equilibrium perspective is an error that should be avoided.

Thus, Pareto thought that the economic system could be analyzed in much the same way that he had once studied molecular structure. Mutual interdependence of parts creates a situation where change in one direction can produce pressures in the opposite direction, with the result that equilibrium phenomena frequently reveal cyclical patterns of change. Thus, in contradiction to the evolutionary theories of his time, Pareto saw economic and social systems as moving equilibria revealing cyclical patterns of change. This metaphor was to be the hallmark of his sociological theory.

The Rise and Fall of the Elites
and *Les systèmes socialistes*

The original title of *The Rise and Fall of the Elites* was "An Application of Sociological Theory," a clear indication of the direction in Pareto's thinking. In this work he sought to identify the major features of society that fluctuate cyclically, to describe the movement of these cycles in equilibrium terms, and to indicate how the structural features and general form of society emerge from operations of the equilibria being described. *The Rise and Fall of the Elites* is an initial statement of Pareto's theory of circulation of elites, for which he became well known. Frequently forgotten in commentaries on this work, however, is that Pareto intended the theory of circulation of elites to be but a single aspect of his more general sociological theory of society and to serve only as a provisional statement and model that could be used in specifying other aspects of his sociological theory.[18] The basic argument in *The Rise and Fall of the Elites* can be stated as follows:

1. Cyclical changes occur in the *sentiments*—that is, values, beliefs, and worldviews—of economic and political elites as well as nonelites.

2. At any time, political processes are dominated by elites, whose members are either *lions* or *foxes*. Lions are strong willed, direct, and conservative. They favor adherence to tradition and show little reluctance to use force. On the other hand, foxes are cunning and devious. Their bravado can be toothless posturing and false imagery. They view government as the art of deceit, misinformation, and secret deals, all cloaked behind a veil of propaganda.

3. At any given time, economic processes are dominated by elites, whose members are either *rentiers* or *speculators*. Rentiers tend to be conservative, are interested in long-term investments, and tend to favor enterprises that produce tangible goods or provide necessary services. Speculators accept risk, are interested in short-term profitability, and tend to engage in intermediary enterprises that make money by transferring things from one set of hands to another without incurring production costs.

4. Because members of elites tend to recruit others like themselves, excluding those who violate their sentiments, political and economic elites tend to become homogeneous over time.

5. Homogeneous elites destroy economic and political vitality and are vulnerable to overthrow by their opposites. Therefore, a country dominated by one kind of elite loses strength and stature. Lions and rentiers are eventually replaced by foxes and speculators, and vice versa.

6. The rate at which change occurs is a dual function of how exploitive elites become and the skill with which elites use force, co-optation, and propaganda to maintain their position.

7. As nonelites become alienated by exploitive activities, their alienation eventually creates pressures that exceed the capacity of elites to use force, thereby resulting in the replacement of one type of elite by another type.

8. The cycles of elites are positively correlated with each other and with economic conditions, with the result that lions and rentiers tend to ascend to elite positions together during times of economic contraction, whereas foxes and speculators tend to ascend to the elite positions during times of economic growth and prosperity.

9. Accompanying, and roughly corresponding to, these political and economic cycles are cycles in ideological beliefs between conservative and liberal tenets.

In these arguments, we can see Pareto's more sociological imagination beginning to assert itself, even though he was still primarily concerned with formal economic models. The concepts of equilibrium and cyclical change have now been extended to embrace sociological variables: elites, mobility, underlying values or sentiments, and more clearly articulated ideologies. Although Pareto had recognized the importance of these variables in his commentaries, they are now part of a more formal equilibrium model of human organization, a model that matured only after his excursion into formal economics.

By the time Pareto wrote *Les systèmes socialistes,* he was completely convinced that people were motivated by nonlogical sentiments disguised in a veneer of ex post facto logic. Thus, to truly understand the dynamics of history, it is necessary to develop a theory of sentiment to explain why people in a given society in a particular period make the choices they do.

In *Les systèmes socialistes,* high priority is given to cultural beliefs and ideology as basic properties of social systems, and Pareto's insights on this subject are an important feature of the *Treatise.* He saw people reacting to events as they are filtered through the prism of their beliefs. Moreover, he saw beliefs as cycling between two poles, one revolving around "faith" and the other around "skepticism." The term *faith* denotes that during some periods beliefs emphasize adherence to tradition and the status quo, whereas during other periods beliefs stress an attempt to assess events logically, even though such "logical assessments" are always an illusion.

One cannot, Pareto argued, understand the nature of a social system unless an assessment of beliefs is made. Of particular importance is determining not only the direction of beliefs, whether toward faith or skepticism, but also their location in the cycle between these two poles.[19] Moreover, in *Treatise* Pareto specifies an inherent dialectic in belief systems, with the dominance of beliefs based on faith setting into motion changes toward those based on skepticism, and vice versa. Thus, as blind conformity to tradition creates tensions between people's beliefs and their actual experiences, they become disillusioned with these beliefs and seek those that "rationally fit" their actual circumstances. As beliefs become dominated by pseudologic, several tensions can be generated. When beliefs are constantly altered to meet changing circumstances, people begin to seek certainty and fixity in their beliefs, thereby setting into motion pressures for beliefs based on faith. Moreover, pseudological beliefs sometimes have less social utility than do beliefs based on faith, and hence, there can be pressure to return to old ways.[20]

Thus, by the end of *Les systèmes socialistes,* beliefs are as prominent as economic and political variables in Pareto's emerging analytical system. Like the circulation of economic and political elites, beliefs reveal a cyclical pattern that is, to some extent, connected to economic and political cycles. With Pareto's retirement from academia in 1907, the stage was set for his most ambitious work, a theory of human social organization.

PARETO'S SOCIOLOGICAL STAGE

During his career, Pareto came to view economic and political events as only parts of more general social processes. Thus, after semiretiring from Lausanne in 1907 (he continued to teach a sociology course until 1916), "the lone thinker of Cèligny" began to work on a purely sociological analysis of social phenomena. This work was the climax of his career, building on all his previous work. The avowed purpose of his *Treatise on General Sociology* is to "discover the form that society assumes by virtue of the forces acting upon it."[21]

The *Treatise* is divided into four volumes, each with its own distinctive emphasis. Volumes 1, 2, and 3 are, in many ways, preliminary and lay the groundwork for sociology. Volume 1 is devoted to establishing the nonrational basis of human behavior and organization. Volume 2 develops Pareto's famous concepts of "sentiments" and "residues." Volume 3 posits a theory of "derivations." Much of the confusion over Pareto's work revolves around these rather unconventional terms, but as we will see, Volume 4, on the "general form of society," employs these concepts in a way that renders their meaning less ambiguous. In our discussion of the *Treatise,* then, we will briefly examine Volumes 1, 2, and 3 and then devote most of our attention to Volume 4, where Pareto finally, at the age of sixty-five, pulled together some of the diverse strands of his theoretical perspective.[22]

Treatise on General Sociology: Volume 1

The basic argument of Volume 1 can be summarized as follows: Most human action is nonrational and is guided by "sentiments" rather than by logic. Pareto had frequently employed the term *sentiments* in his previous work but had never given the concept rigorous definition. Unfortunately, even in this great tome he failed to employ a formal and consistent definition. From the context of his works, however, we can sense the phenomena that he sought to denote with this term. He stressed that humans hold, often unconsciously, basic values and beliefs that guide conduct. We acquire basic standards of evaluation, and they shape our thoughts, mold our perceptions, and guide our actions.[23]

The two most important types of sentiments are values emphasizing the importance of (1) *group persistence* and (2) *combinations*. With these terms Pareto sought to communicate that people's basic value standards cohere around two issues: adherence to tradition (which he alternately phrased "group persistence" or "persistence of aggregates") and desire for change, including the propensity to innovate (which he phrased "combination").

These two basic value standards are somewhat contradictory, and Pareto emphasized that one sentiment or the other tended to dominate at any given point. Over time the aggregate pattern of social sentiments oscillates back and forth from one pole to the other, from insistence on conformity to the compulsion to try new things, and the character of the society changes accordingly.[24]

Treatise on General Sociology: Volume 2

Pareto recognized that many of the forces influencing human behavior, such as instincts and value standards, could not be directly measured. Thus, to discover the operation of these forces, it is necessary to monitor a residual by-product of sentiment—behavior. In this way, Pareto introduced the concept of "residues," by which he meant observable behaviors that serve as empirical indicators of underlying human sentiments.

It is of the utmost importance to understand Pareto's use of the terms *sentiment* and *residue*. Sentiments are underlying value orientations, whereas residues are the behaviors that actors emit in accordance with their orientations. Pareto's meaning has been the subject of some misunderstanding because he often refers to sentiments as residues:

> Returning to the matter of our modes of expression, we must further note that since sentiments are manifest by residues we shall often, for the sake of brevity, use the word "residues" as including the sentiments that they manifest. So we shall say, simply, that residues are among the elements which determine social equilibrium.[25]

What Pareto did, then, was to use behaviors as empirical indicators of value orientations. This is not unlike the approaches employed by other sociologists. However, Pareto regarded the distinction between sentiments and residues as important. Therefore, we will use *sentiment* whenever sentiments were the object of his intent, even if they are referred to as residues in the passages under consideration.

In Pareto's view, all human behavior represents one of six instinctive drives. However, societies differ in the extent to which collective sentiments impede, legitimize, or give rise to behavioral manifestations of instincts. Thus, behavioral expressions, or residues, reflect underlying patterns of sentiment and

Table 20.1 Pareto's Six Classes of Instincts

Instinct	Definition
1. Combinations	Inventive cunning, guile, and creative imagination
2. Group persistence	Stubborn adherence to established ways and vehement defense of tradition
3. Activity	The need to act and express feelings
4. Sociality	The desire for affiliation and acceptance
5. Integrity	Material self-interest and desire for status and self-identity
6. Sex	The urge for carnal gratification

influence the form of society. Pareto's delineation of the six basic types, or "classes," of instincts is set forth in Table 20.1.[26]

Because Pareto assumed that instincts were a constant whereas sentiments varied enormously, variations in behavior will reflect differences in sentiments, for these values will determine the degree to which instincts are allowed to find expression. Thus, Pareto's list of instinct types (which implied the corresponding types of sentiments and residues) accomplished two analytical tasks: (1) It allowed him to view human action as directed along six major axes and thus gave him an exhaustive system of categories for classifying behaviors. (2) It allowed him to classify different populations by varying configurations of value standards. In this way, then, Pareto felt that he had captured both the constancy and variability of human action and organization.

In examining Pareto's actual use of these categories, several themes are evident. First, sentiments vary from population to population. Second, an individual's personality or character can be assessed by enduring patterns of sentiments as they shape and guide the basic instincts. Third, the residues of the instincts for combinations and group persistence are the most important because many of the social system dynamics must be viewed as a result of shifts in the ratio among the corresponding sentiments in the population at large. Sex was of special interest to Pareto because he regarded it as an unambiguous empirical indicator. There is a more tolerant attitude toward sexual expression in times when sentiments of combination are strong relative to sentiments of group persistence.

When we cut through Pareto's awkward terminology and his effort to simultaneously classify instincts, sentiments, and overt behavior with a typology of six classes, we see that a rather simple analytical and empirical point is being emphasized: Human behavior is motivated in basic directions; each line of potential action is circumscribed by corresponding value standards; overt behavior will thus reflect the way in which values have channeled instinctual drives; some classes of residues—behavior shaped by the relative values for combinations and group persistence—are more important than others for understanding social system dynamics.

Treatise on General Sociology: Volume 3

In his commentary phase, Pareto had recognized that humans seek to rationalize and justify their conduct. The products of these efforts are what he termed *derivations,* by which he meant the rationalizations constructed to legitimize a particular line of conduct. As such, derivations rarely reflect actual intentions or the real situation but, rather, efforts to throw into an acceptable light narrow and often destructive interests. Furthermore, if one derivation is discovered to be false, others can easily be created to justify the same behavior. For example, when colonial powers felt, at the turn of the century, that "civilizing" local inhabitants was no longer an appropriate justification for pillaging China, they contented themselves with pillaging to "protect vital interests."

Because derivations are reconstituted at will, without any necessary change in sentiments or interests, they must be analyzed cautiously. As long as their content is not taken literally, derivations can provide clues to which interests in a society are most active and hence most involved in justifying the conduct of particular vested interests. Moreover, derivations can provide an indicator of which sentiments prevail at a given time in a particular society. Again, as long as the accuracy of derivations is not assumed, the general profile of derivations—that is, their emphasis and the nature of their appeal—can provide a rough indicator of underlying value standards because people and groups are likely to attempt to legitimize their actions by appealing to basic and underlying value premises.

Pareto also saw derivations as critical to understanding social system dynamics. The cyclical fluctuations in sentiment that are so important in his model result, in part, from an inherent contradiction between the usefulness of derivations and their correspondence with reality. People want derivations that allow them to do things that might otherwise be questionable, but they also want derivations that seem consistent with the real world. For example, shifts in beliefs from those based on faith to those based on skepticism occur as people search for beliefs that are both useful and in apparent correspondence with their perceptions of reality. There is an inherent dialectic in people's efforts to justify their actions: The more they seek rational accounts, the greater is the likelihood that they will see contradictions and hence be driven to rely on faith and tradition. Conversely, the more people rely on blind faith, the more such faith is contradicted by actual conditions, and thus, the more they will rationalize their accounts of their actions:

> Hence those perpetually recurrent swings of the pendulum, which have been observable for so many centuries, between skepticism and faith, materialism and idealism, logico-experimental science and metaphysics. And so it is, considering for the moment only one or two of such oscillations, that in a little more than a hundred years, and, specifically, from the close of the eighteenth to the beginning of the twentieth century, one witnesses a wave of Voltairean skepticism, and then Rousseau's humanitarianism as a sequel to it; then a religion of Revolution, and then

a return to Christianity; then skepticism once more—Positivism; and finally, in our time, the first stages of a new fluctuation in a mystico-nationalist direction.[27]

By the end of Volume 3, Pareto had performed the preliminary work for his general treatise, but he had not developed, to any great degree, his general analysis of social systems. He had, nevertheless, confirmed for himself that most human behavior was nonlogical; that people constructed symbolic edifices or derivations to justify their conduct; that human action was ultimately guided by value standards, or sentiments, which were reflected in their behavior or residues; and that politics, economics, and value standards revealed both cyclical and equilibrium tendencies. With these initial insights, Pareto then began Volume 4 of his *Treatise.*

The Transformation of Democracy
and *Treatise on General Sociology:* Volume 4

One can sense Pareto's frustration about the time he devoted to Volumes 1, 2, and 3, which in his mind were only preliminary works. As an aging scholar, he appeared to have recognized that he had spent too much time on the early volumes and that only a little time would be left to realize the goal of this last volume on the "general form of society." Perhaps because he was hurried, impatient, and frustrated, this volume lacks complete clarity, and yet, it is sociologically the most important.

Unfortunately, *Treatise* is so long and convoluted that few people have bothered to read it, and even fewer have been able to understand it. Readers were overwhelmed by the case studies Pareto introduced. Consequently, the sociological theory he tried to advance ended up getting lost in a sea of historical detail. Reviewers seem to have regarded *Treatise on General Sociology* as little more than a compendium of awkward terms, without recognizing that Pareto had identified dynamics giving rise to cyclical change in the economy, in politics, and in popular mood.

When the book was finally published, Pareto was sixty-eight years old and in declining health. He did not have the energy to fight many more battles. At the same time, events in Italy and elsewhere were lending credence to his theories. He made one last effort to clarify his ideas for readers and to introduce a few key modifications. The most important of these modifications was to move beyond the crude psychologism of his earlier work by casting his analysis of circulation of elites within a penetrating structural analysis of centralization and decentralization of power. This final effort to clarify and extend his theory is found in a series of articles, which appeared in 1920 and were then published together as *The Transformation of Democracy* in 1921.[28]

Any serious effort to understand Pareto's sociological theory must focus foremost on these two works: *The Transformation of Democracy* and Volume 4 of *Treatise on General Sociology.* With Volume 4, he finally confronted the task of articulating his theory of society. In *The Transformation of Democracy,* he tried to refine and correct that theory by introducing certain pivotal modifications as well as by clarifying his presentation somewhat, to reduce the number of misinterpretations by readers. For these reasons we will treat the two works together. In our opinion, no serious examination of Pareto can neglect either one, nor should it really separate the two.

Given Pareto's career development, one should not be surprised that his general sociology is predicated on the assumption that societies are really *systems* composed of interdependent parts. Hence, any event affecting part of the system is expected to have consequences for the whole. The economy influences public sentiment, which in turn influences politics, and so on. Pareto sought to explain how these patterns of interdependence affect the overall form and character of a society and give rise to predictable patterns of social change. The implication for the social sciences is that the study of economy, politics, and sentiment should be integrated in some way. His general sociology was intended to provide that integration.

If one accepts Pareto's reasoning about systems, anything can affect anything else. Seen in this light, standard causal modeling is rather ridiculous. Instead of selecting a particular event and tracing its causes, Pareto sought to identify the nature of interdependence among societal elements and to generalize about long-term patterns of societal change. In his opinion, history's most striking lesson is that things oscillate. Bad times follow good times, and good times follow bad; his sociology should be seen as a search for the cyclical dynamics that give rise to such patterns of change.

Pareto began with the startling observation that society moves along *three* different cycles simultaneously. One is the business cycle. Another is a cycle between what we might loosely term *liberalism* and *conservatism* in public mood. The third is a cycle between centralization and decentralization of power (see Figure 20.1). It is sometimes difficult to see three separate cycles because each individual cycle influences and tends to synchronize the other two, with the effect that all three tend to move in the same direction at more or less the same time, although Pareto noted that sentiments tend to lag somewhat behind changes in the other two cycles. Economies expand, governments decentralize control, and social constraint is relaxed, all at about the same time. That synchronization results from feedback mechanisms that Pareto hoped to identify.

Until this point, Pareto's sociological work had been essentially empirical. He had been trying to figure out what persistent regularities needed to be explained. Now the interesting theoretical work was to begin. What are the cyclical dynamics that give rise to undulatory change in social sentiment, economic productivity, and political organization? If each cycle is linked to the others, how do those linkages operate? Thus, Pareto's general sociology took form as he attempted to define six separate sets of operating dynamics: dynamics intrinsic to each of the three cycles and dynamics linking each pair of cycles.

For Pareto, discussion of cycles immediately invoked the image of equilibria: Prolonged movement in one direction tends to generate countervailing

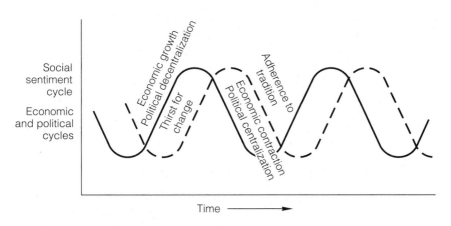

FIGURE 20.1 The Cyclical Patterns of Social Change

pressures, which at first halt and then reverse the direction of change. The same thing happens again later in the cycle, as change in the opposite direction gains momentum until resistance builds to counteract further movement, eventually stimulating a reaction in the opposite direction—what is often termed the "overshoot" problem in some literatures.[29] At this point Pareto's greatness really shows. Most social scientists are content if they can make projections for the near future based on current indicators, but he was far more ambitious and sought to understand what causes the turning points when trends are reversed and events change direction. Let us now explore in detail the theoretical ideas in Volume 4 of *Treatise on General Sociology* and in *The Transformation of Democracy*.

PARETO'S ELEMENTARY THEORIES

In Pareto's view, history displays cyclical patterns of change precisely because the economy, politics, and public sentiment all change rhythmically over time. Each cycle is, in certain respects, autonomous. That is, strictly economic factors account for a considerable portion of business cycle activity, strictly political factors account for much of the oscillation between periods of centralization and decentralization of power, and strictly sociological factors account for a good deal of the swing between periods of liberalism and conservatism in public mood. Pareto's writings suggest that three separate sets of equilibria govern the separate and independent aspects of movement in these three cycles. We term these three equilibria his elementary theories of sentiment, the economy, and politics.

Pareto was convinced that sentiment, economy, and politics were in many ways interdependent. He therefore sought to understand the ways in which movement on one cycle affected movement in the other two cycles. This led him to identify three additional sets of equilibrium dynamics, which govern the interdependence of sentiment with economy, sentiment with politics, and economy with politics.

These linkages will be dealt with later in the chapter. They are important because they unify the various topics of interest to Pareto within a single theoretical framework. This overarching theory is what he called his *general sociology*. We will proceed by describing each of Pareto's elementary theories and will later examine components linking his elementary theories into a general sociology.

An Elementary Theory of Sentiment

Sociology deals with topics that overlap with a number of other fields, such as political science, economics, anthropology, psychology, and religious studies. Public sentiment is important because it is perhaps the one domain of inquiry both absolutely central to sociology and largely unclaimed by any other discipline. Social sentiment nonetheless receives surprisingly little attention. As Pareto made so clear, however, it is impossible to make sense of history without understanding the dynamics that give rise to changes in social sentiment. The study of sentiment was at the very heart of his general sociology.

Sentiments, or underlying value orientations, are never directly observable. What we can do is record the things people say (which Pareto called "derivations") and the things people do ("residues"). It is then possible to triangulate, if you will, to determine social sentiment, and we can take fairly accurate readings of aggregate patterns of change taking place over a period of time across a society at large. This kind of shift in popular sentiment is, after all, what people mean when they talk about the social climate becoming more liberal or more conservative:

> Even a very superficial view of present society reveals streams of opinion
> that manifest underlying patterns of sentiment and interests. These under-
> lying sentiments and interests [rather than opinions about specific issues]
> are the forces at work determining the character of social equilibrium.
> We must therefore avoid becoming overly preoccupied with exactly what
> people say, at the expense of our interest in the underlying sentiments
> which those indicators reflect.

Because we are interested in aggregate patterns of sentiment, we should avoid preoccupation with highly unusual cases.[30]

Pareto's first crucial observation about sentiment is that societies alternate between periods when change is valued and periods when conformity is demanded. Although each society has a different midpoint on the continuum, no society ever remains stationary. All societies move back and forth between times of relative tolerance and relative intolerance of nonnormative behavior, and a clear pattern of oscillation emerges.[31]

Pareto's second observation is that cohort experiences help mold public opinion. Periods of inflation, prosperity, depression, social unrest, or war can leave social sentiments permanently marked. In Pareto's day Italy was one of the last countries to embark on colonial adventures, partly because so many Italians could remember the Austrian domination of northern Italy and could therefore sympathize with the plight of colonized peoples. As 1900 approached and the older generation died off, it was easier for the Italian government to mobilize support for imperial adventure in Africa and the Adriatic Sea.[32]

Pareto's third observation is that changes in sentiment alter the course of history. Every society moves through periods of ascendance and decay, propelled in part by changes in sentiment that encourage or retard innovation, stimulate or impede economic growth, and legitimize or defuse political dissent. As sentiments go, so goes the social order as a whole. A mix of different kinds of people having a variety of talents and dispositions is a necessary precondition for continued prosperity. Bad times are in the offing when, on aggregate, a society becomes either so "liberal" and decadent that all people can think about is personal gratification or so "conservative" and intransigent that any behavior out of the ordinary is punished.

Social sentiments undulate because people want things that are fundamentally incompatible. For example, most people want to live in a world where rules about appropriate behavior are clear and unambiguous, but individuals also want to view norms as flexible guidelines rather than rigid constraints. This presents a dilemma. No society can maximize freedom and at the same time maximize constraint, and every society reacts against excesses of the past by moving in the opposite direction on the freedom-conformity continuum. Discontent builds when a society moves too far in either direction. Such discontent forges itself into a consensus that there is either too much or too little freedom.[33] This consensus does not have to be verbalized. It permeates people's attitudes, affects perceptions, and influences responses. An undulatory pattern of change emerges as a result: "Keeping to surfaces one may say that in history a period of faith will be followed by a period of skepticism, which will in turn be followed by another period of faith, this by another period of skepticism, and so on."[34]

What, then, is Pareto's elementary theory of sentiment? There are times when traditional norms and values seem to be out of date. Speaking in general terms, we can say that a populus sometimes comes to feel, in general, that rules are too restrictive and that the social climate needs to be relaxed. Traditions lose their grip, questioning of authority becomes legitimate, and all kinds of previously punishable behaviors are suddenly tolerated. The society can move in this direction for a long time, with liberalization becoming increasingly pronounced, until a collective sense is reached that too much has been lost. On aggregate, people can reach the conclusion that the society tolerates far more unrestricted ("irresponsible") behavior than it should. At this point, a conservative backlash sets in. Presumption is then on the side of people who want to erect more restrictive rules rather than on the side of the purveyors of unregulated freedom, and the entire climate of the society changes (see Figure 20.2).

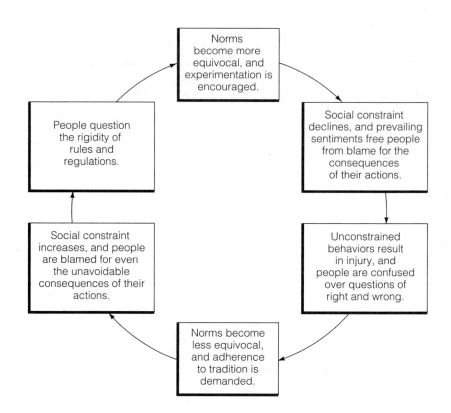

FIGURE 20.2 An Elementary Theory of Sentiment

Pareto was certainly not the only scholar to note that history is marked by cyclical change, but he did more than repeat adages and mirror well-worn generalizations. He suggested a theoretical explanation for changes in sentiment. The important point is that most people do not want to live in either of the worlds that extremes in social sentiment compel us to create. A world devoid of freedom and creativity is sterile and repressive, but a world devoid of rules gives people little protection against the excesses of others. So society oscillates. Pareto's theory reminds us that sentiments compel people to seek relief from one unhappy state of existence by replacing the old regimen with its equally distasteful antithesis.

Pareto emphasized that art, science, philosophy and commerce tended to flourish during periods of liberalization, whereas religion and nationalism tended to do well in conservative periods.[35] For research purposes, he used pornography as an indicator of social change, arguing that pornographic expression was tolerated by a wider audience during periods of liberalization than during periods of conservative backlash, when it was likely to be viewed as a symbol of moral decay. Consequently, changes in the availability of pornographic materials can provide a concrete indicator of shifts in popular sentiment.[36]

The theory of sentiment is the core of Pareto's general sociology. In the next two sections, we will examine his elementary theories of economy and politics. We will then go on to explore the linkages unifying his three elementary theories into a general theory of society.

An Elementary Theory of the Economy

As it was for most economists of his day, the business cycle was a major preoccupation for Pareto. As with other things, he confronted the problem directly. If the economy does cycle between periods of prosperity and depression, it must mean that depressions create the preconditions for economic expansion, whereas prosperity erodes the conditions necessary for sustained growth.

Pareto began his business cycle analysis by focusing on economic fluctuation as a purely economic, rather than sociological, problem. In particular, he identified the availability of capital as the critical economic ingredient in the business cycle. Herein lies a dilemma. It takes a great deal of capital investment to spur periods of economic growth, but, Pareto argued, massive investment over prolonged periods tends to deplete the reservoir of savings available for future use. When money becomes scarce, interest rates rise, investment declines, and the economy slows down.

Pareto's analysis of the economy is similar in important respects to contemporary treatments of accelerator and multiplier effects. The higher the level of net investment, which is defined as investment in excess of depreciation and replacement costs, the greater the number of people put to work building and operating new facilities (the accelerator effect). Those new employees spend money in local shops and further stimulate the economy (the multiplier effect). Investment is the engine that keeps the machine running. Lower investment means fewer people building new plants, which translates into lower sales, with the consequence that wholesale orders decline, and so on. The scarcer capital becomes, the more expensive it is to raise investment funds necessary to fuel increased economic growth, and the more likely it becomes that net investment will decline.

Risk factors also come into play. Entrepreneurs are reluctant to invest in an economy that seems on the decline, especially if the economy is perceived to have nowhere to go but down. A mood of this kind in the business community can seriously aggravate an economic downturn. In contrast, entrepreneurs are eager to invest in an economy that seems on the rise, especially if it is perceived to have nowhere to go but up.

Pareto added to this rather conventional treatment a sophisticated analysis of investment patterns. During depressed periods, many consumer-oriented firms go out of business, with the consequence that the economic infrastructure becomes oriented toward the capital-producing sector. When an economy does expand, people want to satisfy pent-up desires for consumer goods and services. Investment patterns shift as entrepreneurs respond to opportunities for profit in the consumer sector, and over time, the economic infrastructure is transformed. The capital-producing sector begins to shrink in size relative to the rest of the economy. In the long run, Pareto maintained, this shrinkage of the capital-producing sector compounds the difficulty of replacing worn-out equipment and outmoded facilities in a bloated, consumer-oriented economy characterized by high rates of depreciation.[37]

An elementary theory of the economy emerges from this analysis. Investment creates jobs and generates economic activity. The higher is the level of net investment (total investment minus depreciation), the greater will be the level of economic expansion. In the long run, however, expansion can be a factor inhibiting further increases in net investment for three reasons. First, initial investment can erode the pool of available savings, placing upward pressure on interest rates and making future investments more costly. Second, the bigger the economy becomes, the greater is the number of investment dollars needed to offset depreciation each year. Thus, gross investment must increase substantially for net investment to remain at the same level. This also puts upward pressure on interest rates. Finally, after long periods of prosperity, an economic infrastructure tends to be transformed by growth of consumer-oriented businesses. Pareto thought this would reduce the sheer physical availability of capital goods and equipment needed by a large and expanding economy.

The more an economy grows, the greater the probability is that capital available for investment will fall short of investment needs. At that point, the economy enters a downturn, which can be aggravated and prolonged if there is lack of confidence among potential investors. Shrinkage in the economy also means lower levels of annual depreciation, so that relatively low levels of gross investment translate into comparatively high levels of net investment, thus enhancing the prospects for economic recovery (see Figure 20.3).[38]

Below the surface of this elementary theory of the economy, Pareto alluded to sociological dynamics involving sentiment. He entered the domain of sentiment when, for example, he tried to understand consumer spending and when he referred to the level of faith that entrepreneurs have in the economy. At this point, his general sociology comes into play. For Pareto recognized that long-term change in society could be understood only by looking at the interdependence of social and economic, social and political, and economic and political phenomena. Before we delve into his general sociology, we will finish our review of his elementary theories by examining equilibrium dynamics intrinsic to politics.

An Elementary Theory of Politics

Pareto had a lifelong interest in the kind of radical political change in which one regime replaces another. It seemed clear to him that the longer a regime stayed in power, the more decadent it became. As elites become more decadent, they exploit their fellow citizens more and more; as a result, discontent grows. As this trend continues, Pareto felt, it is only a matter of time before a revolutionary cadre leads the masses in a successful uprising in the name of equality and other lofty sounding ideals.

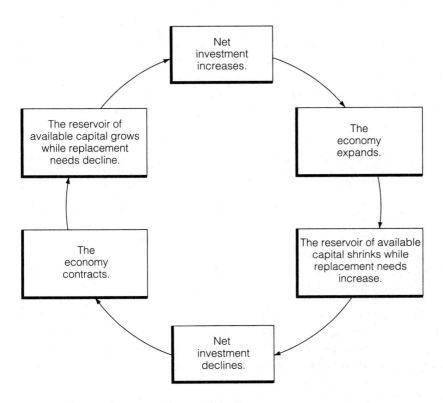

FIGURE 20.3 An Elementary Theory of the Economy

For much of his life Pareto was an idealist who believed that revolutionaries were driven by a desire to ease the plight of the masses. By the turn of the century, however, he had seen enough to conclude that rhetoric and commitment were essentially uncorrelated. He felt that politicians who spouted populist slogans generally did so to enhance their own power rather than out of any compulsion to redress injustice. Over a period of a few generations, energetic new elites develop many of the characteristics of decadent old elites, and the cycle starts over.

This argument is Pareto's famous theory of the circulation of elites.[39] Most sociologists remember him for this theory more than for anything else. But he had other things in mind. Without ever actually rejecting the theory of circulating elites, he subsumed it within a structural analysis of politics in *The Transformation of Democracy,* his final monograph. His mature theory examines a cycle between consolidation and erosion of central power, characterized by changes in political structure.

The structural theory developed in *The Transformation of Democracy* moves away from crude psychologism involving the personalities of leaders and focuses instead on the way systems of political organization change over time.

In essence, Pareto argued that political control could be either centralized or decentralized, with each organizational strategy having its strengths and weaknesses. When erosion of government power reaches dangerous levels, organizational strategies are changed in an effort to consolidate more power.[40]

One holdover from Pareto's theory of circulating elites is his preoccupation with force and co-optation as methods of social control. He began (in *The Rise and Fall of the Elites*) with the observation that some leaders (lions) were adept at the use of force, whereas others (foxes) were proficient in gaining compliance through co-optation. By the 1920s, however, he was moving away from psychologism in favor of a structural analysis of politics. He came to see force as the primary social control mechanism of centralized regimes, and he viewed decentralized regimes as characteristically using patronage to try to co-opt people.

Relying almost exclusively on either force or co-optation can be very dangerous, Pareto believed. Regimes are more likely to maintain their own power when they can use both a carrot and a stick, co-optation and force, to ensure compliance. Even though force can be used to crush opposition for a time, its use also provokes hatred for those in power.

A regime cannot last long when it relies exclusively on force to retain its power.[41] Just as surely, no regime can stay in power by relying exclusively on co-optation. Governments can try a wide range of programs designed to co-opt important sectors of the population—for example, minimum wage laws, social welfare safety nets, free public schools and parks, government contracts and subsidies, protective tariffs, and certification for professionals. But patronage and co-optation are inefficient and expensive.[42]

The Transformation of Democracy is really a case study in the erosion of power that occurs when a government relies too fully on decentralized political organization and the use of patronage to gain compliance. Pareto intended the book to be an examination of one-half of the complete cycle between political centralization and decentralization, a cycle propelled by shifts in the countervailing forces of consolidation and erosion of power.

Authority is relatively consolidated when a regime has the capacity and the will to adjudicate grievances and dispense justice throughout its realm. This capacity is most often present in systems of political organization resting somewhere between the extremes of centralization and decentralization. In contrast, authority erodes when governments lose the capacity for effective and independent action. Erosion of power tends to occur in decentralized systems when a regime loses its ability to solve problems because it panders to special interests or abandons responsibility for activities occurring within its geographic borders. In contrast, erosion of power tends to occur in centralized systems when rulers forbid independent initiative by those in the private sector, stifle public sector initiative by failing to delegate working authority to functionaries, or use force so capriciously that common people come to hate the regime (see Figure 20.4). Readers will note the striking similarity between Pareto's theory and Herbert Spencer's analysis of shifts between militant and industrial organization.[43]

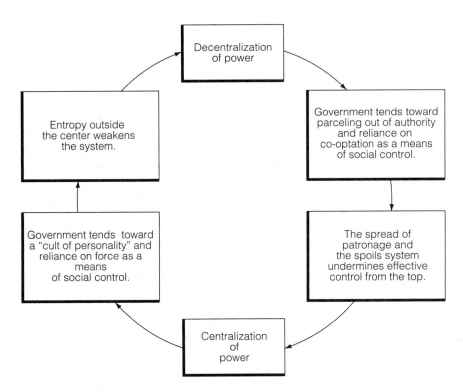

FIGURE 20.4 An Elementary Theory of Politics

The point is that the government's power and authority tend to erode whenever centralized or decentralized organizational strategies are carried to an extreme. The history of any society involves a repetitive cycle of centralization and decentralization of political power. In this sense we can speak of an elementary theory of politics, in the same way we identified an elementary theory of sentiment and economy.[44]

PARETO'S GENERAL SOCIOLOGY: DYNAMIC INTERACTION AMONG CYCLES IN SENTIMENT, POWER, AND THE ECONOMY

There is a clear sense of both mission and progression in Pareto's sociology. He sought to advance our understanding of societal phenomena by recognizing the complexities posed by interdependence among people and events.[45] Most people pay lip service to multiple causality, but he was one of the few to actually identify principles of interdependence linking social, economic, and political realms. These principles of interdependence unify his analysis of economy, polity, and community within a single theoretical framework that constitutes his general sociology.

The Interaction of Social and Economic Phenomena

It is important to remember that Pareto turned to sociology to address fundamental questions he was unable to answer as an economist. He regarded the studies of economics and sociology as inextricably linked because changes in social sentiment dictated the future course of the economy by altering savings and consumption patterns, just as changes in the economy dictated the future course of social sentiment by alternately imbuing people with optimism and fear about the future.

For the economy, changes in social sentiment are tremendously important because of the impact they have on saving and consumption. Consumer saving and consumption are highly elastic, especially for credit buying of durable goods, such as cars and refrigerators. Social values sometimes tend to legitimize hedonism and encourage the pursuit of personal gratification. If people see no end to prosperity in sight, it makes "sense" (in terms of prevailing sentiments) to borrow money to buy consumer goods. Although Pareto did not use the term, he was clearly referring to a *consumer-led* boom.

A consumer-led boom is fueled by two sources, depletion of savings and use of credit. Aggregate shifts in sentiment are critically important because they influence the relative propensities to augment or deplete savings and to extend or retire debt. The gross national product balloons when people busy themselves depleting savings and using credit. But what happens when there are no savings left to deplete and all one's credit has been used? The answer is that spending declines, sending shock waves throughout the economy. Pareto saw some good news, however, even in a dire economic prognosis. Depressions turn people into frugal pessimists who work hard, save their money, and avoid debt. These meager spenders with their careful ways and growing savings accounts provide the backbone for a *business-led* economic recovery. That backbone consists of huge savings reserves that can be borrowed at relatively low rates of interest.[46] Thus, Pareto's great contribution to economics was to provide a sociological explanation for many factors economists tended to treat as givens. People save less and buy more during periods when self-centered pursuit of gratification is deemed legitimate. Conversely, people save more during conservative times when a higher premium is placed on the value of self-denial (see Figure 20–5).

The Interdependence of Social and Political Phenomena

People are implicitly aware of the control strategies employed by a regime, Pareto felt. Each strategy encourages a particular worldview while discouraging certain values and outlooks. For example, the use of co-optation encourages people to view success as a product of whom one knows rather than of what one accomplishes. Moreover, widespread corruption fosters hedonistic

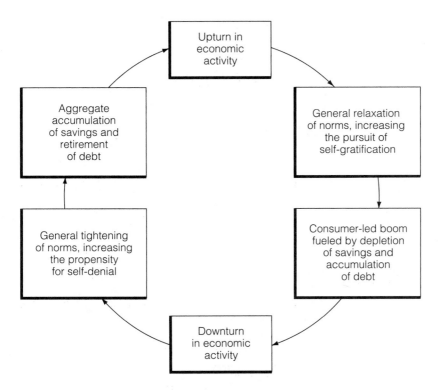

FIGURE 20.5 The Interaction of Social and Economic Phenomena

attitudes at the expense of the work ethic. Thus, politics can have important consequences for the tenor of public sentiment.

Ultimately, people who view the government as a center of bribery and largesse wish to share in the benefits of patronage. As Karl Marx noted, however, human desires are infinitely elastic. The more people get, the more they think they deserve. And the more people see others get, the more likely they are to feel cheated. Thus, governments that practice patronage confront a steady increase in demands for special treatment. This is simply more than most regimes can afford, and discontent rises.

The Transformation of Democracy provides an example: By 1920, the decentralized Italian government was facing a crisis. According to Pareto, it was unable to resist the demands made by unions or corporations. As a result, the government was unable to act decisively. It could not even get railroad workers to adopt uniform schedules based on daylight savings time, which was one of the many reasons the trains had trouble running on time. The government had simply lost power.[47]

Overreliance on force has the opposite effect. Blind, capricious, rigid enforcement of rules creates resistance, which seriously undermines social control. Resistance stimulates repression, which breeds greater resistance, and government power erodes (see Figure 20.6). What emerges from this analysis

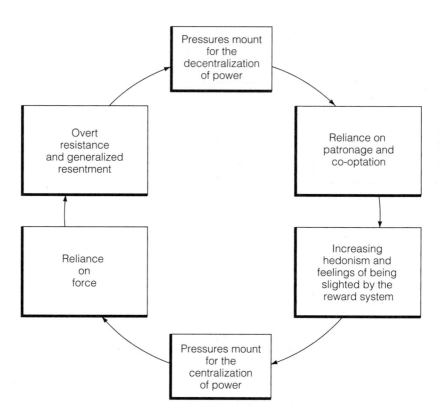

FIGURE 20.6 The Interaction of Social and Political Phenomena

is a close interdependence between Pareto's elementary theories of sentiment and politics.

These ideas illustrate the interdependence of social, economic, and political phenomena. As Pareto pointed out, the nature of this interdependence determines the future direction of change in the overall character of society.

The Interdependence of Economic and Political Phenomena

Pareto did not write very much about the interconnection of the economy and the polity.[48] He seemed to suggest that centralized governments (where decisions are made at the top) discourage activities that are not tightly controlled by the regime, with the result that the government stymies entrepreneurship. On the other hand, decentralized governments tolerate all kinds of behavior, allowing corporations to maximize short-term profits without regard to social costs or long-term consequences. This kind of tolerance undermines confidence in government and is economically unhealthy over the long term. More generally, we can summarize Pareto's argument as follows: Unresponsive governments tend not to create conditions amenable to business expansion. The less responsive the government is, the more likely it is that pressure will build, forcing a decentralization of decision making. But the more responsive to special interests the government becomes, the more business thinks it can manipulate situations and the more product quality tends to deteriorate, with the result that pressure builds for more centralization, coordination, and control (see Figure 20.7).

Once again, Pareto's approach to understanding the world involves equilibrium dynamics generating cyclical change. Business conditions affect the nature of political organization, and vice versa. If we add in the other aspects of Pareto's general sociology, we see the form of society being determined by analytically distinct but functionally interdependent cycles in public sentiment, the economy, and political organization.[49]

CRITICAL CONCLUSIONS

Vilfredo Pareto is discounted by many sociologists who number him among the discipline's minor rather than major intellectual figures.[50] This kind of quick dismissal is unfortunate but understandable. Indeed, Pareto's awkward, laborious, sometimes pedantic writing style almost seems to invite dismissal. His style often makes it difficult to follow his line of argument, which, unfortunately, gives rise to interpretive disagreements about the essence of his theoretical framework.

Perhaps more than with any of sociology's other founding figures, we must ask this: Is Pareto more of a psychologist than sociologist? Those offering the traditional psychologistic interpretations base their conclusions on a more or less literal reading of Pareto's most widely recognized sociological work. At heart, these traditional interpretations focus on Pareto's concern with instinctive drives and his assumption that once personality forms it exerts strong deterministic influence over subsequent behavior. Those arguing in this vein also note Pareto's proviso that the course of a nation's history changes when shifting circumstances create opportunities for new people with different qualities to ascend into positions of leadership. This traditional version of Pareto's sociology is most clearly illustrated in his description of "circulation of elites" suggesting that dominance by leaders who are direct and forceful ("lions") creates conditions for the ascendance of deceitful leaders who like to engage in connivance ("foxes").

Traditional interpretations pay almost no attention to the epistemological framework Pareto employed as an engineer, pay even less attention to his economic manuscripts, and underestimate the importance of *The Transformation of Democracy* because it is a short work written just before Pareto's death. Moreover, it was published in serialized form for broad consumption and was not the kind of exhaustively thorough and academically oriented tome we find in *Treatise on General Sociology*. By privileging *Treatise on General Sociology*, the

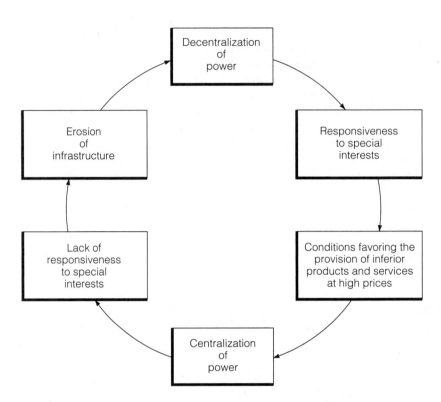

FIGURE 20.7 The Interdependence of Economic and Political Phenomena

laudable goal of traditional interpreters is to try to stay close to Pareto's sociology so that they can most accurately capture its essence. True to their intent, most traditional interpreters are thorough in their study, careful in their writing, and stay close to the literal meaning of the central texts on which they focus. This gives their interpretations of Pareto more than an ample measure of face validity. But there is also strong reason to advance a very different, much more structural version of Pareto's theoretical framework. We feel that structural version of Pareto most genuinely reflects Pareto's actual intentions.

In Pareto's final papers, published as *The Transformation of Democracy*, there are traces of psychologism, but the prevailing flavor puts an undeniably structural cast on themes that had been discussed in more psychologistic terms in his earlier works. Most striking, what had earlier been described as the ascendance of "foxes" was now phrased as the restructuring of government aimed at parceling out of authority by deregulating industry and pandering to special interests. *The Transformation of Democracy* was explicitly intended as a snapshot of the Italian scene in 1920, rather than as the characterization of cyclical change over long periods, as offered in *Treatise on General Sociology*. If we impose this societal snapshot onto Pareto's earlier work analyzing the ebbs and flows of cyclical change over longer periods, we find a powerful social structural theory that is at odds with traditional psychologistic interpretation of Pareto.

A powerful argument can be made that a structuralist interpretation represents the real Pareto, for two reasons. First, and most important, the more structural analysis, which Pareto had developed by the time he wrote *The Transformation of Democracy,* is rather close to what he viewed as good theory from the beginning of his adult life as an engineer. Second, Pareto had a driving need to complete a theory of societal change and to get it right. Given this drive, we find it hard to believe that in his early 70s, confronting what he must have understood could be his last chance to make a statement, he would have written the *Transformation* essays as an intellectually frivolous exercise in political mud slinging or populist entertainment. Pareto would have realistically acknowledged is age and declining health, and as a result, he would have used his time to leave an updated blueprint for the societal theory that he had spent a lifetime trying to build. We believe that is precisely why *The Transformation of Democracy* focuses on themes like the parceling of state authority, complete with some historical references, instead of falling back on psychologistic language of the past.

Our structuralist interpretation of Pareto is not without its difficulties. Chief among these is Pareto's view of political organization and public policy, which is sometimes hard to reconcile with the American experience of stable federalism and a political system with two dominant parties, each of which seems to be responsive to different special interests. As we will explore in the next chapter, however, a more structural interpretation of Pareto offers real insight into societal change and makes it possible to identify a robust set of principles that have contemporary application.

NOTES

1. This chapter draws heavily on Charles H. Powers, *Vilfredo Pareto* (Newbury Park, CA: Sage, 1987).

2. Vilfredo Pareto, "Principi fondamentali della teoria della elasticité dé corpi solidi e ricerche sulla integrazione della equazione differenziali che ne definiscono l'equilibrio," in *Scritti teorici* (Milan: Malfasi, 1952; originally published in 1869), pp. 593–639.

3. Many of the lessons Pareto learned as a manager are alluded to in various passages of Vilfredo Pareto, *Cours d'économie politique* (Geneva: Librairie Droz, 1965; originally published in 1896–1897).

4. The 80/20 rule is an outstanding example. The insight that 80 percent of a company's business revolves around 20 percent of the items in its inventory enables it to make such mundane but important decisions as where to place things to maximize the overall accessibility of merchandise and equipment. This is an incredibly important consideration from the standpoint of maximizing productivity and controlling costs. For example, see *Encyclopedia of Professional Management* (New York: McGraw-Hill, 1978). Another example is Pareto's law of the vital few and the trivial many, reminding managers that with any group of people or events, the best results will be achieved by focusing on the few that really matter. See Robert Goddard, "The Vital Few and the Trivial Many," *Personnel Journal* 66 (1987), pp. 84–94.

5. See, for example, Jacques Gros, *A Paretian Environmental Approach to Power Plant Siting in New England* (New York: Garland, 1979).

6. Chester Barnard, *The Functions of the Executive* (Cambridge, MA: Harvard University Press, 1938).

7. For an excellent selection of articles from Pareto's commentary stage, see Vilfredo Pareto, *La liberté économique et les événements d'Italie* (New York: Burt Franklin, 1968; originally printed in 1898 as a compilation of previously published newspaper and magazine articles).

8. The ardent nature of Pareto's commentary is perhaps most apparent to English-speaking audiences in Vilfredo Pareto, "The Parliamentary Regime in Italy," *Political Science Quarterly* 3 (1893), pp. 677–721.

9. Maffeo Pantaleoni, *Pure Economics* (New York: Macmillan, 1898; originally published in 1889).

10. Pareto, *Cours.*

11. Vilfredo Pareto, *Manual of Political Economy,* 2nd ed., ed. Ann Schwier and Alfred Page and trans. Ann Schwier (New York: August M. Kelley, 1971; originally published in 1909).

12. Vilfredo Pareto, *The Rise and Fall of the Elites,* intro. Hans Zetterberg (Totowa, NJ: Bedminster, 1968; originally published in 1901).

13. Vilfredo Pareto, *Les systèmes socialistes* (Geneva: Librairie Droz, 1965; originally published in 1902–1903).

14. Vilfredo Pareto, *Treatise on General Sociology* (first translated into English as *The Mind and Society*), ed. Arthur Livingston and trans. A. Bongiorno and A. Livingston with J. Rogers (New York: Harcourt Brace Jovanovich, 1935; reprinted under the original title by Dover in 1963 and AMS in 1983), sect. 1.

15. Pareto, *Cours,* sect. 965.

16. See, for example, Vilfredo Pareto, *Escruits sur la courbe de la répartition de la richesse* (Geneva: Librairie Droz, 1965). Pareto's groundbreaking ideas stimulated a significant body of research on welfare economics. See Warren Samuels, *Pareto on Policy* (New York: Elsevier, 1974).

17. Pareto, *Manual,* especially Chapters 1 and 3.

18. Pareto, *Rise and Fall of the Elites,* for instance, pp. 30–31, 36, 40–41, 59–60, and 68–71.

19. Pareto, *Systèmes socialistes,* Chapters 1, 5, and 6. A number of interesting issues relating to sentiment are raised in Brigitte Berger, "Vilfredo Pareto's Sociology as a Contribution to the Sociology of Knowledge," unpublished Ph.D. dissertation, New School for Social Research, 1964.

20. Pareto, *Treatise,* sects. 1678–1690.

21. Ibid., footnote to sect. 1687.

22. Most of vol. 4 appears to have been written during 1913.

23. Pareto, *Treatise,* sect. 888ff.

24. Ibid., sects. 304, 1806–1847, and 2048–2050.

25. Ibid., sect. 1690.

26. Ibid., sects. 885–989 and 992. "Combinations" is Pareto's term for the first instinctive drive listed in the table, the compulsion to be inventive.

27. Ibid., sects. 1680–1681.

28. Vilfredo Pareto, *The Transformation of Democracy,* ed. Charles Powers and trans. Renata Girola (New Brunswick, NJ: Transaction, 1984; originally published in 1921).

29. Overshoot is an important concept in many fields, such as environmental analysis; see, for example, Donella Meadows, Dennis Meadows, and Jorgen Randers, *Beyond Limits* (Post Mills, VT: Chelsea Green, 1992).

30. Pareto, *Transformation of Democracy,* p. 63.

31. Pareto, *Treatise,* sect. 1681.

32. Ibid., sect. 1839.

33. Ibid., sects. 1256–1383.

34. Ibid., sect. 2341.

35. Ibid., sect. 2513.

36. Ibid., sect. 2521.

37. Pareto, *Manual.*

38. Charles Powers, "Sociopolitical Determinants of Economic Cycles:

Vilfredo Pareto's Final Statement,"
Social Science Quarterly 65 (1984),
pp. 988–1001.

39. Pareto, *Rise and Fall of the Elites.*

40. Pareto, *Transformation of Democracy.*

41. Melvin Gurtov and Ray Maghroori,
*Roots of Failure: United States Foreign
Policy in the Third World* (Westport,
CT: Greenwood, 1984).

42. S. E. Finer, "Pareto and Pluto-
Democracy: The Retreat to
Galapagos," *American Political Science
Review* 62 (1968), pp. 440–450. Also
see Suzanne Vromen, "Pareto on the
Inevitability of Revolutions," *American
Behavioral Scientist* 20 (1974), pp.
521–528.

43. See, for example, Jonathan Turner,
Herbert Spencer (Newbury Park, CA:
Sage, 1985).

44. Having moved away from the image
of circulating elites to a theory of
structural change at such a late date
(1920), Pareto can be excused if he
left us with theoretical rough edges to
work out for ourselves. One of the
conceptual problems with his scheme
is that some societies seem to combine the worst characteristics of centralized and decentralized
government. For example, it is possible to have decision making concentrated at the center, leaving
functionaries without the discre-
tionary power they need to operate
government agencies in an effective
manner, and at the same time to rely
on the high levels of co-optation and
patronage Pareto believed to be associated with decentralized systems. It
may be that we should divorce ourselves from the notion that patronage
is more widespread in decentralized
systems than it is in centralized systems.

45. Jean-Martin Rabot, "Le concept
d'équilibre et le philosophie de
Vilfredo Pareto," *Cahiers Vilfredo Pareto*
22 (1984), pp. 117–126.

46. Powers, "Sociopolitical Determinants."

47. Pareto, in *Transformation of Democracy,*
reveals these dynamics.

48. The authors wish to acknowledge a
stimulating exchange of views with
Jürgen Backhaus. See his *Öffentliche
Unternehmen* (Frankfurt: Haag and
Herchen, 1980).

49. Charles Powers and Robert Hanneman,
"Pareto's Theory of Social and Eco-
nomic Cycles: A Formal Model and
Simulation," ed. Randall Collins,
Sociological Theory, vol 1 (1983),
pp. 59–89.

50. Barbara Heyl, "The Harvard 'Pareto
Circle,'" *Journal of the History of the
Behavioral Sciences* 4 (1968),
pp. 316–334; Joseph Lopreato and
Sandra Rusher, "Vilfredo Pareto's
Influence on U.S. Sociology," *Cahiers
Vilfredo Pareto* 69 (1983), pp. 69–122;
Andrzej Kojder, "The Mixed
Reputation of Vilfredo Pareto as a
Classic of Sociology," *Polish Sociological
Bulletin* 10 (1993).

21

✳

Pareto's
Theoretical Legacy

Wilfredo Pareto's theoretical contribution to sociology suffers from the somewhat strange vocabulary he employed, as well as from the complex nature of the theory he developed. As for the vocabulary, terms like "sentiments," "residues," "derivations," "lions," and "foxes" simply do not strike a chord in the modern intellectual world; indeed, they were not even very comfortably received by the intellectual community of Pareto's time. The theory itself is rather complex because it involves three distinct, yet interrelated, sets of forces—cultural, economic, and political—which operate on cycles that are somewhat out of phase with each other. The result is that Pareto's ideas are difficult to understand. Equally significant for our enterprise, they are very hard to translate into models and principles. Undaunted, we will offer a translation in this chapter by, first, using a more modern vocabulary and, second, by developing a complex causal model and a series of propositions that capture the key dynamics in Pareto's theory.

PARETO'S CAUSAL MODEL
OF SOCIETAL DYNAMICS

In Figure 21.1, we have summarized Pareto's model of societal dynamics. The model is complex in two senses: (1) Pareto's theory takes account of many societal forces, and (2) Pareto was alert to the nonlinear nature of causality that is

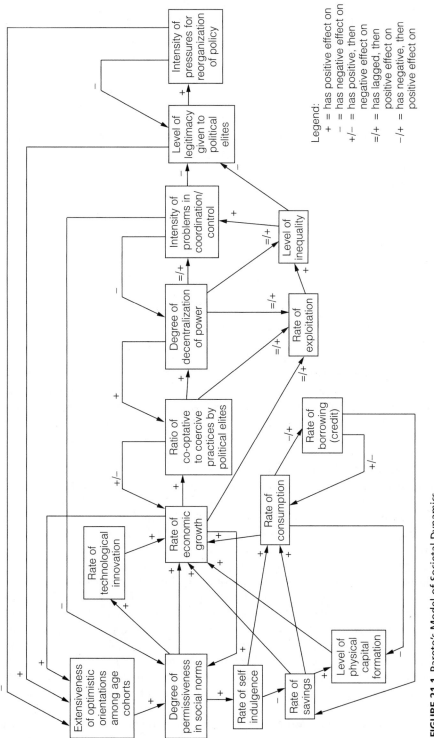

FIGURE 21.1 Pareto's Model of Societal Dynamics

often at work in the social world. If, however, we move through the model with deliberation, this complexity can be mitigated. We must recall that there are three sets of forces operating in a society. First are cultural dynamics reflected by cyclical changes in values, beliefs, and norms. Second are economic forces revolving around production as it is influenced by, and as it influences, both cultural and political forces. Third are political forces that respond to, while causing changes in, cultural and economic processes. The model in Figure 21.1 begins with cultural processes on the far left of the diagram—drawing attention to the pivotal importance that Pareto attributed to varying degrees of optimism among age cohorts and corresponding variability in permissiveness of norms. When diverse age cohorts—young, old, and in-between—are optimistic, these orientations will encourage permissive norms that allow considerable individual choice, experimentation, and innovation. The causal arrows feeding back into these cultural processes, as modeled in Figure 21.1, underscore that optimism and permissiveness vary at subsequent periods, as people react to shifting economic and political events. Economic growth and political decentralization encourage greater optimism and permissiveness, to the point where problems of coordination, control, and exploitation become too intense. At that point, popular outlook shifts in the opposite direction.

Permissive norms will, Pareto argued, cause the rate of technological innovation to increase which, in turn, will increase the rate of economic growth; as the economy grows, people become even more optimistic and tolerant of permissive norms, which in turn encourages even more consumer spending and technological innovation. Other nuances are also important stimulants to economic growth. Permissive norms are associated with toleration of self-indulgence, and this leads people to demand greater quantities and wider varieties of goods and services, thereby fueling a consumer led boom. As people consume, however, they deplete savings; when savings are gone, they begin to borrow on credit. As saving declines and borrowing increases, the aggregate amount of capital available for investment declines, putting a damper on the economy. Pareto noticed that during periods of permissive self-indulgence, investment in capital goods industries slackens. In the end, these forces combine to lower the rate of overall investment and economic growth, and indeed, perhaps even produce a recession. In Figure 21.1, the signs on the causal arrows around this complex of forces on the bottom-left of the model underscore the nonlineal nature of the processes at work. Rates of consumption and borrowing will initially increase production, but this positive relation turns more negative once savings have been depleted. The fundamental dynamic is a simple one. The greater the level of economic investment, the greater is the prospect for additional economic growth. But the more people consume and the less they save, on aggregate, the less money is available to replace capital equipment or to invest in new technologies requiring new kinds of equipment, thereby slowing or reversing economic growth.

Political organization also occupies an important place in Pareto's theory. Economic growth sets into motion political processes that have sweeping consequences. When the economy is growing under conditions of a permissive set of cultural norms and optimistic orientations, the regulatory agencies of government tend to grant considerable freedom and power to powerful economic actors seeking to escape the constraints of regulation. As decentralization proceeds, government tends to parcel out authority to economic actors and increasingly relies on cooptation or manipulation of incentives rather than trying to exercise direct regulatory control. As a consequence, the political system becomes ever-more decentralized, which, in turn, will increase the amount of power economic actors can wield. Eventually, as the lagged positive sign ($=/+$) on the arrow denotes, problems of coordination and control increase by virtue of the government having tolerated the complete or partial deregulation of so much economic activity. When this process is taken to its conclusion, however, the government proves itself unable to coordinate effectively critical activities on which people depend, and its legitimacy is undermined.

This de-legitimation is accelerated by rates of exploitation that tend to increase noticeably when economic elites are given a free reign to do as they please. Exploitation can become rampant as a result of economic actors' desires to increase their wealth at the expense of others; as this process continues, the level of inequality in society increases and institutional and regime legitimacy decline, at least in some quarters. Parceling out authority encourages this process because government compromises its ability to hold special interests in check. As inequality increases, more people become resentful and begin to withdraw legitimacy from a government that cannot stop economic abuse and, more generally, appears to have lost the ability to regulate society. As the legitimacy of government declines, pressures build for a reorganization of polity so that it can better regulate and control actors in society.

The reverse of these causal effects are also an important part of Pareto's argument. As pressures for reorganization mount, optimistic orientations decline. Public sentiment begins to shift toward a more pessimism cast, and individuals become less tolerant of permissive norms. Once these cultural props for economic growth and political decentralization are taken away, the low values for these cultural forces move along the positive arrows in Pareto's model. Technological innovation diminishes, as do attitudes fostering self-indulgence, thereby placing brakes on economic growth. At the same time, rates of savings increase, which lowers consumer demand and borrowing. Over the long term as saving increases, more capital will be available to the economy, and production and technological innovation will increase. This takes time, however.

The sequence of reverse causal arrows moving across the middle of the model emphasize that problems of coordination and control can erode the legitimacy of government and create pressures for reorganization of the polity, encouraging more centralization of power and direct coercive regulation of economicactors. As the ratio of co-option to coercion declines, economic growth will also decrease, at least until savings, capital formation, optimism, and permissive norms return in the next phase of societal dynamics.

A model like the one in Figure 21.1 cannot capture fully the cyclical nature of the dynamics of Pareto's theory, although the nonlinear causal signs on the arrows go a long way toward summarizing Pareto's intent. Still, if we translate Pareto's argument into a series of propositions or principles, we can better see the cyclical nature of the dynamics processes specified in Pareto's theory.

PARETO'S THEORETICAL PRINCIPLES

Elementary Principles of Social Sentiment

Pareto recognized that societal norms change over time, arguing that the general pattern of change is quite predictable because of a tension between humans' desire to be relatively free of social constraint and the simultaneous desire to be safe from the unconstrained actions of others. Briefly stated, Pareto felt that strong norms confer certain benefits on the members of a society, but that individuals eventually rebel against constraint. Initially, the normative liberalization that this rebellion heralds confers recognizable benefits, and these benefits fuel the sentiment that liberalization of norms should continue. The more relaxed norms become, the greater are the mounting costs of living in a society where others are less and less inhibited. Recognition of these costs leads the members of a society to demand more definite lines defining what is appropriate and what is inappropriate. A moderate degree of social constraint offers the members of a society certain benefits, such as some peace of mind, and these benefits legitimate and further propel the trend toward conservative sentiments. Taken far enough, however, the trend toward conservatism engenders more and more social constraint, which people eventually resist. Hence, a pendulous pattern of social change emerges over time.

1. The more equivocal norms become, the more toleration members of a society show for experimentation of all kinds and the less constrained they are in their behavior.

2. The more toleration members of a society show for the behavior of others and the less constrained people feel in their behavior, the more likely they are to be injured or offended by the actions of others.

3. The more injury and offense members of a society sustain because of the actions of others, the more likely they are to seek coherent traditions and unequivocal prescriptions.

4. The less equivocal prescriptions become, the less toleration members of a society show for experimentation of all kinds and the more constrained they are in their behavior.

5. The less toleration members of a society show for the behavior of others and the greater the constraints are on their behavior, the more likely they are to question the rationality of restrictive normative beliefs.

6. The more members of a society question the rationality of normative beliefs, the more equivocal norms become.

Elementary Principles of Economy

Pareto's analysis of the business cycle revolves around the concept of net investment, or total capital investment minus depreciation. During recessionary periods, an economy can spiral downward because low levels of consumer activity discourage capital investment, and without capital investment, an economy has little impetus for growth. Business closures and deterioration of those physical plants still operating act in concert to shrink the economy, while unused physical plants and excess capital drive down the cost of making new capital investments. Thus, as the size of the economy shrinks and the cost of making new investments also declines, the probability that new investment will exceed new depreciation steadily improves. In this manner, economic decline also creates conditions for economic recovery. When new investment does exceed depreciation, this increase in net investment spurs growth that encourages further investment and generates an upward economic spiral by drawing available capital into use. Yet, as the economy grows, it needs increasing amounts of capital investment simply to offset depreciation. Meanwhile, available capital becomes more scarce and, hence, more expensive to secure. As the size of the economy increases and the cost of making new investments increases, it becomes ever more difficult for new investments to offset the massive amounts of aggregate capital depreciation taking place. Hence, cyclical downturn is, at least in Pareto's view, an inevitable outcome.

7. The more an economy grows, the greater is the level of gross capital investment needed to replace depreciated capital stock.

8. The longer an economy has mobilized the capital necessary to replace depreciation and to sustain net investment fueling further growth, the more expensive capital is and the more likely is a decline in net capital investment.

9. The greater the rate of decline is in net capital investment, the greater the expected rate of decline is in overall economic activity.

10. The greater the rate of decline is in overall economic activity, the lower is the level of gross capital investment needed to replace depreciated capital stock.

11. The longer the period of economic contraction is during which an economy fails to mobilize the capital necessary to replace depreciation and to sustain net investment fueling new growth, the less expensive capital is and the more likely net capital investment is to increase.

12. The greater the increase in net investment is, the more likely an economy is to grow.

Elementary Principles of Political Organization

Pareto was intrigued by politics because of the periodic oscillations between periods of consolidation of power, followed by the erosion of effective power. Pareto's analysis centers on the uses of co-optation and exertion force as two main alternative methods of exercising social control. His reading of history led him to believe that leaders consolidate and exercise power most effectively when they are able, skilled, and willing to employ co-optation and force in equal measure. Overreliance on either of these instruments for exerting control leads to the erosion of power. Force can crush opposition for only so long before its excessive use generates hatred for those in power, and long-term reliance on co-optation alone is both unreliable and costly. Thus, in Pareto's view, leaders will overcentralize decision making to combat the erosion of power that occurs when effective sovereignty has been compromised by a decentralized administration and an overreliance on co-optation. Conversely, leaders will decentralize the system of political organization to combat the erosion of power that occurs when resistance to excessive reliance on force escalates.

13. The greater is the tendency toward administrative decentralization, the greater is the reliance by centers of power on co-optation as an instrument of social control, and the more likely centers of power are to extend protection to special interests or to parcel out administrative control to these special interests.

14. The greater is the reliance by centers of power on co-optation as an instrument of social control, and the more administrative action is limited by commitments to special interests or parceled out to these special interests, the greater will be the erosion of power, and, hence, the more problematic will coordination and control become.

15. The more problematic coordination and control become during periods of administrative decentralization, the greater will be the probability of long-term administrative centralization.

16. The greater is the tendency toward administrative centralization, the greater is the reliance on force as an instrument of social control, and the more likely those in power are to constrain productive activities that they do not directly control, thereby increasing political resistance to the regime.

17. The greater is the reliance on force as an instrument of social control, and the greater is the level of political resistance, the greater is the erosion of power and the more problematic coordination and control will become.

18. The more problematic coordination and control become during a period of administrative centralization, the greater is the probability of long-term administrative decentralization.

Pareto's General Sociology:
The Equilibrium of Sentiment, Economics, and Politics

Pareto offers a theoretically sophisticated analysis predicated on a model of a complex, moving, systemic equilibrium. Each element of the system is connected to every other, so that what happens in one part of a system can have far-reaching repercussions throughout the system as a whole. Pareto's "general sociology" examines how progressive changes in social values, in the economy, and in political organization determine the overall character of a society by affecting each other and keeping each other in check. These equilibrating functions connect his elementary theories of sentiment, economy, and politics into a unified theory of social organization.

Principles Linking Sentiment and the Economy Pareto was a famous economist, but left conventional economics to seek a deeper understanding of the connections between the business cycle and other aspects of a social system. Pareto felt that changes in public sentiment exert enormous influence on the economy because of their impact on competing proclivities to save or consume. Conversely, the business cycle influences changes in popular sentiment because economic trends can either encourage optimism or, conversely, induce concerns about the future and aversion to taking risks.

19. The greater is the rate of economic growth, the more likely are members of a society to feel that pursuit of self-gratification is appropriate, causing a shift in sentiments toward more equivocal norms.
20. The more members of a society pursue self-gratification and encourage equivocal norms, the greater are their propensities to spend and consume, leading to the accumulation of long-term consumer debt.
21. The greater is the accumulation of long-term consumer debt, the deeper the economic downturn will be following a period of economic expansion.
22. The deeper is the economic downturn following a period of economic expansion, the less likely people are to feel that pursuit of self-gratification is acceptable, causing a shift in sentiments toward more unequivocal norms.
23. The greater are the constraints on self-gratification and the more unequivocal norms are, the greater are the propensities to be frugal and save, thereby causing an increase of aggregate savings.
24. The greater is the accumulation of aggregate savings, the greater the impetus is for economic recovery.

Principles Linking Sentiment and Political Organization Pareto argued that changes in social sentiment are tied to cyclical changes in political organization, just as they are to periodic movements in the business cycle. During periods of political decentralization, the tendency to placate special interests promotes hedonistic values, while encouraging special interests to seek control over resources and to free themselves from the oversight of governmental agencies. These tendencies erode the effective power of the state, giving rise to a period of administrative centralization whereby the agencies of government attempt to consolidate and reassert their power. With centralization of political control, governmental agencies become less responsive, while social values become more conservative. When political control becomes highly centralized and forceful, the regime is likely to stifle those activities that it does not control directly, and as a result, the vitality of institutions and other domains of activity is diminished. As this dynamism of institutions and activities is constrained, resentments among members of a society increase and, thereby, erode the power wielded by political leaders.

25. The greater is the erosion of power following a period of administrative centralization, the greater is the tendency toward administrative decentralization during which a regime becomes more responsive to demands in general and more apt to co-opt special interests by acquiescing to their demands.

26. The more decentralized is a system of political authority and the more it yields to demands and placates special interests, the more hedonistic social values will become.

27. The more hedonistic are social values during a period of administrative decentralization, the more likely are special interests to exert control over resources and to try to free themselves from effective control by the regime, thereby limiting the sovereignty of political authority and eroding the power of the state.

28. The greater is the erosion of power following a period of administrative decentralization, the greater is the tendency toward administrative centralization during which a regime becomes less responsive to demands and more apt to employ force to exert control.

29. The more centralized is a system of political authority, and the less it yields to demands or placates special interests outside of its hegemonic circle, the more conservative will social values become.

30. The more conservative social values are during a period of administrative centralization and hegemonic repression, the less dynamic are institutional activities and the greater are resentments against constraints, thereby eroding the effective power of the state.

Principles Linking Economic Productivity and Political Organization

As a socioeconomist, Pareto was intrigued by the connections between economic and political cycles. In this respect, his thinking seems remarkably contemporary. In essence, he felt that too much government regulation undermines the dynamism of the economy by stifling business, but he recognized that too little regulation of business undermines the vitality of a society by allowing business interests to divert the resources of the society for their own personal profit while externalizing many of the actual costs of doing business and making profits on the rest of the society. Pareto also felt that economic growth creates an infrastructure that imparts meaningful power to governing agencies, but he believed that growth strains the capacity of government agencies to exert lasting and effective coordination and control.

31. The longer and more pronounced a period of administrative centralization is, the more likely governmental agencies are to obstruct business activities outside the direct centers of power and authority, thereby undermining the dynamism of the economy.

32. The more governmental agencies obstruct business activities and undermine the dynamism of the economy, the more likely popular discontent aimed at the government is to mount and erode effective administrative power.

33. The greater anti-government sentiment is and the greater the erosion of effective administrative power is following periods of administrative centralization, the more likely a political regime is to embark on a program of administrative decentralization.

34. The longer and more pronounced a period of administrative decentralization is, the more businesses and other interests are free to exploit others while imposing many of the real costs of doing business and making profits on the rest of the society.

35. The more exploitive business interests become and the more costs they externalize to the rest of the society, the greater is the number of people who will come to see the government as ineffectual and, hence, the more likely is the erosion of government power.

36. The more people who view the government as ineffectual and the more pronounced the erosion of government power becomes following periods of administrative decentralization, the more likely is a political regime to embark on a program of administrative centralization.

As these propositions reveal, Pareto's work goes beyond offering a description of cyclical change and presents us with a testable theory to explain why change occurs. He offers theoretical explanations that are sufficiently universal to be applied to any society in any period, yet specific enough to encourage operationalization and measurement. In the process, Pareto's work reveals penetrating insights about the systemic ways in which different aspects of society influence one another, giving rise to equilibrated change in socioeconomic and political developments over time.

22

*

The Origin and Context of George Herbert Mead's Thought

In the early decades of this century, sociological theorists understood very little about the micro processes of interaction that connect individuals to the macrostructural dimensions of society. How are society and the individual related? How do individual acts and social structure influence each other? How do societies reproduce themselves through the acts and interactions of individuals? How does society shape people's thoughts and behaviors? These and many related questions remained poorly conceptualized, as can be seen by examining Max Weber's crude categories of action or Émile Durkheim's imprecise attempts to link society, consciousness, ritual, and solidarity. In North America, these questions were given their first definitive answer by a quiet and unassuming philosopher, George Herbert Mead, who made the critical conceptual breakthrough.

Mead's breakthrough was not a blazing insight but, rather, a synthesis of existing concepts into a new perspective, which unlocked the mysteries of how humans interact and, as a consequence, reproduce social structures. Mead did not consider his synthesizing highly original, but as his many students could see and as modern theory now fully appreciates, his work was seminal and changed the course of sociological theorizing. How, then, did such an unpretentious scholar produce such a conceptual breakthrough? The answer resides in Mead's obvious native genius as it was conditioned by his personal experiences and by the prominent intellectual figures who introduced him to crucial ideas.

BIOGRAPHICAL INFLUENCES ON MEAD

George Herbert Mead was born in South Hadley, Massachusetts, in 1863.[1] His father was a minister in a long line of Puritan farmers and clergymen. His mother, who eventually became president of Mount Holyoke College, came from a background similar to her husband's, although perhaps with a more intellectual than religious bent. In 1870, the family moved to Ohio, where Mead's father assumed a position at Oberlin College as a chair of homiletics, or the art of preaching. In 1881, Mead's father died, forcing his mother to sell their house and move into rented rooms. To make ends meet, Mead's mother taught at the college, while he waited on tables to support himself as a student at Oberlin. In 1883, Mead graduated from Oberlin with a major in philosophy, and for the next four years, he appeared to be at loose ends. He taught school for a while, but for the most part, he tutored students and worked as a surveyor on railroad construction in the Northwest. During this period he read voraciously; as was evident throughout his career, he was a broadly read intellectual.

In 1887, Mead decided to enroll in Harvard and pursue further study in philosophy. Here he was exposed to a fuller range of ideas than at Oberlin, which at the time was still a religious school, despite its history of involvement in progressive social affairs. Mead's reading of Charles Darwin brought his growing disenchantment with his father's religion to its culmination, but more important, the substance of Darwin's work had considerable impact on his philosophy and social psychology, as we will explore shortly. Moreover, he read and studied Adam Smith, whose utilitarian position remained an implicit theme in Mead's theorizing. Perhaps an even more important influence was his direct contact with the Harvard psychologist, William James, whose pragmatic philosophy also became a prominent theme in Mead's work.

As was often the custom at this time, Mead went abroad to Leipzig, Germany in his second year of graduate study. There he became familiar with the laboratory work of Wilhelm Wundt, whose psychological experiments and theorizing further moved Mead's philosophical interests toward social psychology. Subsequently, Mead went in 1889 to Berlin, where, Lewis Coser speculates, he might have listened to lectures by Georg Simmel and come to appreciate more completely the importance of status positions and roles in the dynamics of interaction.

In 1891, Mead married and became an instructor of philosophy at the University of Michigan. Here he encountered Charles Horton Cooley and John Dewey, both of whom provided him with critical concepts in his eventual theoretical synthesis. But he did not stay long; in 1894, he followed his friend and colleague Dewey to the new and ambitious University of Chicago. Mead remained there until his death in 1931.

At Chicago, Mead always remained in the shadow of the charismatic Dewey. More significantly, he had great difficulty writing and publishing, and as a result, much of his most important work comes to us as transcriptions of his lectures. These lectures exerted considerable influence among students at Chicago, and the recognition that something important and revolutionary was being said led students to take virtually verbatim notes. In his lifetime, however, Mead never perceived that he had achieved a great theoretical synthesis, one that was to become the conceptual base on which all subsequent theorizing about social interaction would be laid. Indeed, although he was an active and confident man who was very much involved in local efforts at social reform, he saw himself in very modest terms as an intellectual. As we will describe in detail in the next chapter, his modesty was misplaced because he stands as one of the giants of sociological theory.

Before exploring the details of Mead's theory, however, we should pause and further examine the influences on his work. In particular, we will initially explore the various schools of thought, such as utilitarianism, pragmatism, behaviorism, and Darwinism, that shaped his thinking; then we will turn to the key individuals in his intellectual biography and examine the specific concepts he drew from such scholars as Wundt, James, Cooley, and Dewey.

MEAD'S SYNTHESIS
OF SCHOOLS OF THOUGHT

As a philosopher, Mead was attuned to basic philosophical issues and to currents in many diverse intellectual arenas. His broader philosophical scheme reflects this, but, even more significant, his seminal theoretical synthesis on social psychology also pulls together the general metaphors contained within four dominant intellectual perspectives of his time: (1) utilitarianism, (2) Darwinism, (3) pragmatism, and (4) behaviorism.[2]

Utilitarianism

In England during the eighteenth and nineteenth centuries, the economic doctrine that became known as utilitarianism dominated social thought.[3] Mead had read such prominent thinkers as Adam Smith, David Ricardo, John Stuart Mill, Jeremy Bentham, and, to the extent that he can be classified as a utilitarian, Thomas Malthus. Mead absorbed several key ideas from utilitarian doctrines.

First, utilitarians saw human action as being carried out by self-interested actors seeking to maximize their "utility" or benefit in free and openly competitive marketplaces. Although this idea was expressed somewhat differently by various advocates of utilitarianism, Mead appears to have found useful the emphases on (1) actors as seeking rewards, (2) actors as attempting to adjust to a competitive situation, and (3) actors as goal directed and instrumental in their behaviors. Later versions of utilitarianism stressed "pleasure" and "pain" principles, which captured the essence of the behaviorism that emerged in Mead's time and exert considerable influence on his theoretical scheme.

Second, utilitarians often tended to emphasize—indeed, to overemphasize—the rationality of self-seeking actors. From a utilitarian perspective, actors are rational in that they gather all relevant information, weigh various lines of conduct, and select an alternative that will maximize utilities, benefits, or pleasures. Mead never accepted this overly rational view of human action, but his view of the human "mind" as a process of reflective thought in which alternatives are covertly designated, weighed, and rehearsed was, no doubt, partially inspired by the utilitarian position.

Thus, although utilitarianism only marginally influenced Mead, his theoretical scheme corresponded to several of its central points. He probably borrowed these points both directly and indirectly because utilitarianism influenced the other schools of thought that more directly shaped Mead's philosophical scheme. As we will come to see, early behaviorism and pragmatism, although rejecting extreme utilitarianism, nonetheless incorporated some of its basic tenets.

Darwinism

Darwin's formulation of the theory of evolution influenced not only biological theory[4] but also social thought.[5] The view that a species' profile is shaped by the competitive struggle with other species attempting to occupy an environmental niche was highly compatible with utilitarian notions. As a result of this superficial compatibility, Darwinism was carried to absurd extremes in the late nineteenth and early twentieth centuries by a group of thinkers who became known as Social Darwinists.[6] From their viewpoint, social life is a competitive struggle in which the "fittest" will be the best able to "survive" and prosper.[7] Hence, those who enjoy privilege in a society deserve these benefits because they are the "most fit," whereas those who have the least wealth are less fit and worthy. Obviously, Social Darwinism was a gross distortion of the theory of evolution, but its flowering illustrates the extent to which Darwin's theory represented an intellectual bombshell in Europe and America.

Other social theorists borrowed Darwin's ideas more cautiously. Mead used the theory of evolution as a broad metaphor for understanding the processes by which the unique capacities of humans emerge. Mead believed all animals, including humans, must seek to adapt and adjust to an environment; hence, many attributes that an organism reveals are the products of its efforts to adapt to a particular environment. In the distant past, therefore, the unique capacities of humans for language, for mind, for self, and for normatively regulated social organization emerged as a result of selective pressures on the ancestors of humans for these unique capacities.

Mead was not as much interested in the origins of humans as a species as in the infant human's development from an asocial to a social creature. At birth, Mead argued, an infant is not a human. It acquires the unique behavioral capacities of humans only as it adapts to a social environment. Thus, just as the species as a whole acquired its distinctive characteristics through a process of "natural selection," so the infant organism develops its "humanness" through a process of "selection." Because the environment of a person is other people who use language, who possess mind and self, and who live in society, the young must adapt to this environment if they are to survive. As they adapt and adjust, they acquire the capacity to use language, to reveal a mind, to evidence a sense of self, and to participate in society. Thus, Mead borrowed from Darwinian theory the metaphor of adaptation, or adjustment, as the key force shaping the nature of humans.[8] This metaphor was given its most forceful expression in the works of scholars who developed a school of thought known as pragmatism.

Pragmatism

Mead is frequently grouped with pragmatists such as Charles Peirce, William James, and John Dewey. Yet, although Mead was profoundly influenced by James and Dewey, his theoretical scheme is only partially in debt to pragmatism.[9]

Peirce, an American scientist and philosopher, first developed the ideas behind pragmatism in an article titled "How to Make Our Ideas Clear," which appeared in *Popular Science Monthly* in 1878.[10] Pragmatism did not become an acknowledged philosophical school, however, until James delivered a lecture in 1898 titled "Philosophical Conceptions and Practical Results."[11] As Dewey developed his "instrumentalism," pragmatism became a center of philosophical controversy in the United States during the early decades of this century. Pragmatism was primarily concerned with the process of thinking and how it influenced the action of individuals, and vice versa. Although pragmatists carried their banner in different directions, the central thrust of this philosophical school is to view thought as a process that allows humans to adjust, adapt, and achieve goals in their environment.

Thus, pragmatists became concerned with symbols, language, and rational thinking as well as with the way action in the world is influenced by humans' mental capacities. Peirce saw pragmatism as concerned with "self-controlled conduct," which was guided by "adequate deliberation," and hence, pragmatism was based on

> A study of that experience of the phenomena of self-control which is common to all grown men and women; and it seems evident that to some extent, at least, it must always be so based. For it is to conceptions of deliberate conduct that pragmatism would trace the intellectual purpose of symbols; and deliberate conduct is self-controlled conduct.[12]

For Peirce, then, pragmatism stresses the use of symbols and signs in thought and self-control, a point that Mead later adopted. James and Dewey supplemented Peirce's emphasis by stressing that the process of thinking was intimately connected to the process of adaptation and adjustment. James stressed that "truth" was not absolute and enduring; rather, he argued, scientific as well as lay conceptions of truth are only as enduring as their ability to help people adjust and adapt to their circumstances.[13] Truth, in other words, is determined only by its "practical results." Dewey similarly emphasized the significance of thinking for achieving goals and adjusting to the environment. Thought, whether lay or scientific, is an "instrument" that can be used to achieve goals and purposes.[14]

Pragmatism emerged as a reaction to, and an effort to deal with, a number of scientific and philosophical events of the nineteenth century. First, the ascendance of Newtonian mechanics posed the question of whether all aspects of the universe, including human thought and action, could be reduced to invariant and mechanistic laws. To this challenge, pragmatists argued that such laws did not make human action mechanistic and wholly determinative but, rather, that these laws were instruments to be used by humans in achieving goals.[15] Second, the theory of evolution offered the vision of continuity in life processes. Pragmatists added the notion that humans as a species, and as individuals, were engaged in a process of constant adjustment and adaptation to their environment and that thought represented the principal means of achieving such adjustment. Third, the doctrines of utilitarians presented a calculating, rational, and instrumental view of human action. Pragmatists were highly receptive to this perspective, although their concern was with the process of thought and how it is linked to action. Fourth, the ascendance of the scientific method with its emphasis on the verification of conceptual schemes through experienced data presented a consensual view of "proper" modes of investigation. To this point, the pragmatists responded that all action involved an act of verification as people's thoughts and conceptions were "checked" against their experiences in the world. For the pragmatist, human life is a continuous application of the "scientific method" as people seek to cope with the world around them.[16]

Pragmatism thus represents the first distinctly American philosophical system. Mead was personally associated with several of its advocates while being intellectually involved in the debate surrounding the system and its critics. He was clearly influenced by the pragmatists' concern with the process of thinking and with the importance of symbols in thought. He accepted the metaphor that thought and action involved efforts to adjust and adapt to the environment. He embraced the notion that such adaptation involved a continuous process of experiential verification of thought and action. In many ways, utilitarianism and Darwinism came to Mead through pragmatism, and hence to some extent, he must be considered a pragmatist. He was also a behaviorist, and if his social psychology is to be given a label, it is more behavioristic than pragmatic.

Behaviorism

As a psychological perspective, behaviorism began from insights of the Russian physiologist Ivan Petrovich Pavlov (1849–1936), who discovered that experimental dogs associated food with the person bringing the food.[17] He observed, for instance, that dogs would secrete saliva not only when presented with food but also when they heard their feeder's footsteps approaching. After considerable delay and personal agonizing,[18] he undertook a series of experiments on animals to understand such "conditioned responses." From these experiments, he developed several principles that were later incorporated into behaviorism. These include

1. A stimulus consistently associated with another stimulus producing a given physiological response will, by itself, elicit that response.

2. Such conditioned responses can be extinguished when gratifications associated with stimuli are no longer forthcoming.

3. Stimuli that are similar to those producing a conditioned response can also elicit the same response as the original stimulus.

4. Stimuli that increasingly differ from those used to condition a particular response will decreasingly be able to elicit this response.

Thus, Pavlov's experiments exposed the principles of conditioned responses, extinction, response generalization, and response discrimination. Although he clearly recognized the significance of these findings for human behavior, his insights were rediscovered in North America by Edward Lee Thorndike and John B. Watson—the founders of behaviorism.[19]

Thorndike conducted the first laboratory experiments on animals in North America. He observed that animals would retain response patterns for which they were rewarded.[20] For example, in experiments on kittens placed in a puzzle box, he found that they would engage in trial-and-error behavior until they emitted the response allowing them to escape. With each placement in the box, the kittens would engage in less trial-and-error behavior, indicating that the gratifications associated with a response allowing them to escape caused them to learn and retain this response. From these and other studies, which were conducted at the same time as Pavlov's, Thorndike formulated three principles, or laws: (1) the "law of effect" holds that acts in a situation producing gratification will be more likely to occur in the future when that situation recurs; (2) the "law of use" states that the situation-response connection is strengthened with repetition and practice; and (3) the "law of disuse" argues that the connection will weaken when the practice is discontinued.[21]

These laws overlap with those presented by Pavlov, but there is one important difference. Thorndike's experiments were conducted on animals engaged in free trial-and-error behavior, whereas Pavlov's work was on the conditioning of physiological—typically glandular—responses in a tightly controlled laboratory situation. Thorndike's work could thus be seen as more directly relevant to human behavior in natural settings.

Watson was only one of several thinkers to recognize the significance of Pavlov's and Thorndike's work,[22] but he soon became the dominant advocate of what was becoming explicitly known as "behaviorism." His opening shot for the new science of behavior was fired in an article titled "Psychology as the Behaviorist Views It":

> Psychology as the behaviorist views it is a purely objective experimental branch of natural science. Its theoretical goal is the prediction and control of behavior. Introspection forms no essential part of its methods, nor is the scientific value of its data dependent upon the readiness with which they lend themselves to interpretation in terms of consciousness. The behaviorist, in efforts to get a unitary scheme of animal response, recognizes no dividing line between man and brute.[23]

Watson thus became the advocate of the extreme behaviorism against which Mead so vehemently reacted.[24] For Watson, psychology is the study of stimulus-response relations, and the only admissible evidence is overt behavior. Psychologists are to stay out of the "mystery box" of human consciousness and to study only observable behaviors as they are connected to observable stimuli. Mead rejected this assertion and argued that just because an activity such as thinking was not directly observable did not mean that it was not behavior. Mead argued that covert thinking and the capacity to view oneself in situations are nonetheless behaviors and hence subject to the same laws as overt behaviors.

Mead thus rejected extreme behaviorism but accepted its general principle: Behaviors are learned as a result of gratifications associated with them. In accordance with the views of pragmatists, and consistent with Mead's Darwinian metaphor, the gratifications of humans typically involve adjustment to a social environment. Most important, some of the most distinctive behaviors of humans are covert, involving thinking, reflection, and self-awareness. In contrast with Watson's behaviorism, Mead postulated what some have called a *social behaviorism*. From this perspective, both covert and overt behaviors are to be understood through their capacity to produce adjustment to society.

In sum, we can conclude that Mead borrowed the broad assumptions from a number of intellectual perspectives, particularly utilitarianism, Darwinism, pragmatism, and behaviorism. Utilitarians and pragmatists emphasized the process of thinking and rational conduct; utilitarians and Darwinists stressed the importance of competitive struggle and selection of attributes; Darwinists and pragmatists argued for the importance of adaptation and adjustment to an understanding of thought and action; and behaviorists presented a view of learning as the association of behaviors with gratification-producing stimuli. Each of these general ideas became a part of Mead's theoretical scheme, but as he synthesized these ideas, they took on new meaning.

Mead was not only influenced by these general intellectual perspectives, he also borrowed specific concepts from a variety of scholars, only some of whom worked within these general perspectives. By taking specific concepts, reconciling them, and then incorporating them into the metaphors of these four general perspectives, he was able to produce the theoretical breakthrough for which he is deservedly given credit.

WILHELM WUNDT AND MEAD

Even a casual reading of Mead's written work and posthumously published lectures reveals numerous citations to the German psychologist Wilhelm Wundt.[25] Wundt is often given credit for being the father of psychology because by the 1860s he had conducted a series of experiments that could be clearly defined as psychological in nature. In the 1870s, he was one of the first, along with James, to establish a psychological laboratory.

Mead studied briefly in Germany, although not in Heidelberg, where Wundt had established his laboratory and school of loyal followers. Wundt's eminence prompted Mead to read his works carefully.[26] At first glance, it might appear that Mead, the philosopher, would find little of interest in Wundt's voluminous output. Most of Wundt's laboratory work deals with efforts to understand the structure of consciousness, an emphasis not conducive to Mead's insistence on mind as a process. However, Wundt was also a philosopher, social psychologist, and sociologist. Although he would write strictly psychological books like *Physiological Psychology,* he also founded the journal *Philosophical Studies,* in which he published his laboratory studies. Indeed, he saw little reason to distinguish psychology from philosophy. He also devoted many pages in his *Outlines of Psychology* to such issues as gestures, language, self-consciousness, mental communities, customs, myths, and child development, all topics likely to interest Mead. Moreover, Wundt's *Elements of Folk Psychology* was one of the first distinctly social psychological studies, examining the broad evolutionary development of human thought and culture. Thus, he was a scholar of great range and enormous productive energy. Mead would apparently find much in the work of Wundt to stimulate his own thought.

Wundt's View of Gestures

In much of his work, Mead devoted considerable space to Wundt's view of "gestures" and "speech."[27] Mead argued that Wundt had been the first to recognize that gestures represented signs marking the course of ongoing action and that animals used these signs as ways of adjusting to one another. Human language, Wundt argued, represents only an extension of this basic process in lower animals because common and consensual meanings were, over the course of human evolution, given to signs. As human mental capacities grew, such gestures could be used for deliberate communication and interaction.

All these points, although greatly distorted by Wundt's poor ethnographic accounts, were incorporated in altered form in Mead's scheme.[28] Gestures were viewed as the basis for communication and interaction, and language was defined as gestures that carry common meanings. As Wundt implied, humans are unique creatures, and society is possible only by virtue of language and its use to create customs, myths, and other symbol systems.

Wundt's View of "Mental Communities"

Mead did not give Wundt credit for inspiring a more sociological vision of gestures and language. Yet sprinkled throughout Wundt's work are ideas that bear considerable resemblance to those developed by Mead. One such idea is what Wundt termed the *mental community.*[29] Wundt saw the development of speech, self-consciousness, and mental activity in children as emerging from interaction with the social environment. Such interaction, he argued, makes possible identification with a mental community that guides and directs human action and interaction in ways functionally analogous to the regulation of lower animals by instincts. Such mental communities can vary in their nature and extensiveness, producing great variations in human action and patterns of social organization. Just as Durkheim in France had emphasized the significance of the collective conscious, Wundt saw humans as regulated by a variety of mental communities. Mead was, we suspect, to translate this notion of mental community into his vision of "generalized others" or "communities of attitudes" that regulate human action and organization.

In sum, then, Mead appears to have taken from Wundt two critical points. First, interaction is a process of gestural communication, with language being a more developed form of such communication. Second, social organization is more than a process of interaction among people; it is also a process of socialization in which humans acquire the ability to create and use mental communities to regulate their actions and interaction. As we will see, these two points are at the core of Mead's theory of mind, self, and society.

JAMES AND MEAD

By 1890, James was the most prominent psychologist in America, attracting students and worldwide attention. He was also a philosopher who, along with Dewey, became the foremost advocate of pragmatism. Mead borrowed from both James's philosophy and his psychology, incorporating the general thrust of James's philosophy and his specific views on consciousness and self-consciousness.

James's Pragmatism

In many ways James was the most extreme of the pragmatists, advocating that there was no such thing as "absolute truth."[30] Truth is temporary and lasts only as long as it works—that is, only as long as it allows adjustment and adaptation to the environment. James thus rejected the notion that truth involved a search for isomorphism between theoretical principles and empirical reality and that science represented an effort to increase the degree of isomorphism. For James, theories were merely "instruments" to be used for a time in an effort to facilitate adjustment. Hence, objective, permanent, and enduring truth cannot be found.

Mead never completely accepted this extreme position. Indeed, much of his work was directed at discovering some of the fundamental principles describing the basic relationship between individuals and society. He did, however, accept and embrace the pragmatic notion that human life was a constant process of adjustment and that the faculty for consciousness was the key to understanding the nature of this adjustment.

James's View of Consciousness

James defined psychology as the "science of mental life."[31] As a science, psychology aims to understand the nature of mental processes—that is, the nature of "feelings, desires, cognitions, reasons, decisions, and the like."[32] His classic

1890 text, *The Principles of Psychology,* became the most important work in American psychology at the time because it sought to summarize what was then known about mental life. It also contained James's interpretation of mental phenomena, and by far the most important of these interpretations was his conceptualization of consciousness as a process. For James, consciousness is a "stream" and "flow," rather than a structure of elements, as Wundt had proposed.[33] Thus, "mind" is simply a process of thinking, and with this simple fact psychological investigation must begin: "The only thing which psychology has a right to postulate at the outset is the fact of thinking itself, and that must be taken up and analyzed."[34]

James then went on to list five characteristics of thought: (1) thought is personal and always, to some degree, idiosyncratic to each individual; (2) thought is always changing; (3) thought is continuous; (4) thought to the individual appears to deal with objects in an external world; and (5) thought is selective and focuses on some objects to the exclusion of others.[35] Of these characteristics, Mead appears to have been most influenced by the last two. For him, mind is to be seen as a process of selectively denoting objects and of responding to these objects. Although the details of his conceptualization of thinking reflect Dewey's influence more than that of James, this early discussion by James probably shaped Mead's emphasis on selective perception of objects in the environment.

Far more influential on Mead's thought than James's view of consciousness in general was his conceptualization of self-consciousness.[36] Here Mead borrowed much and was directly influenced by James's recognition that one object in the flow of consciousness is oneself.

James's View of Self-Consciousness

James's examination of self began with the assertion that people recognized themselves as objects in empirical situations. He called this process the *empirical self,* or *me*—the latter term being adopted by Mead in his examination of self-images. James went on to describe various types of empirical selves that all people have: (1) the material self, (2) the social self, and (3) the spiritual self. Moreover, James saw each type, or aspect, of self as involving two dimensions: (1) self-feelings (emotions about oneself) and (2) self-seekings (actions prompted by each self). There are types of selves, revealing variations with respect to self-feelings and self-seeking, and there is a hierarchy among the various selves. Thus, James offered an elaborate taxonomy of self-related processes, and although Mead's own conceptualization was sparse by comparison, he selectively borrowed from the entire scheme.

Types of Empirical Selves For James, the material self embraces people's conceptions of their bodies as well as their other possessions because one's actual body and possessions both evoke similar feelings and actions. The social self is, in reality, a series of selves that people have in different situations. Thus, one can have somewhat different self-feelings and action tendencies depending on the type of social situation—whether work, family, club, or community. For Mead, this vision of a social self became most important, for people's self-feelings and actions are, he argued, most influenced by their conception of themselves in various social gatherings. James did not clearly describe the spiritual self, but it appears to embody those most intimate feelings people have about themselves—that is, their worth, talents, strengths, and failings. In *The Principles of Psychology,* James summarized his conceptualization of empirical selves and their constituent dimensions with a table, which is shown in Table 22.1.[37]

The Hierarchy of Empirical Selves James felt that some aspects of different empirical selves were more important than others. As he noted,

> A tolerably unanimous opinion ranges the different selves of which a
> man may be "seized and possessed," and the consequent different orders
> of his self-regard, in an *hierarchical scale, with the bodily Self at the bottom, the*
> *spiritual Self at Top, and the extracorporeal material selves and the various social*
> *selves between.*[38]

Thus, some degree of unity among a person's selves is achieved through their hierarchical ordering, with self-feelings and action tendencies being greatest for those selves high in the hierarchy. A further source of unity comes from the nonempirical self, or what James termed the *pure ego.*

The Pure Ego and Personal Identity Above these empirical selves, James argued, is a unity. People have "a personal identity," or *pure ego,* in that they have a sense of continuity and stability about themselves as objects. Humans take their somewhat diverse empirical selves and integrate them, seeing in them continuity and sameness.[39] Mead was probably greatly influenced by this conception of a stable and unified self-conception. As he argued, humans develop over time, from their experiences in the empirical world, a more "unified" or "complete" self—that is, a stable self-conception. This stable self-conception, Mead emphasized, gives individuals a sense of personal continuity and gives their actions in society a degree of stability and predictability.

In sum, then, Mead's view of self as one of the distinctive features of humans was greatly influenced by James's work. James was not as concerned as Mead was with understanding the emergence of self or its consequences for the social order. But James provided Mead with several critical insights about the nature of self: (1) self is a process of seeing oneself as an object in the stream of conscious awareness; (2) self varies from one empirical situation to another; yet (3) self also reveals unity and stability across situations. Mead never adopted James's taxonomy, but he took the broad contours of James's outline and demonstrated their significance for understanding the nature of human action, interaction, and organization.

Table 22.1   **James's Conceptualization of Empirical Selves**

Dimensions	Types of Empirical Self		
	Material	*Social*	*Spiritual*
Self-seeking	Bodily appetites and instincts	Desire to please, be noticed, admired, and so on	Intellectual, moral and religious aspiration, conscientiousness
	Love of adornment, foppery, acquisitiveness, constructiveness	Sociability, emulation, envy, love, pursuit of honor, ambition, and so on	
	Love of home, and so on		
Self-estimation	Personal vanity, modesty, and so on	Social and family pride, vainglory, snobbery, humility, shame, and so on	Sense of moral or mental superiority, purity, and so on
	Pride of wealth, fear of poverty		Sense of inferiority or of guilt

COOLEY AND MEAD

Mead and Cooley were contemporaries and, as we noted, colleagues in their early careers at the University of Michigan. Their direct interaction was, no doubt, significant, but Cooley's influence extended beyond their period of collegial contact. Indeed, Mead adopted from Cooley several critical insights into the origins and nature of self as well as its significance for social organization.

Cooley's sociology is often vague, excessively mentalistic, and highly moralistic. Yet we can observe several lines of influence on Mead in Cooley's recognition that (1) society is constructed from reciprocal interaction, (2) interaction occurs through the exchange of gestures, (3) self is created from, and allows the maintenance of, patterns of social organization, and (4) social organization is possible by virtue of people's attachment to groups that link them to the larger institutions of society. We should therefore examine these lines of influence in more detail.

Cooley's View of Social Organization

Cooley held the view that society was an organic whole in which specific social processes worked to create, maintain, and change networks of reciprocal activity.[40] Much as Mead later argued, Cooley saw the "vast tissue" of society as constructed from diverse social forms, from small groups to large-scale social institutions. The cement holding these diverse forms together is the capacity of humans to interact and share ideas and conceptions. Such interaction depends on the unique capacities of humans to use gestures and language.

Cooley's View of Interaction

Cooley saw the human ability to assign common meanings and interpretations to gestures—whether words, bodily countenance, facial expressions, or other gestural emissions—as the central mechanism of interaction. In this way humans can communicate, and from this communication they establish social relations: "By communication is here meant the mechanism through which human relations exist and develop—all the symbols of the mind, together with the means of conveying them through space and preserving them in time."[41]

Mead accepted Cooley's view of social organization as constructed gestural communication. More important, however, Cooley helped Mead understand how gestural communication led to interaction and organization. By reading one another's gestures, people are able to see and interpret the dispositions of others. Hence: "Society is an interweaving and interworking of mental selves. I imagine your mind. . . . I dress my mind before yours and expect that you will dress yours before mine."[42]

Mead took this somewhat vague idea and translated it into an explicit view of interaction and social organization as a process of reading gestures, placing oneself mentally into the position of others, and adjusting conduct so as to cooperate with others. Moreover, he accepted Cooley's recognition that self was the critical link in the creation and maintenance of society from patterns of reciprocal communication and interaction.

Cooley's View of Self

Cooley emphasized the human capacity for self-consciousness. This capacity emerges from interaction with others in groups, and once this capacity exists, it allows people to organize themselves into society. Mead adopted the general thrust of this argument, although he made it considerably more explicit and coherent. He appears to have taken three distinct elements from Cooley's somewhat vague and rambling discussion: (1) self as constructed from the *looking glass* of other people's gestures, (2) self as emerging from interaction in groups, and (3) self as a basis for self-control and, hence, social organization.

The "Looking-Glass Self" Cooley adopted James's view of self as the ability to see and recognize oneself as an object. But he added a critical insight: Humans use the gestures of others to see themselves. The images people have of themselves are similar to reflections from a looking glass, or mirror; they are provided by the reactions of others to one's behavior. Thus, by reading the gestures of others, humans see themselves as an object:

> As we see our face, figure, and dress in the glass, and are interested in
> them because they are ours . . . so in imagination we perceive in
> another's mind some thought of our appearance, manners, aims, deeds,
> character, friends, and so on, and are variously affected by it.[43]

As people see themselves in the looking glass of other people's gestures, then, they (1) imagine their appearance in the eyes of others, (2) sense the judgment of others, and (3) have self-feelings about themselves. Thus, during the process of interaction, people develop self-consciousness and self-feelings. Although Cooley did not develop the idea in detail, he implied that humans developed, over time and through repeated glances in the looking glass, a more stable sense of self.

The Emergence of Self Cooley argued that the life history of an individual was evolutionary. Because their ability to read gestures is limited, infants cannot initially see themselves as objects in the looking glass. With time, practice, biological maturation, and exposure to varieties of others, children come to see themselves in the looking glass, and they develop feelings about themselves. Such a process, Cooley felt, is inevitable as long as the young must interact with others because as they act on their environment, others will react, and this reaction will be perceived.[44] Through this process, as it occurs during infancy, childhood, and adolescence, an individual's "personality" is formed. As Mead argued, the existence of a more stable set of self-feelings gives human action stability and predictability, thereby facilitating cooperation with others.

Self and Social Control Cooley saw self as only one aspect of consciousness in general. Thus, he divided consciousness into three aspects: (1) "self-consciousness," or self-awareness of, and feelings about, oneself; (2) "social consciousness," or a person's perceptions of and attitudes toward other people; and (3) "public consciousness," or an individual's view of others as organized in a "communicative group."[45] Cooley saw all three aspects of consciousness as "phases of a single whole."

Cooley never developed these ideas to any great degree, but Mead apparently saw much potential in these distinctions. For Mead, the capacity to see oneself as an object, to perceive the dispositions of others, and to assume the perspective of a broader "public" or "community" gave people a basis for stable action and cooperative interaction. Because of these capacities, then, society is possible.

Cooley's View of Primary Groups

Cooley argued that the most basic unit of society was the "primary group," which he defined as those associations characterized by "intimate face-to-face association and cooperation":

> They are primary in several senses but chiefly in that they are fundamental in forming the social nature and ideals of individuals. The result of intimate association, psychologically, is a certain fusion of individualities in a common whole, so that one's very self, for many purposes at least, is the common life and purpose of the group.[46]

Thus, the looking glass of gestures emitted by those in one's primary group are the most important in the emergence and maintenance of self-feelings.

Moreover, the link between individuals and the broader institutional structure of society is the primary group. Institutions cannot, Cooley stressed, be maintained unless past traditions and public morals are given immediate relevance to individuals through the intimacy characteristic of primary groups. Indeed, for Cooley, primary groups "are the springs of life, not only for the individual but for social institutions."[47]

Cooley's concept of the primary group appears to have influenced Mead in two ways. First, Mead retained Cooley's position that self emerges, in large part, by virtue of an individual's participation in face-to-face, organized activity. Second, Mead implicitly argued that one of the bridges between the individual and broader institutional structure of society was the small group, although Cooley's emphasis on this point was much greater than was Mead's.

Thus, Mead was enormously influenced by the work of Cooley. Mead saw the full implications of Cooley's ideas for understanding the nature of the relationship between the individual and society. As we will appreciate in the next chapter, Mead borrowed, extended, and integrated into a more coherent theory Cooley's views on gestures, interaction, self and its emergence, and social organization.

DEWEY AND MEAD

Dewey and Mead were initially young colleagues at the University of Michigan, and when Dewey moved to the new University of Chicago in 1894, he invited Mead to join him in the Department of Philosophy and Psychology. Mead and Dewey were thus colleagues until 1905, when Dewey left for Columbia University. They engaged in much dialogue; therefore, it is not surprising that their thoughts reveal many similarities. Yet as we emphasized earlier, Dewey was the intellectual star of Chicago, and he wrote in many diverse areas and generated much attention, inside and outside the academic world.[48] In contrast, the retiring Mead, who had great difficulty writing, was constantly in Dewey's shadow. Ironically, however, Mead made the more important, long-term intellectual contribution to sociology.

Mead accepted the broad contours of Dewey's pragmatism, but a more important influence on Mead's thought was Dewey's conceptualization of thought and thinking. Thus, Dewey extended his brilliance into philosophy, morals, education, methodology, the history of science, psychology, and virtually any area of inquiry that caught his interest.[49] Mead selectively borrowed several key ideas from Dewey's wide-ranging inquiries and incorporated them into a vision of what Mead termed *mind*.

Dewey's Pragmatism

Dewey's pragmatism attacked the traditional dualisms of philosophy: knower and known, objects and thought of objects, and mind and external world. He thought that this dualism was false, that the act of knowing and the "things" to be known were interdependent. Objects can exist in a world external to an individual, but their existence and properties are determined by the process of acting toward these objects. For Dewey, the basic process of human life consisted of organisms acting on, and making inquiry into, their environment. Indeed, the history of human thought has been a quest for greater certainty about the consequences of action. In a scheme that resembles Auguste Comte's law of the three stages, Dewey argued that religion had represented the primitive way to achieve certainty, that the classical world had classified experience to achieve certainty, and that the modern world sought to control nature through the discovery of its laws of operation.[50] With the development of modern science, Dewey argued, a new problem emerges: Moral values can no longer be legitimized by God or by appeals to the natural order.[51]

Such is the philosophical dilemma Dewey proposed and then resolved with his pragmatism. Values, morality, and other evaluational ideas are to be found in the action of people as they cope with their environment. Value is discovered and found as people try to adjust and adapt to a problematic situation. Thus, when humans do not know the morality of a situation, they will construct one and use it as an "instrument" to facilitate their adjustment to that situation. Value-oriented action is like all other thought and action: It emerges from people's acts in a problematic situation. Mead never absorbed the details of Dewey's "instrumentalism" or Dewey's almost frantic efforts to create the "good society,"[52] but he did adopt the view that thinking and thought arose from the process of dealing with problematic situations.

Dewey's View of Thinking

Dewey's pragmatism led him to view thinking as a process involving (1) blockage of impulses, (2) selective perception of the environment, (3) rehearsal of alternatives, (4) overt action, and (5) assessment of consequences. Then if a situation is still problematic, a new sequence is carried out until the problematic situation is eliminated.[53]

Mead adopted two aspects of this vision. First, he saw thinking as part of a larger process of action. Thinking occurs when an organism's impulses are blocked and when it is in maladjustment with its environment.[54] This argument became part of Mead's theory of motivation and "stages of the act." Second, thinking, for Mead, involves selective perception of objects, covert and imaginative rehearsal of alternatives, anticipation of the consequences of alternatives, and selection of a line of conduct. He termed these behavioral capacities *mind,* and they were clearly adapted from Dewey's discussion of human nature and conduct.[55]

MEAD'S SYNTHESIS

We can now appreciate the intellectual world as Mead encountered it. The convergence of utilitarianism, pragmatism, Darwinism, and behaviorism gave him a general set of assumptions for understanding human behavior. The specific concepts of Wundt, James, Cooley, and Dewey gave Mead the necessary intellectual tools to understand that humans were unique by virtue of their behavioral capacities for mind and self. Conversely, mind and self emerge from gestural interaction in society. Once they emerge, however, mind and self make a distinctive form of gestural interaction and an entirely revolutionary creation: symbolically regulated patterns of social organization. In broad strokes, such is the nature of Mead's synthesis. We can now examine the details of this synthesis in the next chapter. We should remember that Mead, like most great intellects, built his synthesis on "the shoulders of giants."

NOTES

1. As Lewis Coser notes, biographical materials on Mead are rather scarce, so much of the information in this chapter relies primarily on his excellent review in Lewis A. Coser, *Masters of Sociological Thought* (New York: Harcourt Brace Jovanovich, 1977).

2. See also Jonathan H. Turner, *The Structure of Sociological Theory,* 6th ed. (Belmont, CA: Wadsworth, 1998).

3. See Chapter 2 of this book.

4. Charles Darwin, *On the Origin of Species* (London: Murry, 1859).

5. See, for example, William G. Sumner, *What Social Classes Owe Each Other* (New York: Harper & Row, 1883).

6. Richard Hofstadter, *Social Darwinism in American Thought, 1860–1915* (Philadelphia: University of Pennsylvania Press, 1945).

7. Spencer first used the phrase "survival of the fittest," which apparently influenced Darwin, as he acknowledged in *On the Origin of Species.* Other early American sociologists, such as William Graham Sumner, took this idea to extremes. Spencer's utilitarianism was recessive in his sociological works, however, so it is unfair to count Spencer as a Social Darwinist.

8. Mead also reacted to Darwin's later efforts to understand emotions in animals. See Charles Darwin, *The Expression of Emotions in Man and Animals* (London: Murry, 1872). Mead used this analysis as his straw man in developing his own theory of gestures and interaction.

9. For relevant summaries of pragmatism, see Charles Morris, *The Pragmatic Movement in American Philosophy* (New York: George Braziller, 1970), and Edward C. Moore, *American Pragmatism: Peirce, James, and Dewey* (New York: Columbia University Press, 1961).

10. For Peirce's general works, see Charles Sanders Peirce, *The Collected Papers of Charles Sanders Peirce,* 8 vols. (Cambridge, MA: Harvard University Press, 1931–1958). Peirce's "How to Make Our Ideas Clear" is in vol. 5, pp. 248–271.

11. This lecture was delivered at Berkeley, California. See also William James, *Pragmatism* (Cambridge, MA: Harvard University Press, 1975).

12. Morris, *Pragmatic Movement,* p. 11.

13. William James, *The Meaning of Truth: A Sequel to "Pragmatism"* (New York: Longmans, Green, 1909).

14. John Dewey, *Human Nature and Conduct* (New York: Holt, Rinehart & Winston, 1922).

15. In particular, see John Dewey, *The Quest for Certainty* (New York: Minton, Balch, 1925), or James, *Meaning of Truth.*

16. See Morris, *Pragmatic Movement,* pp. 5–11.

17. See, for relevant articles, lectures, and references, I. P. Pavlov, *Selected Works,* ed. K. S. Kostoyants, trans. S. Belsky (Moscow: Foreign Languages Publishing House, 1955), and *Lectures on Conditioned Reflexes,* 3d ed., trans.

W. H. Gantt (New York: International, 1928).

18. I. P. Pavlov, "Autobiography," in *Selected Works,* pp. 41–44.

19. For an excellent summary of their ideas, see Robert I. Watson, *The Great Psychologists,* 3d ed. (Philadelphia: Lippincott, 1971), pp. 417–446.

20. Edward L. Thorndike, "Animal Intelligence: An Experimental Study of the Associative Processes in Animals," *Psychological Review Monograph,* Supplement 2 (1898).

21. See Edward L. Thorndike, *The Elements of Psychology* (New York: Seiler, 1905), *The Fundamentals of Learning* (New York: Teachers College Press, 1932), and *The Psychology of Wants, Interests, and Attitudes* (New York: Appleton-Century-Crofts, 1935).

22. The others included Max F. Meyer, *Psychology of the Other-One* (Columbus: Missouri Books, 1921), and Albert P. Weiss, *A Theoretical Basis of Human Behavior* (Columbus: Adams, 1925).

23. J. B. Watson, "Psychology as the Behaviorist Views It," *Psychological Review* 20 (1913), pp. 158–177. For other basic works by Watson, see *Psychology from the Standpoint of a Behaviorist,* 3rd ed. (Philadelphia: Lippincott, 1929), and *Behavior: An Introduction to Comparative Psychology* (New York: Holt, Rinehart & Winston, 1914).

24. For example, in his *Mind, Self, and Society* (Chicago: University of Chicago Press, 1934), Mead has eighteen references to Watson's work and is highly critical of the latter's extreme methodological position. Mead considered himself a behaviorist nonetheless. For further documentation of this conclusion, see Jonathan H. Turner, "A Note on G. H. Mead's Behavioristic Theory of Social Structure," *Journal for the Theory of Social Behavior* 12 (July 1982), pp. 213–222, and John D. Baldwin, *George Herbert Mead* (Beverly Hills, CA: Sage, 1986).

25. *Mind, Self, and Society* alone contains more than twenty references to Wundt's ideas.

26. For basic references on Wundt's work, see Wilhelm Wundt, *Principles of Physiological Psychology* (New York: Macmillan, 1904; originally published in 1874), *Lectures on Human and Animal Psychology,* 2nd ed. (New York: Macmillan, 1894; originally published in 1892), *Outlines of Psychology,* 7th ed. (Leipzig: Engleman, 1907; originally published in 1896), and *Elements of Folk Psychology: Outlines of a Psychological History of the Development of Mankind* (London: George Allen, 1916).

27. See, for a more complete discussion, Wilhelm Wundt, *The Language of Gestures* (The Hague: Mouton, 1973).

28. See, for example, Wundt, *Folk Psychology.*

29. Wundt, *Outlines of Psychology,* pp. 296–298.

30. James, *Meaning of Truth.*

31. William James, *The Principles of Psychology* (New York: Holt, Rinehart & Winston, 1890), p. 1.

32. Ibid., p. 1.

33. James had also developed the notion of a "stream of consciousness" in his *The Varieties of Religious Experience* (New York: Longmans, Green, 1902).

34. James, *Principles of Psychology,* p. 224.

35. Ibid., pp. 225–290.

36. Ibid., pp. 291–401.

37. Ibid., p. 329.

38. Ibid., p. 313 (emphasis in original).

39. Ibid., p. 334.

40. Charles Horton Cooley, *Social Process* (New York: Scribner's, 1918), p. 28.

41. Charles Horton Cooley, *Social Organization: A Study of the Larger Mind* (New York: Scribner's, 1916), p. 61.

42. Charles Horton Cooley, *Life and the Student* (New York: Knopf, 1927), p. 200.

43. Charles Horton Cooley, *Human Nature and the Social Order* (New York: Scribner's, 1902), p. 184.

44. Ibid., pp. 137–211.

45. Cooley, *Social Organization,* p. 12.

46. Ibid., p. 23.

47. Ibid., p. 27.

48. Dewey's bibliography is more than seventy-five pages long, indicating his incredible productivity.

49. Dewey's most important works, in addition to those cited earlier, include *Outlines of a Critical Theory of Ethics* (Ann Arbor, MI: Register, 1891); *The Study of Ethics* (Ann Arbor, MI: Register, 1894); *The School and Society* (Chicago: University of Chicago Press, 1900); *Studies in Logical Theory* (Chicago: University of Chicago Press, 1903); with James H. Tufts, *Ethics* (New York: Holt, Rinehart & Winston, 1908); *How We Think* (Boston: D. C. Heath, 1910); *The Influence of Darwin on Philosophy* (New York: Holt, Rinehart & Winston, 1910); *Democracy and Education* (New York: Macmillan, 1916); *Essays in Experimental Logic* (Chicago: University of Chicago Press, 1916); *Reconstruction in Philosophy* (New York: Holt, Rinehart & Winston, 1920); *Experience and Nature* (La Salle, IL: Open Court, 1925); *Philosophy and Civilization* (New York: Minton, Balch, 1931); *Art as Experience* (New York: Minton, Balch, 1934); *A Common Faith* (New Haven, CT: Yale University Press, 1934); *Logic: The Theory of Inquiry* (New York: Holt, Rinehart & Winston, 1938); *Theory of Valuation* (Chicago: University of Chicago Press, 1939); *Problems of Men* (New York: Philosophical Library, 1946); and with A. F. Bentley, *Knowing and the Known* (Boston: Beacon, 1949).

50. Dewey, *Quest for Certainty.*

51. Dewey, *Influence of Darwin,* p. 22.

52. However, Mead's ideals paralleled those of Dewey, and he was occasionally drawn into various reform causes.

53. See, for example, Dewey, *Human Nature and Conduct* and *How We Think.*

54. Indeed, Mead appears to have borrowed Dewey's exact terms in *Human Nature and Conduct.*

55. Dewey, *Human Nature and Conduct.*

23

✳

The Sociology of George Herbert Mead

Because Mead wrote relatively little in his lifetime, his major works are found in the published lecture notes of his students. As a result, the four posthumous books that constitute the core of his thought are somewhat long and rambling. Moreover, with the exception of *Mind, Self, and Society,* his ideas are distinctly philosophical rather than sociological in tone.[1] Our goal in this chapter, therefore, is to pull from his philosophical works key sociological insights, while devoting most of our analysis to the explicitly social-psychological work, *Mind, Self, and Society.*

MEAD'S BROADER PHILOSOPHY

Much of Mead's sociology is only a part of a much broader philosophical view. This view was never fully articulated, nor was it well integrated, but two posthumous works, *Movements of Thought in the Nineteenth Century* and *The Philosophy of the Present,* provide a glimpse of his broader vision.

Many fascinating themes are contained in these works, but one of the most persistent is that all human activity represents an adjustment and adaptation to the social environment. In *Movements of Thought,* Mead traced the development of social thought from its early, pre-scientific phases to the contemporary, scientific stage. In a way reminiscent of Auguste Comte's law of the three stages, Mead saw the great ideas of history as moving toward an ever more rational, or scientific, profile, because with the emergence of scientific thought, a better level of adaptation and adjustment to the world could be achieved.

The Philosophy of the Present contains a somewhat disjointed series of essays that represent a more philosophical treatment of ideas contained in Mead's social psychology, particularly in *Mind, Self, and Society*. Here again, he emphasized that what was uniquely human is nothing but a series of particular behavioral capacities that have evolved from adaptations to the ongoing life process. Much of the discussion addresses purely philosophical topics about the ontological status of consciousnesses in the past, present, and future, but between the lines he stressed that the capacities of humans for thought and self-reflection do not necessitate a dualism between mind and body because all the unique mental abilities of humans were behaviors directed toward facilitating their adjustment to the environment as it is encountered in the present.

MIND, SELF, AND SOCIETY

Mead's "book," *Mind, Self, and Society*, consists of verbatim transcripts from his famous course on social psychology at the University of Chicago. Although the notes come from the 1927 and 1930 versions of the course, the basic ideas on social interaction, personality, and social organization had been developed a decade earlier.

At the time Mead was addressing his students, behaviorists like J. B. Watson had simply abandoned any serious effort to understand consciousness, personality, and other variables in the "black box" of human cognition. Mead felt that such a "solution" to studying psychological processes was unacceptable.[2] He also felt that the opposite philosophical tendency to view "mind," "spirit," "will," and other psychological states as a kind of spiritual entity was untenable. What is required, he argued, is for mind and self, as the two most distinctive aspects of human personality, and for "society" as maintained by mind and self, to be viewed as part of ongoing social processes.

Mead's View of the "Life Process"

The Darwinian theory of evolution provided Mead with a view of life as a process of adaptation to environmental conditions. The attributes of a species, therefore, are the result of selection for those characteristics allowing for adaptation to the conditions in which a species finds itself. This theory provided Mead with a general metaphor for viewing life in general and thus with a broad perspective for analyzing humans. Pragmatism, as a philosophical doctrine developed in great detail by John Dewey, represented one way of translating the Darwinian metaphor into principles for understanding human behavior: Humans are "pragmatic" creatures who use their facilities for achieving "adjustment" to the world; conversely, much of what is unique to any individual arises from making adjustments to the world. John Dewey's pragmatism, termed *instrumentalism,* stressed the importance of critical and rational thought in making life adjustments to the world; it gave Mead a view of thinking as the basic adjustment by which humans survive. Behaviorism, as a prominent psychological school of thought, converged with this element in pragmatism because it emphasized that all animals tend to retain those responses to environmental stimuli that are rewarded, or reinforced. Although the processes of thinking were regarded as too "psychical" by behaviorists like Watson, the stress on the retention of reinforced behaviors was consistent with Darwinian notions of adaptation and survival as well as with pragmatist ideas of response and adjustment. Even utilitarianism—especially that of such thinkers as Jeremy Bentham, who emphasized the pleasure and pain principles—could be seen by Mead as compatible with the theories of evolution, behaviorism, and pragmatism. The utilitarian emphasis on "utility," "pleasure," and "pain" was certainly compatible with behaviorist notions of reinforcement; the utilitarian concern with rational thought and the weighing of alternatives was compatible with Dewey's instrumentalism and its concept of critical thinking, and the utilitarian view that order emerges from competition among free individuals seemed to parallel Darwinian notions of struggle as the underlying principle of the biotic order.

Thus, the unique attributes of humans, such as their capacity to use language, their ability to view themselves as objects, and their capacity to reason, must all be viewed as emerging from the life processes of adaptation and adjustment. Mind and self cannot be ignored, as behaviorists often sought to do, nor can they be seen as a kind of mystical and spiritual force that elevated humans out of the basic life processes influencing all species. Humans as a species evolved like other life forms, and, hence, their most distinctive attributes—mind, self, and society—must be viewed as emerging from the basic process of adaptation. Further, each individual member of the human species is like the individuals of other species: What they are is the result of the common biological heritage of their species as well as of their adjustment to the particulars of a given environment.

Mead's Social Behaviorism

Mead did not define his work as *social* behaviorism, but subsequent commentators have used this term to distinguish his work from Watsonian behaviorism. In contrast with Watson, who simply denied the distinctiveness of subjective consciousness, Mead felt that it was possible to use broad behavioristic principles to understand "subjective behavior":

> Watson apparently assumes that to deny the existence of mind or consciousness as a psychical stuff, substance, or entity is to deny its existence altogether, and that a naturalistic or behavioristic account of it as such is out of the question. But, on the contrary, we may deny its existence as a psychical entity without denying its existence in some other sense at all; and if we then conceive of it functionally, and as a natural rather than a transcendental phenomenon, it becomes possible to deal with it in behavioristic terms.[3]

If subjective experiences in humans are viewed as behaviors, it is possible to understand them in behavioristic terms. The unique mental capacities of humans are a model of behavior that arises from reinforcement processes that explain nonsubjective and overt behavior. Of particular importance for understanding the attributes of humans, then, is the reinforcement that comes from adaptation and adjustment to environmental conditions. At some point in the distant past, the unique mental capacities of humans, and the creation of society employing these capacities, emerged by the process of natural selection under natural environmental conditions. Once the unique patterns of human organization are created, however, the "environment" for any individual is social; that is, it is an environment of other people to whom an individual must adapt and adjust.

Thus, social behaviorism stresses the processes by which individuals acquire a certain behavioral repertoire by virtue of their adjustments to ongoing patterns of social organization. Analysis must begin with the observable fact that organized activity occurs and then attempt to understand the particular actions of individuals as they adjust to cooperative activity:

> We are not, in social psychology, building up the behavior of the social group in terms of the behavior of separate individuals composing it; rather, we are starting out with a given social whole of complex group activity, into which we analyze (as elements) the behavior of each of the separate individuals composing it. We attempt, that is, to explain the conduct of the individual in terms of the organized conduct of the social group, rather than to account for the organized conduct of the social group in terms of the conduct of the separate individuals belonging to it.[4]

The behavior of individuals—not just their observable actions but also their internal behaviors of thinking, assessing, and evaluating—must be analyzed within a social context. For what is distinctively human emerges from adjustment to ongoing social activity, or "society." Thus, Mead's social behaviorism must be distinguished from the behavioristic approach of Watson[5] in two ways. First, the existence of inner subjective experiences is not denied or viewed as methodologically irrelevant[6]; rather, these experiences are viewed as behavior. Second, the behaviors of humans, including those distinctly human behaviors that Mead called *mind* and *self*, arise from adaptation and adjustment to ongoing and organized social activity. Reinforcement is thus equated with the degree of adjustment and adaptation to society.

Mead's Behavioristic View of Mind

For Mead, "mind" is a type of behavioral response that emerges from interaction with others in a social context. Without interaction, mind could not exist:

> We must regard mind, then, arising and developing within the social process, within the empirical matrix of social interactions. We must, that is, get an inner individual experience from the standpoint of social acts which include the experiences of separate individuals in a social context wherein

those individuals interact. The processes of experience which the human brain makes possible are made possible only for a group of interacting individuals: only for individual organisms which are members of a society; not for the organism in isolation from other individual organisms.[7]

Gestures and Mind The social process in which mind emerges is one of communication with gestures. Mead gave the German psychologist Wilhelm Wundt credit for understanding the central significance of the gesture to communication and interaction. In contrast with Charles Darwin, who had viewed gestures as expressions of emotions, Wundt recognized gestures as that part of the ongoing behavior of one organism that stimulates the behavior of another organism.[8] Mead took this basic idea and extended it in ways that became the basis not only for the emergence of mind and self but also for the creation, maintenance, and change of society.

Mead formulated the concept of the "conversation of gestures" to describe the simplest form of interaction. One organism emits gestures that stimulate a response from a second organism. In turn, the second organism emits gestures that stimulate an "adjusted response" from the first organism. Then, if interaction continues, the adjusted response of the first organism involves emitting gestures that result in yet another adjustment of behavior by the second organism, and so on, as long as the two organisms continue to interact. Mead frequently termed this conversation of gestures the *triadic matrix,* because it involves three interrelated elements:

1. Gestural emission by one organism as it acts on its environment

2. A response by another organism that becomes a gestural stimulus to the acting organism

3. An adjusted response by the acting organism that takes into account the gestural stimuli of the responding organism

This triadic matrix constitutes the simplest form of communication and interaction among organisms. This form of interaction, Mead felt, typifies "lower animals" and human infants. For example, if one dog growls, indicating to another dog that it is about to attack, the other will react, perhaps by running away, requiring the growling dog to adjust its response by chasing the fleeing dog or by turning elsewhere to vent its aggressive impulses. Or, to take another example, a hungry infant cries, which in turn arouses a response in its mother (for instance, the mother feeds the infant), which in turn results in an adjusted response by the infant.

Much of the significance of Mead's discussion of the triadic matrix is that the mentalistic concept of "meaning" is lodged in the interaction process rather than in "ideas" or other mentalistic notions that might reside outside interaction. If a gesture "indicates to another organism the subsequent behavior of a given organism, then it has meaning."[9] Thus, if a dog growls and another dog uses this gesture to predict an attack, this gesture of growling has meaning. Meaning is thus given a behavioristic definition: It is a kind of behavior—a gesture—of one organism that signals to another subsequent behavior of this organism. Meaning, therefore, need not involve complex cognitive activity. A dog that runs away from another growling dog, Mead would assert, is reacting without "ideas" or "elaborate deliberation"; yet the growl has meaning to the dog because it uses the growl as an early indicator of what will follow. Thus meaning is

> not to be conceived, fundamentally, as a state of consciousness, or as a set of organized relations existing or subsisting mentally outside the field of experience into which they enter; on the contrary, it should be conceived objectively, as having its existence entirely within this field itself.[10]

The significance of the conversation of gestures for ongoing activity resides in the triadic matrix, and associated meanings, allowing organisms to adjust their responses to one another. Thus, as organisms use one another's gestures as a means for adjusting their respective responses, they become increasingly capable of organized and concerted conduct. Yet, such gestural conversations limit the capacity of organisms to organize themselves and to cooperate. Among humans, Mead asserted, a qualitatively different form of communication evolved. This is communication involving *significant symbols.* He felt that the development of the capacity to use significant symbols distinguished humans from other species. *Mind* arises in a maturing human infant from the development of the capacity to use significant symbols. In turn, as we will show, the existence of mind ensures the development of self and the perpetuation and change of society.

Significant Symbols and Mind   The gestures of "lower organisms," Mead felt, do not evoke the same response in the organism emitting a gesture and the one interpreting the gesture. As he observed, the roar of the lion does not mean the same thing to the lion and its potential victim. When organisms become capable of using gestures that evoke the same response in each other, then they are employing what he termed *significant,* or *conventional,* gestures. As he illustrated, if a person shouts "Fire!" in a movie theater, this gesture evokes the same response tendency (escape, fleeing, and so on) in the person emitting the gesture and in those receiving it. Such gestures, he felt, are unique to humans and make possible their capacities for mind, self, and society.[11]

Significant symbols are, as Mead emphasized, the basis for language. Of particular significance are vocal significant symbols because sounds can be readily heard by both sender and receiver, thus evoking a similar behavioral tendency. Other nonvocal gestures, however, are also significant in that they can come to mobilize similar tendencies to act. A frown, glare, clenched fist, rigid stance, and the like can all become significant because they serve as a stimulus to similar responses by senders and receivers. Thus, humans' capacity for language—that is, communication by significant symbols—allows the emergence of their unique capacities for mind and self. An infant of the species cannot have a mind until it acquires the rudimentary capacity for language.

In what ways, then, does language make mind possible? Mead borrowed Dewey's vision of "reflective" and "critical" thinking, as well as the utilitarian's vision of "rational choice," in formulating his conceptualization of mind. For Mead, mind involves the behavioral capacities to

1. Denote objects in the environment with significant symbols
2. Use these symbols as a stimulus to one's own response
3. Read and interpret the gestures of others and use these as a stimulus for one's response
4. Suspend temporarily or inhibit overt behavioral responses to one's own gestural denotations or those of others
5. Engage in "imaginative rehearsal" of alternative lines of conduct, visualize their consequences, and select the response that will facilitate adjustment to the environment

Mind is thus a behavior, not a substance or entity. It is a behavior that involves using significant symbols to stimulate responses but, at the same time, to inhibit or delay overt behavior so that potential responses can be covertly rehearsed and assessed. Mind is thus an "internal conversation of gestures" using significant symbols because an individual with mind talks to itself. The individual uses significant symbols to stimulate a line of response; the individual visualizes the consequences of this response; if necessary, the individual inhibits the response and uses another set of symbols to stimulate alternative responses; and the individual persists until it is satisfied with its response and overtly pursues a given line of conduct.

This capacity for mind, Mead stressed, is not inborn; it depends on interaction with others and the acquisition of the ability to interpret and use their significant symbols (as well as biological maturation). As Mead noted, feral children who are raised without significant symbols do not seem "human" because they have not had to adjust to an environment mediated by significant symbols and, hence, have not acquired the behavioral capacities for mind.

Role-Taking and Mind Mind emerges in an individual because human infants, if they are to survive, must adjust and adapt to a social environment—that is, to a world of organized activity. At first, an infant is like a "lower animal" in that it responds reflexively to the gestures of others and emits gestures that do not evoke similar responses in it and those in the environment. Such a level of adjustment, Mead implied, is neither efficient nor adaptive. A baby's cry does not indicate what it wants, whether food, water, warmth, or whatever, and by not reading accurately the vocal and other gestures emitted by others in their environment, the young can frequently create adjustment problems for themselves. Thus, in a metaphor that is both Darwinian and behavioristic, there is "selective pressure" for acquiring the ability to use and interpret significant gestures. Hence, those gestures that bring reinforcement—that is, adjustment to the environment—are likely to be retained in the response repertoire of the infant.

A critical process in using and interpreting significant gestures is what Mead termed "taking the role of the other," or *role-taking*. An ability to use significant symbols means that the gestures emitted by others in the environment allow a person to read or interpret the dispositions of these others. For example, an infant who has acquired the rudimentary ability to interpret significant symbols can use its mother's tone of voice, facial expressions, and words to imagine her feelings and potential actions—that is, to "take on" her role or perspective. Role-taking is critical to the emergence of mind, for unless the gestures of others, and the disposition to act that these gestures reveal, can become a part of the stimuli used to covertly rehearse alternative lines of conduct, overt behavior will often produce maladjustment to the environment. For without the ability to assume the perspective of others with whom one must deal, it is difficult to adjust to, and coordinate responses with, these others.

The Genesis of Mind Mead saw mind as developing in a sequence of phases, as is represented in Figure 23.1. Because an infant depends on others and, in turn, these others depend on society for their survival, mind develops from the forced dependency of an infant on society. Because society is held together by actors who use language and who can role-take, the infant must seek to meet its needs in a world mediated by symbols. Through conscious coaching by others, and through simple trial and error, the infant comes to use significant symbols to denote objects relevant to satisfying its needs (such as food, mother, and so on). To consummate other impulses, the infant eventually must acquire greater capacities to use and understand language; once an infant can use language, it can begin to read the gestures of others and call out in itself the dispositions of others. When a young child can role-take, it can soon begin to consciously think, reflect, and rehearse responses. In other words, it reveals the rudimentary behavioral abilities Mead termed *mind*.[12]

The causal arrows in Figure 23.1 actually represent a series of preconditions for the next stage of development. The model is "value added" in that certain conditions must be met before subsequent events can occur. Underlying these conditions is Mead's implicit vision of "social selection," which represents his reconciliation of learning-theory principles with pragmatism and Darwinism. The development of abilities for language, role-taking, and mind are selected for as the infant seeks to consummate impulses in society. If the infant is to adjust and adapt to society, it must acquire the ability for minded behavior. Thus, as the infant lives in a social environment, it learns those behaviors—first significant symbols, then role-taking, and eventually mind—that facilitate, to ever-increasing degrees, its adjustment to the social environment.

The model presented in Figure 23.1 underscores Mead's view that there is nothing mysterious or mystical about the human mind. It is a behavior acquired like other behavioral tendencies as a human organism attempts to adapt to its surroundings. Mind is a behavioral capacity acquired in stages, with each stage setting the conditions for the next. As mind emerges, so does self-awareness. In many respects, the emergence of mind is a precondition for the genesis of self. Yet, the rudiments of self begin with an organism's ability to role-take, for the organism can then derive self-images or see itself as an object.

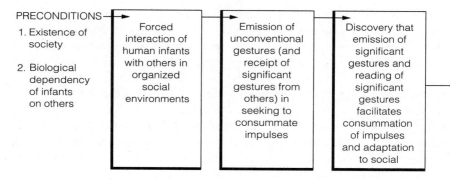

FIGURE 23.1 Mead's Model of the Genesis of Mind

Mead's Behavioristic View of Self

The Social Nature of Self As a "social behaviorist," Mead emphasized that the capacity to view oneself as an object in the field of experience was a type of learned behavior. This behavior is learned through interaction with others:

> The self is something which has a development; it is not initially there, at birth, but arises in the process of social experience and activity, that is, develops in the given individual as a result of his relations to that process as a whole and to other individuals within that process.[13]

Self emerges from the capacity to use language and to take the role of the other. Borrowing the essentials of Charles Horton Cooley's "looking-glass self,"[14] Mead viewed the social self as emerging from a process in which individuals read the gestures of others, or "take their attitudes," and derive an image, or picture, of themselves as a certain type of object in a situation. This image of oneself then acts as a behavioral stimulus, calling out certain responses in the individual. In turn, these responses of an individual cause further reactions by others, resulting in the emission of gestures that make possible role-taking by an individual, who then derives new self-images and new behavioral stimuli. Thus, like mind, self arises from the triadic matrix of people interacting and adjusting their responses to one another. The individual does not experience self directly, only through reading the gestures of others:

> The individual experiences himself, not directly, but only indirectly, from the particular standpoints of other individual members of the same social group, or from the generalized standpoint of the social group as a whole to which he belongs . . . and he becomes an object to himself only by taking the attitudes of other individuals toward himself within a social environment or context of experience and behavior in which both he and they are involved.[15]

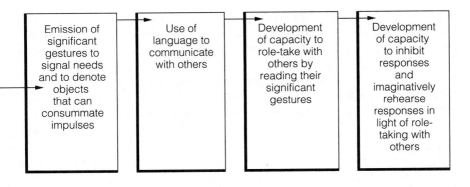

The Structure of Self Mead appeared to use the notion of "self" in two different ways. One usage involves viewing self as a "transitory image" of oneself as an object in a particular situation. Thus, as people interact, they role-take and derive self-images of themselves in that situation. Second, in contrast with this conceptualization, Mead also viewed self as a structure, or configuration of typical habitual meanings toward self, that people carry to all situations. For "after a self has arisen, it in a certain sense provides for itself its social experiences."[16]

These views are not, of course, contradictory. The process of deriving self-images in situations leads, over time, to the crystallization of a more permanent, trans–situational set of attitudes toward oneself as a certain type of object. Humans begin to interpret selectively the gestures of others in light of their attitudes toward themselves, and thus, their behaviors take on a consistency. For if the view of oneself as a certain type of object is relatively stable, and if we use self like all other environmental objects as a stimulus for behavior, overt behavior will reveal a degree of consistency across social situations.

Mead sometimes termed this development of stable attitudes toward oneself as an object the *complete,* or *unified,* self. Yet, he recognized that this complete self was not a rigid structure and that it was not imperviously and inflexibly imposed on diverse interactions. Rather, in different social contexts various aspects of the complete self are more evident. Depending on one's audience, then, different "elementary selves" will be evident:

> The unity and structure of the complete self reflects the unity and structure of the social process as a whole; and each of the elementary selves of which it is composed reflects the unity and structure of one of the various aspects of that process in which the individual is implicated. In other words, the various elementary selves that constitute, or are organized into, a complete self are the various aspects of the structure of that complete self answering to the various aspects of the structure of the social process as a whole; the structure of the complete self is thus a reflection of the complete social process.[17]

In this passage a further insight into the structure of self is evident: Although elementary selves are unified by a complete self, people who experience a highly contradictory social environment with *dis*unity in the social process will also experience difficulty in developing a complete self, or a relatively stable and consistent set of attitudes toward themselves as a certain type of object. To some extent, then, people present different aspects of their more complete and unified selves to different audiences, but when these audiences demand radically contradictory actions, the development of a unified self-conception becomes problematic.

In sum, then, Mead's conceptualization is behavioristic in that he viewed seeing oneself as an object as a behavior unique to humans. Moreover, like other objects in one's environment, the self is a stimulus to behavior. Thus, as people develop a consistent view of themselves as a type of object—that is, as their self reveals a structure—their responses to this stable stimulus take on a consistency. Mead's conceptualization of the structure of self involves the recognition that the stability of self is largely a consequence of the unity and stability in the social processes from which the self arises.

Phases of the Self Mead wanted to avoid connoting that the structure of self limited a person's repertoire of potential responses. Although a unified self-conception lends considerable stability and predictability to overt behaviors, there is always an element of spontaneity and unpredictability to action. This is inherent in the "phases of self," which were conceptualized by Mead in terms of the *I* and *me.*

The image that a person derives from his or her behavior in a situation is what Mead termed the *me.* As such, the "me" represents the attitudes of others and the broader community as these influence an individual's retrospective interpretation of his or her behavior. For example, if we talk too loudly in a crowd of strangers, we see the startled looks of others and will become cognizant of general norms about voice levels and inflections when among strangers. These "me" images are received by reading the gestures of specific others in a situation and by role-taking, or assuming the attitude of the broader community. In contrast with the "me" is the "I," which is the actual emission of behavior. If a person speaks too loudly, this is "I," and when this person reacts to his or her loudness, the "me" phase of action is initiated. Mead emphasized that the "I" can only be known in experience because we must wait for "me" images to know just what the "I" did. People cannot know until after they have acted ("I") just how the expectations of others ("me") are actually carried out.

Mead's conceptualization of the "I" and "me" allowed him to conceptualize the self as a constant process of behavior and self-image. People act; they view themselves as objects; they assess the consequences of their action; they interpret others' reaction to their action; and they resolve how to act next. Then, they act again, calling forth new self-images of their actions. This conceptualization of the "I" and "me" phases of self enabled Mead to accomplish several conceptual tasks. First, he left room for spontaneity in human action; if the "I" can be known only in experience, or through the "me," one's actions are never completely circumscribed. Second, as we will explore in more detail later, it gave Mead a way of visualizing the process of self-control. Humans are, in his view, cybernetic organisms who respond, receive feedback and make adjustment, and then respond again. In this way, he emphasized, self, like mind, is a process of adaptation; it is a behavior in which an organism successively responds to itself as an object as it adjusts to its environment. Third, the "I" and "me" phases of self gave Mead a way to conceptualize variations in the extent to which the expectations of others and the broader community constrain action. The *relative values* of the "I" and "me," as he phrased the matter,[18] are a positive function of people's status in a particular situation. The more involved they are in a group, the greater the values of "me" images and the greater is the control of "I" impulses. Conversely, the less the involvement of a person in a situation, the less salient are "me" images, and hence, the greater is the variation in that person's overt behavior.

The Genesis of Self Mead devoted considerable attention to the emergence of self and self-conceptions in humans. This attention allowed him to emphasize again that the self was a social product and a type of behavior that emerged from the efforts of the human organism to adjust and adapt to its environment. Self arises from the same processes that lead to the development of "mind," while depending on the behavioral capacities of mind.

For self to develop, a human infant must acquire the ability to use significant symbols. Without this ability, it is not possible to role-take with others and thereby develop an image of oneself by interpreting the gestures of others. Self also depends on the mind's capacities because people must be able to designate themselves linguistically as an object in their field of experience and to organize responses toward themselves as an object. Thus, the use of significant symbols, the ability to role-take, and the behavioral capacities of mind are all preconditions for the development of self, particularly a more stable self-conception, or "unified" self.

Mead visualized self as developing in three stages, each marked by an increased capacity to role-take with a wider audience of others. The first stage is *play,* which is marked by a very limited capacity to role-take. A child can assume the perspective of only one or two others at a time, and play frequently involves little more than discourse and interaction with "imaginary companions" to whom the child talks in enacting a particular role. Thus, a child who plays "mother" can also, at the same time, assume the role of "baby," and the child might move back and forth between the mother's and infant's roles. The play stage is thus typified by the ability to assume the perspective of only a few others at a time.

With biological maturation and with practice at assuming the perspectives of others, a child eventually acquires the capacity to take the role of multiple others engaged in ongoing and organized activity. The second stage is what Mead termed the *game* in which individuals can role-take with multiple others at the same time. Perhaps the most prototypical form of such the role-taking is to be a participant in a game, such as baseball, where the child must assume the role of other players, anticipate how they will act, and coordinate responses with their likely course of action. Thus, children begin to see themselves as objects in an organized field, and they begin to control and regulate their responses to themselves and to others to facilitate the coordination of activity. During this stage in the development of self, the number and variety of such game situations expand:

> There are all sorts of social organizations, some of which are fairly last-ing, some temporary, into which the child is entering, and he is playing a sort of social game in them. It is a period in which he likes "to belong," and he gets into organizations which come into existence and pass out of existence. He becomes something which can function in the organized whole, and thus tends to determine himself in his relationship with the group to which he belongs.[19]

In both the play and game situations, individuals view themselves in rela-tion to specific others. By role-taking with specific others lodged in particular roles, individuals derive images of themselves from the viewpoint of these oth-ers. Yet the self, Mead contended, is incomplete until a third stage is realized: role-taking with *the generalized other.* He saw the generalized other as a "com-munity of attitudes" among members of an ongoing social collective. When individuals can view themselves in relation to this community of attitudes and then adjust their conduct in accordance with the expectations of these atti-tudes, they have reached the third stage in the development of self. They can now role-take with the generalized other. For Mead, the play and game rep-resent the initial stages in the development of self, but in the final stage, indi-viduals can generalize the varied attitudes of others and see themselves and regulate their actions from a broader perspective.

Without this capacity to view oneself as an object in relation to the gener-alized other, behavior could only be situation-specific. Unless people can see themselves as objects implicated in a broader social process, their actions can-not reveal continuity across situations. Moreover, humans could not create larger societies, composed of multiple groupings, without the members of the society viewing themselves, and controlling their responses, in accordance with the expectations of the generalized other.[20]

Mead recognized that in complex social systems there could be multiple generalized others. Individuals can view themselves and control their behav-iors from a variety of broader perspectives. Moreover, a generalized other can represent the embodiment of collective attitudes of concrete and func-tioning groups, or it can be more abstract, pertaining to broad social classes and categories:

> In the most highly developed, organized, and complicated human social communities . . . , [the] various socially functional classes or subgroups of individuals to which any given individual belongs . . . are of two kinds. Some of them are concrete social classes or subgroups, such as political parties, clubs, corporations, which are all actually functional social units,

in terms of which their individual members are directly related to one another. The others are abstract social classes or subgroups, such as the class of debtors and the class of creditors, in terms of which their individual members are related to one another only more or less indirectly, and which only more or less indirectly function as social units, but which afford or represent unlimited possibilities for the widening and ramifying and enriching of the social relations among all the individual members of the given society as an organized and unified whole.[21]

The capacity to take the role of multiple and diverse generalized others—from the perspective of a small group to that of an entire society—enables individuals to engage in the processes of self-evaluation, self-criticisms, and self-control from the perspective of what Mead termed *society*. Thus, by virtue of self-images derived from role-taking with specific others in concrete groups as well as from role-taking with generalized others personifying varying communities of attitudes, people come to see themselves as a particular type of object, with certain strengths, weaknesses, and other attributes.

Moreover, people become capable of regulating their responses to sustain this vision of themselves as a certain type of object. As people come to see themselves, and consistently respond to themselves, through their particular configuration of specific and generalized attitudes of others, they come to possess what Mead termed a *complete* and *unified* self.

Figure 23.2 attempts to summarize the dynamic processes involved in creating a self. It is more complex than Figure 23.1 because Mead used the concept of "self" in several interrelated ways. Thus, for purposes of interpreting the model portrayed in Figure 23.2, let us recapitulate his various notions about self. First, he saw the development of self as a process of role-taking with increasingly varied and generalized "others." This facet of self is represented across the top of Figure 23.2, because increasing acuity at role-taking influences the other aspects of self (this is emphasized by the vertical arrows connecting the boxes at each stage in the emergence of self). Second, as is shown in the middle row of Figure 23.2, Mead visualized self as a process of self-control. As he emphasized in his notion of the "I" and "me" phases of self and in his view of "mind," this facet of self involves the growing ability to read the gestures of others, to inhibit inappropriate responses in relation to these others, and to adjust responses in a way that will facilitate interaction. In its more advanced stages, self-control also includes the capacity to assume the "general" perspective, or "community of attitudes," of specific groups and, eventually, of the broader community. The process of self-control thus represents the extensions of the capacities for mind, and for this reason the precondition for self—that is, the "incipient capacities for mind" at the left of the model—is seen to tie almost directly into the self-control aspect of self. Third, as shown along the bottom row in Figure 23.2, Mead also saw self as involving the emergence of a self-conception, or stable disposition to act toward oneself as a certain type of object. Such a stable self-conception evolves out of the accumulation of self-images and self-evaluations with reference to specific, and then increasingly generalized, others.

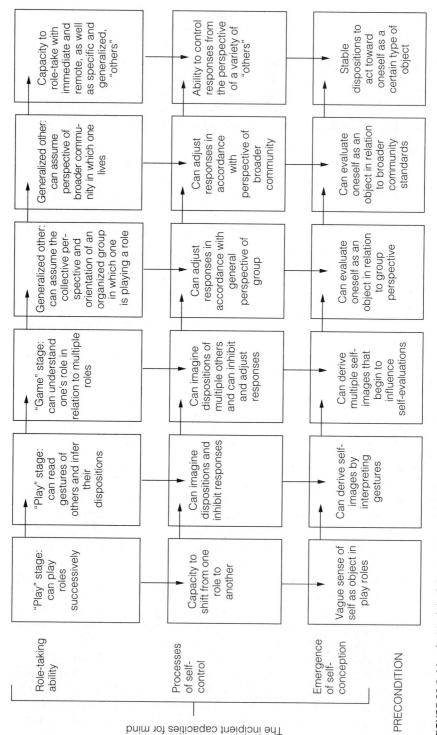

FIGURE 23.2 Mead's Model of the Genesis of Self

Thus, in reading Mead's model, the arrows that move from left to right denote the development of each aspect of self. The arrows that move down the columns stress his emphasis on the role-taking process and on how developments in the ability to role-take influence the etiology of self-control processes and self-conceptions. Of course, we might also draw arrows back up the columns because to some extent, self-control processes and self-conceptions influence role-taking abilities. But we feel that the arrows, as currently drawn, best capture Mead's vision of causal processes in the initial emergence of those multiple behavioral capacities he subsumed under the label *self.* These capacities are, in turn, vital to the production and reproduction of society, as will be explored.

Mead's Conception of Society

The Behavioral Basis of Society Mead labeled as *mind* those behavioral capacities in organisms that allow the use of symbols to denote objects and to role-take, to use objects as stimuli for various behaviors, to inhibit responses, to imaginatively rehearse alternative responses, and to select a line of conduct. Thus, mind allows cooperation among individuals as they attempt to select behaviors that will facilitate cooperation. *Self* is the term Mead used to describe the behavioral capacity to see oneself as an object in the environment and to use a stable conception of oneself as a certain type of object as a major stimulus for organizing behavior. The capacity for mind and self arises from, and continues to depend on, the process of role-taking because one's view of oneself as an object and one's capacity to select among alternative behaviors are possible through reading the gestures of others and determining their attitudes and dispositions.

In many ways mind is the capacity for denoting alternatives, whereas self involves the capacity for ordering choices in a consistent framework. An organism with only mind could visualize alternatives but could not readily select among them. The capacity for self allows the selection of behaviors among alternatives. In so doing, self provides a source of stability and consistency in a person's behavior, while integrating that behavior into the social fabric, or society.

Mead saw several ways in which self provides for the integration of behavior into society. First, the capacity to see oneself as an object in a field of objects allows individuals to see themselves in relation to other individuals. They can see their place in the field of perception and, hence, adjust their responses (through the capacity for mind) to coordinate their activities.

Second, the emergence of a unified and complete self, or stable self-conception, means that individuals consistently place into their perceptual field a view of themselves as a certain type of object. This object becomes a stimulus to subsequent behaviors that reveal consistency because they are responses to the self as a type of object that has certain stable attributes. People's behavior across widely divergent situations thus reveals consistency because they interject, to some degree, a stable self-conception of themselves as a certain type of object. This object, as much as any of the objects peculiar to a situation, serves as a stimulus to the organization for behavior. The more rigid the self-conception, the more the gestures of others are selectively interpreted and used to organize responses consistent with one's self-conception. The consequence for society of these self-related processes is that as people's actions take on consistency from situation to situation or from time to time in the same situation, their behaviors become predictable, thereby making it easier for individuals to adjust to, and cooperate with, one another.

Third, the process of role-taking allows individuals to see themselves not only in relation to specific others in particular situations but also in relation to varieties of generalized others. Thus, if one evaluates one's actions in terms of a generalized other, behaviors will take on consistency from situation to situation and from time to time. Moreover, to the degree that all participants to an interaction role-take with the same generalized other, they will approach and perceive situations within "common meanings," and they will be prepared to act in terms of the same perspective. By viewing themselves as objects relative to the same set of expectations, people approach situations with common understandings that will facilitate their adjustment to one another.

A fourth—and related—point is that the capacity to role-take with varieties of generalized others allows individuals to elaborate patterns of social organization. Individuals are now liberated from the need for face-to-face interaction as the basis for coordinating their activities. Once they can role-take with varieties of generalized others, some of whom are abstract conceptions, they can guide their conduct from a common perspective without directly role-taking with one another. Thus, the capacity to view oneself as an object and to adjust responses in relation to the perspective of an abstract generalized other greatly extends the potential scope of patterns of social organization.

Fifth, in addition to providing behavioral consistency and individual integration into extended networks of interaction, self also serves as a vehicle of social change. The phases of self—the "*I*" and "*me*," as Mead termed them—ensure that individual behaviors will, to some degree, alter the flow of the social process. Even if "me" images reflect perfectly the expectations in a situation, and even if one's view of oneself as a certain type of object is totally congruent with these expectations, actual behavior—that is, the "I"—can deviate from what is anticipated in "me" images. This deviation, however small or great, forces others in the situation to adjust their behaviors, providing new "me" images to guide subsequent behaviors ("I")—and so on, in the course of interaction that moves in and out of "I" and "me" phases. Of course, when expectations are not clear and when one's self-conception is at odds with the expectations of others, "I" behaviors are likely to be less predictable, requiring greater adjustments by others. Or when the capacity to develop "me" images dictates changes in a situation for an individual—and this is often the case among individuals whose self-conception or generalized others are at odds—even greater behavioral variance and social change can be expected as the "I" phase of action occurs. Thus, the inherent phases of self—the "I" and "me"—make inevitable change in patterns of interaction. Sometimes these changes are small and imperceptible, and only after the long accumulation of small adjustments is the change noticeable.[22] At other times, the change is great, as when a person in political power initiates a new course of activity. In either case, Mead went to great lengths to emphasize, self not only provides a source of continuity and integration for human behavior, but also is a source of change in society.[23]

In his analysis of "society," or patterns of social organization, then, Mead attempted to visualize how society was created, maintained, and changed through the processes of interaction among humans with minds and selves. In emphasizing this connection between personality and society, Mead provided a valuable supplement to the macrostructural analyses of European sociology. The result is often a rather vague portrayal of society, however, because Mead had little interest in developing a coherent or detailed view of social structure. Thus, one does not find in Mead's works the sense of substructures and superstructures evident in Karl Marx's works, nor does one find Max Weber's passion for constructing ideal types of structural relations. To some extent, Mead and Émile Durkheim converge, in that both were vitally concerned with the symbolic bases for social integration. They diverge, however, in that Durkheim tended to view integration as cultural and social structures, whereas Mead saw integration as the result of the behavioral capacities of mind and self. Mead and Georg Simmel reveal some affinity, in that both were concerned with interaction, roles, and self, but Simmel never sought to develop a systematic theory of micro-social processes, whereas Mead focused on these processes.

What emerges from Mead's view of society, then, is not a vision of social structure and the emergent properties of these structures. Rather, he reaffirmed that patterns of social organization, whatever their form and profile, were mediated by human behavioral capacities for language, role-taking, mind, and self. Aside from a general view stressed by all thinkers of his time, that societies are becoming more differentiated and complex, he offered only a few clues about the properties of social structures in human societies.

Mead's analysis of society, therefore, is actually a series of statements on the underlying processes that make coordination among individuals possible. As long as this is recognized, we can avoid severe criticism of his fragmentary and superficial discussion of social evolution and morphology. His real contribution resides in his understanding of the behavioral mechanism—role-taking by language-using organisms with minds and selves—by which humans are able to coordinate their activities and construct elaborate patterns of social organization.

The Process of Society For Mead, the term *society* is simply a way of denoting that interactive processes can reveal stability and that humans act within a framework imposed by stabilized social relations. The key to understanding society lies in the use of language and the practice of role-taking by individuals with mind and self. By means of the capacity to use and read significant gestures, individuals can role-take and use their mind and self to articulate their actions to specific others in a situation and to a variety of generalized others.

Because generalized others embody the broader groups—organizations, institutions, and communities—that mark the structure of society, they provide a common frame of reference for individuals to use in adjusting their conduct.

Society is thus maintained by virtue of humans' ability to role-take and to assume the perspective of generalized others. Mead implicitly argued that society as presented to any given individual represents a series of perspectives, or "attitudes," which the individual assumes in regulating behavior. Some attitudes are those of others in one's immediate field; other perspectives are those of less immediate groups; still other attitudes come from more remote social collectives; additional perspectives come from the abstract categories used as a frame of reference; and ultimately, the entire population using a common set of symbols and meanings constitutes the most remote generalized other. Thus, at any given time an individual is role-taking with some combination of specific and generalized others. The attitudes embodied by these others are then used in the processes of mind and self to construct lines of conduct.

Mead believed, then, that the structure and dynamics of society concern those variables that influence the number, salience, scope, and proximity of generalized others. Thus, by implication, Mead argued that to the degree individuals could accurately take the role of one another and assume the perspective of common generalized other(s), patterns of interaction would be stable and cooperative. Conversely, to the degree that role-taking is inaccurate and occurs relative to divergent generalized other(s), interaction will be disrupted and perhaps conflictual.[24]

From this perspective, the theoretical key to explaining patterns of social organization involves isolating those variables that influence (1) the accuracy of role-taking and (2) the convergence of generalized others. What might some of these variables be? Mead did not discuss them in detail because he was not interested in building formal sociological theory. Rather, his concerns were more philosophical, and hence he stressed recognizing the general nature of the processes underlying the maintenance of the social order. In a number of places, however, he offered some clues about what variables influence the capacity of actors to role-take with the same generalized other.

One barrier to role-taking with the same generalized other is social differentiation.[25] In complex societies, people play different roles, and often, the immediate generalized others for these roles will vary. This is, of course, a somewhat different way of stating Comte's and Durkheim's concerns about the mal-integrative effects of differentiation. Mead recognized that when individuals' immediate generalized others vary, it is possible to have a more general or abstract generalized other with which they can mutually role-take. As a result, despite their differences, people can role-take with a common perspective and use it to guide their conduct. Durkheim's similar conceptualization emphasized the "enfeeblement" or abstractness of the collective conscience (or culture) and the resulting anomie and egoism.

Mead's view, however, offers the recognition that although the community of attitudes of two individuals' immediate groups might diverge somewhat, they can at the same time assume the perspective of a more remote, or abstract, generalized other and use this community of attitudes as a common perspective for guiding their conduct. Unlike Durkheim, who saw structural units like "occupational groups" as necessary mediators between the "collective conscience" and the individual, Mead's formulation of mind and self implicitly argues that through the capacity to role-take with multiple and remote others, diversely located individuals can become integrated into a common social fabric. Thus, structural differentiation will tend, Mead appears to have argued, to force role-taking with more remote and abstract generalized others. Thus, the dimensions of a society can be greatly extended because people's interactions are mediated and regulated by reference to a common community of attitudes rather than by face-to-face interaction.

Also related to differentiation—indeed, it is a type of differentiation—is stratification.[26] Class barriers increase the likelihood that individuals in different classes will not share the same community of attitudes. To the degree that a system of hierarchical differentiation is to be integrated, role-taking with a more distant generalized other will supplement the community of attitudes peculiar to a particular social class.

Another aspect of differentiation is population.[27] As populations increase in size, it becomes increasingly likely that any two individuals will role-take with somewhat different perspectives in their interaction with specific others in their immediate groups. If a large population is to remain integrated, Mead argued, individuals will supplement their immediate communities of attitudes by role-taking with more abstract generalized others. Hence, as the size of interaction networks increases, these networks will be integrated by role-taking with an ever-more-abstract perspective or community of attitudes.

In sum, then, Mead's view of society is dominated by a concern with the social-psychological mechanisms by which social structures are integrated. For Mead, *society* is just a term for the processes of role-taking with varieties of specific and generalized others and the consequent coordination of action made possible by the behavioral capacities of mind and self. By emphasizing the processes underlying social structures, he presented a highly dynamic view of society. Not only is society created by role-taking, it can be changed by these same processes. Thus, as diverse individuals come into contact, role-take, and adjust their responses, they create a community of attitudes, which they then use to regulate their subsequent actions. As more actors are implicated, or as their roles become more differentiated, they generate additional perspectives to guide their actions. Similarly, because actors possess unique self-conceptions and because they role-take with potentially diverse perspectives, they often must restructure existing patterns as they come to adjust to one another.

Thus, we get little feeling in Mead's work for the majesty of social structure. His conceptualization can perhaps be seen as a de-mystification of society because society is nothing more than a process of role-taking by individuals who possess mind and self and who seek to make adjustments to one another. We should note, however, that Mead did offer some partial views of social morphology—that is, of the structural forms created by role-taking. We now briefly examine these more morphological or structural conceptualizations of society.

The Morphology of Society Mead frequently used terms that carry structural connotations, with notions of *group, community, institution,* and *society* being the most common. To some degree, he used these terms interchangeably to denote regularity in patterns of interaction among individuals. Yet at times he appears to have had an image of basic structural units that compose a total society.

Mead used the term *society* in two senses: (1) society simply refers to ongoing, organized activity, and (2) society pertains to geopolitical units, such as nation-states. The former usage, however, is the most frequent, and thus, we will retain the view that *society* is the term for ongoing and organized activity among pluralities of actors, whether this activity is that of a small group or of a total society.

Mead's use of *community* was ambiguous, and he often appeared to equate it with society. His most general usage referred to a plurality of actors who share a common set of significant symbols, who perceive that they constitute a distinguishable entity, and who share a common generalized other, or community of attitudes. As such, a community can be quite small or large, depending on whether people perceive that they constitute an entity. Mead typically employed the concept of community to denote large pluralities of actors, and thus other structural units were seen to operate within communities.

Within every community, there are certain general ways in which people are supposed to act. These are what Mead defined as *institutions:*

> There are, then, whole series of such common responses in the community in which we live, and such responses are what we term "institutions." The institution represents a common response on the part of all members of the community to a particular situation.[28]

Institutions, Mead argued, are related, and thus, when people act in one institutional context, they implicitly invoke responses to others. As Mead emphasized,

> Institutions . . . present in a certain sense the life-habits of the community as such; and when an individual acts toward others in, say, economic terms, he is calling out not simply a single response but a whole group of related responses.[29]

Institutions represent only general lines of response to varying life situations, whether economic, political, familial, religious, or educational. People take the role of the generalized other for each institution, and because institutions are interrelated, they tend to call out appropriate responses for other institutions. In this way, people can move readily from situation to situation within a broader community, calling out appropriate responses and inhibiting inappropriate ones. One moves smoothly, for example, from economic to familial situations because responses for both are evoked in the individual during role-taking with one or the other.

Mead recognized that institutions, and the attendant generalized other, provide only a broad framework guiding people's actions. People belong to a wide variety of smaller units that Mead tended to call *groups*. Economic activity, for example, is conducted by different individuals in varying economic groups. Familial actions occur within family groups, and so on for all institutional activity. Groups reveal their own generalized others, which are both unique and yet consistent within the community of attitudes of social institutions or of the broader community. Groups can vary enormously in size, differentiation, longevity, and restrictiveness, but Mead's general point is that activity of individuals involves simultaneous role-taking with the generalized other in groups, clusters of interrelated institutions, and broad community perspectives.

The Culture of Society Mead never used the concept of *culture* in the modern sense of the term. Yet his view of social organization as mediated by generalized others is consistent with the view that culture is a system of symbols by which human thought, perception, and action are mobilized and regulated. As with social structure or morphology, however, Mead was not interested in analyzing in detail the varieties of symbol systems humans create and use to organize their affairs. Rather, he was primarily concerned with the more general insight that humans use significant symbols, or language, to create communities of attitudes. And by virtue of the capacity for role-taking, humans regulate their conduct not only in relation to the attitudes of specific others but also relative to generalized others who embody these communities of attitudes.

The concept of *generalized other* is Mead's term for what would now be seen as those symbol systems of a broader cultural system that regulate perception, thought, and action. His generalized other is thus composed of norms, values, beliefs, and other such regulatory systems of symbols. He never made careful distinctions, for example, among values, beliefs, and norms, for he was interested only in isolating the basic processes of society: Individuals with mind and self role-take with varieties of generalized others to regulate their conduct and, thus, coordinate their actions.

Mead's conception of society, therefore, emphasizes the basic nature of the processes underlying ongoing social activity. He was not concerned, to any great degree, with the details of social structure or the components of culture. His great insight was that regardless of the specific structure of society, the processes by which society is created, maintained, and changed are the same. Social organization is the result of behavioral capacities for mind and self as these allow actors to role-take with varieties of others and, thus, to regulate and coordinate their actions. This insight into the fundamental relationship between the individual and society marks Mead's great contribution in *Mind, Self, and Society*. Before his synthesis, we should emphasize, the nature of this relationship had not been conceptualized, as can best be illustrated by comparing his analysis with those of the theorists examined in previous chapters.[30]

THE PHILOSOPHY OF THE ACT

Mead left numerous unpublished papers, many of which were published posthumously in *The Philosophy of the Act*.[31] Much of this work is not of great interest to sociologists; in the first essay, however, one on which the editors imposed the unfortunate title "Stages of the Act," Mead offered new insights that cannot be found in his other essays or in his lectures. In this piece, he presented a theory of human motivation that should be viewed as supplemental to his conceptualization of mind, self, and society.

Mead did not present his argument as the concept of *motivation,* but his intent was to understand why and how human action was initiated and given direction. For Mead, the most basic unit of behavior is "the act," and much of *The Philosophy of the Act* concerns understanding the nature of this fundamental unit. The behavior of an individual is ultimately nothing more than a series of acts, sometimes enacted singularly but more often emitted simultaneously. Thus, if we are to gain insight into the nature of human behavior, we must comprehend the constituent components of behavior—that is, "acts."

In his analysis of the act, Mead retained his basic assumptions. Acts are part of a larger life process of organisms adjusting to the environmental conditions in which they find themselves. Moreover, human acts are unique because of people's capacities for mind and self. Thus, Mead's theory of motivation revolves around understanding how the behavior of organisms with mind and self and operating within society is initiated and directed. He visualized the act as composed of four "stages," although he emphasized that humans could simultaneously be involved in different stages of different acts. He also recognized that acts vary in length, degree of overlap, consistency, intensity, and other variable states, but In his analysis of the stages of the act, he was more interested in isolating the basic nature of the act than in developing propositions about its variable properties.

Mead saw acts as consisting of four stages: (1) impulse, (2) perception, (3) manipulation, and (4) consummation.[32] These are not entirely discrete, for they often blend into one another, but they constitute distinctive phases involving somewhat different behavioral capacities. Our discussion will focus on each stage separately, but we must emphasize that Mead did not view the stages of a given act as separable or as isolated from the stages of other acts.

Impulse

For Mead, an *impulse* represents a state of disequilibrium, or tension, between an organism and its environment. Although he was not concerned with varying states of impulses—that is, their direction, type, and intensity—he did offer two implicit propositions: (1) The greater the degree of disequilibrium between an organism and its environment, the stronger its impulse and the more likely behavior is to reflect this. (2) The longer an impulse persists, the more it will direct behavior until it is consummated.

The source of disequilibrium for an organism can vary. Some impulses come from organic needs that are unfulfilled, whereas others come from interpersonal maladjustments.[33] Still others stem from self-inflicted reflections, and many are a combination of organic, interpersonal, and intrapsychic sources of tension. The key point is that impulses initiate efforts at their consummation, while giving the behavior of an organism a general direction. Mead was quick to point out, however, that a state of disequilibrium could be eliminated in many different ways and that the specific direction of behavior would be determined by the conditions of the environment. For Mead, humans are not pushed and pulled around by impulses. On the contrary, an impulse is defined as the degree of harmony with the environment, and the precise ways it is consummated are influenced by the manner in which an organism is prepared to adjust to its environment.

For example, even seemingly organic drives such as hunger and thirst are seen as arising from behavioral adaptations to the environment. Hunger is often defined by cultural standards for when meals are to be eaten, and it arises when the organism has not secured food from the environment. The way in which this disequilibrium will be eliminated is greatly constrained by the individual's social world. The types of foods considered edible, the way they are eaten, and when they can be eaten will all be shaped by environmental forces as they impinge on actors with mind and self. Thus, for Mead, an impulse initiates behavior and gives it only a general direction. The next stage of the act—perception—will determine what aspects of the environment are relevant for eliminating the impulse.

Perception

What humans see in their environment, Mead argued, is highly selective. One basis for selective perception is the impulse: People become attuned to those objects in their environment perceived relevant to the elimination of an impulse. Even here, past socialization, self-conceptions, and expectations from specific and generalized others all constrain what objects are seen as relevant to eliminating a given impulse. For example, a hungry person in India will not see a cow as a relevant object of food but rather will become sensitized to other potential food objects.

The process of *perception* thus sensitizes an individual to certain objects in the environment. These objects become stimuli for repertoires of behavioral responses. Thus, as an individual becomes sensitized to certain objects, he or she is prepared to behave in certain ways toward those objects. Mead believed, then, that perception is simply the arousal of potential responses to stimuli; that is, as the organism becomes aware of relevant objects, it also is prepared to act in certain ways. Humans thus approach objects with a series of hypotheses, or notions, about how certain responses toward objects can eliminate their state of disequilibrium.

Manipulation

The testing of these hypotheses—that is, the emission of behaviors toward objects—is termed *manipulation.* Because humans have mind and self, they can engage in covert as well as overt manipulation. A human can often covertly imagine the consequences of action toward objects for eliminating an impulse. Hence, humans frequently manipulate their world mentally, and only after imagining the consequences of various actions do they emit an overt line of behavior. At other times, humans manipulate their environment without deliberate or delayed thinking; they simply emit a behavior perceived as likely to eliminate an impulse.

What determines whether manipulation will be covert before it is overt? The key condition is what Mead saw as *blockage,* a condition where the consummation of an impulse is inhibited or delayed. Blockage produces imagery and initiates the process of thinking. For example, breaking a pencil while writing (creating impulse or disequilibrium with the environment) leads to efforts at manipulation: One actor may immediately perceive a pile of sharpened pencils next to the writing pad, pick up a new pencil, and continue writing without a moment's reflection. Another writer, who did not prepare a stack of pencils, might initially become attuned to the drawer of the desk, open it, search for a pencil, and generally start searching "blindly" for a pencil. At some point, frequently after a person has "wandered around unconsciously" for a while, the blockage of the impulse begins to generate conscious imagery, and a person's manipulations become covert. Images of where one last left a pencil are now consciously evoked, or the probable location of a pencil sharpener is anticipated. Thus, when the impulse, perception, and overt manipulation stages of the act do not lead to consummation, thinking occurs, and manipulation becomes covert, using the behavioral capacities of mind and self.

Thinking can also be initiated earlier in the act. For example, if perception does not yield a field of relevant objects, blockage occurs at this stage, with the result that by virtue of the capacities for mind, an actor immediately begins covert thinking. Thus, thinking is a behavioral adaptation of an organism experiencing disequilibrium with its environment and unable to perceive objects or manipulate behaviors in ways leading to consummation of an impulse.

In the process of thinking, then, an actor comes to perceive relevant objects; the actor might even role-take with the object if it is another individual or a group; a self-image may be derived, and one may see self as yet another object; and then various lines of conduct are imaginatively rehearsed until a proper line of conduct is selected and emitted. Of course, if the selected behavior does not eliminate the impulse, the process starts over and continues until the organism's behavior allows it to achieve a state of equilibrium with its environment.

The stage of manipulation is thus "cybernetic" in that it involves behavior, feed-back, readjustment of behavior, feedback, readjustment, and so on until an impulse is eliminated.[34] Mead's vision of thinking as "imaginative rehearsal" and his conceptualization of the "I" and "me" fit into this more general cybernetic view of the act. Thinking involves imagining a behavior and then giving oneself the feedback about the probable consequences of the behavior. The "I" and "me" phases of self involve deriving "me" images (feedback) from behaviors ("I") and then using these images to adjust subsequent behaviors. Unlike many theorists of motivation, Mead saw acts as constructed from a succession of manipulations that yield feedback, which, in turn, is used to make subsequent manipulations. Thus, motivation is a process of constant adjustment and readjustment of behaviors to restore equilibrium with the environment.

Although Mead did not develop any formal propositions on the manipulatory stage of the act, he implicitly assumed that the more often an impulse was blocked, the more it grew in intensity and the more it consumed the process of thinking and the phases of self. Thus, individuals who have not eliminated a strong impulse through successful manipulation will have a considerable amount of their thinking and self-reflection consumed by imagery pertaining to objects and behaviors that might eliminate the impulse. For example, people who cannot satiate their hunger or sexual appetites or who cannot achieve the recognition they feel they deserve are likely to devote a considerable, and ever-increasing, amount of their time in covert and overt manipulations in an effort to control their impulses.

Consummation

The *consummation* stage of the act simply denotes the act's completion through the elimination of the disequilibrium between an organism and its environment. As a behaviorist, Mead emphasized that successful consummation of acts by the emission of behaviors in relation to certain objects led to the development of stable behavior patterns. Thus, general classes, or types, of impulses will tend to elicit particular responses from an individual if these responses have been successful in the past in restoring equilibrium. Individuals will tend to perceive the same or similar objects as relevant to the elimination of the impulse, and they will tend to use these objects as stimuli for eliciting certain behaviors. In this way, people develop stable behavioral tendencies to act on their environment.

Figure 23.3 represents Mead's conceptualization of these phases of the act. For any person, of course, multiple impulses are operating, each at various stages of consummation and at potential points of blockage. For humans, perception involves seeing not only physical objects but also oneself, others, and various generalized others as part of the environment. Manipulation for humans with the capacities for mind and self involves both overt behavior and covert deliberations where individuals weigh alternatives and assess their consequences with reference to their self-conception, the expectations of specific others, and various generalized others. Consummation for humans, who must live and survive in social groups, almost always revolves around adaptation to, and cooperation with, others in ongoing collective enterprises. As the feedback arrows denoting blockage emphasize, the point of blockage influences the salience of any phase in the flow of an act. Moreover, this process of blockage determines the strength of the causal arrows connecting stages in the act. Intense impulses are typically those that have been blocked, thereby causing heightened perception. In turn, heightened perception generates greater overt and covert manipulation; if blockage occurs, perception is further heightened, as are impulses. If manipulation is unsuccessful, escalated covert manipulation ensues, thereby heightening perception and the impulse (via the feedback arrows at the top of Figure 23.3).

This model of the act allows an understanding of how individuals can be "driven" to seemingly irrational or excessively emotional behavior, and it can provide insight into the dynamics of compulsive behavior. These behaviors would result from the blockage of powerful impulses that persist and escalate in intensity, thereby distorting an individual's perceptions, covert thinking, and overt behavior. For example, individuals who were rejected by significant others in their early years might have a powerful series of unconsummated impulses that distort their perceptions and manipulations to abnormal extremes. Given that the unstable or abnormal self-conceptions of such individuals can distort the process of perception, as well as covert and overt manipulation, they might never be able to perceive that they have consummated their impulses in interpersonal relations.

Unlike Freud or other clinicians and psychologists of his time, Mead was not interested in types of abnormal behavior. He was more concerned with constructing a model that would denote the fundamental properties of human action, whether normal or abnormal. His critics often portray Mead's social behaviorism as overly rational, but this view does not consider his model of the act. This model contains the elements for emotional as well as rational action, and although he was not interested in assessing the consequences of various weights among the arrows in Figure 23.3, the model provides a valuable tool for those who are concerned with how various types of impulses, when coupled with different patterns of impulse blockage, will produce varying forms of covert and overt behavior.

Mead's analysis of the stages of the act thus provides a useful supplement to his discussion of *Mind, Self, and Society*. We now have a better vision of why people initiate action and why behaviors take a certain direction. In many ways, Mead's conception of motivation represents a synthesis of diverse schools of thought. The stimulus-response, or reflex-arc, approach of J. B. Watson and, more recently, of B. F. Skinner is retained without the restrictive tendency to avoid the "black box" of human cognition. The Gestalt psychology of Mead's time is evident through the emphasis on behavior as initiated by a desire to maintain harmony within a perceptual field of objects and relations. The psychoanalytic view of behavior as a reconciliation of impulses and ego processes is maintained in the emphasis of biosocial sources of disequilibrium and the stages of perception and covert manipulation by actors with mind and self. Moreover, Mead's emphasis on blockage, and how blockage increases the intensity of impulses, is consistent with the psychoanalytic view of the sources of disruptive emotional states.

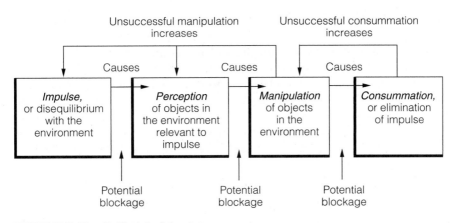

FIGURE 23.3 Mead's Model of the Act

In sum, then, Mead's theory of motivation, much like his view of mind, self, and society, represents a synthesis of diverse and often contradictory viewpoints. The biological individual is not ignored, the internal psychological processes of individuals are highlighted, and the relation of acts to the ongoing processes of society is still prominent. Mead's view of motivation is thus distinctly sociological, emphasizing the relationship of individuals to one another and to the social as well as physical environments. What drives actors and shapes the course of their behaviors is the relationship of the organism to its environment. For human actors, who by virtue of mind and self are able to live and participate in society, this environment is decidedly social. Therefore, humans initiate and direct their actions in an effort to achieve integration into the ongoing social process. Mead's *social* behaviorism marked a synthesis of utilitarian, pragmatist, behaviorist, and even Darwinian notions. Mead's basic premise is this: Behaviors that facilitate the adjustment and adaptation of organisms to their environment will be retained.

The environment of any individual organism is society. Thus, the infant must adjust to society, developing the behavioral capacities for language, role-taking, mind, and self. The continued use of these fundamental capacities is essential for the mature individual's ongoing adjustment and adaptation to society. Mead's critical insight is that the capacities for mind and self are behaviors. Moreover, these capacities ensure that among humans much action will be covert and involve role-taking, reflective thinking, self-criticism, and self-assessment.

These behavioral capacities for mind and self make humans distinctive. Only from interaction among actors with mind and self is society possible. For a species not organized by instincts, as are ants and bees, the ability to role-take becomes crucial in such interactions. Indeed, impulses are both caused and constrained by the capacity to role-take, not only with one another but also with broader perspectives, values, beliefs, and norms. All these constraints on impulses allow the organization of the species into society. This behavioral ability also makes human social organization flexible, thereby facilitating the adjustment and adaptation of the species as a whole to the environment.

Thus, the acquisition of mind and self enables individuals to adapt to their social environment, and the flexible interactive abilities of individuals with mind and self facilitate the species' adaptation to the environment through the creation, maintenance, and change of society. Mead's ideas must be viewed within this basic framework. As we saw in Chapter 22, he borrowed much from other scholars, but he combined their ideas in new ways into an approach that unraveled the basic nature of the relationships between the individual and society.

CRITICAL CONCLUSIONS

Before Mead, the process of interaction was not well understood. Various thinkers had captured a portion of the process, but Mead synthesized various lines of thinking into a coherent conceptual framework. The strength of Mead's analysis resides in his understanding of the relationship among ongoing patterns of social organization, or society, and the behavioral capacities that arise from human needs to adapt to these patterns and that, as a result, sustain society. Using conventional gestures, role taking, mind, and self are, in Mead's eyes, behaviors rather than entities or things, and they are learned like all behaviors because they provide reinforcement to individuals or, alternatively in pragmatist terminology, because they allow for adaptation to society. Thus, for Mead, society always stands above the individual in the sense that it exists before a person is born and, consequently, is the environment to which individuals must adjust and adapt. Yet, without learning conventional gestures and role taking, and without acquiring the ability to engage in minded deliberations or self reflection and appraisal from the perspective of society and its various generalized others, society would not be possible.

It is difficult to criticize the thinker who, in essence, unlocked the mysteries of micro social processes, but we can offer several criticisms. One is that Mead never developed a very clear conception of society or culture. He saw "institutions" as ongoing patterns of cooperative behavior, and he viewed culture in terms of various generalized others. Yet, this is a rather minimal conception of macrostructures that are sustained by microprocesses; so, even though Mead saw society as standing above the individual, his theory is really about how people acquire the behavioral capacities to adjust to society and culture. We are not, however, given a theory of society or culture.

One result of this failure is that many contemporary theorists assume that a separate theory, or set of theories, about the dynamics of society and culture is not necessary. Instead, all we need is a theory of interpersonal behavior to explain institutional and cultural systems. Such theories become, however, little more than pronouncements that, for example, assert that "society is symbolic interaction"[35] which says very little and explains virtually nothing about society beyond the interpersonal processes necessary to sustain it. Thus, Mead's sociology is decidedly micro, which is fine as long as we realize this limitation; many contemporary sociologists unfortunately forget this fact.

We can even criticize Mead on the more micro level of analysis. Probably his greatest failing is the lack of a theory of emotions. One of the most critical aspects of interaction is its emotional content, and when individuals role-take, engage in minded deliberations, or make self appraisals, they are being emotional. Mead even had a Freudian-looking theory of "the act" that could easily have been used to address the emotions involved as impulses go unconsummated or as they are consummated. But we are never given a word. Moreover, Mead had used Darwinian metaphors in all his work, and he was certainly aware of Darwin's book[36] on expressions and emotions in animals, and yet, he did not pick up this lead. Thus, because Mead was considered the key figure in microsociology for most of the century, his lapse became the discipline's gap in knowledge. For, not until the late 1970s did the sociology of emotions emerge as a field of inquiry in interactionist theorizing. The only explanation for this late interest is the slavish conformity to Mead's lead, which, as profound as it was, did not tell a complete story of microsocial processes.

NOTES

1. The philosophical tone of Mead's posthumously published lectures is revealed in the titles of the four books: *The Philosophy of the Present* (La Salle, IL: Open Court, 1959; originally published in 1932); *Mind, Self, and Society* (Chicago: University of Chicago Press, 1934); *Movements of Thought in the Nineteenth Century* (Chicago: University of Chicago Press, 1936); and *The Philosophy of the Act* (Chicago: University of Chicago Press, 1938). *Mind, Self, and Society* contains a bibliography of Mead's published work (pp. 390–392).

2. As he observed with respect to Watson's efforts to deal with subjective experience: "John B. Watson's attitude was that of the Queen in *Alice in Wonderland*—Off with their heads!—there were no such things." Mead, *Mind, Self, and Society*, pp. 2–3.

3. Mead, *Mind, Self, and Society*, p. 10.

4. Ibid., p. 7.

5. And, of course, the more recent version of B. F. Skinner and others of this stripe.

6. That is, because they cannot be directly observed, they cannot be studied.

7. Mead, *Mind, Self, and Society*, p. 133.

8. As Mead observed, "The term gesture may be identified with these beginnings of social acts which are stimuli for the response of other forms." *Mind, Self, and Society*, p. 43.

9. Mead, *Mind, Self, and Society*, p. 76.

10. Ibid., p. 78.

11. However, the evidence is now clear that other higher primates can use such "significant gestures."

12. For other published statements by Mead on the nature and operation of mind, see "Image and Sensation," *Journal of Philosophy* 1 (1904), pp. 604–607; "Social Consciousness and the Consciousness of Meaning," *Psychological Bulletin* 7 (1910), pp. 397–405; "The Mechanisms of Social Consciousness," *Journal of Philosophy* 9 (1912), pp. 401–406; "Scientific Method and Individual Thinker," in *Creative Intelligence* (New

York: Holt, Rinehart & Winston, 1917), pp. 176–227; and "A Behavioristic Account of the Significant Symbols," *Journal of Philosophy* 19 (1922), pp. 157–163.

13. Mead, *Mind, Self, and Society,* p. 135.

14. Mead did reject many of the specifics in Cooley's argument about "the looking-glass self." See, for example, *Mind, Self, and Society,* p. 173, "Cooley's Contribution to American Social Thought," *American Journal of Sociology* 35 (1929–1930), pp. 385–407, and "Smashing the Looking Glass," *Survey* 35 (1915–1916), pp. 349–361.

15. Mead, *Mind, Self, and Society,* p. 138.

16. Ibid., p. 140.

17. Ibid., p. 144.

18. Ibid., p. 199.

19. Ibid., p. 160.

20. The similarity between Durkheim's notion of the collective conscience and Mead's conception of generalized other should be immediately apparent. But in contrast with Durkheim, Mead provided the mechanism—role-taking and self-related behaviors—by which individuals become capable of viewing and controlling their actions in the perspective of the collectivity. For more details along this line, see Jonathan H. Turner, "A Note on G. H. Mead's Behavioristic Theory of Social Structure," *Journal for the Theory of Social Behavior* 12 (July 1982), pp. 213–222, and *A Theory of Social Interaction* (Stanford, CA: Stanford University Press, 1988), Chapter 10.

21. Mead, *Mind, Self, and Society,* p. 157.

22. Ibid., pp. 180, 202, and 216 for the relevant statements.

23. For Mead's explicitly published works on self, see "The Social Self," *Journal of Philosophy* 10 (1913), pp. 374–380, "The Genesis of the Self and Social Control," *International Journal of Ethics* 35 (1924–1925), pp. 251–277, and "Cooley's Contribution."

24. Mead, *Mind, Self, and Society,* pp. 321–322.

25. Ibid., pp. 321–322.

26. Ibid., p. 327.

27. Ibid., p. 326.

28. Ibid., p. 261.

29. Ibid., p. 264.

30. See, in particular, Weber's and Durkheim's analyses to appreciate how crudely the interactive basis of social structure had been conceptualized before Mead's synthesis.

31. Mead, *Philosophy of the Act.*

32. For an excellent secondary discussion of Mead's stages of the act, see Tamotsu Shibutani, "A Cybernetic Approach to Motivation," in *Modern Systems Research for the Behavioral Scientist,* ed. Walter Buckley (Hawthorne, NY: Aldine, 1968), and *Society and Personality, An Interactionist Approach to Social Psychology* (Englewood Cliffs, NJ: Prentice-Hall, 1961), pp. 63–93.

33. For Mead's conceptualization of biologic needs, see the supplementary essays in *Mind, Self, and Society,* particularly essay 2.

34. See Shibutani, "Cybernetic Approach," for a more detailed discussion.

35. Herbert Blumer, *Symbolic Interactionism: Perspective and Method* (Englewood Cliff, NJ: Prentice-Hall, 1969).

36. Charles Darwin, *The Expressions of Emotions in Man and Animals* (London, UK: Wats, 1982).

24

✳

Mead's
Theoretical Legacy

George Herbert Mead unlocked the basic mechanisms of human action and interaction. Mead saw humans as acquiring the basic behavioral capacities necessary for participation in ongoing patterns of social organization. By virtue of these capacities for reading and using conventional gestures, role-taking, minded deliberations, and awareness of self as an object in relation to generalized others, people are able to cooperate and, thereby, sustain society. The basic process of interaction, then, involves individuals (a) mutually role-taking as they read the conventional gestures emitted by others, (b) deriving "me" images of oneself as an object while determining the likely course of action of others, (c) deriving additional "me" image from the point of view of various generalized others attached to the structure of the situation, (d) deliberating about the best course of action in light of these "me" images, and (e) finally behaving towards others (the "I" phase of interaction). As soon as this "I" phase is complete, Mead argued, the entire process in repeated again.

Mead added to this model of interaction an analysis of the "phases of the act" which has not, in our opinion, been given sufficient consideration in sociological discussions of Mead's work. In his view, humans always have a configuration of "impulses" at various states of consummation. Each of these impulses represents a point of disequilibrium with the environment. Some are long-term and chronic; others are in the process of being consummated; and still others are presently guiding perception and thought. The point that Mead emphasized is that interaction is motivated, but this process is complicated by the simultaneous operation of configurations of impulses at various phases of

485

the act. As impulses or states of disequilibrium with the environment are activated, "perception" of relevant objects that can consummate these impulses ensues. This perception is highly selective, oriented to those objects that can allow consummation of any impulse. The next phase of the act is "manipulation" in which individuals overtly behave in their environment or covertly engage in minded deliberations about how best to consummate the impulse. The final stage involves the consummation of the impulse, or the failure to do so. When impulses go unconsummated, they increase in intensity and begin to distort the perception and manipulation phases of the act.

Acts and interaction all occur in cultural and social contexts, but as we noted at the close of the last chapter, these issues are rather under-theorized by Mead. He conceptualized culture in terms of generalized others, which can be temporary and immediate to a particular episode of interaction, or attached to various levels of social structure, from the group through community and organizations to a whole society. The phrase "community of attitudes" used to describe the generalized other is rather vague, but it connotes the perspective of collectivities as represented by values, beliefs, norms, and perhaps cognitive processes such as categories of thought and perception. Social structure, or society, is even more imprecisely analyzed. At the level of acts and interaction, social structure is the pattern of activity in which individuals are cooperating, but beyond this more micro view, social structure at the macro level is not theorized beyond statements that institutions exist and constrain micro patterns of ongoing cooperative activity. In many ways, Mead is more a cultural theorist because the generalized other appears to be the way that social structure influences acts and interactions. For each level of social structure—from groups to the society as a whole—there is a generalized other, individuals role-take with these generalized others, and thus, social structure influences acts and interaction through individuals' efforts at role-taking with generalized others.

MEAD'S CAUSAL MODEL
OF SOCIAL PROCESSES

With these considerations behind us, we can construct a model of social process contained in Mead's sociology. As is shown in Figure 24.1, the model tries to reconcile Mead's analysis of the phases of the act with the analysis of interaction. As the left portion of the model emphasizes, behavior is initiated by impulses, and the more intense are the impulses, the more active are both overt and covert behaviors. Most impulses come from levels of adjustment and adaptation to social environments. Hence, when individuals feel out of synchronization with ongoing patterns of social organization, or sense that such might be the case, they become motivated to find a way to cooperate. Their perceptions of relevant objects are heightened and selective, and as the model emphasizes, two of the most important objects to be perceived are self and generalized others. Perceptions can lead to immediate behavioral responses,

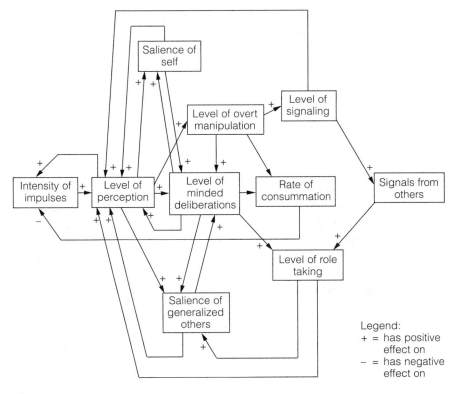

FIGURE 24.1 Mead's Model of Social Processes

what is termed "level of overt manipulation" in the model, but if impulses are powerful or have failed to be consummated in the past, then covert deliberations occur, and these are labeled "minded deliberations" in the model. The more mind processes are activated, the more salient will self and generalized others become, and the more they will influence the overt emission of gestures or signals to others.

Another important object in individuals' perceptions and manipulations are other people, whose dispositions to act are discovered through role-taking of their conventional gestures. The more active is role-taking, the more others become objects in the perceptual and manipulative phases of the act. Again, the more mind intervenes in the deliberation about what the gestures of others mean, the more overt behavior will reflect an effort to reconcile conceptions of self and dictates of generalized others with expectations and courses of actions of others in the situation. Mead also suggested that individuals can role-take with others not present in the situation, using these others' expectations and "actions" as a frame of reference in minded deliberations and overt behaviors.

Human interaction constantly activates these processes. Even before an interaction occurs, these processes can be in full operation. During the course of one episode of interaction, many iterations of these processes ensue as impulses, perceptions, minded deliberations involving self, generalized others, and others are constantly readjusted. As anyone who has thought about an interaction that has been completed knows, these processes can operate long after the interaction as people play out alternative scenarios of how they could or should have acted. Mead's great contribution to general social theory is this model of action and interaction.

Most of Mead's work, however, examines how the capacities for using conventional gestures, for role-taking, for mind, and for self are acquired through learning. We have already reviewed these processes in detail in the last chapter and have summarized their dynamics in models. Keeping in mind the models in Figures 23.1 on page 462, 23.2 on page 468, and 23.3 on the act on page 481, along with the general model in figure 24.1, we can now examine the elementary principles that describe how humans acquire the behavioral capacities that allow them to engage in interactions sustaining the viability of society.

MEAD'S THEORETICAL PRINCIPLES

Although Mead never stated his ideas as theoretical principles, it is clear that he thought in terms of basic and fundamental relationships among phenomena. Mead is read, and then re-read today, because he articulated some universal features of interaction processes, and we would do justice to his genius by reviewing the principles that can be readily extracted from his lectures on social psychology and philosophy in general.

We can best appreciate Mead's theoretical principles by recognizing that Mead sought to show how human action, interaction, and organization are qualitatively unique, yet extensions of behavioral processes evident in other species. Mead initially postulated principles of action and interaction in general and then attempted to show how the emergence of the behavioral capacities for mind and self make human action and interaction unique and how this uniqueness arises from, and at the same time allows, the creation, maintenance, and change of society. Thus, we will first examine Mead's general principles of action and interaction and, then, explore his general principles of *human* action, interaction, and social organization.

Principles of Animal Action

In *The Philosophy of the Act,* Mead presented several general principles of action—that is, of what is involved in initiating and giving direction to behavior in all animals. As we saw, his ideas are expressed as "stages" and can be modeled. These ideas can also be expressed as a series of basic principles of action:

1. The greater is the degree of maladjustment of an organism to its environment, the stronger are the organism's impulses.

2. The greater is the intensity of an organism's impulse, the greater is the organism's perceptual awareness of objects that can potentially consummate the impulse and the more the organism manipulates objects in the environment.
   a. The more maladjustment stems from unconsummated organic needs, the more intense is the impulse.
   b. The longer an impulse has gone unconsummated, the more intense is the impulse.
3. The more impulses have been consummated by the perception and manipulation of certain classes of objects in the environment, the more perceptual and behavioral responses will be directed at these and similar classes of objects when similar impulses arise.

These three principles summarize Mead's social behaviorism. Action emerges from adjustment and adaptation problems encountered by an organism. Behavior is directed at restoring equilibrium between the organism and the environment. The essence of behavior involves perception and manipulation of objects. Successful perception and manipulations are retained in an organism's behavioral repertoire. As Mead argued, human action emerged from this behavioral base, but the capacities for mind, self, and society require, as we will document shortly, additional theoretical principles if the distinctive qualities of human action are to be understood. To appreciate fully these additional principles, however, we need first to summarize Mead's general principles of interaction among animals without mind, self, and society.

Principles of Animal Interaction

Mead's view of interaction can be modeled as a causal sequence of events over time, as was done in Chapter 3, or it can be expressed by the following two principles:

4. The more organisms seek to manipulate objects in their environment in an effort to consummate impulses, the greater is the visibility of gestures marking their course of action.
5. The greater are the number and visibility of gestures emitted by acting organisms, the more these organisms adjust their responses to each other.

These two principles underscore Mead's view that the essence of interaction involves (a) an organism emitting gestures as acts on the world, (b) another organism responding to these gestures and, hence, emitting its own gestures as it seeks to consummate its impulses, and (c) readjustments of responses by each organism in response to the gestures emitted. Mead termed the process the *triadic matrix,* and as we emphasized in the last chapter, it can occur without cognitive manipulations and without the development of common meanings. Indeed, as Mead argued, only among humans with mind and self, living in society, does this fundamental interactive process involve cognitive manipulations, normative regulation, and shared meanings.

Principles of Human Action, Interaction, and Organization

Critical to understanding Mead's view of human action, interaction, and organization is the recognition that humans develop mind and self from their participation in society. We first review Mead's formulation of principles on the development of mind and self, then we can see how these two behavioral capacities alter the principles of human action, interaction, and organization.

Principles on the Emergence of Mind Any particular individual is born into a society of actors with mind and self. Mind, self, and human society were thus seen by Mead as intimately connected because mind and self are learned as a result of having to participate in society while society is reproduced by virtue of mind and self. We must jump into this cycle of interconnections at some point, so we begin by isolating Mead's view of socialization and the emergence of mind. This view assumes the prior existence of society and adult actors with mind and self; young infants who do not possess these behavioral capacities must adapt and adjust into this social milieu. In attempting to understand how infants adjust to adult actors and to society, Mead offered a series of important principles describing the fundamental properties of the socialization process:

6. The more an infant must adapt to an environment composed of organized collectivities of actors, the more is its exposure to significant gestures.

7. The more an infant seeks to consummate its impulses in an organized social collectivity, the greater is the selective value for consummating impulses of leaning how to read and use significant gestures.

8. The more an infant can use and read significant gestures, the greater is its ability to role-take with others in its environment, and hence, the greater is its capacity to communicate its needs and to anticipate the responses of others on whom it depends for the consummation of impulses.

9. The greater is the capacity of an infant to role-take and use significant gestures, the greater is its capacity to communicate with itself.

10. The greater is the capacity of an infant to communicate with itself, the greater is its ability to covertly designate objects in its environment, inhibit inappropriate responses, and select a response that will consummate its impulses and thereby facilitate its adjustment.

11. The more an infant reveals such minded behavior, the greater is its ability to control its responses and, hence, to cooperate with others in ongoing and organized collectivities.

These principles should be read in two ways. First, each principle, by itself, expresses a fundamental relationship in the nature of human development. For example, principle 6 states that human infants are, by virtue of being born into society, inevitably exposed to a collage of significant gestures; or principle 7 states that because infants must consummate impulses in a world of significant gestures, they will learn to read and use these gestures as a means of increasing their adjustment. Thus, each principle states that one variable condition, stated in the first clause of the principle, will lead to the development of another capacity, which is stated in the second clause of the principle, in the maturing human infant. Second, this sequence of six principles should be viewed as marking "stages" in the genesis of a critical behavioral capacity, mind.

Principles on the Emergence of Self As the capacities for mind begin to emerge, self also becomes evident; however, the full development of self is, like the development of mind, the result of a series of fundamental processes. These processes are summarized by the following principles:

12. The more a young actor engages in minded behavior, the more it can read significant gestures, role-take, and communicate with itself.

13. The more a young actor reads significant gestures, role-takes, and communicates with itself, the more it can see itself as an object in any given situation.

14. The more diverse the specific others with whom a young actor can role-take, the more the actor sees itself as an object in relation to the dispositions of multiple others.

15. The more generalized is the perspective of others with whom a young infant can role-take, the more the infant sees itself as an object in relation to general values, beliefs, and norms of increasingly larger collectivities.

16. The greater has been the stability of a young actor's images of itself as an object in relation to both specific others and generalized perspectives, the more reflexive is the actor's role-taking and the more consistent are its behavioral responses.
    a. The more first self-images derived from role-taking with others have been consistent and noncontradictory, the more reflexive is role-taking and the more consistent are behaviors.
    b. The more self-images derived from role-taking with generalized perspectives are consistent and noncontradictory, the more reflexive is role-taking and the more consistent are behaviors.

17. The more a young actor reveals stability in its responses to itself as an object, and the more it sees itself as an object in relation to specific others as well as generalized perspectives, the greater its capacity to control responses and, hence, to cooperate with others in ongoing and organized collectivities.

These principles document Mead's view of certain fundamental relation-ships among role-taking acuity, images of the self as an object, and capacities for social control. These principles also summarize Mead's conceptualization of the sequence of events involved in generating a "unified" self in which an indi-vidual adjusts its responses in relation to (a) a stable self-conception, (b) spe-cific expectations of others, and (c) general values, beliefs, and norms.

As the consecutive numbering of the principles underscores, the develop-ment of mind and self is a continuous process. Once the behavioral abilities for mind and self in individual human organisms are evident, action and interac-tion as well as patterns of social organization among humans become qualita-tively different from that of nonhuman organisms. Yet, Mead emphasized that there is nothing mysterious or mystical about this qualitative difference. Indeed, even though human action, interaction, and organization are distinct by virtue of the capacity for mind, self, and symbolically mediated organiza-tion into society, this distinctiveness has been built on a base common to all acting organisms.

Principles of Human Action and Interaction The emergence of mind and self somewhat complicate Mead's view of the act as involving impulse, perception, and manipulation as well as his notion of the triadic matrix as a simple process of actors emitting and reading gestures as they adjust their responses to each other. Indeed, the complications introduced into the processes of action and interaction make an entirely new way to organize a species into society.

Concerning action, Mead noted one additional principle to account for the distinctive features of human acts:

18. The greater the intensity of impulses of humans with mind and self,
    (a) the more heightened is perceptual awareness of objects that can
    potentially consummate the impulse to be selective, (b) the more likely is
    manipulation to be covert, and (c) the more likely are both perception
    and manipulation to be circumscribed by self-conceptions, expectations
    of specific others, and generalized perspectives of organized collectivities.

When action, as described in principle 18, occurs in a social context with others, it then becomes overt interaction. Even isolated acts, where others are not physically present, involve interaction with symbolically invoked others and generalized perspectives. The capacities for mind and self, Mead argued, ensure that humans will invoke the dispositions of others and broader com-munities of attitudes to guide behavior during the course of their acts, even if specific others are not physically present and even if others do not directly react to one's behaviors. When others are present, the use of significant symbols and role-taking becomes more direct and immediate, requiring several supplemen-tary principles on *inter*action:

19. The more humans with mind and self seek to consummate impulses in
    the presence of others, the more they emit overt significant gestures and
    read the significant gestures of others, and hence, the greater is their
    role-taking activity.

20. The more humans role-take with each other, the more their interaction is guided by specific disposition of others present in a situation, by images of the self as a certain type of object, and by generalized perspectives of the organized collectivities.

When stated as principles, the key relationships among impulses, significant gestures, role-taking, self-conceptions, expectations of others, and generalized perspectives are highlighted. For Mead, impulses drive action, but most action occurs in a context of others and, hence, becomes interaction. Interaction among humans depends on role-taking, which, in turn, produces self-images, awareness of the dispositions of others, and cognizance of generalized perspectives. Such is the nature of human action and interaction among organisms with mind and self who must consummate their impulses within the framework imposed by ongoing patterns of social organization. Thus, principles 18, 19, and 20 can be interpreted as Mead's "laws" of action and interaction among humans.

By comparing these laws with his general principles of action and interaction among nonhuman organisms (principles 1 through 5), we can see that these principles on humans represent extensions of those on nonhumans. We have now summarized Mead's principles on the emergence of mind and self as well as those on action and interaction. These principles place into theoretical context Mead's vision of how society is created, maintained, and changed. As a philosopher and social psychologist, rather than as a structural sociologist, Mead had a distinctively social-psychological view of social organization. This vision supplements and complements the structural perspective developed in Europe by such figures as Herbert Spencer, Karl Marx, Max Weber, and Émile Durkheim. Although each of these scholars sought to uncover some of the social-psychological dynamics of macro-social structures, none was able to achieve the insights that Mead developed on the fundamental social-psychological properties underlying patterns of social organization.

Principles of Human Social Organization Because interaction among humans is possible by virtue of role-taking abilities, and because society involves stabilized patterns of interaction, society for Mead is ultimately a process of (a) role-taking with various "others" and (b) using the dispositions and perspectives of these others for self-evaluation and self-control. The nature and scope of society, Mead implicitly argued, are a dual function of the number of specific others and the abstractness of the generalized others with whom individuals can role-take. In many ways, Mead viewed society as a "capacity" for various types of role-taking. If actors can role-take with only one other at a time, then the scale of society is limited, but once they can role-take with multiple others, and then, with generalized others, the scale of society is greatly extended. These fundamental relationships are summarized in Mead's two basic principles on the dynamics underlying society:

21. The more actors role-take with pluralities of others and use the dispositions of multiple others as a source of self-evaluation and self-control, the greater is their capacity to create and maintain patterns of social organization.

22. The more actors role-take with the generalized perspective of organized collectivities and use this perspective as a source of self-evaluation and self-control, the greater is their capacity to create and maintain patterns of social organization.

If actors cannot meet the conditions specified in these two "laws" of social organization, then instability and change in patterns of interaction are likely. Actors who cannot role-take with multiple others at a time and use the dispositions of these others to see themselves as objects and to control their responses will not be able to coordinate their responses as well as actors who can perform such role-taking. Actors who cannot role-take with the general norms, beliefs, values, and other symbol systems of organized groups and use these to view themselves and to regulate their actions will not be able to extend patterns of social organization beyond immediate face-to-face contact. Only when actors can role-take with a broader "community of attitudes" and use a common set of expectations to guide their conduct do extended patterns of social organization become possible.

In addition to these two basic principles, Mead elaborated several propositions on role-taking with generalized others. Because the scope of society is ultimately a positive function of role-taking abilities with generalized others, Mead apparently felt it necessary to specify some of the variables influencing the relations among role-taking, generalized others, and the nature of society. Three variables are most prominent in Mead's scheme: (1) the degree to which actors can hold a *common* generalized other, (2) the degree of *consistency* among multiple generalized others, and (3) the degree of *integration* among different types and layers of generalized others. Mead implicitly incorporated these variables into additional principles of social organization:

23. The more actors role-take with a common generalized perspective and use this common perspective as a source of self-evaluation and self-control, the greater is their capacity to create and maintain cohesive patterns of social organization
    a. The more similar are the positions of actors, the more they will role-take with a common generalized perspective.
    b. The smaller is the population of actors, the more they will role-take with a common generalized perspective.

In this principle, the ability to role-take with a common perspective (norms, values, beliefs, and other symbolic components) is linked to the degree of cohesiveness in patterns of social organization. Thus, Mead argued, a common collective perspective maintains unified and cohesive patterns of organization. On this score, Mead came close to Durkheim's emphasis on the need for a "common" or "collective" conscience. In contrast with Durkheim, however, Mead was able to tie this point to the theory of human action and interaction, and hence, he was in a position to specify the mechanisms by which individual conduct is regulated by a "generalized other" or "collective conscience."

Much like Durkheim, Mead also recognized that the size of a population and its differentiation into roles influence the degree of commonality of the generalized other. Naturally, the converse of principle 23 could signal difficulties in achieving unified and cohesive patterns of social organization. If the members of a population cannot role-take with a common generalized other, then cohesive social organization will be more problematic. If a population is large or highly differentiated, role-taking with a common generalized other will be more difficult.

Like Durkheim, Mead recognized that a common generalized other becomes increasingly tenuous with growing size and differentiation of a population. For large differentiated populations, there are multiple "generalized others" because people participate in many different organized collectivities. These considerations led Mead to view consistency of generalized others as related to how extensive differentiation of social structure could become.

24. The more actors role-take with multiple but consistent generalized perspectives and use these perspectives as a source of self-evaluation and self-control, the greater is their capacity to differentiate roles and extend the scope of social organization.

In this principle, Mead argued that if the basic profile of norms, values, and beliefs of different groupings in which individuals participate are not contradictory, then differentiation does not lead to conflict and degeneration of social organization. On the contrary, multiple and consistent generalized others allow functional differentiation of roles and groups which, in turn, expands the scope (size, territory, and other such variables) of society. Of course, if generalized others are contradictory, then conflictual relations are likely, thereby limiting the extent of social organization.

Mead's analysis also resembled Durkheim's who recognized that symbolic components of culture exist at different levels of generality. Some are highly abstract and cut across diverse groupings, whereas others are tied to specific groups and organizations. Mead distinguished between "abstract" generalized others and concrete "organized" others to denote this facet of symbolic organization. Like Durkheim, Mead saw that the scope of social organization is limited by how well "abstract others" (values and beliefs, for example) are integrated with more concrete "organized others" (particular norms and doctrines of concrete groups, social classes, organizations, and regions, for example). Large-scale social organization, Mead felt, depends on common and highly abstract values and beliefs that are integrated with the specific perspectives of differentiated collectivities. The concept of integration, Mead appeared to argue, involves more than consistency and lack of contradiction; it denotes that the generalized others of particular organized collectivities represent concrete applications of the abstract generalized other. The abstract generalized other sets the parameters for less-abstract perspectives, thereby ensuring not just consistency between the two but also integration where the tenets of the specific follow from the abstract.

Mead's argument came close to that of Durkheim in this recognition. We can visualize this similarity in the following principle:

25. The more actors role-take with specific perspectives of particular collectivities that follow from more general perspectives, and the more these integrated perspectives are a source of self-evaluation and self-control, the greater is their capacity to extend the scope of social organization.

As with principles 1 through 24, the converse of this 25th principle can point to some of the conditions producing conflict and change. To the degree that abstract and specific perspectives are not integrated, actors will potentially have different interpretations of situations, and to the degree that they come into contact, the probability for conflict will be increased.

Principles 21 to 25 summarize Mead's vision of the basic properties of social organization. For society to exist at all, actors must be able to role-take with multiple others and with generalized others (principles 21 and 22). For a highly cohesive organization to exist and persist, actors must be able to role-take a common generalized other (principle 23). For somewhat less-cohesive but more differentiated and extensive patterns of social organization to be viable, actors must be able to role-take with multiple, but nevertheless non-contradictory, generalized others (principle 24). For large-scale and highly extensive patterns of organization, actors must role-take with well-integrated abstract and specific generalized others (principle 25).

In sum, we see that Mead's ideas can be converted into workable models and propositions. The propositions take time to delineate because they address the emergence of those behavioral capacities—use of significant gestures, role-taking, mind, and self—that make it possible for individuals to cooperate and, hence, for society to persist. Mead was a philosopher, but he gave sociology some of its most important ideas about the nature of human social interaction and about how micro-level processes arise from and, at the same time perpetuate, the macro social order.

Name Index

Annenkov, Paul, 105

Bacon, Sir Francis, 2
Bakunin, Mikhail, 104
Barnard, Chester, 381
Bauer, Bruno, 103, 104, 111–113
Baumgarten, Herman, 175
Baumgarten, Ida, 175
Baxter, Richard, 224
Bendix, Reinhard, 143, 211
Bentham, Jeremy, 436, 456
Böhm-Bawerk, Eugen, 182
Bonald, Louis de, 10, 18
Boutroux, Émile, 307
Brentano, Lujo, 181
Bukunin, Dina, 382

Calvin, John, 223
Charles X, 8
Comte, Auguste, 3, 4, 6–36, 38, 42, 51, 55, 60,
    85, 101, 267, 308–311, 318–321, 323,
    330, 337–340, 343, 351, 354, 360, 372,
    384, 386, 388, 454, 456, 472. *See also* sub-
    ject index
Condorcet, Jean, 3, 10–12, 14, 15, 20
Cooley, Charles Horton, 435, 436, 446–449, 451,
    462
Coser, Lewis, 251, 308, 435

Darwin, Charles, 46, 48–50, 255, 257, 335, 366,
    368, 371, 435–437, 441, 455, 456, 458,
    483
Dewey, John, 435, 436, 438, 443, 444,
    449–451, 455, 456, 459
Diderot, Denis, 3
Dilthey, Wilhelm, 173, 177, 178, 185–189, 254
Durkheim, André, 309
Durkheim, Émile, 9–11, 18, 24, 26, 27, 29, 31,
    34, 36–39, 45, 61, 101, 181, 231, 252,
    261, 267, 288, 307–378, 434, 443,
    471–473, 493–496. *See also* subject index

Einstein, Albert, 38
Elliot, George, 51
Engels, Friedrich, 104–107, 112, 114, 118–123,
    125, 126, 152, 153, 163. *See also* subject
    index

Feuerbach, Ludwig, 103, 104, 107, 111–114
Freud, 480, 483

Gandhi, 204
Gerth, Hans, 178
Goethe, Johann, 102, 174, 254

Harvey, William, 49
Hegel, Georg Wilhelm Friedrich, 102–104,
    107–114, 118, 127, 130, 132, 181, 256,
    257, 259, 385. *See also* subject index
Heine, Heinrich, 104
Hildebrand, Bruno, 181
Hitler, Adolf, 204, 252
Hobbes, Thomas, 51, 310, 315
Hume, David, 256, 355

James, William, 435, 436, 438, 443–447, 451
Jaspers, Karl, 177
Jevons, W. S., 182

Kant, Immanuel, 174, 181, 252, 254–259, 265,
    355
Knies, Karl, 176, 181

Lewes, George, 51
Lipset, Seymour Martin, 225
Locke, John, 256, 310, 315
Louis Phillippe, duke of Orleans, 8
Louis XVIII, 8

Machiavelli, Niccolo, 386
Maistre, Joseph-Marie de, 10, 18
Malthus, Thomas, 48, 49, 436
Marcus, Steven, 121
Marx, Heinrich, 102
Marx, Henrietta, 102
Marx, Karl, 13, 14, 34–37, 39, 40, 45, 101–172,
    178–180, 184, 189, 193–195, 208, 210,
    213, 214, 218, 228, 231, 242, 244, 252,
    254, 255, 259–261, 275, 280, 288, 292,
    299, 300, 304, 305, 310, 325, 338–341,
    359, 360, 377, 385, 388, 389, 417, 471,
    493. *See also* subject index
Mattenier, Marie, 380
Mauss, Marcel, 356
Mazzini, Giuseppe, 379
Mead, George Herbert, 34, 36, 37, 261, 280,
    434–496. *See also* subject index
Menger, Karl, 182–184
Michels, Robert, 177
Mill, John Stuart, 9, 436
Mills, C. Wright, 178
Milne-Edward, Henri, 49
Mommsen, Theodor, 173
Montesquieu, Charles, 3, 10, 11, 13–15, 20, 29,
    309–316, 318, 319, 321–323, 326, 327,
    337, 340, 374. *See also* subject index

Napoleon, 7, 8, 379
Napoleon III, 8

Subject Index

Act, Mead's model of, 481*f*
Action
  animal, Mead's principles of, 488
  human
    dynamics of, Simmel's image of, 281*f*
    Mead's principles of, 490, 492
  and interaction, Mead's theory of, 485
  Weber's conception of, 201*f*
Analytical scheme, defined, 39
Anomie, Durkheim's view of, 338, 369, 375

Bias, Spencerian theory of, 58, 60
Bourgeoisie
  Engels's analysis of, 122
  Marx's theory of, 133

Calvin, John, *Westminster Confession of 1647* by, 223
Capitalism
  collapse of, Marx's views of, 150*t*
  demise of, Marx's theory of, 147
  emergence of, Weber's causal argument for, 226*f*
  in historical context, Marx's view of, 139, 150
  Marx's analysis of, 106
  in nineteenth-century England, 47
  Weber's view on, 217
Causal model
  complex, defined, 40
  simple, defined, 39
  simple and complex, comparison of, 41*f*
Centralization, and conflict, Simmel's view of, 276
Chance, Weber's view of, 215*t*
Class, Weber's theory of, 207
Class structure
  Marx's model of, 214*t*
  Weber's model of, 212, 214*f*
Coalitions, and conflict and group formation, Simmel's view of, 277
Collective conscience, Durkheim's view of, 331
Communist League, 105, 132
Communist party, Marx's description of, 139
Communists
  literature of, Marx's view of, 138
  Marx's theory of, 136
Community
  of attitudes, Mead's theory of, 486
  Mead's definition of, 474
Competition, Simmel's view of, 275
Comte, Auguste
  as advocate of sociology, 29
  analytic mode of, 25
  and Condorcet, 14
  The Course of Positive Philosophy by, 9, 19, 21, 29
  and Durkheim, 318
  early essays by, 20
  as early master, 34
  early scientific stage of, 19
  and emergence of sociological theory, 7–33

experimentation by, 24
formulation of sociological methods by, 23
functionalism of, 25
historical analysis by, 24
law of the three stages of, 20, 27*f*
laws of social organization and change of, 22
and Montesquieu, 10
moralistic and quasi-religious stage of, 19
observation by, 24
organization of sociology by, 25
and Pareto, 386
and religion, 19
and Saint-Simon, 8
and Saint-Simon's early work, 15
and Saint-Simon's later work, 17
and social dynamics, 27
social physics of, 11, 20
and social statics, 25
  implicit model of, 27*f*
sociological goals of, 23
and sociological theory, 21
sociology of, 19
and Spencer, 51
strange biography of, 7
Systems of Positive Polity by, 9, 19
thought of
  conservative trends in, 18
  intellectual origins of, 10
  liberal trends in, 18
and Turgot, 12
Concept, defined, 37
Conflict
  among recognized and accepted opponents, Simmel's view of, 275
  among those with common personal qualities, Simmel's view of, 274
  between groups, Simmel's view of, 276
  and centralization, Simmel's view of, 276
  and coalitions and group formation, Simmel's view of, 277
  and de-legitimation, Weber's theoretical principles of, 247
  Marx's model of, 164*f*
  Simmel's view of, 272, 304
  social, Simmel's image of, 279*f*
  and social differentiation and societal integration, Simmel's model of, 300, 301*f*
  as social form, Simmel's view of, 273
  and solidarity and intolerance, Simmel's view of, 277
  and stratification, Weber's model of, 242, 243*f*
  as threat to group, Simmel's view of, 275
  within groups, Simmel's view of, 274
Confucianism, Weber's view on, 227
Consciousness, James's view of, 443
Consummation, Mead's view of, 479
Cooley, Charles Horton, and Mead, 446